**Imaging X-Ray Astronomy**
**A Decade of Einstein Observatory Achievements**

# Imaging X-Ray Astronomy

## *A Decade of Einstein Observatory Achievements*

**Editor**

**MARTIN ELVIS**

*Harvard-Smithsonian Center for Astrophysics*

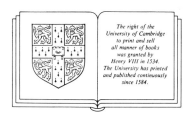

The right of the
University of Cambridge
to print and sell
all manner of books
was granted by
Henry VIII in 1534.
The University has printed
and published continuously
since 1584.

CAMBRIDGE UNIVERSITY PRESS
*Cambridge*
*New York   Port Chester*
*Melbourne   Sydney*

Published by the Press Syndicate of the University of Cambridge
The Pitt Building, Trumpington Street, Cambridge CB2 1RP
40 West 20th Street, New York, NY 10011, USA
10 Stamford Road, Oakleigh, Melbourne 3166, Australia

© Cambridge University Press 1990

First published 1990

Printed in Great Britain at the University Press, Cambridge

*British Library Cataloguing in Publication Data*

*Library of Congress cataloguing in publication data*

ISBN 0 521 38105 3

# Contents

# *Preface*

When the HEAO-B satellite was launched just after midnight on November 13th 1978 I was the most junior of junior scientists who was fortunate enough to be present at Cape Canaveral. We all knew that this mission, if only it was successfully launched, would be one of the most obviously revolutionary steps taken in astrophysics this century. How often, after all, could there be a single mission that improved sensitivity, angular resolution and spectroscopic resolution by over two orders of magnitude for a whole field of astrophysics? This was the promise of imaging X-ray astronomy, in its first incarnation as the HEAO-B. That this promise was fulfilled is by now part of history. This book tries to record that achievement.

At launch the NASA acronym (from "High Energy Astrophysical Observatory") had added to it the more appealing name of the "Einstein Observatory". This was an appropriate name for a mission, launched in the centennial of the birth of Albert Einstein, to investigate high energy phenomena around relativistic objects, neutron stars and black holes, and, through studying the cosmic X-ray background, attempt to understand the structure of the universe. The launch into the clear night sky was spectacular. The Atlas-Centaur rocket was visible to the naked eye a hundred miles downrange. We learned at the 1 am launch breakfast that the satellite had successfully reached its orbit. This was the engineers' triumphal banquet. Scientists, especially opportunistic latecomers like myself, kept a low profile.

Until four days after launch we did not know if the machine would deliver as expected. We could only be sure of this when we saw the first X-ray *image* up on the TV screen. It was impressive that the software to make such an image was working only a few days after launch. Standing at the back of the crowd watching the image appear block by block I was tense, but that can have been only a shadow of what was felt by those who had already spent a decade on the project. Gradually the picture of Cygnus-1 (seen on the cover of this book) appeared. The streak of light shows the path of the source across the detector as the satellite slewed onto the target position... and there it stopped, on a dime. The revolution was on. After that it was daily revelations, even during the test phase, when unexpected intense O-star X-ray emission was discovered.

No-one at the launch would have thought that we would still be analyzing data from *Einstein* 10 years later, or that we would still be waiting for AXAF as the next US X-ray astronomy mission. Yet also perhaps no-one would have though that the Einstein data would still have within it enough richness to be producing novel science a decade later.

The 10th anniversary of the launch of *Einstein* was celebrated with a Symposium held in Cambridge, Massachusetts (USA) on 13–15th November 1988. This book grew out of the talks given at this symposium.

The book is in three parts: the first part demonstrates that *Einstein* results have redefined X-ray astronomy; the second part shows that the *Einstein* data continue to be a source of innovation in the field; and the third part explains how the whole of the *Einstein* data is being made available to all astronomers through a series of data bases.

**Part 1** is a series of reviews of almost all the areas of astrophysics on which the *Einstein Observatory* had an impact. Inspecting the table of contents shows that this was almost every part of astrophysics. The series begins with an overview of the *Einstein* mission by Harvey Tananbaum, and shows how the questions it raised can be addressed with the next US X-ray astronomy mission, AXAF. Unfortunately not all invited speakers were able to contribute to this volume. This means that, for example, supernova remnants are covered primarily in the chapter on "High Resolution X-ray Spectroscopy of Thermal Plasmas" by Claude Canizares, and active nuclei and quasars are covered for the most part by the chapter by Steve Holt on "Spectra of Non-thermal X-ray Sources". This part of the book ends with a paper by Riccardo Giacconi who looks beyond AXAF and its European counterpart XMM, toward future X-ray astronomy missions of new design that might make similar sized steps to those taken by *Einstein*. **Part 2** is a series of short papers, selected from the contributed papers at the symposium, that illustrate the wide-ranging used being made of the *Einstein* data today; **Part 3** describes each of the data bases – including the IPC source catalog and the CD-ROM preparation of all the processed imaging data – now being prepared at the *Einstein* Data Centre at the Smithsonian Astrophysical Observatory. In this way the book gives a complete view of the *Einstein Observatory* in its historical position and in its many current uses.

In the preparation of the symposium and of this book many people had a role. In particular I would like to thank Polly Sullivan for her energy and attention to detail in arranging the symposium logistics, Pepi Fabbiano for much assistance especially in arranging the banquet, Elizabeth Bohlen for the poster graphics (which make the cover illustration for this book), and Karen Modestino for bringing the various manuscripts into a uniform and handsome format. Thanks must also go to the other members of the local organising committee: Tommaso Maccacaro, Belinda Wilkes, Isabella Gioia and Bill Forman; and to the scientific organising committee: Riccardo Giacconi (chair), Claude Canizares, David Helfand, Steve Holt, Harvey Tananbaum and Tommaso Maccacaro. I am particulary grateful to the organisations that sponsored the symposium: NASA (through the Astrophysics Division of OSSA, division chief C. Pellerin), the Smithsonian Astrophysical Observatory, and the Massachusetts Institute of Technology. Thanks also go to all those people who have worked on the Einstein project, from its earliest days to the present, and who made its tremendous success possible.

The Symposium included a Banquet held in the Fogg Museum of Harvard University just a short walk across Harvard Yard from the conference location. Harvey Tananbaum arranged for almost all the key scientists involved in the beginnings of the Einstein project to give short reminiscences of this time. These Banquet talks, edited by George Clark, open our book (Prologue) giving it a historical perspective and a glimpse of the human side of this large undertaking.

Martin Elvis

*Cambridge, Mass.,*
*August 1989*

# PROLOGUE
## The Origin of
## The Einstein Observatory:
## Personal Recollections

# PERSONAL RECOLLECTIONS
# ON THE ORIGINS OF
# THE EINSTEIN OBSERVATORY

## Bruno Rossi

I wish to give some reminiscences that go back thirty years to the end of the 50's.

At that time, the only known celestial source of x-rays was the Sun. Some of us (and in these I include particularly Riccardo Giacconi, George Clark, and Stan Olbert) thought that it would be interesting to see if one could find other, more distant sources of x-rays in the sky, sources located outside the solar system. Fortunately, none of us was an astronomer — we were all physicists — otherwise, we would have known that our idea was not promising and would probably have dropped it after a little reflection on the known facts of astronomy. Martin Annis was interested and put at our disposal the facilities of the company he had recently started — American Science and Engineering, Inc.

It was immediately recognized how difficult it would be to do what we had in mind. To detect an x-ray source similar to the Sun, placed at a distance on the order of the distance to the nearest star, we would need a detector a billion times more sensitive than those that had been used in solar x-ray observations. Of course, we hoped there might be sources of celestial x-rays more powerful than the Sun. Still it was clearly necessary to develop detectors of the greatest attainable sensitivity if we were to have any hope of success.

To simply increase the sensitive area of the detector was not the best solution. The sensitivity limit of a detector in a given amount of observing time is set by the cosmic-ray background. By increasing the size of a detector you increase the flux of x-ray incident on the sensitive area, but, at the same time, you also increase the cosmic-ray background. What was needed was an instrument capable of concentrating on a very small detector the x-ray incident over a large area of collection — some sort of a concentrator.

Riccardo suggested the use of grazing incidence single reflection from a parabolic mirror to achieve concentration. He then came across the article written by H. Wolter in 1952 on x-ray optics and the concept of an x-ray microscope employing two grazing incidence reflections. He immediately suggested to build an image-forming grazing-incidence mirror system for an x-ray telescope. The idea sounded promising and Riccardo and his associates began right away to work on it. The final result, nearly twenty years later and after an important success with the solar x-ray telescope on the Apollo mission, was the *Einstein* stellar x-ray telescope the tenth birthday of which we are here to celebrate.

However, in 1960 it was clear that it would take many years to reach that goal, and we thought that it might be worthwhile, in the meantime, to pursue a rocket-borne x-ray exploration of the sky with instruments easier to construct than a grazing-incidence telescope but, of course, much less sensitive. I was confident that, without resorting to any new technology, it would be possible to produce x-ray detectors a few orders of magnitude more sensitive than those in existence.

NASA was asked to support our program; but NASA had no interest in our project which was judged to have no chance of success. Fortunately, at that time, the Cambridge Research Laboratory of the U.S. Air Force was engaged in studies of the moon. John Salisbury, Chief of the Lunar and Planetary Exploration Branch, obtained support for an AS&E proposal to detect lunar x-rays in hopes of obtaining information on the chemical composition of the lunar surface (it seemed likely that the moon would prove to be, by virtue of scattering solar x-rays, the strongest source of celestial x-rays other than the sun itself). Thus our experiment was nominally an attempt to detect solar-induced x-ray emission by the moon.

Two rocket flights failed. In preparation for a third, Herb Gursky and Frank Paolini took on the task of developing large, flat Geiger-Mü counters partially protected against the cosmic-ray background by an anti-coincidence system. Their sensitivity was about 100 times greater than that of the detectors used previously in the x-ray observations of the sun.

Three of these detectors were flown on a rocket on June 18, 1962. To everybody's astonishment we detected something that was clearly not the moon. Nature had been so kind as to place the strongest x-ray source in the sky, Sco X-1, within the field of view of the sensors. The source could be seen very clearly in the telemetry record, and so, on that date, extra-solar x-ray astronomy was born.

*****************************************************

## Herbert Gursky

I wish to relate three events that illustrate the genesis of *Einstein*.

These are events that have stayed with me over the years and I go back over them in my mind from time-to-time and think of their significance.

The first of these is the letter from Riccardo Giacconi to Homer Newell written in July 1967. The letter is described in the Annual Reviews monograph, "Telescopes for the 1980's", but is not discussed in "The X-ray Universe" by Tucker and Giacconi. As I remember it, the letter was an attempt to set the stage for an explorer-class follow-on to *UHURU* (not yet flown and known at that time only as SAS-1) but involving a focusing x-ray telescope. It was felt that such a mission would be more appealing if it could be shown to be part of a larger program. So the explorer-scale mission was embedded in a program, culminating in a facility-class observatory using meter size optics. You all know what happened—NASA showed no enthusiasm for the explorer, but bought the facility.

I have oversimplified and compressed the events—there was several year period of planning, studying and lobbying—but I distinctly recall a specific moment of being stunned by the realization that NASA had accepted the idea of a major mission involving a focusing x-ray telescope, as opposed to a more modest explorer-class mission.

The second event relates to the order of the HEAO missions. You will recall that *Einstein* was initially LOXT and was the third of the HEAO series. After the program was restructured in 1973, the scientific panel that reviewed the program recommended that the telescope mission be moved to number 2, ahead of the cosmic ray mission. I recall being surprised that the review committee would agree to change the mission order but we thought it to be terribly important, since, among other things, there was concern that the final mission of a series is easy to cancel. That did not occur, but the several years headstart has obviously been very important to the U.S. program of x-ray astronomy.

Finally I point out how ill-prepared we were for the development of *Einstein*. Although we had done a fair amount of telescope development, there had been virtually no detector development work carried out prior to beginning the satellite project. In fact we had only one successful rocket flight involving an *Einstein*-like focusing telescope and a high-resolution imaging detector looking at a cosmic x-ray source. It was conducted around 1970 while we were still at AS&E. The three scientists, principally responsible for the payload were Ed Kellogg, Steve Murray and Leon Van Speybroeck. The results from that flight revealed three detected photons coming from Sco X-1! That was it, three photons. Their significance was very high. They had arrived within the blur circle of the mirror and the number of background events was virtually zero, but still prior to *Einstein*'s turn on in orbit to view Cygnus X-1 these were the only photons detected using a similar system.

On thing is clear, the emergence of *Einstein* as a mission is counter to the standard paradigm for the development of a space discipline—for which x-ray astronomy is held up as an example—of an orderly, sequenced development beginning with sounding rockets leading to moderate space missions, on to facility class missions. Rather, the facility class mission, *Einstein*, emerged full grown. The proper sequence of development had taken place in the case of solar x-ray telescope missions, but that was just barely relevant except for demonstrating that efficient, high quality grazing incidence telescopes could be built.

My own opinion as to what happened was that NASA had this enormous hole that needed to be filled based on the expectation that its funding would continue at the Apollo-like levels and was looking for big missions, like HEAO, rather than explorers like SAS. And this bureaucratic imperative took precedence over any theoretical ideas relating to orderly development.

But I think there was genuine wisdom displayed; first on the part of NASA administrators like John Naugle in pushing major facilities in spite of the limited technical developments. Then on the part of the HEAO review committee—that moved the mission up one notch—I recall that Al Cameron chaired it, but I don't remember who else was on it.

Finally I have to extend my greatest admiration to the individuals—Harvey Tananbaum, Steve Murray, Leon Van Speybroeck, others—who actually brought to fruition what I and others had committed them to.

*****************************************************

# Elihu Boldt

My first opportunity for participating in the planning of an orbiting x-ray telescope observatory came via Nancy Roman's invitation to attend a conference at NASA headquarters on October 9-10, 1967; this meeting was to survey what we expected to accomplish with such a mission from the point of view of specific objectives in astronomy. Resolving and studying sources of the x-ray background was (and remains) an issue of considerable interest to me and was the basis of my particular contribution, which was entitled "The Local Supergalaxy as the Structured Aspect of a Universal X-ray Background" (GSFC X-611-67-486).

The next meeting on this program that I attended took place on August 7, 1968 at the Astrionics Laboratory of the Marshall Space Flight Center and was concerned with the question of what sort of instrumentation would be appropriate for placement at the focal plane of a large grazing-incidence x-ray telescope. The general consensus was that high resolution imaging and high resolution dispersive spectroscopy were absolutely vital and that polarimetry would be very desirable. I found that this situation disturbed me in two rather separate ways. First, I realized that our main x-ray hardware program at Goddard was inappropriate in that it emphasized large area *mechanically collimated* gas proportional chambers; these were of the type developed by Peter Serlemitsos for OSO-8 and ultimately adapted for our HEAO-1 broadband all-sky study of the x-ray background and foreground sources. Second, I was concerned about the apparently inherent inefficiency of the experiments being considered for the focal plane; somehow, the photon collecting power of the telescope was not being fully exploited. Then I remembered that back at Goddard there was a relatively young newcomer to our group by the name of

Steve Holt who was fascinated by the prospect of using a cooled solid-state device for detecting x-rays and was working on a balloon-borne experiment for observing the Crab Nebula with a Ge(Li) crystal. This sort of detector had high quantum efficiency and much better energy resolution than gas proportional counters but was intrinsically small in area. It was immediately obvious to me that such a high performance non-dispersive spectrometer at the focus of a large grazing-incidence x-ray telescope could obviate both of my concerns and would be an ideal complement to the other focal plane instruments. When I returned to Goddard Steve agreed with this and pointed out that a Si(Li) crystal would be more suitable than Ge for the energies to be covered by the telescope.

Although a Si(Li) crystal would detect essentially every photon (>0.5 keV) focused onto it by the telescope the device by itself could not be used for locating the image differentially within its overall field of view of several arc minutes. The real imaging capability of the telescope was not being utilized in such a situation, and this bothered me. However, we noted that the front surface of the detector could serve as a mirror for light and devised a rather complex scheme for exploiting this to correlate visible-light and x-ray images. We estimated that the optical and x-ray images could be correlated to a precision of about ten arc-seconds by using a rotating modulation wire-collimator wheel in the focal plane (in front of the detector). It was with a strong feeling of pride that I presented a description of this elaborate invention at a meeting held in Huntsville during the summer of 1969; after all, this device could do essentially *everything* (short of polarization). Before I finished, though, Bob Novick stood up and announced, "That's a giant step backwards!". I was sufficiently intimidated by this to drop the idea of image location. Perhaps we were thereby saved from an excessively complicated experiment likely to fail, but I sometimes wonder.

A short while before the launch of HEAO-2 Riccardo Giacconi visited us at Goddard to get some first-hand information about preliminary results emerging from our all-sky HEAO-1 experiment, particularly as regards the question of extragalactic sources that contribute to the previously unresolved background. It was clear that this was an issue of strong mutual interest and that the HEAO-2 telescope mission should go a long way towards providing us with the answer. He told us of his desire to give a *real* name to the observatory and mentioned several candidates; the name Einstein was on the list because of the centennial celebration of his birth. I gave my endorsement to the choice of Einstein by reminding Riccardo that the tools of general relativity would be of particular importance to us for this mission and would be used rather routinely in the analysis of extragalactic results.

Steve took on the tremendous challenge of preparing the SSS (solid-state spectrometer) experiment for the *Einstein Observatory*. Refrigerator issues (*e.g.*, cryogen lifetime) and microphonics isolation were just two of the many state-of-art problems which had to be solved and apparently were by the time of launch. Early in the actual mission, however, there was a very tense moment of concern that Steve has reminded me about. Standing by in the control room at Goddard late at night, he was ready for a hefty signal from Cyg X-3, one of the first objects to be observed with the SSS and one of the brightest sources in the *UHURU* catalog. In fact, the counting rate actually obtained was alarmingly small and Steve was dismayed. What could possibly have gone wrong? Conferring with each other by phone, we managed to allay our anxiety by consulting a graph of the Cyg X-3 spectrum previously obtained from one of our rocket-borne experiments and noting that most of the incident photon flux resided at energies above the 4 keV telescope cutoff (*i.e.*, the Cyg X-3 spectrum exhibited pronounced absorption). Fortunately, we had not done anything to "fix" the experiment.

In August 1982 Frank McDonald, who was then head of the GSFC Laboratory for High Energy Astrophysics, called a meeting of our x-ray astronomy group for a discussion of our plans for possible AXAF proposals. I summarized the situation by noting that we were probably too late to make a competitive contribution to imaging or to high resolution dispersive spectroscopy. Concerning polarimetry and high resolution non-dispersive spectroscopy I suggested that what was called for were some bright new ideas. In particular, I emphasized the importance of greatly improving the resolution of non-dispersive spectrometers. Frank pointed out that Goddard was involved in making significant advances in infra-red bolometers and that we might want to check into this. Fortunately Rich Mushotzky, then a junior staff member, took this challenge seriously and subsequently walked down the hall to confer with Harvey Moseley of the IR group; the concept of an x-ray microcalorimeter emerged from this encounter. We now have a most effective team working on this; under Steve's impetus they are well along in developing an instrument that provides the breakthrough in AXAF spectroscopy that we called for. However, this team has a warning to pass on. Exposing the AXAF microcalorimeter to Cyg X-3 could present us with a problem. With the extended bandwidth of AXAF, the counting rate is likely to be too *high* this time.

*****************************************************

# George W. Clark

At this symposium Claude Canizares has described results obtained from the analysis of cosmic plasmas by high-resolution x-ray spectroscopy with the Focal Plane Crystal Spectrometer (the FPCS) on the *Einstein Observatory*. Claude guided the development and us of that instrument since he came to MIT in 1972 and took on the tasks of Project Scientist for the development of the flight instrument.

Seven years earlier I had decided to focus a part of the work in x-ray astronomy at MIT toward the goal of high-resolution x-ray spectrometry. It was an area not yet occupied by our friends at American Science and Engineering, Inc. where observational extra-solar x-ray astronomy began in the early 60's with the invention of the grazing incidence x-ray telescope and the discovery of Sco X-1. I remember with gratitude the generous help that Riccardo Giacconi and the AS&E president, Martin Annis, gave us at MIT as we began our work in x-ray astronomy, specially in the area of the rocket experiments of Hale Bradt.

Using an x-ray generator acquired with funds provided by the new MIT Center for Space Research, I began some experiments in 1965 with grazing incidence reflection gratings in the MIT Laboratory for Nuclear Science. During that period I met Herb Schnopper, then an x-ray physicist at Cornell interested in the new field of x-ray astronomy. Herb accepted an offer to join the MIT faculty, and during the spring of 1966, before he officially arrived, we submitted to NASA a proposal for development of a crystal spectrometer, which was turned down. Soon after he settled at MIT Herb turned our attention to curved crystal focusing x-ray spectrometers. His first engineering model of an FPCS was a Tinker-Toy model illustrating how a mechanical linkage could keep slit, curved crystal and detector aligned on the Rowland circle. In the summer of '67 we proposed the focusing spectrometer concept to Joseph Chamberlain at the Kitt Peak National Observatory where it was incorporated in the KPNO proposal for a "low-budget" x-ray observatory submitted a year later. Then KPNO dropped its space astronomy program. Later in '67 we submitted a proposal to the Marshall Space Flight Center for "A Curved Crystal Focussing Cosmic X-ray Spectrometer" which also wasn't funded.

Meanwhile, Riccardo, with the support of Martin Annis and Bruno Rossi, was urging NASA to adopt his plan for a dedicated exploratory x-ray satellite mission that eventually materialized as *UHURU*, and for a program for development of a much larger mission with an imaging x-ray telescope and focal plane instruments. In early '68 he got Bob Novick from Columbia, who was specially interested in astronomical x-ray polarimetry, and me to join him in promoting the idea of a scientifically integrated x-ray astrophysics mission based on an imaging telescope. As a self-styled "Principal Investigator Group" we sent a letter to Richard Halpern at NASA proposing a "Large Orbiting X-ray Telescope". That spring NASA issued an AO for x-ray and gamma-ray astronomy, and Herb and I submitted another curved crystal spectrometer proposal from MIT. This time we got a development study contract and an invitation to attend a meeting at the Marshall Space Flight Center on August 7, 1968 to discuss plans for a Large Orbiting X-ray Telescope. I think it was this meeting that set the wheels of government in motion that led finally to *Einstein* in orbit.

In January of '69 Giacconi, Novick and I sent a report to Halpern urging that the scientific package for LOXT be "planned as an integrated system for the analysis of the structures, spectra, polarizations, and time variations of X-ray sources,—." On March 19 there was another meeting at MSFC to discuss the stellar grazing incidence x-ray telescope and possible focal plane instruments, which now included ideas for a solid state spectrometer introduced by Elihu Boldt and Steve Holt of the Goddard Space Flight Center. Following that meeting the MSFC sent out proposal requests including one to MIT for a design study of a "Curved Crystal Focussing X-ray Spectrometer" to which we responded successfully. And still in March I circulated a memo among Boldt, Giacconi, Novick and the key NASA people, Halpern, John Downey, and Nancy Roman confirming our "intention to work together during the next six months in developing an overall plan for the scientific package of the projected x-ray telescope facility", which we did in a series of meetings during the spring and summer of 1969.

Around this time the concept of a series of "Super Explorers" emerged at NASA as a part of the effort to plan a post-Apollo future for the Agency. In June of '69 Giacconi, Novick, Boldt nad I wrote to John Naugle, Associate Director for Space Science and Applications, urging the start of a x-ray telescope project, possibly within the SEX series. In the August meeting of the planning group the idea of a scientific consortium for LOXT with a single PI emerged. It matured by January of '70 to the point of a Memorandum of Agreement among Giacconi, Gursky, Clark, Novick, and Boldt to work together on the development of an integrated scientific instrument package and to establish the areas of scientific responsibility — image analysis, aspect determination and transmission grating spectrometry at AS&E, Bragg spectrometry at MIT, polarimetry and objective grating spectrometry at Columbia, and solid state cryogenic non-dispersive spectrometry at Goddard Space Flight Center.

Throughout this early phase of the *Einstein* Observatory Riccardo maintained the principle of an integrated scientific package for the LOXT and impressed on NASA the scientific urgency of putting the LOXT as number 2 in the projected line of what had come to be called the High Energy Astrophysical Observatories.

In '72 I sent a proposal for the formal phase B study (design definition) of a Cosmic X-ray Spectrometer to AS&E where Riccardo was coordinating the consortium effort to initiate the preliminary work necessary to prepare for the LOXT flight program that was now definitely going to happen. With my graduate student, Doran Bardas, I explored the properties of crystals suitable for the projected wavelength range of the proposed LOXT mirrors, including artificial lead stearate multilayers for use in the ultra-long wavelength region in the manner pioneered by Bert Henke at the University of Hawaii. At this point Claude Canizares left particle physics at Harvard for x-ray astronomy at MIT and joined our phase B spectrometer project. Intense work of a flight program ensured, leading to one terrible moment on January 5, 1973 when the HEAO program was partially terminated on account of a NASA budget crisis. Called to an emergency meeting at NASA Headquarters by Jesse Mitchell, Naugle's successor as Associate Directory for Space Science and Applications, we were told that, to save the sinking ship, some passengers would be thrown overboard. He urged those in the water to swim along and push. Actually, most of the swimmers found ways to climb back on board as in the case of the Columbia group who lost their polarimeter project but retained their major role in the scientific use of HEAO-2, the *Einstein Observatory*. Nevertheless, it was a bitter moment for many.

Five more years went by during which our group at MIT developed the instruments for the SAS-3 X-ray Observatory and shared responsibility for two experiments on the HEAO-1 mission. Then on November 13, 1978 came the thundering launch of HEAO-2, celebrated at a Coco Beach restaurant where a toast was raised to Riccardo Giacconi and Bruno Rossi, whose vision had inspired the 20-year progress from scientific speculation to the realities of extra-solar x-ray astronomy.

**********************************************************

# Riccardo Giacconi

It is a pleasure to share with so many colleagues and friends some of the moments in the *Einstein* program which remain most vividly in my mind.

In 1963 I had prepared with Herb Gursky a document outlining a long term program of development of x-ray instrumentation culminating with the launch in 1968 off a 1.2 meter x-ray telescope. I had an opportunity to defend the need for focusing x-ray optics at the Woodshole summer study by the National Academy of Sciences in 1967, which was so influential in getting the HEAO program started. The first real high moment in the program came with approval by a NASA review committee of the LOXT in September 1970. Remember that this was before the successful launch of *UHURU*.

The sheer magnitude of what was proposed (a 1.2 meter focusing telescope and a Baez-like high-throughput concentrator of $1 \times 1$ m with a 13 meter focal length) and approved is still dazzling today. It will only be after the launch of AXAF and XMM that the full potential of that mission will be realized.

Then came the "Friday massacre" of 1973. LOXT was reduced to what became "*Einstein*". Bob Novick and polarimetry were the innocent victims of a drastic reduction in capabilities and cost. Though the Columbia group fully shared with the other Consortium institutions in data rights, I am sure it was a grievous personal blow to Bob. The science that got lost still awaits to be done.

In 1977 following the computer-aided test plans of Leon Van Speybroeck, Harvey Tananbaum and I were sharing the vigil at MSFC, with our colleagues from other institutions. I remember that Leon, whose queasy stomach seems to prevent him from personally attending crucial tests or launch activities, was not there when, after turning on the X-ray calibration source, no signal was received. The accidental placement of an occulting ribbon in the light path took only a few minutes to be discovered but what Harvey and I shared at that moment is hard to describe.

I remember the launch in 1978 with Harvey and I passing through all gamuts of emotions, finally exultant for a perfect orbital insertion. I remember sharing my joy with my young son, who thought launching satellites was almost as exciting as Disney Land.

Finally, I remember a particular moment with Ethan Schreier and Steve Murray when we first pointed Einstein to Cyg X-1. As the telescope slewed slowly to the required celestial position, a streak of x-ray light appeared on the screen, finally to concentrate in a brilliant point source. The power of focusing x-ray optics to study the sky never before privileged, my companions and I, to experience together the joy and excitement of a new field of science opening up for the first time.

It has been a privilege to share some of these moments with you tonight.

# PART 1
# IMAGING
# X-RAY ASTRONOMY:
# EINSTEIN ACHIEVEMENTS

# FROM EINSTEIN TO AXAF

**Harvey Tananbaum**

Harvard-Smithsonian Center for Astrophysics

## 1. INTRODUCTION

The presentations at the 10th Anniversary *Einstein* Symposium and the articles in this book cover a wide variety of scientific topics describing some of the important advances and discoveries made with the *Einstein* Observatory. The breadth and depth of science carried out with *Einstein* has made it essentially impossible to cover fully individual subdisciplines in single review talks and papers; nonetheless, this has not deterred me from attempting to summarize some of the major *Einstein* highlights and then to assess scientific prospects for AXAF.

To appreciate better the significance of the *Einstein* Observatory, let me start with an historical perspective, beginning with the June 1962 rocket flight which discovered the first extrasolar x-ray source (Sco X−1) and first observed the all-sky x-ray background (Giacconi, Gursky, Paolini, and Rossi 1962). As an aside, I note how fitting it is that three of the four principals involved in this landmark flight were able to participate in this Symposium. During most of the first decade, the emphasis in x-ray astronomy was on source detection and existence — determining which signals corresponded to real celestial sources. Rocket and balloon surveys were accompanied by efforts to identify the brightest sources: the identifications of Sco X−1 with a relatively faint, blue star and of the bright x-ray source in Taurus with the Crab Nebula represent two major successes. This period also saw the detection of x-rays from M87 — the first discovery of an extragalactic x-ray source (Byram, Chubb, and Friedman 1966).

The situation changed dramatically with the launch of the first Small Astronomy Satellite, *UHURU*, in 1970. With *UHURU* we carried out the first sensitive all-sky survey, reaffirming the presence of the 30 or so known brightest sources, concentrated primarily in the Galactic plane, and discovered more than 300 new x-ray sources, mostly extragalactic (Giacconi *et al.* 1974, and Forman *et al.* 1978). During the early 1970's the emphasis in x-ray astronomy shifted from demonstrating source existence to understanding how the x-rays are produced in at least some of the sources.

Two major break-throughs achieved with *UHURU* were:

1. The discovery of extended x-ray emission from clusters of galaxies (Gursky *et al.* 1971) suggesting that the x-rays are generated *via* thermal bremsstrahlung emission from hot gas filling the cluster [subsequently confirmed *via* detection of iron line emission with the Ariel-5 (Mitchell *et al.* 1976) and OSO-8 satellites (Serlemitsos et al. 1977)], and

2. The discovery that some Galactic x-ray sources are x-ray pulsars, in binary systems with detailed observations demonstrating that the energy source is gravitational. X-rays are generated when accreting material is accelerated and heated while falling onto a compact object (Giacconi *et al.* 1971; Schreier *et al.* 1972).

Although the pulsing x-ray sources were generally accepted to be neutron stars, strong evidence was developed that, at least for Cygnus X−1, the compact x-ray star was most likely a black hole (Oda *et al.* 1971; for a full discussion and more complete set of references, see Giacconi, 1981).

While the *UHURU* observations and analysis were taking place, the technology of grazing incidence telescopes for x-ray astronomy was being developed (originally for solar observations with rockets and the Skylab Apollo Telescope Mount) by G. Vaiana, L. VanSpeybroeck, and their colleagues under the overall guidance of R. Giacconi. It is in this context that we come to the *Einstein* telescope and its ability to image celestial x-ray sources.

## 2. THE EINSTEIN OBSERVATORY

The second High Energy Astronomy Observatory (HEAO-B) was launched into orbit by an Atlas-Centaur rocket on November 13, 1978. Renamed the *Einstein* Observatory, it operated successfully (with one significant interruption) until April 1981, carrying out over 5,000 targeted observations. Table 1 summarizes the capabilities of the *Einstein* instruments.

On the basis of the more than 1000 *Einstein*-related papers written over the past decade and with new papers still being written, it is my assessment that the major impact of *Einstein* has been to bring x-ray astronomy into the mainstream of current astronomical research. *Einstein* has demonstrated the importance of sensitive x-ray studies for understanding essentially all types of astronomical systems.

In this section, I summarize some of the *Einstein* highlights and discoveries and describe some new areas of research opened by *Einstein*. Many of these topics are reviewed in much greater detail elsewhere in these Proceedings. While we have learned much with *Einstein*, we have also raised many new questions to be explored with the next generation of x-ray satellites especially AXAF. The capabilities of AXAF are discussed at appropriate points in this section and in the concluding section of this paper.

### 2.1 Stars

One of the earliest discoveries with *Einstein* was the pervasive x-ray emission from stars of all types. Figure 1 shows a High Resolution Imager (HRI) observation of the nearby binary star system α-Centauri (Golub *et al.* 1982). The significant

---

### Table 1
### The Einstein (HEAO-2) Mission

Einstein operated for 2.5 years from its launch on 13 Nov 1978 until April 1981. During that time it was pointed toward some 5000 celestial targets, most of which were detected. Einstein also discovered several thousand 'serendipitous' sources that fell within its field of view.

Einstein had four focal plane instruments:

**Imaging Proportional Counter (IPC)**, PI institution SAO, with CAL collaboration. The IPC imaged a $\sim 1$ degree square field of view with $\sim 1$ arcminute angular resolution. It also had modest spectral resolution ($E/\Delta E \sim 1$).

**High Resolution Imager (HRI)** PI institution SAO. This instrument had the highest spatial resolution ($\sim 2$ arcseconds) over a 30′ dia. field of view, but had no intrinsic spectral resolution.

**Solid State Spectrometer (SSS)** - supplied by GSFC. The SSS was a high quantum efficiency, non-imaging device with a 6 arcminute diameter field-of-view that gave good spectral resolution ($\Delta E \sim 200$ eV) over a broad band . Due to its limited cryogen supply the SSS operated only during the first 10 months of the mission.

**Focal Plane Crystal Spectrometer (FPCS)** - supplied by MIT. This was the highest spectral resolution device on Einstein ($E/\Delta E \sim 100 - 1000$ ),but observed only a narrow energy range at any one time.

In addition there were two auxiliary instruments:

**Objective Grating Spectrometer (OGS)** supplied by Univ. of Utrecht. These gratings dispersed spectra onto the HRI where they gave a high resolution ($E/\Delta E \sim 50$) spectrum over a broad band, but with relatively low throughput.

**Monitor Proportional Counter (MPC)** supplied by SAO. To give the Einstein mission some sensitivity in the more traditional 2–20 keV band a collimated proportional counter, the MPC, was co-aligned with the telescope. Its field of view was essentially the same as that of the IPC, although of course the MPC was non-imaging. *All* targets were observed with the MPC, regardless of the instrument at the focal plane.

All the Einstein data have now been in a public Data Bank or archive for several years.

result here is the finding that the K1 star is almost twice as bright as the G2 star, contrary to the predictions of simple acoustical heating models developed to explain x-ray emission from the solar corona. *Einstein* detected substantial coronal x-ray emission from all classes of later-type stars (F,G,K, and M) with luminosities ranging from $10^{26}$ to $10^{31}$ ergs s$^{-1}$ (Vaiana *et al.* 1981). Models to explain these observations invoke dynamo mechanisms to describe how magnetic fields can channel energy from the stellar interior to the corona while confining the hot plasma. Such models are discussed in this volume by Linsky and by Vaiana, who indicate the importance of stellar rotation and age.

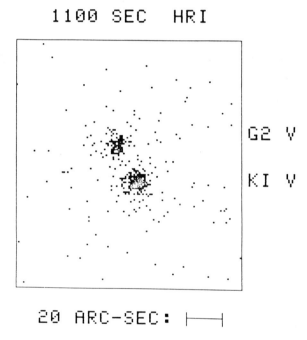

Figure 1. High Resolution image of nearby binary system, Alpha Centauri. The brighter x-ray source corresponds to the K-star and the fainter to the G-star.

The systematic understanding of coronal x-ray emission has been substantially advanced by samplings of star clusters which contain many stars of essentially the same age, all at the same distance from us. Figure 2 shows a superposition of the *Einstein* Imaging Proportional Counter (IPC) observations of the central region of the Hyades Cluster. For the covered fields, *Einstein* detected 66 of 121 observed cluster members, including 6 of 9 dA stars, 16 of 18 dF, 13 of 15 dG, 16 of 30 dK, and 14 of 46 dM. From data such as these, we can use the detected x-ray fluxes and upper limits to determine the mean x-ray luminosity and distribution for various spectral classes. By comparing results for various clusters of different ages and for field stars, we find that coronal x-ray emission depends on age: generally decreasing as stars become older (probably due in large part to decreasing rate of rotation), but with the dependence of x-ray luminosity on age varying as a function of spectral class (Micela *et al.* 1988).

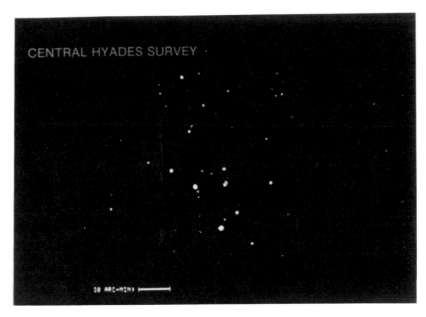

Figure 2. Mosaic of IPC images showing Einstein survey of central region of Hyades cluster.

*Einstein* observations extended these studies to still younger stars with the discovery of substantial x-ray emission from pre-main sequence stars (Ku and Chanan 1979; Gahm 1980; Feigelson and DeCampli 1981). This situation is dramatically illustrated in Figure 3 which is a superposition of *Einstein* IPC images covering a $2° \times 2°$ region in the $\rho$ Oph cloud (Montmerle *et al.* 1983). Approximately 50 x-ray sources are seen, mostly identified with pre-main sequence stars, or very young stellar objects. Many of these x-ray sources are highly variable over a day, suggesting that the x-ray emission is dominated by a succession of stellar flares. The observed x-ray luminosities range from $5 \times 10^{29}$ to $3 \times 10^{31}$ ergs s$^{-1}$. Since the column density within molecular clouds is quite high, we are most likely seeing only sources near the periphery of the $\rho$ Oph cloud in these *Einstein* images. AXAF will extend to significantly higher energies and should be able to probe more deeply into such star-forming regions, providing us with a sensitive tool for detecting T Tauri and naked T Tauri stars (*c.f.*, Walter, 1986). These objects may be relatively bright in x-rays because they are rapidly rotating and still in a state of collapse, not yet stabilized by internal nuclear fusion and therefore highly convective or turbulent throughout.

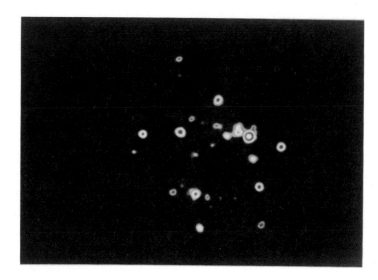

Figure 3.   Mosaic of IPC images covering 2° × 2° region in ρ Oph Cloud (Courtesy of Montmerle *et al.* 1983).

*Einstein* also discovered unexpected, and still unexplained, x-ray emission from early-type stars (B and O). Figure 4 shows an IPC image of the field around η Carinae, the variable star which erupted so spectacularly in the 1800's and which might have been expected to be an x-ray emitter. Also seen in this image is x-ray emission from several hot, massive O stars in the field, radiating $\sim 10^{33}$ ergs s$^{-1}$ in x-rays (Seward *et al.* 1979). O stars such as these are thought to be too hot to have convective zones, so simple dynamo models extrapolated from later-type stars can not be applied here. While stellar winds and shocks may play an important role, a satisfactory explanation may await high-resolution spectroscopy and imaging with AXAF.

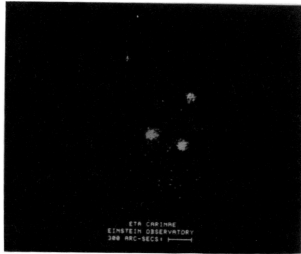

Figure 4. IPC image of Eta Carinae region.

## 2.2 Supernova Remnants

*Einstein* also provided a wealth of information about the later stages of stellar evolution. Figure 5 is an HRI image of the remnant of the supernova observed with the naked eye by Tycho Brahe in 1572. The appearance of the remnant in x-rays is an almost circular shell of diameter 8 arc minutes. The x-ray shell can be resolved in this HRI image and has a thickness which averages 30% of the radius. The outermost emission is thought to correspond to the interaction of the expanding blast wave with the interstellar medium, while the brighter regions slightly inside the outermost shell may be associated with ejected material heated *via* the "reverse" shock (Seward, Gorenstein, and Tucker, 1983). There is no evidence for a point-source, stellar remnant within this shell; the relative youth of the system suggests that any neutron star which had been left would not yet have had time to cool below the limit of detectability of this observation. These circumstances strongly suggest that no neutron star was created in this explosion.

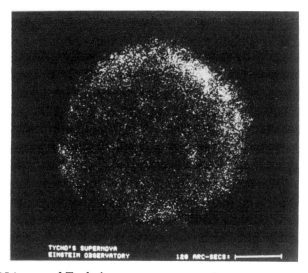

Figure 5. HRI image of Tycho's supernova remnant.

Figure 6 shows the x-ray spectrum obtained by Becker *et al.* (1980) with the *Einstein* Solid State Spectrometer (SSS) for Tycho's supernova remnant. The x-ray emission is dominated by lines, with more than 50% of the photons in the *Einstein* band coming from helium-like ions of silicon and sulfur. Data such as these have been used to deduce elemental abundances and to infer properties of the progenitor star and the supernova explosion itself. A major step with *Einstein* has been the recognition that for younger supernova remnants ionization
indexionization equilibrium equilibrium does not apply — insufficient time has elapsed for the ions to reach equilibrium.

The optical light curve recorded by Tycho Brahe is that expected for a Type I supernova explosion, and most theoretical models predict substantial production

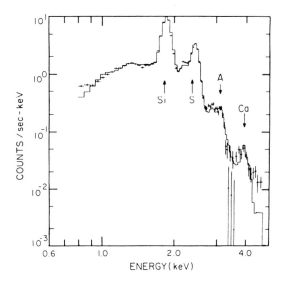

Figure 6. Pulse height spectrum for Tycho's SNR as observed with the Einstein SSS (Courtesy of Becker *et al.* 1980).

of iron — up to several tenths of a solar mass. However, the *Einstein* spectrum does not show much iron emission, and therefore raises a number of questions.

- Is there substantial iron emission at 6-7 keV beyond the *Einstein* band?

- Was the iron actually produced?

- Does the iron exist preferentially in the innermost region due to stratification in the pre-supernova star and due to the absence of substantial mixing, and is it therefore still too cold to emit x-rays because it has not yet been heated by the reverse shock? or,

- Has the iron not yet been shock heated and fully ionized because the gas density in that region is relatively low?

Spatially resolved spectra, extending above 7 keV, with AXAF should provide the information needed to address these questions and to resolve the debate. It will also be interesting to compare data for Tycho with AXAF observations of SN1987A in the Large Magellanic Cloud (LMC). Here mixing has already been invoked to explain the optical, x-ray, and $\gamma$-ray light curves. The detected neutrino burst (and the recently reported detection of optical pulsations) strongly indicate that a neutron-star remnant is present. Of course, for SN1987A the progenitor is known to be a high mass star, which is unlikely for Tycho.

Figure 7 is a mosaic of HRI images of Puppis A, a much older supernova remnant (Petre *et al.* 1982). Some of the irregular structure may be related to irregularities in the surrounding interstellar medium. The bright region near the eastern edge may result from the collision of the expanding remnant blastwave with an interstellar cloud of size $\sim 1$ pc diameter and mass $\sim 1 M_\odot$.

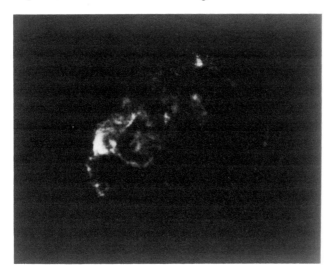

Figure 7. Mosaic of HRI images for Puppis A (Courtesy of Petre *et al.* 1982).

Winkler *et al.* (1981) have obtained a high-resolution spectrum for this bright region of Puppis A using the *Einstein* Focal Plane Crystal Spectrometer (FPCS). Figure 8 shows the observed counting rate from 700 to 1100 eV. Iron, oxygen, and neon lines are identified and have been used to deduce electron temperature, ionization state, and selective elemental abundances (Canizares, this volume).

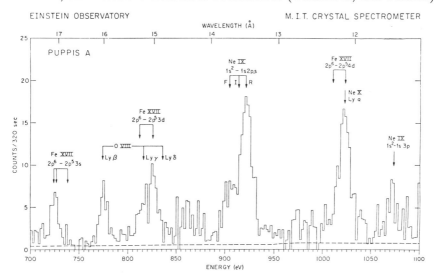

Figure 8. X-ray spectrum (700-1100 eV) of Puppis A as observed by the Einstein FPCS. Data combined from 14 scans of TAP crystal. The dashed line indicates the background level (Courtesy of Winkler *et al.* 1981).

The ratio of O VIII Ly $\alpha$ (seen at $\sim$ 650 eV) to O VIII Ly $\beta$ provides the electron temperature. Ratios of forbidden, intersystem, and resonance lines for Ne IX provide information on departure from ionization equilibrium. Ratios of O VIII to O VII (seen at lower energies) indicate that electron collisional excitation is the dominant process for this region. After corrections for interstellar absorption, neon to oxygen ratios are computed as $\sim$ twice solar and oxygen to iron ratios as $\sim$ five times solar. An estimated distance of 1 kpc leads to estimates for the progenitor star of $\geq 2M_\odot$ of oxygen and $\geq 25M_\odot$ total mass.

The IPC image of SS 433, an even more complicated object, is shown in Figure 9 (Watson *et al.* 1983). The emphasis in this image is on the large scale x-ray emission in jets emanating from the central, accreting binary star — possibly a neutron star, possibly a black hole. Radio maps show that this system is probably a supernova remnant, and optical observations show a pair of precessing jets moving at $\sim$ 1/4 the speed of light. Arguably, the most bizarre astronomical object presently under study, SS 433 was first studied in detail on the basis of its association with a faint x-ray source (*c.f.*, Margon *et al.* 1979). This situation points to an important aspect of x-ray emission in selecting for detailed investigation some of the most remarkable astronomical systems from the multitude of optical objects in the heavens.

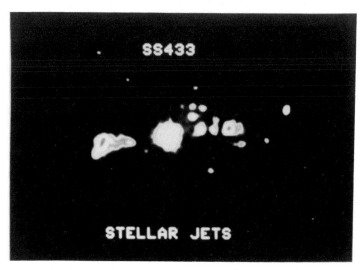

Figure 9. Mosaic of 3 IPC images showing central source and jet emission from SS 433.

## 2.3 Normal Galaxies

*Einstein* images have been used to study the x-ray contents of entire galaxies. Figure 10 is a mosaic of IPC images of the LMC assembled by our colleagues at Columbia (Wang *et al.* 1989). These observations reach a limiting sensitivity of $3 \times 10^{35}$ ergs s$^{-1}$ over the entire LMC and as much as 10 times fainter in selected regions. The image shows a few very bright, classical x-ray binaries known before the *Einstein* mission, many x-ray emitting supernova remnants — some discovered with *Einstein* through their x-ray emission, a number of still unidentified sources, and a large-scale, patchy, diffuse emission associated with the interstellar medium of the LMC. With AXAF, a $10^5$ sec exposure will be able to reach $10^{32}$ ergs s$^{-1}$ for an LMC field, allowing us to detect all of the individual O stars and many of the cataclysmic variables for regions of the LMC observed this deeply.

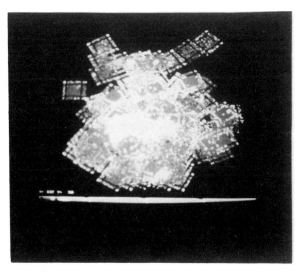

Figure 10. Mosaic of IPC images depicting point sources, supernova remnants, and diffuse interstellar emission from LMC (Courtesy of Wang *et al.* 1989).

*Einstein* studies of Andromeda (M31)(Trinchieri, Van Speybroeck *et al.* 1989) have found ∼120 individual x-ray sources and have demonstrated significant differences compared to our own Galaxy. For example, the central 2′ bulge of M31 contains at least 20 bright x-ray sources, compared to 3 in the same 400 pc-sized region of our Galaxy. This suggests different stellar populations as well as evolutionary differences for the two galaxies. (see Fabbiano, this volume, and references therein for further details.)

## 2.4 Active Galactic Nuclei and Quasars

Observations of the BL Lac object PKS 2155-304 with the *Einstein* Objective Grating Spectrometer (OGS) by Canizares and Kruper (1984) have detected a broad absorption feature near ∼ 0.6 keV. This feature results from absorption by highly ionized oxygen (O VIII), either in the intergalactic medium (at a density approaching that needed to close the universe) or in a region local to the BL Lac object (in a cloud of matter with velocities of $2 - 3 \times 10^4$ km s$^{-1}$, sufficient to Doppler broaden an absorption feature to the width observed). While these observations are unique among the *Einstein* results, they suggest a wealth of opportunities to be pursued with the increased sensitivity and spectral resolution which will be available with AXAF.

*Einstein* studies have shown that as a class quasars are x-ray loud (Tananbaum *et al.* 1979; Ku, Helfand, and Lucy 1980; Zamorani *et al.* 1981; Avni and Tananbaum 1986). Only three quasars had been detected as x-ray sources before the launch of *Einstein*; now hundreds of optically and radio-selected quasars have reasonably well determined 0.5 - 3.5 keV x-ray fluxes and luminosities. Moreover, the *Einstein* medium and deep surveys (Maccacaro *et al.* 1982; Gioia *et al.* 1989; Primini *et al.* 1989) have discovered hundreds of new quasars *via* their x-ray emission.

The relationship between summed x-ray emission from individual quasars and the x-ray background continues to be one of the more hotly debated issues. The limit set by the x-ray background for the summed total x-ray emission from quasars was used by Setti and Woltjer (1979) and Zamorani *et al.* (1981) to predict a turnover in the optical counts of quasars for magnitudes fainter than B $\sim 20^m$; this turnover has now been observed (Koo and Kron 1988; Boyle *et al.* 1987; Marano, Zamorani, and Zitelli 1988). The best available data on quasar optical number counts, x-ray to optical luminosity ratios, luminosity and evolution functions, fluctuations in the x-ray background (Hamilton and Helfand 1987), and the intensity and spectral shape of the x-ray background in the 0.5 -3.5 keV band have sufficient uncertainties (and conflicts) that plausible calculations have been published "showing" that quasars (and their lower luminosity relations) contribute from 20% to 100% of the all-sky background. Here again the greater sensitivity and larger bandwidth of AXAF will provide significant light (to accompany the present heat and sound), possibly enough to resolve this question.

## 2.5 Hot Gas in Galaxies and Clusters of Galaxies

*Einstein* observations have shown that x-ray data can provide unique insights into galaxies, tracing their evolution, revealing their interactions with their environment, and measuring their underlying mass distribution. Figure 11 is a visible light image showing several bright galaxies in the Virgo cluster. Of particular interest here are M86 and M84, two large elliptical galaxies about 1° away from M87. The *Einstein* x-ray contours first reported by Forman *et al.* (1979) are superimposed on the optical image. It is immediately apparent that in spite of their optical similarity, M86 and M84 have very different x-ray properties. M86 is nearly an order of magnitude more luminous in x-rays and has a large, extended, asymmetric plume of x-ray emission trailing behind it. On the basis of these x-ray observations, plus optical radial velocities, a scenario has been developed in which M86 spends much of its lifetime away from the core of the Virgo cluster. Over some $10^9$ years, stellar evolution in M86 leads to the production of interstellar gas - perhaps at a rate of $\sim 1 M_\odot$/yr. While M86 is away from the cluster core, its gravitational hold is sufficient to bind the gas to the galaxy. When M86 enters the cluster core, it encounters the dense cluster gas already there, and the ram pressure force is sufficient to strip away the accumulated gas producing the plume we observe. By contrast, M84 spends most or all of its time in the cluster core, and gas produced by stellar evolution is continuously being stripped from the galaxy. Without the accumulation of a reservoir of gas, there will be no sudden stripping to produce a luminous x-ray plume. This scenario exemplifies how x-ray data can be used to trace a galaxy's history and interpret its interaction with its environment.

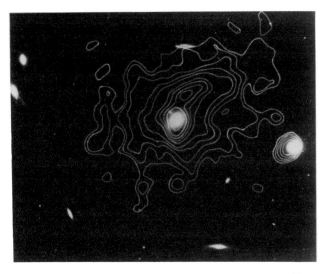

Figure 11. IPC contours superimposed on optical photograph (from Kitt Peak 4m telescope) containing galaxies M86 and M84 in the Virgo cluster.

These data also illustrate another major discovery of the *Einstein* Observatory — the presence of substantial amounts of hot gas in many early-type galaxies. An important step in this process was the finding by Forman, Jones, and Tucker (1985) that such hot gas is present in early-type field galaxies as well as in cluster galaxies. While there has been some debate over the maximum sizes of the extended x-ray emission, and while there are only limited data on the temperature profiles of the hot gas, there is no question about the existence of this hot gas. This detection already tells us much about ongoing stellar evolution and about the current absence of strong galactic winds in early-type galaxies. There is also a strong suggestion that these early-type galaxies have massive dark halos (Forman, Jones, and Tucker 1985), although there are limitations in the *Einstein* data for carrying out accurate determinations of masses for many of these galaxies (Trinchieri, Fabbiano, and Canizares 1986). In this regard, observations with AXAF (and possibly earlier with ROSAT, ASTRO-D, and SPEC-X-GAMMA) will allow us to measure with great precision the density and temperature profiles of the hot gas and thereby to trace the underlying gravitational potential, possibly providing new insight into the composition of the gravitating material.

The presence of this hot gas and the availability of this observational technique for measuring mass provide astronomers with a critical tool for studying early-type galaxies (analogous to rotation curves used to study mass distributions in spiral galaxies; Burstein and Rubin, 1985, and references therein). The utilization of x-ray observations to determine mass distributions has been most successfully applied to M87, and a full description of the assumptions and equations are provided by Fabricant, Lecar, and Gorenstein (1980); and Fabricant and Gorenstein (1983). In summary, the ideal gas law and the equation for hydrostatic equilibrium (balance between gravitational and thermal pressure) allow one to determine the total mass as a function of radius from the observed gas temperature and radial gradients of gas density and temperature. AXAF will have the energy bandwidth, field of view, angular resolution, low background, and ability to obtain the spatially resolved spectra needed to apply this method to nearby and distant galaxies and clusters of galaxies. For M87, from which photons are plentiful, AXAF will measure the mass to better than 10%.

*Einstein* observations have also advanced our understanding of clusters of galaxies in many ways. X-ray images have been used by Jones and Forman (1981) to classify clusters, thereby facilitating comparisons to theoretical models for cluster formation and evolution. Figure 12 organizes clusters into two groups: those with

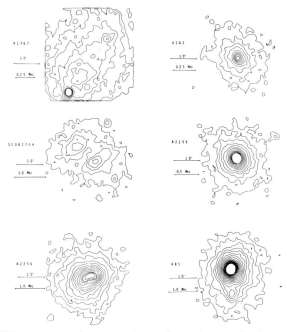

Figure 12. IPC contours of x-ray emission for 6 clusters depicting the evolution (from top to bottom) of clusters with central x-ray bright galaxies (on right) and of clusters without such central galaxies (on left).

a centrally located, x-ray bright galaxy (on the right), and those without such a central galaxy (on the left). For those with a central galaxy, cooling flows may play a particularly important role in the evolution of both the central galaxy and the cluster (see Fabian, these Proceedings). For both types of clusters, the scenario depicts evolution proceeding along the lines suggested by hierarchical models (*c.f.*, Peebles 1970; Aarseth 1969; and White 1976), with galaxies forming first and then interacting gravitationally to generate cluster structure. In the simplest description, with the passage of time, the cluster potential deepens and becomes smoother, eventually dominating the potential associated with individual galaxies. In this picture, x-ray luminosity and temperature increase with time, as does the velocity dispersion of the galaxies comprising a cluster. The fact that the large majority of relatively nearby clusters without x-ray bright central galaxies resemble A1367 in their x-ray emission and not A2256 or the Coma cluster indicates that most clusters are still relatively young and unevolved.

Further support for this scenario was provided by the discovery with *Einstein* of several clusters with double x-ray structure as illustrated by Figure 13 (Forman *et al.* 1981). Clusters in this image show two well-defined clumps of x-ray emitting gas, remarkably similar to the picture produced for an intermediate state of evolution by White's (1976) numerical simulations.

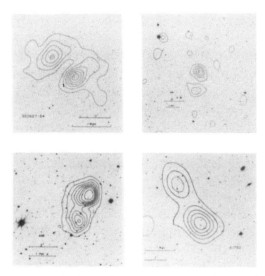

Figure 13. IPC x-ray iso-intensity contours superimposed on Palomar Sky Survey prints for four clusters showing double-peaked x-ray structure.

With AXAF, we should be able to detect (and discover) clusters out to redshifts between 1 and 2, perhaps reaching the epoch at which cluster formation is predominantly taking place, and at the least, spanning the time over which very substantial cluster evolution does occur. Models for cluster evolution will be tested (and modified) on the basis of statistically well-defined data on cluster x-ray luminosity and temperature versus redshift. Individual cluster maps should prove most informative as we search for individual x-ray bright member galaxies with large gaseous halos at times before the cluster potential became dominant. The ability to measure elemental abundances, particularly iron and oxygen, as a function of redshift may provide essential information on the fraction of primordial cluster gas and the fraction processed through galaxies. These studies will also be enhanced by spatially resolved spectroscopy, for example, of the plume of gas associated with M86 to determine elemental abundances for gas currently being stripped from galaxies.

## 3. AXAF

*Einstein* has shown that essentially all known astronomical objects (from Jupiter in our solar system to disant quasars near the edge of the observable universe) are candidates for detailed x-ray studies. For example, x-ray observations can tell us much about the behavior of matter under extreme physical conditions, about energy transport in stellar interiors, about evolution and explosions of massive stars, about acceleration of particles *via* shock waves, about the formation and evolution of galaxies and clusters, about the rate of expansion of the universe, and about the distribution and possible nature of dark matter. As I have already stated, AXAF will be at the center of these future studies of astronomical objects and fundamental physical processes. Some of the essential features of AXAF which will enable us to carry out these ambitious objectives are described in this section.

AXAF will be more sensitive than *Einstein* by nearly a factor of 100 for faint source detection and by up to a factor of 1000 for high resolution spectroscopy. One major contributor to this increase in sensitivity is the higher angular resolution resulting from improvements in mirror figure and surface smoothness. A second contributor is improved detector performances — particularly higher quantum efficiency, lower background, and increased energy resolution. A third contributor is an increase of $\sim 4$ in the area at lower energies accompanied by an extension of the energy bandwidth to beyond 8 keV.

Figure 14 shows the predicted encircled energy fraction as a function of encircling radius for AXAF at 2.98 keV, with Figure 14a covering radii from 0 to 2″ and Figure 14b covering radii from 0 - 15″. The encircled energy fraction for a given radius $r$ is the fraction of imaged photons from a point source which fall within a circle of radius $r$. The current AXAF specification (labelled 2) is based on actual accomplishments with the AXAF Technology Mirror Assembly. The AXAF design goal (labelled 1) is within reach on the basis of our current understanding of the x-ray optics, but requires modest advances in the metrology used to characterize the mirror surfaces in the mid-frequency spatial domain. Within a circle of radius of 0.″5 (appropriate

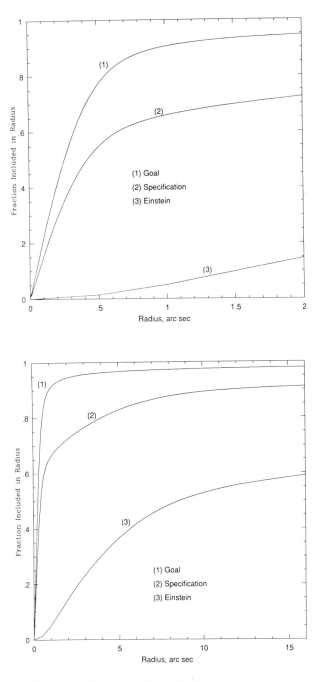

Figure 14. Predicted performance for AXAF x-ray optics and actual performance of Einstein telescope. Fraction of imaged 2.98 keV x-rays from a point source contained within a circle of a given radius plotted versus radius. (a) Covers radii from 0 to 2″; (b) covers radii from 0 to 15″.

for high resolution imaging), AXAF will contain at least 50% of the 2.98 keV photons and achievement of the goal would increase this number to 75%. In contrast,
*Einstein* (labelled 3) included only a few percent of the energy within an 0."5 radius. This concentration of signal means that for point sources the AXAF detection
cell can be as small as 1" × 1". This, in turn, reduces the total background associated with a detection cell to ∼0.1 counts per day, leading to a great increase in
sensitivity for faint source detection relative to *Einstein*. For imaging of extended
sources and for some spectroscopy it is relevant to consider the power contained
in a somewhat larger region. AXAF will contain 80% of the 2.98 keV photons in
a circle of 5" radius (goal 96%) compared to 35% for *Einstein*, again indicating a
substantial improvement.

Figure 15 shows the effective area for the primary AXAF imaging detectors as a
function of energy. The two curves illustrate the complementarity of the two instruments, with the High Resolution Camera (using microchannel plates) having the
highest throughput at lower energies and the AXAF CCD Imaging Spectrometer
(using charge-coupled devices) superior above ∼ 1 keV. The performance summarized in this figure encompasses the benefits derived from larger mirror area, larger
energy bandwidth, and higher detector quantum efficiency relative to *Einstein*.

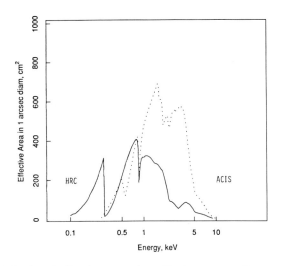

Figure 15. Calculated effective area for AXAF imagers (HRC and ACIS) plotted versus energy.

Figure 16 shows the energy resolution versus energy for point sources to be
observed with the AXAF spectrometers. The two sets of gratings are labelled LETG
and HETG; individual crystals of the Bragg Crystal Spectrometer are labelled BCS;
the calorimeter is labelled XRS with two curves shown: one for 5 eV and one for 10
eV resolution (with ongoing development work, the actual resolution of the XRS
is still to be determined, but is expected to be of this order); and the CCD array

is labelled ACIS. At various different energies, different spectrometers provide the highest resolution which provides one basis for choosing an instrument with which to observe. The ability to achieve resolutions $(E/\Delta E) \geq 1000$ over much of the AXAF band is an essential aspect of the increased capability of AXAF relative to *Einstein* for carrying out high resolution spectroscopy.

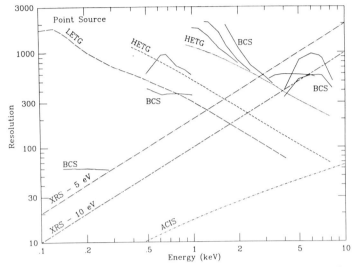

Figure 16. Energy resolution $(E/\Delta E)$ of AXAF spectrometers plotted as a function of energy. See text for identification of labels. Note the resolution shown is for point sources of emission.

As already stated, the resolutions in Figure 16 are for point sources. The gratings can also be used to study multiple point sources in a single field of view, but cannot readily deal with extended sources. Over a limited field of view, the XRS is divided into a mosaic of smaller cells and can provide data on a scale of $\sim 10'' \times 10''$ arc sec over a region of size $1' \times 1'$. For very extended sources reaching $20'$ or $30'$ in size the BCS and the ACIS provide the best capabilities for AXAF spectroscopic studies.

Another relevant parameter for spectroscopic observations is "grasp" which is a measure of throughput or average quantum efficiency obtained by folding a typical source spectrum through the mirror and spectrometer response, taking into account the fraction of the total energy band covered by a spectrometer at any one time. Figure 17 plots maximum resolving power for four AXAF spectrometers versus grasp. The ACIS and XRS have the highest throughput or grasp, while the LETGS and BCS (not shown in this figure) have the highest energy resolution. A formal computation for the BCS determines a grasp of $\sim 10^{-4}$, but this primarily serves to illustrate that the BCS is not the appropriate instrument for determining continuum spectral properties or surveying the overall spectrum to search for lines. For studies of specific lines such as helium-like silicon and sulfur, BCS has a grasp of several percent to accompany its high spectral resolution. We also note that at 1 keV the AXAF LETGS has more than 20 times the throughput of the *Einstein* OGS.

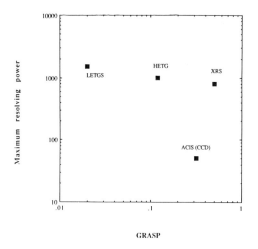

Figure 17. Energy resolution plotted versus grasp (or throughput) for AXAF spectrometers.

A final perspective comparing AXAF to *Einstein* can be visualized in Figure 18. The left panel shows a segment $\sim 0.°1 \times 0.°1$ of the deepest *Einstein* exposure of $\sim$300,000 s with the HRI in Ursa Minor. There are two faint sources detected (and possibly not even apparent on this reproduction). Sources at this limiting sensitivity may constitute 20-30% of the all-sky background. In a similar exposure with AXAF we expect to detect more than 100 times as many sources. The actual number of sources will depend on the slope of the number counts-flux relationship. If the *logN-logS* slope remains steep, we will actually resolve the background with AXAF and more or less run out of sources. The AXAF simulation shown in the right hand panel assumes a rather flat *logN-logS* relationship with integral slope = 1.15, in which case the discrete sources, numerous as they are, will not quite fully resolve the background.

Whether or not AXAF fully resolves the background, it will provide essential information on the formation epoch, evolution, and luminosity functions of quasars, BL Lacs, clusters of galaxies, and possibly lower luminosity active galactic nuclei. The large samples of faint, distant extragalactic sources provided by AXAF will complement superbly the ROSAT All-Sky Survey. Between now and the launch of AXAF in 1996, other powerful x-ray missions are also scheduled to be flown. One should anticipate further discoveries from these observatories, providing still more exciting prospects for AXAF to pursue with its high resolution imaging and spectroscopic capabilities.

Figure 18. Einstein deep survey (left) with two very faint sources in 0°.1 × 0°.1 region compared to simulated AXAF deep survey for same region.

In summary, *Einstein* has shown us where to look and what questions to pursue. Future x-ray observatories, particularly AXAF, are intended to carry out specific measurements and to advance our understanding of the objects and fundamental physical processes which comprise our Universe.

## ACKNOWLEDGMENTS

I wish to thank my many colleagues at CFA, MIT, Columbia, Goddard Space Flight Center, and Marshall Space Flight Center, as well as the members of the wider astronomical community who have contributed so much to the success of the *Einstein* program. I apologize for my inability to acknowledge them all individually or to reference all (or even a substantial fraction) of the important papers which have advanced our knowledge and understanding. I also thank Leon Van Speybroeck and the other members of the AXAF Science Working Group for providing me with specific AXAF information.

This work was supported by NASA Contract NAS 8-30751.

## REFERENCES

Aarseth, S. 1969, *M.N.R.A.S.*, **144**, 537.
Avni, Y. and Tananbaum, H. 1986, *Ap. J.*, **305**, 83.
Becker, R. H., Holt, S.S., Smith, B. W., White, N.E., Boldt, E. A., Mushotzky, R. F., and Serlemitsos, P.J. 1980, *Ap. J.(Letters)*, **235**, L5.

Boyle, B.J., Fong, R., Shanks, T., and Peterson, B.A. 1987, *M.N.R.A.S.*, **227**, 717.

Burstein, D., and Rubin, V.C. 1985, *Ap. J.*, **297**, 423.

Byram, E.T., Chubb, T.A., and Friedman, H. 1966, *SCIENCE*, **152**, 66.

Canizares, C.R. and Kruper, J. 1984, *Ap. J. (Letters)*, **278**, L99.

Fabricant, D., Lecar, M. and Gorenstein, P. 1980, *Ap. J.*, **241**, 552.

Fabricant, D. and Gorenstein, P. 1983, *Ap. J.*, **267**, 535.

Feigelson, E.D. and DeCampli, W.M. 1981, *Ap. J. (Letters)*, **243**, L89.

Forman, W., Jones, C., and Tucker, W. 1985, *Ap. J.*, **293**, 102.

Forman, W., Schwarz, J., Jones,C., Liller, W., and Fabian, A.C. 1979 *Ap. J. (Letters)*, **234**, L27.

Forman, W., Bechtold, J., Blair, W., Giacconi, R., VanSpeybroeck, L., and Jones, C. 1981, *Ap.J. (Letters)* **243**, L133.

Forman, W. *et al.* 1978. *Ap. J. (Suppl.)*, **38**, 357.

Gahm, G.F. 1980, *Ap. J. (Letters)*, **242**, L163.

Giacconi, R., Gursky, H., Paolini, F., and Rossi, B.B. 1962, *Phys. Rev. Letters*, **9**, 439.

Giacconi, R., *et al.* 1974, *Ap. J. (Suppl.)*, **27**, 37.

Giacconi, R., Gursky, H., Kellogg, E., Schreier, E., and Tananbaum, H., 1971, *Ap. J. (Letters)*, **167**, L67.

Giacconi, R. 1981, in *Proc. of the UHURU Memorial Symposium*, J. of the Washington Acad. of Sciences, **71**, 1.

Gioia, I. *et al.* 1989, *Ap. J. (Suppl.)*, submitted.

Golub, L., Harnden, F. R., Pallavicini, R., Rosner, R., and Vaiana, G. S., 1982, *Ap. J.*, **253**, 242.

Gursky, H., Kellogg, E., Murray, S., Leong, C., Tananbaum, H. and Giacconi, R. 1971, *Ap. J. (Letters)*, **167**, L81.

Hamilton,T. T. and Helfand, D. H. 1987, *Ap. J.*, **318**, 93.

Jones, C. and Forman, W. 1981, *Ap. J.* **276**, 38.

Koo, D.C. and Kron, R. G. 1988, *Ap. J.*, **325**, 92.

Ku, W.H.-M and Chanan, G. A. 1979, *Ap. J. (Letters)*, **234**, L59.

Ku, W.H.-M, Helfand, D.J. and Lucy, L.B. 1980, *Nature*, **288**, 323.

Maccacaro, T. *et al.* 1982, *Ap. J.*, **253**, 504.

Marano, B., Zamorani, G., and Zitelli, V. 1988, *M.N.R.A.S.*, **232**, 111.

Margon, B., Ford, H.C., Katz, J.I., Kwitter, K.B., Ulrich, R.K., Stone, R.P.S., and Klemola, A. 1979, *Ap. J. (Letters)*, **230**, L41.

Micela, G., Sciortino, S., Vaiana, G., Schmitt, J., Stein, R., Harnden, F. R., and Rosner, R. 1988, *Ap. J.*, **325**, 798.

Mitchell, R.J., Culhane, J.L., Davison, P.J.N. and Ives, J.C. 1976, *M.N.R.A.S.*, **176**, 29.

Montmerle, L., Koch-Maramond, L., Falgarone, E., and Grindlay, J. 1983, *Ap. J.*, **269**, 182.

Oda, M., Gorenstein, P., Gursky, H., Kellogg, E., Schreier, E., Tananbaum, H., and Giacconi, R. 1971, *Ap. J. (Letters)*, **166**, L1.

Peebles, P.J.F. 1970, *Ap. J.*, **75**, 13.

Petre, R., Canizares, C.R., Kriss, G.A., and Winkler, P.F. 1982, *Ap. J.*, **258**, 22.

Primini, F. *et al.* 1989, in preparation

Schreier, E., Levinson, R., Gursky, H., Kellogg, E., Tananbaum, H., and Giacconi, R. 1972, *Ap.J. (Letters)*, **172**, L79.

Serlemitsos, P.J., Smith, B.W., Boldt, E.A., Holt, S.S. and Swank, J.H. 1977, *Ap. J. (Letters)*, **211**, L63.

Setti, G. and Woltjer, L. 1979, *Astro. Astrophys.*, **76**, L-1.

Seward, F.D., Forman, W. R., Giacconi, R., Griffiths, R. E., Harnden, F. R., Jones, C., and Pye, J. P. 1979, *Ap. J. (Letters)*, **234**, L55.

Seward, F. D., Gorenstein, P., and Tucker, W. 1983, *Ap. J.*, **266**, 287.

Tananbaum, H. *et al.* 1979, *Ap. J. (Letters)*, **234**, L9.

Trinchieri, G., VanSpeybroeck, L., *et al.* 1989, in preparation.

Trinchieri, G., Fabbiano, G., and Canizares, C. R. 1986, *Ap. J.*, **310**, 637.

Vaiana, G.S. *et al.* 1981, *Ap. J.*, **245**, 163.

Wang, Q. *et al.* 1989, in preparation.

Walter, F.M. 1986 *Ap. J.*, **306**, 573.

Watson, M.G., Willingale, R., Grindlay, J.E., and Seward, F. D. 1983, *Ap. J.*, **273**, 688.

White, S.D. M. 1976 *M.N.R.A.S.*, **177**, 717.

Winkler, P. F., Canizares, C.R., Clark, G.W., Markert, T.H., Kalata, K. and Schnopper, H.W. 1981, *Ap. J. (Letters)*, 246, L27.

Zamorani, G. *et al.* 1981, *Ap. J.*, **245**, 357.

# EINSTEIN AND STELLAR SOURCES

**Jeffrey L. Linsky[1]**
Joint Institute for Laboratory Astrophysics
National Institute of Standards and Technology and
University of Colorado, Boulder, CO

"There is something fascinating about science. One gets such wholesale returns of conjecture out of such a trifling investment of fact."

– Mark Twain, *Life on the Mississippi*, 1874


## 1. PERSPECTIVE

Only 10 years ago the initial *Einstein* observations excited the astronomical world with discoveries of new phenomena and the confrontation of theory with the real world. Stellar astronomy was especially blessed by the pleasant but unanticipated discovery that nearly all classes of stars are x-ray sources. The conventional wisdom then was that stellar coronae are heated by acoustic waves generated in stellar convective zones. According to this theory the O,B and A-type stars should not have hot coronae as these stars lack convective zones, the K and M dwarfs should be extremely weak x-ray sources as their predicted acoustic wave fluxes were small, and the x-ray surface flux of a star should increase with bolometric luminosity (*e.g.* Mewe 1979; Renzini *et al.* 1977).

Reviews of the early stellar results by Linsky (1981a,b), Vaiana (1981), Rosner, Golub, and Vaiana (1985), and others emphasized the demise of the acoustic heating theory as none of its predictions were confirmed by the new data. Instead they argued that magnetic fields must play a dominant role in heating stellar coronae at least for the dwarf stars of spectral types F-M. Our present recognition of the importance of magnetic processes in the heating of stellar coronae should have preceded *Einstein* by at least 5 years as the Skylab x-ray observations of the solar corona conclusively showed the spatial correlation of closed coronal magnetic structures with bright x-ray emission above sunspots and active regions (*e.g.* Poletto *et al.* 1975; Rosner, Tucker, and Vaiana 1978). In general, the coronal plasma is hotter and denser where the coronal field is predicted to be large on the basis of potential field extrapolations from the measured photospheric fields. Nevertheless, no magnetic field measurements were then available for stars cooler than the Sun and the Sun was viewed as a particular star that might not be representative. In fact, *Einstein* has shown us just the opposite, the Sun has a very low luminosity corona down by 3 orders of magnitude compared to the most active stars in which magnetic heating processes occur.

---

[1]Staff Member, Quantum Physics Division, National Institute of Standards and Technology

Since the early flurry of observations and their instant analysis, many in-depth studies and detailed surveys have been published. (I include in Table 1 a selected list of important stellar surveys obtained with the *Einstein* Observatory.) Also, re-examinations of archival images and spectra have provided new discoveries and revised the hasty early conclusions based on only a few detections and inadequate consideration of upper limits and volume-limited samples. In particular, I will challenge the popular view that the coronae of hot stars and solar-type stars are fundamentally different. It is no longer feasible to cite the full range of the *Einstein* stellar results and literature in one review (see Pallavicini 1988 for a recent review of this topic); instead, I will summarize some of the major conclusions about stellar coronae that we have obtained from the *Einstein* data and what major questions remain for AXAF.

Rather than showing a conventional Hertzsprung-Russell diagram of detected x-ray stellar sources, I will introduce two H-R diagrams that I believe are in fact more fundamental to understanding the physics responsible for the x-ray emission detected by *Einstein*. The first (Figure 1) is an H-R diagram showing the types of stars with detected or inferred magnetic fields. These fields have been measured using the Robinson (1980) line broadening technique for some 30 late-type dwarf stars with spectral types G0V-M5V by Saar and coworkers (*e.g.* Saar 1987a,b), and using polarization techniques for some 20 B and early A-type dwarf stars (Borra, Landstreet, and Thompson 1983). In addition, the presence of dark spots indicates that at least the cooler components of the RS CVn, W UMa, and Algol-type spectroscopic binary systems must have large magnetic fluxes in their atmospheres.

Since it is difficult to measure photospheric magnetic fields in stars with broad line profiles and no model-independent techniques are available for measuring coronal fields in any star other than the Sun, I must appeal to another H-R diagram for indirect evidence of stellar magnetic fields. The H-R diagram of stars with nonthermal microwave emission (Figure 2), typically gyrosynchrotron emission or a coherent process during flares, includes the dMe stars (*e.g.* Bastian and Bookbinder 1988), RS CVn binaries (Drake, Simon, and Linsky 1989), O stars (Abbott *et al.* 1985), B and A-type stars with peculiar abundances (Drake *et al.* 1987a), and with some uncertainty the M giants (see Drake *et al.* 1987b). The gyrosynchrotron emission process is indicated by circular polarization, high inferred brightness temperatures ($10^9 - 10^{12}$ K), negative radio spectral indices, and/or by appreciable temporal variability. The presence of coronal magnetic fields is required for the emission process, but is also implied by the presence of relativistic electrons.

| | Table 1. | | |
|---|---|---|---|
| | Selected *Einstein* Surveys of Nondegenerate Stars and Binary Systems | | |
| Types of Stars in Survey | Number of Detections | Range in log $L_x$ | References |
| O-M Field star survey | 38 | 26.1 - 33.6 | Vaiana *et al.* (1981) |
| O3 - K2 III field star survey | 69 | 26.9 - 33.3 | Pallavacini *et al.* (1981) |
| Flux-limited field star survey | 79 | | Helfand and Caillault (1982) |
| Stellar sample within 25 pc. | 54 | 26.8 - 30.5 | Johnson (1986) |
| Of and O&B supergiants | 10 | 31.0 - 32.8 | Cassinelli *et al.* (1981) |
| Wolf-Rayet stars | 4 | 31.9 - 33.0 | White and Long (1986) |
| Ap and Am stars | 8 | 28.2 - 28.8 | Cash and Snow (1982) |
| A, Ap, and Am stars | 21 | 26.6 - 30.7 | Golub *et al.* (1983) |
| Late-A to early-F stars | 62 | 27.6 - 30.0 | Schmitt *et al.* (1985) |
| Magnitude-limited F star sample | 33 | 27.3 - 30.7 | Topka *et al.* (1982) |
| F dwarfs | 13 | | Walter (1983) |
| Late-F and G dwarfs | 33 | 26.7 - 29.4 | Maggio *et al.* (1987) |
| G0-K5 dwarfs | 18 | | Walter (1981) |
| F6 - M2 dwarfs-supergiants | 22 | | Ayres *et al.* (1981) |
| G4 - M2 giants and supergiants | 14 | | Haisch and Simon (1982) |
| F - M giants and supergiants | 87 | 26.8 - 29.6 | Maggio *et al.* (1989) |
| F0 - M5 stars in *Einstein* MSS | 128 | 27.8 - 32.6 | Fleming *et al.* (1989) |
| K and M dwarfs within 10 pc. | 128 | 26.1 - 29.9 | Bookbinder *et al.* (1989) |
| M dwarfs | 104 | 27.7 - 29.8 | Caillault *et al.* (1986) |
| M dwarfs in *Einstein* MSS | 25 | 27.8 - 30.1 | Fleming *et al.* (1988) |
| PMS stars in Tau-Aur and Per | 59 | 29.5 - 31.5 | Feigelson *et al.* (1987) |
| Naked T Tau stars in Tau-Aur | 32 | | Walter *et al.* (1988) |
| Orion cluster (log T= 6.5) | 58 | 30.4 - 32.8 | Ku *et al.* (1982) |
| Post-T Tauri G stars in Orion | 13 | 30.2 - 31.3 | Smith *et al.* (1983) |
| Rho Oph cluster survey | 60 | 29.7 - 31.4 | Montmerle *et al.* (1983) |
| Pleiades cluster (log T= 7.8) | 44 | 29.4 - 31.2 | Caillault and Helfand (1985) |
| Ursa Major cluster (log T= 8.5) | 7 | 28.9 - 29.6 | Walter *et al.* (1984) |
| Hyades cluster (log T= 8.8) | 48 | 28.4 - 30.1 | Stern *et al.* (1981) |
| Hyades cluster survey | 66 | 28.4 - 30.1 | Micela *et al.* (1988) |
| RS CVn systems (SSS data) | 8 | 30.2 - 31.4 | Swank *et al.* (1981) |
| RS CVn systems | 47 | 29.4 - 31.5 | Walter and Bowyer (1981) |
| RS CVn systems | 24 | 29.0 - 31.6 | Majer *et al.* (1986) |
| Algol systems | 7 | 30.1 - 31.3 | White and Marshall (1983) |

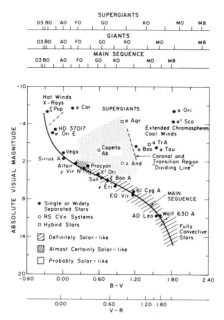

Figure 1. An H-R diagram showing the location of different types of stars discussed in this review. Regions of the diagram where magnetic fields have definitely been measured are labelled "definitely solar-like." Also two of the Bp stars with both measured fields and nonthermal radio emission are HD 37017 and σ Ori E. Stars for which magnetic fields are probably present but not yet measured are labeled "almost certainly solar-like" or "probably solar-like." From Linsky (1985).

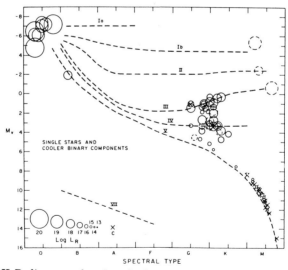

Figure 2. An H-R diagram showing the location of stars with 6-cm nonthermal emission known in 1984. The size of each bubble symbol is proportional to the radio luminosity in $ergs\,s^{-1}\,Hz^{-1}$. Coherent emitters are indicated by x symbols. From Gibson (1985).

The existence of magnetic fields in a diverse range of stars, as indicated by the measurement of photospheric fields or the detection of gyrosynchrotron radio emission, provides a new perspective on the study of soft coronal x-ray emission which is produced by bound-free and free-free processes in a hot thermal plasma. The spatial correlation of strong x-ray emission with magnetic flux in the underlying photosphere is well-documented for solar active regions, and the same relationship has recently been established for late-type stars as well (Saar and Schrijver 1987). This relationship is almost certainly a consequence of magnetic heating processes and the confinement of dense, hot plasma within closed magnetic structures, at least for the late-type stars without massive winds. I will now discuss each type of star detected by *Einstein* and comment on the probable relations between magnetic fields, hot thermal plasma, and nonthermal processes. I will also call attention to the similarities and differences among the coronae of the different types of stars.

## 2. O-TYPE AND EARLY B-TYPE STARS

Prior to the first observations with the *Einstein Observatory*, no x-rays had been detected from O stars and the general opinion was that none should be detected, because optical and ultraviolet observations indicated that these stars have massive winds with temperatures in the range 20-40,000 K. Also, most theories of coronal heating then presumed that the energy source is damping of acoustic waves and these stars do not have significant convective zones with which to generate such waves. A small minority, however, predicted x-ray emission from O stars on the basis of the "minimum flux corona" theory (Hearn 1975) or the anomalously strong resonance lines of O VI. Thus it was a surprise to most astronomers that the O stars are the most x-ray luminous class of nondegenerate stars with log $L_x$ in the range 31-33 (erg s$^{-1}$) for O dwarfs and 31.9-33.6 (erg s$^{-1}$) for O supergiants.

The first large surveys of these stars by Pallavicini *et al.* (1981) and by Cassinelli *et al.* (1981) led to the surprising result that the mean $L_x/L_{bol}$ ratio is $1.5 \times 10^{-7}$ with scatter less than a factor of 3, independent of stellar luminosity and rotational velocity (*cf.* Chlebowski *et al.* 1989). The same ratio holds for the B stars in the young Pleiades cluster (Caillault and Helfand 1985). Pallavicini *et al.* (1981) argued that this relation holds for the full range in spectral type from O3 to A5 on the basis of a few detections of A stars and the stars Canopus (F0 Ib) and $\alpha$ Hyi (F0 V or A9 III), but Cassinelli (1985) showed that the $L_x/L_{bol}$ ratio decreases rapidly below $10^{-7}$ for cooler stars beginning at spectral type B2, just where stellar mass loss rates also decrease rapidly. Most of the A star x-ray detections have subsequently turned out to be spurious (see below).

The very presence of detected x-rays from these stars argues that the hot emitting plasma must be distributed throughout the stellar wind as overlying material would completely absorb x-rays emitted from a corona located only at the base of the outflow. To get around this problem, Lucy and White (1980) and subsequent investigators have proposed that the hot plasma is located in shocks throughout the wind produced by the radiative instability inherent in a radiatively-accelerated wind. The long term variability detected in the soft energy band for four O stars by Collura *et al.* (1989) is consistent with this picture of x-ray emission distributed throughout a turbulent wind.

Spectroscopic studies with the *Einstein* SSS have provided valuable additional information. For example, the absence of detectable absorption at the 0.6 keV K shell absorption of oxygen does not completely rule out a base corona if the wind geometry is highly asymmetric or if the x-rays from a base corona completely ionize the wind as in the recombination wind model of Waldron (1984). Cassinelli (1985) and Swank (1985) showed that the SSS spectra of $\zeta$ Ori (O9.5 Ia) and $\tau$ Sco (B0 V) cannot be fit with a one- temperature plasma, but instead require at least two thermal components with a hotter component temperature above $15 \times 10^6$ K. Since plasma hotter than about $7 \times 10^6$ K cannot be confined by gravity alone, Cassinelli (1985) suggested that the plasma may be confined by magnetic fields of less that 100 Gauss. This is very interesting because Abbott, Bieging, and Churchwell (1984) identified the radio emission from five O-type or Wolf-Rayet stars as definitely nonthermal on the basis of their measured spectral indices and high radio luminosities. White (1985) proposed that this nonthermal emission is synchrotron emission from electrons accelerated to relativistic energies by the shocks present in these winds. This explanation leads to a picture of O stars somewhat similar to the solar-type stars in which both thermal x-ray and nonthermal radio emission occur in a magnetic envelope with the heating and accelerating mechanism being radiatively-accelerated shocks for the O stars and magnetically-driven phenomena in the solar-type stars.

## 3. MAGNETIC B-TYPE AND A-TYPE STARS

Does the solar-hot star connection extend to the next cooler class of stars in the H-R diagram? Beginning with Babcock (1947), strong fields with apparently simple dipole geometries have been measured in B and A-type stars with peculiar chemical compositions. Cash and Snow (1982) used the *Einstein* IPC to search for x-ray emission from a sample of eight such stars. They detected two of these stars ($\omega$ Oph and $\beta$ Scl) which are single with log $L_x = 28.0 - 28.8$ (erg s$^{-1}$). Using the VLA, Drake *et al.* (1987a) discovered nonthermal radio emission, which they interpret as gyrosynchrotron emission from continuously injected, mildly relativistic electrons, from three helium-strong early-Bp stars and two helium-weak, silicon-strong stars with spectral types near A0p. These stars all have strong magnetic fields, and in particular GL Lac = Babcock's star (B8p) has a field of 16.9 kilogauss, the strongest magnetic field for a nondegenerate star. Two of these stars ($\sigma$ Ori E and HD 37017) have measured x-ray fluxes, which correspond to log $L_x = 30.5$ and $L_x/L_{bol} = 1 \times 10^{-7}$. This agreement with the $L_x/L_{bol}$ ratio characteristic of the hotter stars is astonishing given that the Bp stars show no evidence for winds and the hot stars show no evidence for kilogauss strength magnetic fields. Nevertheless, both classes of stars show strong thermal x-ray emission and nonthermal radio emission. Why?

## 4. A-TYPE AND EARLY F-TYPE STARS: WHERE DOES STELLAR ACTIVITY BEGIN?

We now recognize that the magnetic A and B stars have coronae with both a hot thermal plasma and a permanent population of nonthermal electrons which are contained, heated and accelerated by strong magnetic fields in a manner that appears similar to the Sun. Do stars with spectral types intermediate between these stars and the Sun share similar properties and, in particular, at what spectral type do coronae begin to appear for stars with convective envelopes in which the production of magnetic fields is likely by dynamo processes and the activity indices depend upon rotational velocity? The chemically peculiar stars, which are all slow rotators, are the only stars hotter than the Sun for which magnetic fields have been measured. The chemically normal stars are typically more rapidly rotating than the Sun and are thus difficult stars in which to measure magnetic fields by line-broadening or polarization techniques. *Einstein* x-ray measurements have been particularly useful in providing further insight into this problem as the photospheric x-ray flux is negligible, whereas the photospheric flux in the optical and ultraviolet rises rapidly with increasing effective temperature. This effect makes it nearly impossible to measure such indicators of nonradiatively-heated plasma in stellar chromospheres and transition regions as emission in the CaII H and K lines, the MgII h and k lines and the ultraviolet resonance lines of CII and CIV for stars hotter than about spectral type F0V.

Early in the mission Vega (A0V) and Sirius A (A1V) were detected by the HRI but not the IPC. Golub *et al.* (1983) identified these stars as low level x-ray sources with $\log L_x = 27$ and $\log L_x/L_{bol} = -7$, consistent with the $L_x/L_{bol}$ ratio for the hotter stars. They identified the more luminous x-ray emission from a few other A-type stars as x-rays from known or previously unknown K or M dwarf companions. The reliable detection of x-rays from A stars is particularly difficult, because most stars are members of binary systems and most companions are late-type dwarfs which are luminous in x-rays but very faint optically. To make matters worse, the *Einstein* HRI had some sensitivity to ultraviolet light, and in an erratum, Golub *et al.* (1983) concluded that both Vega and Sirius A are not x-ray sources.

So, where do coronae in stars with convective envelopes begin to appear in the H-R diagram? The most comprehensive study to date is the survey of A and F field stars by Schmitt *et al.* (1985). They found that the statistics for the detection of x-rays from M dwarf stars in the interval $0.1 < B-V < 0.3$ are very different from those in the interval $0.3 < B-V < 0.5$. They concluded that the onset of detectable stellar coronae begins somewhere in the range $\sim 0.26 < B-V < 0.30$ (roughly spectral type

F0V or effective temperature 7200 K), and that the mean value of log $L_x$ for F stars is 28.3. This is consistent with the spectral type at which other spectral indicators for nonradiatively-heated atmospheres first appear in the H-R diagram, including HeI 5876Å absorption (Wolff, Boesgaard, and Simon 1986) and emission in the CII 1335Å and CIV 1549Å transition region lines (Walter and Linsky 1986). An apparent exception to this rule is the star Altair ($\alpha$ Aql, A7 IV-V) for which B-V = 0.22. Golub *et al.* (1983) find this star to be a very soft x-ray source (coronal temperature of about 800,000 K) with log $L_x$ = 27.14 and log $L_x/L_{bol}$ = $-7.5$. Simon and Landsman (1987) detected Lyman $\alpha$ emission from Altair, and Walter and Linsky (1986) detected CII and CIV emission from Altair with IUE. Since no hotter A-type stars show detectable x-ray, Lyman $\alpha$, CII, or CII emission or HeI absorption, I conclude that Altair marks the location in the H-R diagram where coronae begin for stars with convective envelopes.

There is ample evidence that magnetic fields on the Sun and late-type stars in general are amplified by dynamo processes involving the interaction of differential rotation and deep convective zones. Thus we anticipate that such activity indicators as the x-ray luminosity should depend upon the stellar rotation rate. Schmitt *et al.* (1985) found only weak evidence for a correlation of $L_x$ with the projected rotational velocity $v \sin i$ for their large sample of F stars. With a smaller sample of 14 F stars, Walter (1983) argued that such a correlation is not present for stars hotter than B$-$V = 0.45 (about spectral type F4 V) but is definitely present in cooler stars. From this one might conclude that coronae for stars hotter than B$-$V = 0.45 are not heated by a magnetic mechanism. In fact the situation may be more complicated with two heating processes occurring in these stars. Walter and Schrijver (1987) argue that a nonmagnetic heating process, perhaps simple acoustic waves, is increasingly more important in the hotter F stars until it disappears rapidly for stars just hotter than B$-$V = 0.30. They find that when the level of nonmagnetic heating is subtracted from the x-ray luminosity of F stars, the remaining component has the same functional dependence upon $v \sin i$ for stars as hot as B$-$V = 0.30.

## 5. PRE-MAIN SEQUENCE AND ACTIVE MAIN SEQUENCE STARS

The initial *Einstein* images of star forming regions were surprisingly rich in x-ray sources. Comparisons of the same fields at time intervals as short as one day apart give the appearance of blinking lights that Montmerle *et al.* (1983) likened to a Christmas tree. The very large amplitude of the x-ray variability suggests that there may be no "quiescent" coronae for these stars, but rather a distribution of flares with a wide range of luminosities. These stars have x-ray luminosities typically 1,000 to 10,000 times that of the quiet Sun, but only a few are coincident with the optical positions of known classical T Tauri stars (CTTS). Investigations of the other bright x-ray sources led Walter (1987) to propose a new category of pre-main smaller sequence stars which he called Naked T Tauri stars (NTTS) on the basis that their ages (log t = 6 – 7) and kinematical properties are the same as the CTTS, but they lack the enhanced infrared continuum and strong Hα emission that are formed in disks that surround the CTTS. The NTTS have apparently lost their disks and are thus "naked." In many cases this loss of a disk may be a simple consequence of these stars wandering out of their nascent molecular cloud. Some 200 NTTS have now been identified in or near the Tau-Ori (Feigelson *et al.* 1987), Orion (Ku *et al.* 1982), Sco-Oph (Walter 1987), and CrA star-forming regions. NTTS are generally low mass stars with spectral types Late-F to mid-M, and are probably the first new class of stars identified on the basis of *Einstein* observations.

Comparison of pre-main sequence stars (age in years, log t = 6 – 7) with the young main sequence stars in the Pleiades (log t = 7.8), Ursa Major (log t = 8.5), and Hyades (log t = 8.8) clusters reveals some interesting correlations. One is that for each spectral type the mean value of $L_x$ decreases with age but not according to the simple $t^{-1/2}$ law originally proposed by Skumanich (1972). Instead, the decrease in $L_x$ is very gradual until about log t = 9 and then much more rapid than the Skumanich relation (Caillault and Helfand 1985). The estimated x-ray luminosity of the Sun uniformly covered with active regions is log $L_x$ = 29.3, about 5 times smaller than the brightest Pleiades G stars. This comparison led Caillault and Helfand (1985) to propose that young stars have different coronal structures than the Sun perhaps with larger loops, but there is an alternative explanation described below.

Several groups have searched for variations of x-ray flux with rotational phase in young stars without success, leading to the hypothesis that their coronae are nearly completely covered with bright active regions. Also no obvious dependence of $L_x$ on rotational parameters has been identified for young stars, contrary to what is seen when studying stars with restricted mass and age ranges (*e.g.* Fleming *et al.* 1988). No explanation has surfaced for this anomalous behavior, but the pronounced time variability observed for the pre-main sequence stars may mask such a dependence and the magnetic fields in these stars may be partly remnant rather than dynamo-generated. The presence of strong nonthermal radio emission from a number of pre-main sequence stars indicates the presence of magnetic fields in their coronae.

Vilhu (1987) has introduced the concept of saturation that is helpful in understanding coronal emission on a macroscopic scale. He showed that the ratio of the x-ray flux to the bolometric flux ($F_x/F_{bol} = L_x/L_{bol}$) has a clearly defined upper limit (called the saturation level) as a function of B–V for all single and binary stars except the CTTS (see Figure 3). This upper limit is determined by stars that are rapid rotators either because of their youth or because they are members of tidally-synchronous binary systems such as the RS CVn systems. The total of the saturation radiative flux in x-rays and in spectral features formed in the chromosphere and transition region is only a factor of 10 below the estimated mechanical flux available in the subphotospheric convective zone. Thus saturation may indicate the maximum efficiency with which a stellar atmosphere can convert mechanical energy into heat in its outer atmospheric layers. Vilhu argues that since the heating occurs where the magnetic fields are strongest, saturation likely occurs when a star is completely covered with the strongest magnetic fields, a condition given by the equality of magnetic and gas pressure in the photosphere (equipartition). solar active regions lie well below saturation and the fraction of their area covered with equipartition fields is also well below unity. An interesting point made in Figure 3 is that the Sun would lie at the saturation level if it were covered completely with active regions scaled up in brightness such that equipartition fields completely cover the surface. Thus the coronae of the active Pleiades G stars may differ from the Sun only in that equipartition fields completely cover the surface. The one missing element in this picture is the CTTS which lie above the saturation levels. This could be explained if much of the heating for these stars comes from accretion of the disk material as has been proposed by several investigators.

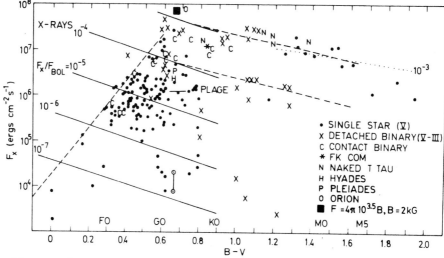

Figure 3. X-ray surface fluxes (0.2-4 keV) determined for stars are indicated by the symbols according to type of star or binary system. The filled square indicates where the Sun would lie if it were covered completely with active regions scaled up in x-ray flux such that equipartition magnetic fields completely cover the solar surface. From Vilhu (1987).

Additional supporting evidence for saturation comes from the analysis of correlations of $L_x$ with rotation. Using an optically-selected sample of 34 stars later than spectral type F7, which included stars with a wide range of coronal activity, Pallavicini *et al.* (1981) showed that $L_x$ is correlated with projected rotational velocity with the relation $L_x \sim 10^{27}(v \sin i)^2$. Other authors using different stellar samples have subsequently obtained other functional relations. These correlations are consistent with the picture of coronal heating by dynamo-amplified magnetic fields. However, using a x-ray selected sample from the *Einstein* Medium Sensitivity Survey, Fleming, Gioia, and Maccacaro (1989) recently found that $L_x$ increases with angular rotation until $L_x$ reaches a maximum value proportional to the square of the stellar radius. Thus for this stellar sample, which is strongly biased toward stars with strong coronal activity, the x-ray surface fluxes reach a limiting saturation value that does not depend on rotation.

## 6. M-TYPE STARS

I will discuss the class of main sequence stars with the lowest effective temperatures separately from the hotter stars, since the M dwarfs possess several distinctive properties. Beginning abruptly at spectral type M0 V, these stars are easily detected as quiescent radio sources, generally interpreted as gyrosynchrotron emission, and flaring radio sources, generally interpreted as emission by the cyclotron maser or plasma radiation mechanisms (Dulk 1985). Also, optical and x-ray flaring begins rather abruptly at spectral type M0 V, except for the pre-main sequence stars where flaring is observed in earlier spectral types. The rarity of optical flares in the hotter stars can be explained as a lack of contrast with respect to the bright photosphere, but the rapid onset of radio flaring at spectral type M0 V may be due to some unknown property of magnetic fields in these stars.

An important question is whether the coronae of M dwarfs change character near spectral type M5 V where stellar interior models predict that cooler stars become fully convective rather than have radiative cores and convective envelopes. This is important because "shell" type dynamos that are thought to amplify magnetic fields in solar-type stars operate in the interface region between the base of the convective zone and the radiative core. If a shift to a "distributed" dynamo operating throughout the convective zone occurs near spectral type M5 V, then there may be some observable consequence in the coronal heating rate and x-ray luminosity. Rucinski (1984) noted that $L_x$ trends downward with decreasing effective temperature from spectral types M0 V to M6 V, whereas the $L_x/L_{bol}$ ratio appears to be roughly independent of spectral type but with considerable scatter. Bookbinder *et al.* (1989) confirm Rucinski's first finding but suggest that $L_x/L_{bol}$ increases with later spectral type. The answer to the question of whether the coronae of M dwarfs change character near spectral type M5 V must therefore await observations with ROSAT and AXAF.

The *Einstein Observatory* did measure significant age effects in the coronal emission from M dwarfs. In his unbiased survey of stars within 25 pc, Johnson (1986) detected 23 M dwarfs. The most luminous of the kinematically young disk stars had $\log L_x = 29.0$ (erg s$^{-1}$), whereas the most luminous of the older disk stars was nearly a factor of 10 fainter with $\log L_x = 28.2$ (erg s$^{-1}$). Bookbinder, Golub, and Rosner (1989) show that the $L_x$ integral luminosity function for young disk M dwarfs is systematically displaced toward larger $L_x$ also by nearly a factor of 10 relative to the old disk stars. This difference is valid for both early and late-M spectral types separately. Fleming *et al.* (1988) state that $L_x/L_{bol}$ probably increases with stellar rotation rate, but their data are inadequate to determine a functional dependence.

Since the surface areas of M dwarfs are factors of 2 to 10 smaller than the Sun, the x-ray surface fluxes are between 10 and 1,000 times larger. Also, x-ray flares can be 10,000 times more luminous than large solar flares. There is no time to discuss the extensive literature on flares (*cf.* Haisch 1983; Pallavicini *et al.* 1988) and coronal heating mechanisms, but I do wish to mention two interesting clues that the *Einstein Observatory* has provided us concerning the difference in the coronae between M dwarfs and the Sun. First, M dwarf coronae are considerably hotter than the nonflaring solar corona. The SSS data for Wolf 630 AB (Swank and Johnson 1982) require a two-temperature fit with roughly equal emission measure at log t = 6.8 and roughly 7.6. Pallavicini *et al.* (1988) have also fit two-temperature thermal spectra to EXOSAT observations of 6 M dwarfs with log t typically 6.6 and 7.2. Second, Ambruster, Sciortino, and Golub (1987) conclude that the x-ray emission from the mostly M dwarf stars in their sample are variable on time scales of minutes to hours with an amplitude of roughly 30% independent of spectral type. Since most of this variability is not in the form of large flares, they suggest, as have others, that a microflaring heating mechanism may be important for these stars. Pallavicini *et al.* (1988) has reviewed the EXOSAT data relevant to this question and concludes that these data do not support the hypothesis that the quiescent emission from dMe or other stars originates with low-amplitude flaring activity. I believe that the role of flaring as a coronal heating process is another question that requires a more sensitive x-ray observatory like AXAF.

## 7. ACTIVE POST-MAIN SEQUENCE BINARIES

In this survey of coronal x-ray emission from main sequence and pre-main sequence stars, I have identified a pattern in which both hot thermal and nonthermal electrons exist in the outer atmospheres of most types of stars, with magnetic fields likely playing major roles in both heating the thermal plasma and in accelerating the nonthermal plasma. Does this pattern persist as stars evolve off the main sequence? I first consider those close binary systems which include a late-type evolved star.

Binary systems are divided into different classes according to the spectral types of the component stars, orbital periods, spectral feature discriminants, and whether the Roche lobes are filled by both stars (contact systems), one star (semi-detached systems), or neither star (detached systems). *Einstein* observed a number of RS CVn (usually detached) systems, W UMa (generally contact) systems, and Algol (usually semi-detached) systems, but I will concentrate here on the RS CVn-type systems as *Einstein* has contributed most to our understanding of their coronae. The RS CVn systems have periods of 1-14 days, and usually consist of a K0 IV primary and a late G dwarf secondary. Hall (1976, 1981) has reviewed the properties of these systems as well as the related longer period systems with giant star components and the contact W Uma systems, and Popper and Ulrich (1977) have discussed their evolutionary status. The most striking peculiarity of the RS CVn systems is a migrating quasi-sinusoidal distortion in their optical light curves (Hall 1981) that is generally explained by an uneven distribution of dark, cool photospheric spots (*cf.* Eaton and Hall 1979). Rodono *et al.* (1987) showed that a bright active region seen in ultraviolet emission lines overlies a dark starspot seen in optical photometry for the well-studied example II Peg. Linsky (1984) has reviewed the ultraviolet and x-ray observations of these systems.

Even before the launch of the *Einstein Observatory*, Walter *et al.* (1980) were able to detect with HEAO-1 some 15 RS CVn systems, because of their high x-ray luminosity and high space density. Walter and Bowyer (1981) then used the *Einstein* IPC to detect 47 systems with log $L_x$ = 29.4 − 31.5 (erg s$^{-1}$). Thus the RS CVn systems are as x-ray luminous as the pre-main sequence stars; only the O stars and Wolf-Rayet stars are more luminous nondegenerate stellar x-ray sources. Walter and Bowyer (1981) also determined that both the RS CVn systems and the single G-type stars follow the same rotation law $L_x/L_{bol} \sim 1/P \sim \Omega$, indicating that rotation rather than binarity *per se* controls the coronal heating rates. Of course, the rapid rotation rates for RS CVn systems with orbital periods less than about 20 days are tidally-induced and thus a consequence of binarity. Nevertheless, the identification of rotation rather than binarity or flows between the stars as the most important stellar parameter characterizing the x-ray emission is critical, because it provides evidence for the presence of large magnetic fluxes generated in the deep convective zones of these rapidly-rotating stars by a dynamo process. Such fields have not been measured directly but are inferred from the presence of large dark starspots (*e.g.* Vogt and Penrod 1983; Rodono *et al.* 1986) and nonthermal radio emission (Morris and Mutel 1988; Drake *et al.* 1989), which is usually explained as gyrosynchrotron emission from mildly-relativistic electrons. I will return to this point later.

These systems are sufficiently bright that Swank *et al.* (1981) were able to obtain SSS spectra of seven RS CVns and Algol itself. The spectra cannot be fit well by an isothermal plasma, but each can be fit by a two-temperature plasma with a warm component in the range log t = 6.6 − 6.9 and a hot component in the range log t = 7.3 − 8.0, although the SSS was not able to determine the temperature of the hot components accurately. The luminosity ratio of the warm to hot components range from 10 for Capella to about unity for UX Ari. Also, the hot component luminosity is time variable, whereas the warm component luminosity is nearly constant. Swank (1985) argued that the bimodal distribution of temperatures is not an artifact of the limited spectral resolution of the SSS, because fits to the spectra with emission measure distributions including appreciable plasma at intermediate temperatures are poorer than the fits with bimodal distributions. This conclusion is confirmed by Lemen *et al.* (1989) in their analysis of EXOSAT TGS spectra of Capella, $\sigma$ CrB and Procyon, but AXAF spectra are needed to infer accurate stellar emission measure distributions.

What is the meaning of this unanticipated bimodal temperature distribution, and how does it fit into a comprehensive picture of these systems? *Einstein* Observatory observations of the eclipsing system AR Lac (G2 IV + K0 IV) provide some valuable clues. Walter *et al.* (1983) observed AR Lac with the IPC and MPC as a function of orbital phase during both primary eclipse (K0 IV star in front) and secondary eclipse (G2 IV star in front). Their analysis of the x-ray light curve indicated three major components to the corona – geometrically-thin (scale height $\sim 0.02$ $R_*$) components surrounding each star, and a geometrically-extended component (scale height about $1R_*$) above the K0 IV star. All of these components are asymmetric in longitude and lie close to the equator. They proposed that the extended component contains the hot plasma indicated by both the IPC and SSS data, and that the hot and warm plasma therefore are located in different size coronal loop structures. Using the Rosner *et al.* (1978) coronal loop scaling law and estimates of the emitting volume, they derived the pressures and number of loop structures in each component. The pressures in the compact loops are similar to solar flares and the number of extended loops is of order 10, which is consistent with the large time variability detected by Swank *et al.* (1981) in the hot component of the x-ray emission from RS CVn systems.

Subsequent EXOSAT observations of AR Lac have provided a beautiful confirmation of this picture. White *et al.* (1989) observed AR Lac continuously for a complete orbital period (2.0 days). They found a dip in the LE (0.05-2.0 keV) flux during secondary eclipse but no corresponding dip in the ME flux (1.0-6.0 keV), indicating that the warm and hot components of the coronal plasma are geometrically separate. Also, the absence of any dip in the ME flux at the time of either eclipse is strong evidence that the hot component is geometrically extended as Walter *et al.* (1983) hypothesized. This picture is consistent with the VLBI radio observations. For example, Mutel *et al.* (1985) determined that the 6 cm emission from UX Ari and other RS CVn systems consists of an extended component with the dimensions of the separation of the stars in the binary system and a compact component that is located near the surface of one star and is associated with flaring.

Drake *et al.* (1989) have tied this all together by showing that the extended low-level radio emission seen in the VLBI data can be explained as synchrotron emission by the same thermal electrons that are detected as the hot component by the SSS and seen as extended x-ray emission from AR Lac. These electrons emit in an extended region of 200 Gauss magnetic fields. Uchida and Sakurai (1983) computed the magnetic field distribution in a typical RS CVn system, showing that strong field lines can fill the space between the stars. Differential rotation and photospheric motions should stretch and twist the field lines between the stars such that heating over a large volume can occur by many possible magnetic modes.

## 8. POST-MAIN SEQUENCE SINGLE STARS

I now consider those evolved stars that do not have a nearby companion to force rapid rotation. The list of such stars detected by *Einstein* is about 14 (Ayres *et al.* 1981; Haisch and Simon 1982), consisting of one supergiant (Canopus, F0 Ib-II), three of the Hyades K0 III giants (Stern *et al.* 1981), and the rest mostly giants with spectral types G5 III to K2 III. Maggio *et al.* (1989) have extended the list of detections to 18 single stars and 69 binaries and have also measured upper limits for 293 additional stars. EXOSAT has added a few stars to the list of detected sources (Gondoin, Mangenay, and Praderie 1987) but with no surprises.

Ayres *et al.* (1981) called attention to the absence of x-ray emission from any evolved single star cooler than a roughly vertical line in the H-R diagram passing through spectral type K2 III. This coronal "dividing line" lies at the same location in the H-R diagram as the transition region dividing line, discovered by Linsky and Haisch (1979), that separates the F0-K2 giants, which typically show evidence for $10^5$ K plasma, from the giants cooler than K2 III, which typically show no evidence for $10^5$ K plasma as indicated by the absence of CIV emission in IUE spectra. The so-called hybrid chromosphere stars, generally G Ib or K II stars which show evidence for both cool winds and CIV emission, do not have detected x-ray emission. The only exception is $\alpha$ TrA (K2 IIb-IIIa), which has detected x-rays (Brown 1986) and lies to the right of the coronal dividing line, but has a companion star that could be the x-ray source. Considering both detections and upper limits from the full *Einstein* data set, Maggio *et al.* (1989) established the drop in x-ray detections for stars to the right of the x-ray dividing line on a firm statistical basis.

How does this information fit in with the previously described picture of both hot thermal plasma and nonthermal electrons in the magnetic envelopes of the stars discussed so far? Haisch (1987) has summarized the recent observations of x-rays from evolved single stars and has compared the coronal and transition region dividing lines with boundaries for the rapid decrease in rotational velocity and the disappearance of HeI 10830Å absorption or emission. The rapid decrease in rotational velocity for stars cooler than spectral types G0 IV (Gray and Nagar 1985) and G5 III (Gray 1982) provides a natural explanation for the absence of x-ray emission from much cooler stars as the magnetic fields should be weak. More luminous stars do not show a rapid decrease in rotational velocity at any spectral type (Gray and Toner 1986), however, which cannot explain the absence of x-ray emission from these stars.

I believe that the measurement of x-rays in the warmer evolved stars but not in the cooler stars can be understood as a confluence of several effects. First, magnetic fields have not been measured in any of these stars and should be very difficult to measure, because photospheric gas pressures decrease systematically with decreasing stellar gravity and increasing luminosity. Thus the balance of magnetic and gas pressure in stellar photospheres observed by Saar (1987b) predicts photospheric fields of only 100 Gauss for K giants and 10 Gauss for M supergiants but much smaller fields higher in their atmospheres. The absence of nonthermal microwave emission from any evolved single late-type star is consistent with the predicted weak fields. The reported detection of radio flare emission in certain cool luminous stars to the right of the coronal dividing line by Slee *et al.* (1989), however, is inconsistent with this picture and should be checked by observations with radio arrays.

A second consideration is that the critical point for thermally-driven winds in the absence of magnetic fields decreases with decreasing stellar gravity. Thus one predicts lower coronal temperatures toward the upper right portion of the H-R diagram. In particular, $\beta$ Dra (G2 Ib-II), which is an x-ray source and lies to the left of the coronal dividing line, should have a critical temperature of about $1 \times 10^6$ K, whereas Arcturus (K2 III) and Betelgeuse (M2 Iab), which do not have detected x-ray emission and lie to the right of the coronal dividing line, should have critical temperatures of a few times $10^5$ K. This argument is very simplistic, however, as shock waves may be responsible for heating these atmospheres. In this case the temperature structure may be transient and far different from that predicted by steady-state solutions to the hydrodynamic equations (*cf.* Cuntz 1987). Thus portions of the atmospheres of cool luminous stars may be heated occasionally to $10^6$ K, in which case AXAF may detect transient x-ray emission. At present, the *Einstein* upper limits on $L_x/L_{bol}$ for Arcturus, Betelgeuse, and Antares (M1 Ib) are 30 times smaller than for solar coronal holes, regions on the Sun where the x-ray flux is smallest.

A final consideration is the presence of HeI 10830Å absorption or emission in stars well to the right of the coronal dividing line. This is of concern because the population of the lower state of the HeI transition in the Sun and late-type stars is usually interpreted as due to recombination following photoionization of HeI by x-rays shining down from a hot corona (Zirin 1975). We do not know whether the 10830Å line in the cool luminous stars indicates photoionization by very soft photons from a cool corona or by some other mechanism. In any case, whether the absence of x-rays from the cool luminous stars confirms or refutes the general picture presented here is one of many difficult problems that *Einstein* has bequeathed to AXAF.

## 9. AXAF AND THE FUTURE

Within 10 years, if all goes well, we anticipate beginning a new era in x-ray astronomy with AXAF. By that time the *Einstein* data set will have been mined extensively and the new perspectives on stellar astronomy inherent in that data should be published and understood. Also by that time we should have digested the ROSAT and EUVE all-sky surveys and the ROSAT pointed images, which should go roughly a factor of 5 deeper than the *Einstein* IPC and HRI images.

What then will be the stellar astronomy agenda for AXAF? I see three major areas in which AXAF will be able to break new ground:

(1) AXAF will be able to image far more deeply than either *Einstein* or ROSAT with far better angular resolution (0.5-1.0 arcsec) and much crisper images. Thus AXAF will be able to detect each of the known types of stellar x- ray sources at much larger distances where the chemical compositions and environments may be much different. As an example of the sensitivity of AXAF, I include in Table 2 the distances to which AXAF will be able to detect the known types of stellar x-ray sources. Many of these sources will be detectable over large volumes in our galaxy including the halo. The O stars should be easily detected in the Magellenic Clouds. Also, AXAF will be able to study crowded fields and resolve visual binaries. With its great sensitivity, AXAF may identify whole new classes of x-ray sources.

Table 2.
Distance to Which AXAF/ACIS Can Detect Different Types of Stars

| Star Class | Example | log $L_x$ | Distance (kpc) in $10^4 s^1$ | in $10^5 s^1$ |
|---|---|---|---|---|
| Old M dwarf | Barnard's star | 26.1 | 0.020 | 0.060 |
| Old K giant | $\alpha$ Boo | <27.3 | <0.070 | <0.220 |
| Middle aged G dwarf | Quiet Sun | 27.7 | 0.115 | 0.360 |
| dMe star | YY Gem | 29.6 | 1.0 | 3.3 |
| Dwarf star (log T = 8.8) | Hyades G dwarf | 29.6 | 1.0 | 3.3 |
| Dwarf star (log T = 7.8) | Pleiades G dwarf | 30.2 | 2.1 | 6.5 |
| Bright T Tauri star | GW Ori | 31.7 | 11. | 36. |
| Bright RS CVn system | RS CVn | 31.8 | 13. | 41. |
| Bright O supergiant | HD 93129A | 33.6 | 100. | 320. |

(2) The AXAF spectrometers will be capable of obtaining moderate resolution spectra of nearly all point sources that *Einstein* could image. Thus AXAF will revolutionize our ability to understand the physical properties and processes occurring in stellar coronae, an area where the *Einstein* Observatory had very limited capabilities. The AXAF gratings and the XRS will have resolutions in excess of several hundred over most of the AXAF energy range, which is sufficient to resolve most strong spectral lines and thus to derive the temperature and emission measure distributions of coronal plasmas (see Linsky 1987,1989).

(3) Finally, AXAF will begin the era of high resolution x-ray spectroscopy. The AXAF gratings and the XRS have resolutions of 1000-1500 over limited energy ranges. This resolution provides modest capability to study flows in stellar coronae

and to map the location of extended coronal structures in the coronae of rapidly-rotating stars (such as RS CVn systems) by the Doppler imaging technique as Walter *et al.* (1987) and Neff *et al.* (1989) have pioneered in the ultraviolet.

## ACKNOWLEDGMENTS

This work is supported by several NASA grants, including a grant to the National Institute of Standards and Technology to cover my work as an Interdisciplinary Scientist on AXAF. I would like to thank Jay Bookbinder and Alec Brown for their comments, and Steve Holt and Rich Mushotzky for their hospitality at Goddard where much of this paper was finally written. Finally, thanks are due to Leslie Haas and Philip Judge for struggling to create an attractive $T_EX$ manuscript.

## REFERENCES

Abbott, D.C., Bieging, J.H., and Churchwell, E. 1984, *Ap. J.*, **280**, 671.

Abbott, D.C., Bieging, J.H., and Churchwell, E. 1985, in *The Origin of Nonradiative Heating/Momentum in Hot Stars*, NASA CP 2358, p. 47.

Ambruster, C.W., Sciortino, S. and Golub, L. 1987, *Ap. J. (Suppl.)*, **65**, 273.

Ayres, T.R., Linsky, J.L., Vaiana, G.S., Golub, L. and Rosner, R. 1981, *Ap. J.*, **250**, 293.

Babcock, H.W. 1947, *Ap. J.*, **105**, 105.

Bastian, T. and Bookbinder, J. 1988, *Nature* **326**, 678.

Bookbinder, J., Golub, L. and Rosner, R. 1989, *Ap. J.*, . (submitted).

Borra, E.F., Landstreet, J.D., and Thompson, I. 1983, *Ap. J. (Suppl.)*, **52**,151.

Brown, A. 1986, *Space Science Reviews*, **6**, No. 8, p. 195.

Caillault, J. P. and Helfand, D. J. 1985, *Ap. J.*, **289**, 279.

Caillault, J. P., Helfand, D. J., Nousek, J.A. and Takalo, L.D. 1986, *Ap. J.*, **304**, 318.

Cash, W. Jr and Snow, T.P. Jr. 1982, *Ap. J.*, bf 263, L59.

Cassinelli, J.P. 1985, in *Origin of Nonradiative Heating/Momentum in Hot Stars*, ed. A.B. Underhill and A. Michalitsianos, NASA CP 2358, p. 2.

Cassinelli, J. P., Waldron, W.L., Sanders, W. T., Harnden, R.F. Jr, Rosner, R. and Vaiana, G.S. 1981, *Ap. J.*, **250**, 677.

Chelebowski, T., Harnden, F.R. Jr., and Sciortino, S. 1989, *Ap. J.*, , submitted.

Collura, A., Sciortino, S., Serio, S., Vaiana, G.S., Harnden, F.R. Jr., and Rosner, R. 1989, *Ap. J.*, **338**, 296.

Cuntz, M. 1987, *Astr. Ap.*, **188**, L5.

Drake, S.A., Abbott, D.C., Bastian, T.S., Bieging, J.H., Churchwell, E., Dulk, G., and Linsky, J.L. 1987a, *Ap. J.*, **322**, 902.

Drake, S.A., Linsky, J.L., and Elitzur, M. 1987b, *A. J.*, **94**, 1280.

Drake, S.A., Simon, T., and Linsky, J.L. 1989, *Ap. J. (Suppl.)*, (in press).

Dulk, G.A. 1985, *Ann. Rev. Astron. Astrophys.* **23**, 103.

Eaton, J.A. and Hall, D.S. 1979, *Ap. J.*, **227**, 907.

Feigelson, E.D., Jackson, J.M., Mathieu, R.D., Myers, P.C. and Walter, F.M. 1987, *A. J.*, **94**, 1251.

Fleming, T.A., Gioia, I.M., and Maccacaro, T. 1989, *Ap. J.*, **340**, 1011.

Fleming, T.A., Liebert, J., Gioia, I.M., Maccacaro, T. 1988, *Ap. J.*, **331**, 958.

Gibson, D.M. 1985, in *The Origin of Nonradiative Heating/Momentum in Hot Stars*, NASA CP 2358, p. 70.

Golub, L., Harnden, F.R. Jr, Maxson, C.W., Rosner, R., Vaiana, G.S., Cash, W. Jr and Snow, T.P. Jr 1983, *Ap. J.*, **271**, 264 [Erratum 278, 456].

Gondoin, P., Mangenay, A., and Praderie, F. 1987, *Astr. Ap.*, **174**, 187.

Gray, D.F. 1982, *Ap. J.*, **262**, 682.

Gray, D.F. and Nagar, P. 1985, *Ap. J.*, **298**, 756.

Gray, D.F. and Toner, C.G. 1986, *Ap. J.*, 310, 277.

Haisch, B.M. 1983, in *Activity in Red-Dwarf Stars*, ed. P.B. Byrne and M. Rodono (Dordrecht: Reidel), p. 255.

Haisch, B.M. 1987, in J.L. Linsky and R.E. Stencel (eds.), in *Cool Stars, Stellar Systems, and the Sun*, Springer-Verlag Publ. Co., Berlin, Germany, p. 269.

Haisch, B.M. and Simon, T. 1982, *Ap. J.*, **263**, 252.

Hall, D.S. 1976, in *Multiple Periodic Variable Stars*, ed. W.S. Fitch (Dordrecht: Reidel), p. 287.

Hall, D.S. 1981, in *Solar Phenomena in Stars and Stellar Systems*, ed. R.M. Bonnet and A.K. Dupree (Dordrecht: Reidel), p. 431.

Hearn, A.G. 1975, *Astr. Ap.*, **40**, 355.

Helfand, D.J. and Caillault, J.P. 1982, *Ap. J.*, **253**, 760.

Johnson, H.M. 1986, *Ap. J.*, **303**, 470.

Ku, W.H.M., Righini-Cohen, G. and Simon, M. 1982, *Science* **215**, 61.

Lemen, J.R., Mewe, R., and Schrijver, C.J. 1989, *Ap. J.*, (in press).

Linsky, J. L., 1981a, in *X-ray Astronomy with the Einstein Satellite*, ed. R. Giacconi (Dordrecht: Reidel), p. 19.

Linsky, J. L. 1981b, in *X-ray Astronomy in the 1980s*, NASA TM 83848, p. 17.

Linsky, J.L. 1984, in *Cool Stars, Stellar Systems, and the Sun*, ed. S.L. Baliunas and L. Hartmann (Berlin: Springer-Verlag), p. 244.

Linsky, J.L. 1985, *Sol. Phys* **100**, 333.

Linsky, J.L. 1987 *Astrophysics Letters* **26**, 21.

Linsky, J.L. 1989, Proceedings of IAU Colloquium 115, *High Resolution X-ray Spectroscopy of Cosmic Plasmas*, to appear.

Linsky, J.L. and Haisch, B.M. 1979, *Ap. J.*, **229**, L27.

Lucy, L.B. and White, R.L. 1980, *Ap. J.*, **241**, 300.

Maggio, A. *et al.* 1987, *Ap. J.*, **315**, 687.

Maggio, A. *et al.* 1989, *Ap. J.*, , submitted.

Majer, P., Schmitt, J.H.M.M., Golub, L., Harnden, F.R. Jr, and Rosner, R. 1986, *Ap. J.*, **300**, 360.

Mewe, R. W. 1979, *Space Sci. Rev.*, **24**, 101.

Micela, G. *et al.* 1988, *Ap. J.*, **325**, 798.

Montmerle, T., Koch-Miramond, L., Falgarone, E. and Grindlay, J. 1983, *Ap. J.*, **269**, 182.

Morris, D.H. and Mutel, R.L. 1988, *A. J.*, **95**, 204.

Mutel, R.L., Lestrade, J.F., Preston, R.A., and Phillips, R.B. 1985, *Ap. J.*, **289**, 262.

Neff, J. *et al.* 1989, *Astr. Ap.*, (in press).

Palavicini, R. 1988, in Proceedings of the Ninth Sacramento Peak Summer Workshop on *Solar and Stellar Coronal Structures and Dynamics*, to appear.

Pallavicini, R. *et al.* 1981, *Ap. J.*, **248**, 279.

Pallavicini, R. *et al.* 1988, *Astr. Ap.*, **191**, 109.

Poletto, G., Vaiana, G.S., Zombeck, M.V., Krieger, A.S., and Timothy, A.F. 1975, *Sol. Phys.* **44**, 83.

Popper, D.M. and Ulrich, R.K. 1977, *Ap. J.*, **212**, L31.

Renzini, A., Cacciari, C., Ulmschneider, P., and Schmitz, F. 1977, *Astr. Ap.*, **61**, 39.

Robinson, R. D. 1980, *Ap. J.*, **239**, 961.

Rodono, M. *et al.* 1986, *Astr. Ap.*, **165**, 135.

Rodono, M. *et al.* 1987, *Astr. Ap.*, **176**, 267.

Rosner, R., Golub, L., and Vaiana, G.S. 1985, *Ann. Rev. Astron. Ap.* **23**, 413.

Rosner, R., Tucker, W.H., and Vaiana, G.S. 1978, *Ap. J.*, **220**, 643.

Rucinski, S.M. 1984, *Astr. Ap.*, **132**, L9.

Saar, S.H.: 1987a, Ph. D. thesis, University of Colorado.

Saar, S.H.: 1987b, in J.L. Linsky and R.E. Stencel (eds.), *Cool Stars, Stellar Systems, and the Sun*, Springer-Verlag Publ. Co., Berlin, Germany, p. 10.

Saar, S.H. and Schrijver, C.J. 1987, in *Cool Stars, Stellar Systems and the Sun*, ed. J.L. Linsky and R.E. Stencel (Dordrecht: Reidel), p. 38.

Schmitt, J.H.M.M., Golub, L., Harnden, F.R. Jr, Maxson, C.W. and Vaiana, G.S. 1985, *Ap. J.*, **290**, 307.

Simon, T. and Landsman, W. 1987, in *Cool Stars, Stellar Systems and the Sun*, ed. J.L. Linsky and R.E. Stencel (Springer-Verlag: Berlin), p. 265.

Skumanich, A. 1972, *Ap. J.*, **171**, 565.

Slee, O.B. *et al.* 1989, preprint.

Smith, M.A., Pravdo, S.H. and Ku, W.H.M. 1983, *Ap. J.*, **272**, 163.

Stern, R. A. *et al.* 1981, *Ap. J.*, **249**, 647.

Swank, J.H. 1985, in *Origin of Nonradiative Heating/Momentum in Hot Stars*, ed. A.B. Underhill and A. Michalitsianos, NASA CP 2358, p. 86.

Swank, J.H. and Johnson, H.M. 1982, *Ap. J.*, **259**, L67.

Swank, J. H., White, N. E., Holt, S. S. and Becker, R. H. 1981, *Ap. J.*, **246**, 208.

Topka, K. *et al.* 1982, *Ap. J.*, **259**, 677.

Uchida, Y. and Sakurai, T. 1983, in *Activity in Red-Dwarf Stars*, ed. P.B. Byrne and M. Rodono (Dordrecht: Reidel), p. 629.

Vaiana, G. S. 1981, in *X-ray Asatronomy with the Einstein Satellite*, ed. R. Giacconi (Dordrecht: Reidel), p. 1.

Vaiana, G. S. *et al.* 1981, *Ap. J.*, **245**, 163.

Vilhu, O. 1987, in *Cool Stars, Stellar Systems and the Sun*, ed. J.L. Linsky and R.E. Stencel (Dordrecht: Reidel), p. 110.

Vogt, S.S. and Penrod, G.D. 1983, *Publ. Astr. Soc. Pacific* **95**, 565.

Waldron, W.L. 1984, *Ap. J.*, **282**, 256.

Walter, F.M. 1981, *Ap. J.*, **245**, 677.

Walter, F.M. 1983, *Ap. J.*, **274**, 794.

Walter, F.M. 1987, in *Cool Stars, Stellar Systems and the Sun*, ed. J.L. Linsky and R.E. Stencel (Dordrecht: Reidel), p. 422.

Walter, F.M. and Bowyer, S. 1981, *Ap. J.*, **245**, 671.

Walter, F.M., Brown, A., Mathieu, R.D., Myers, P.C. and Vrba, F.J. 1988, *A. J.*, **96**, 297.

Walter, F.M., Cash, W., Charles, P.A., and Bowyer, C.S. 1980, *Ap. J.*, **236**, 212.

Walter, F.M., Gibson, D.M., and Basri, G.S. 1983, *Ap. J.*, **267**, 665.

Walter, F.M. and Linsky, J.L. 1986, in *New Insights in Astrophysics: Eight Years of UV Astronomy with IUE*, ESA SP-263, p. 103.

Walter, F.M., Linsky, J.L., Simon, T., Golub, L. and Vaiana, G.S. 1984, *Ap. J.*, **281**, 815.

Walter, F.M. and Schrijver, C. 1987, in *Cool Stars, Stellar Systems and the Sun*, ed. J.L. Linsky and R.E. Stencel (Springer-Verlag: Berlin), p. 262.

Walter, F.M. *et al.* 1987, *Astr. Ap.*, **186**, 241.

White, N.E. and Marshall, F.E. 1983, *Ap. J.*, **268**, L117.

White, N.E. *et al.* 1989, *Ap. J.*, (in press).

White, R.L. 1985, *Ap. J.*, **289**, 698.

White, R.L. and Long, K.S. 1986, *Ap. J.*, **310**, 832.

Wolff, S.C., Boesgaard, A.M. and Simon, T. 1986, *Ap. J.*, **310**, 360.

Zirin, H. 1975, *Ap. J.*, **199**, L63.

# X-RAY EMISSION FROM STARS: A SHARPER AND DEEPER VIEW OF OUR GALAXY

**G.S. Vaiana**

Osservatorio Astronomico di Palermo, Italy

Harvard-Smithsonian Center for Astrophysics, Cambridge, USA

## 1. INTRODUCTION: FROM EARLY DAYS TO THE PRESENT

The *Einstein Observatory* (Giacconi *et al.* 1979) was the first imaging telescope capable of systematically studying the x-ray properties of normal stars. Prior to *Einstein*, the domain of stellar x-ray astronomy was limited to a few rather special types of stars: Luminous compact binaries, dwarf novae, cataclysmic variables, and RS CVn's; but the thousand-fold increase in sensitivity of this observatory over previous observations was expected to make accessible ordinary stellar x-ray sources of solar luminosity out to distances of 10 pc.

This expectation was more than fulfilled. From the very first *Einstein* images, several surprises emerged, and in consequence, the early results reported from the observatory were largely of a discovery nature, as x-ray emission was observed from virtually every type of "normal" star in the HR diagram. However, subsequent effort by both a large number of *Einstein* guest investigators and the stellar survey groups at the four PI institutions focused on systematic determinations of the x-ray properties of stars, based on the then-current data reduction software.

Their results showed: that x-ray emission is characteristic of most stars, with the possible exception of late-B and A dwarfs, and late spectral type giants and supergiants; that for late-type stars, the level of x-ray emission correlates with rotation rate, and spans a range of roughly 3 orders of magnitude; that for early-type stars the level of x-ray emission correlates with the underlying star's bolometric luminosity, and spans a range of roughly 4 orders of magnitude; and that the temperature of the emitting matter in the case of late-type stars lies dominantly in the 0.2 - 1.0 keV range.

I would like to focus here on an aspect of the x-ray stellar results which will became more completely addressed as we enter the second decade of the *Einstein* data reduction, as new observations finally become available, and as new satellites are being planned for the future: The x-ray stars as a subclass of all galactic and extragalactic x-ray sources. Thus, one ultimately would like to answer technical questions such as: Can ordinary stars be distinguished in any way from the remaining sources, based solely on a limited set of optical and x-ray characteristics (say, only the $f_x/f_v$ and hardness ratios), or, in the extreme case, based solely on x-ray emission properties alone? On a less technical, and more physical level, one wishes to know what is the stellar contribution to the x-ray background; the make-up of the galaxy's x-ray emission; the stellar contribution to the x-ray emission of extra-galactic objects – such questions are all part of a new focus in the analysis which is made possible by access to the total data set, and by a more clear understanding of x-ray emission from the various categories of stars.

Our present effort is a step along this route, and is aimed at establishing a reference stellar x-ray list. This survey work is expected to be completed next year, and has already required several years of effort by a collaborative team of researchers, who are primarily at the CfA and the Osservatorio Astronomico di Palermo, but have also included researchers at the MPE, the Univ. of Chicago, and other European and U.S. institutions. In the process of assembling, testing, and providing the ground rules required, we are learning a great deal about the totality of the data set and the stellar data in particular.

## 2. THE X-RAY SELECTED SOURCE DATA BASE

Our group has been working for several years on studying stellar x-ray sources observed with the *Einstein Observatory*; these observations constitute the largest stellar x-ray data base to date (*cf.* Table 1); and we are presently preparing a stellar catalog (Vaiana *et al.* 1989) based on a uniform analysis and validation process of all Imaging Proportional Counter (IPC; Gorenstein, Harnden and Fabricant 1981) data obtained during *Einstein*'s operational life. This catalog has been implemented on a relational data base, using the $Ingres^{TM}$ data base language, so that statistical analyses of the properties of the full source catalog are relatively easily and flexibly accomplished (Sciortino *et al.* 1988; Harnden *et al.* 1989a,b).

| Table 1 | |
|---|---|
| Identified Stellar Detection | ∼ 2000 |
| Expected Stellar from Unidentified Sources | ∼ 1000-2000 |
| Upper Bounds from several Optical Catalogs | ∼ 30,000 |

During the operational lifetime of *Einstein*, a total of approximately 4,000 IPC images were collected, with a coverage of approximately 10% of the entire sky. The sky coverage was inhomogeneous since both the Galactic plane region was more extensively surveyed than the other regions (*cf.* Figure 1), and because the obtained images span a wide range of exposure times (*cf.* Figure 2). However, approximately half of the available images have exposure time of 2-3 ksec, with limiting sensitivities ranging between $10^{-13} - 10^{-12} erg\ cm^{-2}\ s^{-1}$ (with a conversion factor of $2 \times 10^{-11} erg\ cm^{-2}\ ct^{-1}$) in the 0.16 - 3.5 keV energy band (*cf.* Figure 3). These characteristics are determined by a variety of factors, which include the exposure time, the intensity of the background, telescope vignetting and local properties of detector and gain.

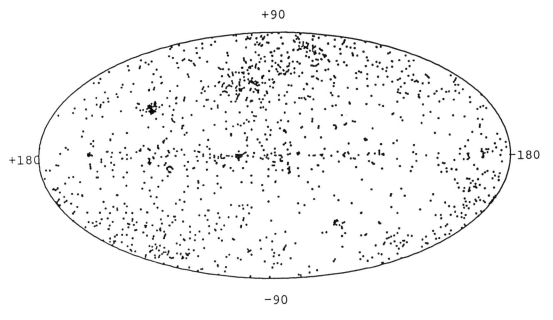

Figure 1. Sky coverage and distribution of the ∼ 4,000 *Einstein* IPC exposures. They cover approximately 10% of the sky. Note the non-uniformity of sky coverage and, in particular, the oversampling of the Galactic plane region.

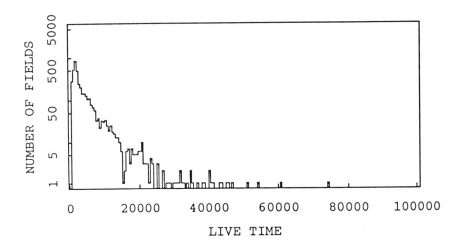

Figure 2. Live time coverage ranges from few hundreds seconds to the longest deep survey exposures. Approximately 50% of the images have exposure times of 2-3 ksec.

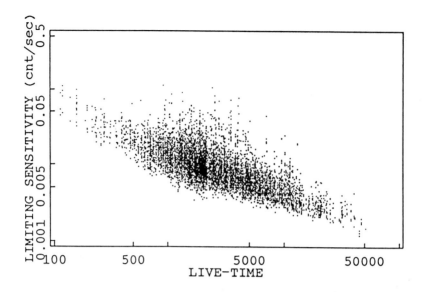

Figure 3. Scatter plot of individual local $3\sigma$ sensitivity evaluated at the position of undetected SAO star positions to determine upper limits to the surveyed more than 10 thousand SAO stars. Position partially shadowed by the detector window entrance support structure have been screened out. The combined effect of background, vignetting, and gain gives a range of thresholds for each exposure time.

In the final data processing (Rev-1; Harnden *et al.* 1984), the x-ray sources are detected in either of 3 (not entirely independent) energy bands [soft (0.16 - 0.8 keV), hard (0.8 - 3.5 keV), and broad (0.16 - 3.5 keV)], based on the use of one of two detection algorithms; in the first, source detection is based on a *Local* threshold determined by a local determination of the background,while the second is based on a *Map* threshold determined from a background reference map (properly scaled to the given image). The latter method is more sensitive than the first, especially in crowded fields such as open clusters and stellar associations, but finds more spurious sources in regions with weak diffuse emission; for this reason, it has not been applied to all available images.

As results of the refined source recognition system implemented in the final processing of the IPC data, a total of $\sim 16,000$ detections (based on the use of both detection methods, and counting all 3 energy bands) have emerged above the thresholds. On the basis of an analysis of simulated data by T. Maccacaro (private communication), it is expected that the standard threshold setting of the final processing results in an expected number of 0.3 spurious source per field for each detection method and for each band. Hence, of the $\sim 9,000$ *Map* broad detections found in the $\sim 3,200$ fields where the *Map* detection algorithm has been applied, we expect that $\sim 1,000$ are spurious, *i.e.*, are indeed local fluctuation of the image background.

Can one do better? The IPC does not discriminate between real detections and spurious local fluctuations; but one can investigate how the number of sources changes as one increases the acceptance threshold by a fixed amount (we do not intend to adopt a fixed value threshold for all exposures, since we want to retain the expected dependence of the threshold on the background counts); we have found that if we increase the S/N threshold by 0.6 (in order to obtain essentially no spurious detections), we lose $\sim 40\%$ of the detections, and that an increase in the threshold by 0.3 will include only a few spurious sources and lead to a loss of only 25% of the detections (*cf.* Figure 4a). A simple test of all this is based on the idea that a detection of the same source in several distinct images is an independent verification of its "reality"; a close look at Figures 4a and 4b shows that, as expected on statistical grounds, approximately half of the low S/N sources are indeed real.

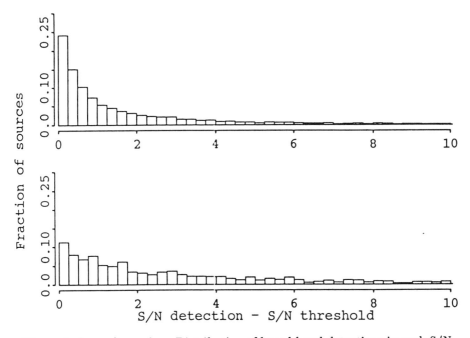

Figure 4. (upper) panel a - Distribution of broad band detections in each S/N bin above the detection threshold for the images in which they lie. (lower) panel b - As in panel a, but only for the x-ray sources detected more than once.

## 3. THE STELLAR X-RAY SOURCE IDENTIFICATIONS

The stellar x-ray catalog project requires an input list of optical stars in order to allow determination of the x-ray spectra and luminosities of selected classes of stars. In Figure 5 are shown the HR diagrams of the surveyed Woolley stars and of the surveyed Bright Star Catalog, from which have been drawn the volume-limited samples of main-sequence and giant stars; these samples have allowed us to clarify the role of various stellar parameters in determining the x-ray emission level. The principal optical catalogs on which we have based this projects are given in Table 2.

| Table 2: Optical Catalogs used for X-ray Stellar Survey | |
| --- | --- |
| Woolley Catalog | Stars with distances smaller than 25 pc |
| Bright Star Catalog (BSC) | Complete for V < 6, color indexes and distances for a large fraction of stars |
| Open Clusters Catalogs | Hyades, Pleiades, UMa, etc. |
| Kukarkin Catalog | Variable Stars |
| SAO Catalog | Complete for V < 8.5 |
| OB Star Catalog | Stars up to $m_v \sim 12$ |
| | Complete up to 2.5 kpc |

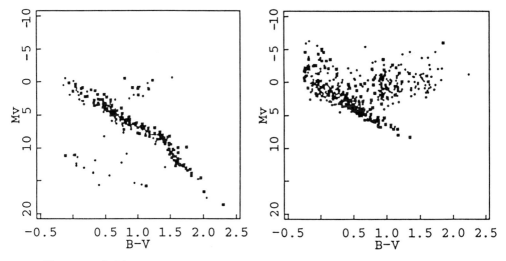

Figure 5. (left) H-R diagram of the *Einstein*-surveyed Woolley sample; bold symbols indicate x-ray detections and small symbols indicate stars for which only upper limits to x-ray luminosity have been measured. Note the great number of late dK and dM stars; (right) H-R diagram of *Einstein* surveyed BSC sample, symbols as on left panel. Note the great number of dB, dA and late-type giant stars.

## 4. SORTING THE X-RAY SOURCES BY OPTICAL CATALOG

### 4.1 The Late-type Stars

The collection of all the x-ray data for the nearby dwarfs in the color range $0.1 < B-V < 2.3$ pose a puzzle because there is no evidence for systematic patterns, but rather a large scatter of more than 3 orders of magnitude, which dominates both in $L_x$ and $L_x/L_v$ (*cf.* Figures 6, 7).

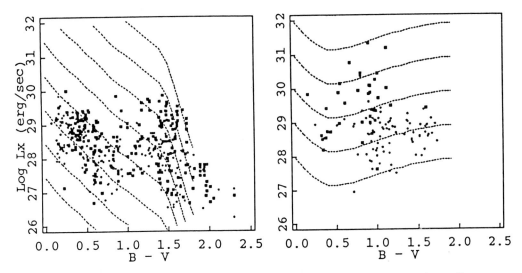

Figure 6. (left) - Scatter plot of $\log(L_x)$ *vs.* $B - V$ for all the surveyed Woolley stars (bold symbols indicate x-ray detections, and small symbols indicate stars for which only upper limits to x-ray luminosity have been measured). Dotted curves at constant $F_x/F_v$ ratios are superimposed. (right) As on the left panel for all the surveyed F-M giants and super-giants listed in the BSC.

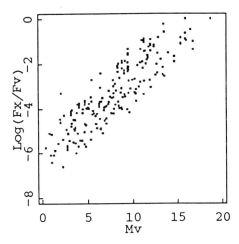

Figure 7. Plot of $\log(F_x/F_v)$ *vs.* $M_v$ for detected Woolley stars. Note that the fainter stars have $\log(F_x/F_v)$ between -1 and 0.

At this point, it is very useful to take advantage of our previous knowledge of solar x-ray emission in order to suggest what stellar parameters might account for the observed scatter of the data. That is, we expect that the relevant parameters might be related in a general sense to stellar dynamo activity. The search for possible correlations must start from an appropriate unbiased sample (from the optical point of view); that is, instead of dealing with x-ray flux-limited samples (to which we shall return in the following section), we must deal with volume-limited samples (or rather, statistical approximation to such samples). Furthermore, we need to apply appropriate statistical techniques, which take proper account of upper limits as well as detections. The techniques adopted make use of survival analysis as adapted to astronomical needs (Avni *et al.* 1980; Feigelson and Nelson 1985; Schmitt 1985; Isobe, Feigelson and Nelson 1986) to build maximum likelihood integral x-ray luminosity functions for various categories of stars, to evaluate non-parametric sample mean x-ray luminosities, and to estimate errors on means by using bootstrap techniques. A generalization of linear regression analysis to account for the presence of upper bounds has been applied to establish trends present in the data in a quantitative manner, and to compute the statistical significance of these trends. Extensive applications of these techniques can be found in the analysis of the nearby volume-limited stellar x-ray survey (Schmitt *et al.* 1985a; Bookbinder 1985; Maggio *et al.* 1987; Maggio *et al.* 1989a,b), as well as in the surveys of early-type stars (Sciortino *et al.* 1989a,b) and of young open clusters and moving groups (Micela *et al.* 1985, 1988, 1989a; Schmitt *et al.* 1989a; Feigelson and Kriss 1989).

A systematic study using these techniques confirms the importance of magnetic field-activity indicators such as stellar rotation, stellar age, and stellar Rossby number, which are presumably all proxies of solar-type magnetic activity, as will become evident in the following.

### 4.1.1  Main Sequence Stars

That x-ray luminosity levels scale with stellar rotation rate is expected on very general grounds based on stellar magnetic dynamo theory; however, there is no general consensus whether the specific quantities to use are the x-ray luminosity or the x-ray surface flux, and the rotation rate or the Rossby number. Several analysis have shown that the x-ray luminosity levels of field and cluster stars scale with the stellar rotation rate $L_x \propto v \sin i^2$ (Pallavicini *et al.* 1981; Walter 1982, 1983; Micela *et al.* 1985; Maggio *et al.* 1987). A good correlation has also been found between x-ray luminosity and the Rossby number (Micela, Sciortino, Serio 1984; Vihlu 1984; Mangenay and Praderie 1984; Schmitt *et al.* 1985a; Maggio *et al.* 1987), following the original findings of Noyes *et al.* (1984) of a correlation between the intensity of CaII H and K lines and the Rossby number.

However, we note that for the rapidly-rotating K Pleaides stars the observed x-ray luminosity is significantly lower than that expected on the basis of the above scaling law (*cf.* Figure 8). This result has been interpreted in two different ways: (i) The x-ray luminosity level is associated with evolutionary changes of the stellar rotational structure, *i.e.*, strong radial gradients are need to generate intense x-ray luminosity (Micela, Sciortino and Serio 1984; Micela *et al.* 1985) and these are not present in the still rapidly-rotating K Pleiades stars; (ii) The x-ray luminosity level cannot increase above a maximum value fixed by the efficiency of the mechanism transporting energy from the top of the convection zone to heat the corona, *i.e.*, the K Pleiades stars are as active as they can be (Stepien 1988). We will return to this subject.

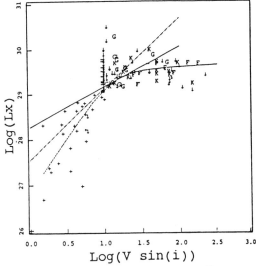

Figure 8. Scatter plot of x-ray luminosity *vs.* stellar rotation rate for the sample of nearby G and K stars (crosses), and for the sample of Pleiades stars (the letter code indicates spectral type). The solid line is the best-fit correlation to the data (including upper bounds), while the dashed-dotted line is the $L_x \propto v \sin i^2$ scaling law. The inclusion of the Pleiades stars changes the slope from 2 to 1.2. The dashed line is a robust local fit to all data. This fit clearly indicates a trend of decreasing slope for the fast rotating Pleiades K stars.

Several samples of coeval stars have been surveyed with the *Einstein Observatory* allowing the determination of the x-ray luminosity functions at several ages for given spectral type ranges (*cf.* Figure 9). These studies (Stern *et al.* 1981; Walter *et al.* 1984; Micela *et al.* 1985, 1988, 1989a; Caillault and Helfand 1985; Maggio *et al.* 1987; Feigelson and Kriss 1989) allow us to draw the conclusion that the x-ray emission in each given spectral type range decreases with increasing stellar age (*cf.* Figure 10). Micela *et al.* (1989a) have also shown that in the Pleiades G and K stars there is no evidence for x-ray emission variations similar in amplitude (up to a factor 2-3) and frequency to that observed from young stellar objects in $\rho$ Oph; such variations could have been detected if present, and therefore they concluded that already at the age of the Pleiades( log *age* $\sim$ 7.7 years) the occurrence rate of such amplitude variations is drastically reduced from that seen at an age of log *age* $\sim$ 6.5 years (Montmerle *et al.* 1983).

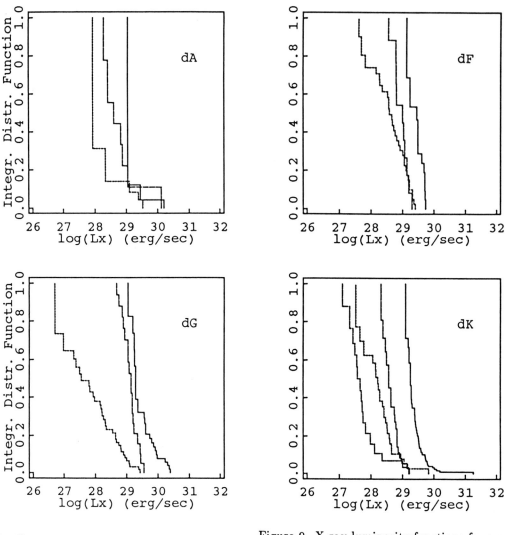

Figure 9. X-ray luminosity functions for several samples of coeval stars in 5 distinct ranges of spectral types. Solid lines indicate the Pleiades, long-dashed lines the Hyades, and short-dashed lines the field-stars; for the dK and dM stars, the short-dashed lines refer to the kinematically-young field stars, and the dotted-dashed lines to the kinematically-old field stars. A decrease of the mean of the x-ray luminosity with increasing stellar age has been shown to exist for the dG and the dK stars, while for the dM stars the available cluster data do not allow to us to probe for this phenomenon.

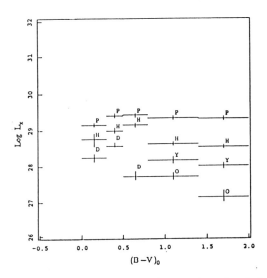

Figure 10. Mean values of $\log L_x$ versus $(B - V)_o$ color index, for different groups of stars, as labelled: P - Pleiades (Micela *et al.* 1989a), H - Hyades (Micela *et al.* 1988), D - disk field stars (dA and dF from Schmitt *et al.* 1985a, dG from Maggio *et al.* 1987), Y - young (and O - old) dK and dM disk stars (Bookbinder 1985). Horizontal bars indicates the range of color index over which the mean value was computed, and vertical bars indicate formal bootstrap errors on the means.

Walter (1983) and Schmitt *et al.* (1985a) have studied the x-ray emission from late A and early F stars; the expected onset of convection (hence of any related magnetic dynamo activity) makes this spectral type range essential to test the hypothesis that the observed x-ray emission from late-type stars is magnetic in its origin. Schmitt *et al.* (1985a) have shown that the mean x-ray luminosity of single stars in that color range increases from late dA to early dF. Similar results have been obtained for the Hyades and the Pleiades members in the same color index range (Caillault and Helfand 1985; Micela *et al.* 1988, 1989a). However, there are a few A stars which are not known members of binary systems, that show evidence for intense x-ray emission, and in some cases so intense as to not be explainable in term of x-ray emission from an unseen late-type companion. Furthermore, x-ray emission is common among Ap stars with strong magnetic fields (Cash and Snow 1982).

### 4.1.2  Giant Stars

Using the available IUE observations, Linsky and Haisch (1979) proposed the existence of a "dividing line" separating stars (with $B - V \leq 1$) with evidence for closed magnetic solar-like atmospheric structure, from stars (with $B - V \geq 1$) with evidence only of winds.

Vaiana *et al.* (1981), and Ayres *et al.* (1981) reported the detection of soft x-ray emission from late F through early K giants, but only upper bounds from cooler single giants or from supergiants, with the exception of $\alpha$ Car. Using a sample of about 30 giants and supergiants Ayres *et al.* (1981) noted that the stars to the left of the "boundary" line showed an emission level range of more than two orders of magnitude. These results were confirmed by Haisch and Simon (1982), who added observations of 8 more G-K giants.

Maggio *et al.* (1989a,b) have completed the survey of all the 380 giants and supergiants listed in the Bright Star Catalogue which were observed with the *Einstein* IPC. They have built x-ray luminosity functions (*cf.* Maggio *et al.* this volume) and $F_x/F_v$ integral distribution functions for a volume-limited sample of F, G, K giants and for the F-M supergiants, and have found a statistically significant drop in the number of x-ray-detected giants at spectral type K ($B - V \sim 1.1$). This drop seems to be associated with the ascending phase of the red giant branch (*cf.* Maggio *et al.* this volume). The detection of x-ray emission from G giants with masses $M \geq 2M_\odot$ whose likely precursors are dA stars suggests that the onset of convective envelopes may trigger an efficient dynamo mechanism during the approach to the red giant branch. They find a poor correlation between stellar rotation rate and x-ray luminosity in the available data, and argue that additional parameters, probably the radial gradient of rotational rate and the depth of the convection zone, are needed to explain stellar activity levels of giants on the active side of the HR diagram.

### 4.1.3  Coronal Temperature of Late-Type Stars

The IPC provides rather coarse x-ray spectra with $\Delta E/E \sim 1$ at 1 keV; this resolution is insufficient to resolve lines in the incident spectra and therefore to perform detailed plasma diagnostics of the kind possible for the handful of stars observed with the *Einstein* and *EXOSAT* Objective Grating Spectrographs (Mewe *et al.* 1982; Brinkman *et al.* 1985). A somewhat larger sample of stellar x-ray spectra has been taken with the *Einstein* SSS, but with a strong bias towards RS CVn's and with no information below the carbon edge (Swank *et al.* 1981; Swank 1984). Therefore the general characterization of the stellar coronal x-ray spectra can only be done with the IPC, which provides a sufficiently large data sample to carry out statistical studies of stellar x-ray spectra. Several authors have attempted to fit thermal x-ray spectra (Raymond and Smith 1977; Raymond 1988) to the IPC spectra (Vaiana *et al.* 1983; Vaiana 1983; Schrijver *et al.* 1984; Majer *et al.* 1986), concluding that the spectra show evidence for multitemperature plasmas;

Majer *et al.* (1986) suggested that these data are consistent with a continuous emission measure distribution in temperature, and so might be best modeled within the framework of loop model analysis (*cf.* Schmitt *et al.* 1985b; Stern *et al.* 1986).

Recently a complete homogeneous analysis of all stars whose IPC spectra contain more than 200 counts and which are identified with an entry in the BSC (Hoffleit and Jaskeck 1984) or in the Woolley Catalog (Woolley *et al.* 1970) has been undertaken (*cf.* Schmitt 1988; Collura *et al.* 1989a; Schmitt *et al.* 1989b). In figure 11 we show a summary of the resulting coronal temperatures derived from the successful single temperature fits. Main sequence and subgiant stars in the color range $0.2 < B - V < 0.9$ have coronal temperatures in a narrow range around $\log T \sim 6.4$; giants are usually hotter, and show temperature at $\log T \sim 7.0$; and RS CVn's have extremely high temperatures, peaking at about $\log T \sim 7.2$. A sizable fraction of stellar x-ray spectra cannot be described with single temperature fits; in such cases (which generally tend to be the higher S/N spectra), a two-temperature fits is usually a good description of observed spectra. The majority of two-components fits are required for stars later than G5 (Schmitt *et al.* 1989b). The two derived temperatures peak usually at $\log T_{low} \sim 6.4$ and $\log T_{high} \sim 7.2$, as noted by Schmitt (1984), Majer *et al.* (1986), and Schmitt *et al.* (1987).

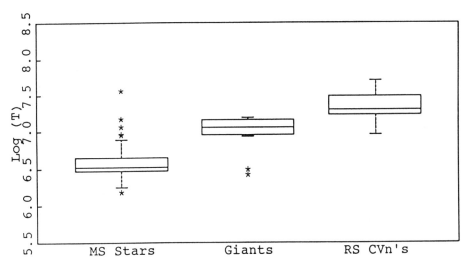

Figure 11. Box-plot of most-likely temperature derived by single-temperature fits for the sample of main sequence/subgiant stars, giant stars, and RS CVn's.

What is the physical meaning of the two-component fits? To answer this question, two points must be considered: (i) two-component fits are usually required when the data quality is better (*i.e.*, whenever the number of counts in the spectrum is high), with the exception of the high-quality spectra for giants (which do not require a two-component description, thus suggesting that their coronal emission is intrinsically different from that of main-sequence stars); (ii) the derived $T_{low}$

and $T_{high}$ can be interpreted as "effective" temperatures associated with the two well-separated transmittance windows (lying below and above the carbon edge) where the IPC is most sensitive to incoming x-rays. With these considerations in mind, the spectral data have been interpreted as evidence for a multi-temperature coronal plasma.

Schmitt *et al.* (1987), upon comparing IPC spectra and EXOSAT measurements taken with various large band-pass filters, argued that a description in terms of a differential emission measure distribution of the form of equation $Q(T) = EM_{max} (T/T_{max})^{\alpha}$ provides another form of parameterizations of the observed x-ray spectra somewhat more related to a physical description. This description has been applied to the sample of nearby and bright stars surveyed with *Einstein* (Schmitt 1988; Collura *et al.* 1989a; Schmitt *et al.* 1989b); while main sequence stars appear to cluster at values of $\alpha \sim 0.8$, with small dispersion, rather larger values of $\alpha$ are found in low gravity objects, such as giants and RS CVn's.

### 4.1.4 Non-Parametric Characterization of Spectra

Substantial efforts have been made in the last few years to use the limited spectral capability of the IPC to categorize x-ray sources (*cf.* Cordova *et al.* 1989). Our group is presently working in this area, and some interesting results have already been obtained for the stellar sources. Sciortino *et al.* (1989c) have shown on the basis of all available IPC detections of known stars that there is a trend of increasing stellar x-ray luminosity with hardening of the x-ray spectrum. To pursue their investigation, they have defined a hardness-ratio, $HR$,

$$HR = \frac{[Count(Hard) - Count(Soft)]}{[Count(Hard) + Count(Soft)]} \tag{1}$$

where $Count(Hard)$ are IPC counts in the [0.8-3.5 keV] energy band and $Count(Soft)$ are those in the [0.16-0.8 keV] energy band. To prevent spurious values of HR, they have eliminated all x-ray sources from their analysis which were partially obscured by field edges or entrance window supports.

The detected Bright Star Catalog (BSC) stars show a clear trend of increasing x-ray luminosity with the hardening of spectrum (*cf.* Figure 12); and an analogous trend is present in the Woolley sample, notwithstanding the more restricted range of $L_x$ values spanned by these nearby stars. It is noteworthy that a detailed fit procedure assuming an isothermal Raymond spectral model furnishes, for a smaller size sample, a similar result, and *does not support* an explanation of hardening of the spectrum due to interstellar absorption (Schmitt *et al.* 1989b).

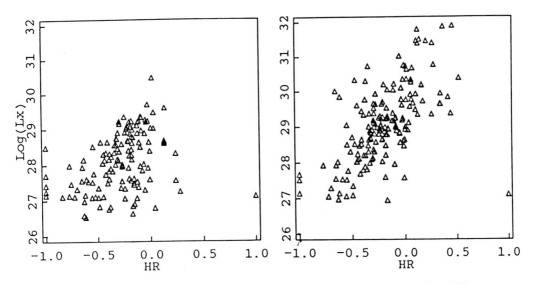

Figure 12. (left) Scatter plot of $L_x$ *vs.* Hardness ratio for the detected Woolley stars; (right) analogous plot for the detected BSC stars.

Since the sample of detected BSC stars is largely composed of giants (*cf.* Figure 5), Sciortino *et al.* (1989c) concluded that detected giant stars (*i.e.*, those with B-V $\sim 1$, *cf.* Maggio *et al.* 1989a,b) generally have a spectrum which is harder than that of main sequence stars of similar spectral type (*cf.* also Schmitt *et al.* 1989b).

To further investigate the spectral properties of selected samples of stars, we can consider the star distribution in the $\log(F_x/F_v)$, $HR$ plane (*cf.* Figure 13). The Woolley stars show two major concentrations in this diagram, at $HR \sim -0.2$, $\log(F_x/F_v) \sim -4$ and at $HR \sim -0.1$, $\log(F_x/F_v) \sim -2.5$ respectively; the second concentration is essentially entirely composed of dM stars. There are no BSC stars with $\log(F_x/F_v) \sim -3.0$ (due to the lack of faint stars), but there is a large number of stars with $HR \sim 0.3$, $\log(F_x/F_v) \sim -4.0$. We note that in this range of $F_x/F_v$ we can exclude the presence of RS-CVn's stars, as recently show by Maggio *et al.* (1989b), so that again these stars have to be largely "normal" giants.

If we increase the limiting magnitude of the stellar sample from $V \leq 6$ (BSC) to $V \leq 8.5$ (SAO), many stars with $\log(F_x/F_v)$ between -3 and -2, and $HR > 0$ will show up as a characteristic feature of the SAO star distribution (*cf.* Figure 14). These stars can be most likely explained as members of the RS CVn population, or as members of a population of young active dF-dG stars. Part of these stars can be associated to binary systems where an A stars (responsible for the optical luminosity) is associated with an K-M stars (responsible for the x-ray luminosity).

For those systems the deduced $(F_x/F_v)$ values severely underestimate the true ratios of the emitting x-ray sources.

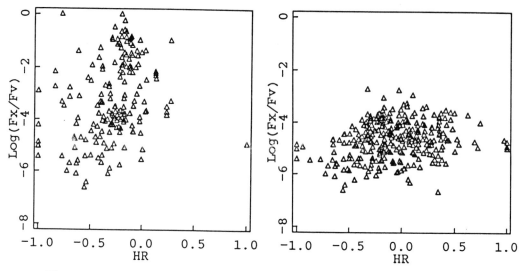

Figure 13. Scatter plot of $\log(F_x/F_v)$ *vs.* HR for detected Woolley (left) and for detected BSC (right) stars.

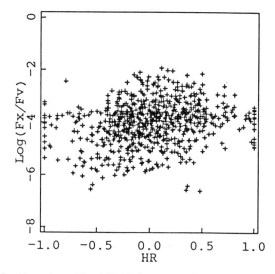

Figure 14. Scatter plot of $\log(F_x/F_v)$ *vs.* HR for detected SAO stars.

## 4.2 The Early-Type Stars

The first stellar x-ray observations performed with *Einstein* were of the Cyg OB2 association (Harnden *et al.* 1979), and it was very surprising to find that "normal" OB stars emit at x-ray luminosity levels between $10^{32} - 10^{33} erg\ s^{-1}$. A series of follow-up observations confirmed the ubiquity and the high level of x-ray emission in the entire O-B3 spectral type range (Seward *et al.* 1979; Long and White 1980; Cassinelli *et al.* 1981; Pallavicini *et al.* 1981; Seward and Chlebowski 1982) and indicated a striking correlation between x-ray and bolometric luminosity.

Several analyses have attempted to account for the physical mechanism(s) responsible for the observed x-ray emission. Since the spectral analysis of the IPC data for a few early-type stars (Cassinelli and Swank 1983) does not show evidence for the expected amount of absorption below 0.5 keV by the stellar winds known to be flowing from these stars, the slab corona plus cool wind model (which predated these observations, and at first provided the only theoretical account for explaining these data; *cf.* Hearn 1975; Cassinelli and Olson 1979; Waldron 1984) cannot be reconciled with the experimental evidence. The long-term variations in x-ray emission levels (Snow, Cash and Grady 1981; Collura *et al.* 1989b), especially in the soft region of the spectrum, indicate that likely changes in the absorption properties of an O star's wind occurs on time scale of months. The upper limits on the amplitude of short-term variations (Collura *et al.* 1989b) are compatible with the short-term x-ray emission changes predicted by the chaotic wind model of Lucy (1982). This model, with an *ad hoc* choice of shock strength distribution parameters, can explain the principal features of the observed spectra. Hence at this moment, this class of models (Lucy and White 1980; Lucy 1982; Owocki and Rybicki 1984, 1985) is generally favored; however, we note that Cassinelli (1985) has pointed out the presence of very hot plasma ($T > 1.5 \times 10^7 K$), suggesting that this plasma requires some sort of (magnetic?) confinement.

More recently, the entire set of IPC and HRI images has been reanalyzed in order to compile the complete catalog of all O stars surveyed with *Einstein* (Chlebowski, Harnden, and Sciortino 1989). As a result of this effort, the x-ray luminosities and the ratio between x-ray and bolometric luminosities (or the upper limits to these quantities) have been derived for 289 O stars, merging available data and taking into account the effect of interstellar absorption. This catalog has been adopted as a starting point for several follow-up studies. Chlebowski (1989) has suggested that the $L_x/L_{bol}$ ratios of single run-away O stars is smaller than that of "normal" single O stars, and has hypothesized that the run-away stars emit the "minimal" x-rays generated in accordance with Lucy's (1982) chaotic wind model, while the other O stars (which usually lie in associations) emit at higher level than this minimal value, the excess produced by the bow-shocks generated by the wind hitting the remnant-condensations still around the new-born, slow-moving, O stars.

Sciortino *et al.* (1989a,b) have analyzed all the available data on O stars, building the first x-ray luminosity function for a volume-limited statistically representative subsample extracted from the Chlebowski, Harnden and Sciortino catalogue (*cf.* Sciortino *et al.* 1989a, this volume). They have confirmed that the x-ray and bolometric luminosity are well correlated, have found no evidence for a dependence of x-ray luminosity on Galactocentric or heliocentric distance, and have found a correlation between $\dot{M}V_\infty$ and x-ray luminosity that is stronger than the correlation with $V_\infty$ or $\dot{M}$ alone. They suggest that the former correlation may be primary, and argue that this finding while compatible with the x-rays produced by shocks generated in a chaotic wind (Lucy 1982), it is not likely to be reconciled with the "coronal" model. Moreover they have reconsider Chlebowski's (1989) hypothesized mechanism, comparing for the same data set the $L_x/L_{bol}$ distribution of single field/run-away O stars with that of the single O stars in associations, and have found that these two distributions cannot statistically be distinguished at a sufficient level.

In summary, the optically selected surveys have shown that the initially-puzzling wide range in observed stellar x-ray emission levels can be accounted for in terms of stellar parameters which play a central role in stellar activity: For the late-type stars, these parameters are related to the stellar rotation rate, the stellar age, and the onset of convection, all somehow involved in stellar dynamo action. The x-ray luminosity of early-type stars through the correlations with stellar bolometric luminosity, $\dot{M}$, and $V_\infty$ seem to suggest a mechanism related to the instability of radiation-driven winds from these stars.

## 5. X-RAY FLUX LIMITED SURVEYS: THEIR IMPACT ON STELLAR ASTRONOMY

Several x-ray flux-limited surveys have had a substantial impact on the knowledge and understanding of stellar x-ray emission: These surveys have furnished a large fraction of presently available data on pre-main sequence stars, and have allowed us to test conclusions based on volume-limited surveys.

### 5.1  The Stellar content of the EMSS

The Extended Medium Sensitivity Survey (EMSS; Gioia, Maccacaro and Wolter 1987; Maccacaro *et al.* 1988; Gioia *et al.* 1989, this volume) has serendipitously detected a large number of x-ray selected late-type stars (*cf.* Fleming 1988; Fleming *et al.* 1988, 1989), detecting among others 14 new RS CVn's, 5 new W Uma binaries, and 8 previously unknown M dwarfs within 25 pc. All together the EMSS stars represent an unbiased sample of the brightest stellar x-ray sources at high Galactic latitude; however, since the EMSS considers only the fields above $|b| > 20°$ this sample cannot be used to study OB and pre-main sequence stars, which tend to lie closer to the Galactic plane.

Fleming *et al.* (1989) have shown that the x-ray luminosity of EMSS F-M stars scales with $v \sin i$, and argue that this is a plausible argument for rotational saturation of the emission from the coronae of late-type stars. A detailed study of the x-ray selected M stars (Fleming *et al.* 1988) shows that their luminosity distribution is compatible with the tail of the luminosity function derived from the volume-limited sample of M stars (Bookbinder 1985). The former authors have also shown a good correlation between $L_x$, and $H_\alpha$, and Ca II K emission, the total chromospheric luminosity being about half of the x-ray luminosity. They argue that these findings are consistent with the hypothesis that the chromospheres of M-dwarfs are heated by the x-rays coming from the overlying corona (Fleming *et al.* 1988).

Favata *et al.* (1988), on the basis of the x-ray luminosity functions of the nearby volume-limited stellar samples and of a detailed numerical model of stellar content of the Galaxy (Bahcall and Soneira 1980), have shown the presence of an excess of x-ray detected yellow stars in the MSS survey (Gioia *et al.* 1984) with respect to the expected number of detected late-type main-sequence and giant stars and have concluded that the excess is likely explained in terms of a population of RS CVn x-ray emitters. These tentative identifications have been recently confirmed by the optical identification reported by Fleming (1988) for the stars in the EMSS survey.

## 5.2 The Pre-Main Sequence Stars

As already noted the EMSS cannot provide information on the class of pre-main sequence stars. However since the early days of *Einstein* observations several specialized surveys have investigated star-formation regions such as the T Tauri associations in the Taurus-Auriga complex (Ku and Chanan 1979; Gahm 1980; Feigelson and Kriss 1981; Feigelson and DeCampli 1981; Walter and Kuhi 1981; Walter *et al.* 1987; Feigelson *et al.* 1987), the $\rho$ Ophiuchi cloud (Montmerle *et al.* 1983), the Orion Nebula (Ku *et al.* 1982; Caillault and Zoonematkermani 1986; Zoonematkermani and Caillault 1987), NGC 2264 (Simon, Cash and Snow 1985), and the Chamaleon I cloud (Feigelson and Kriss 1989). These observations have shown that pre-main sequence stars constitute a new class of soft x-ray sources and that their emission could be up to 1000 times more intense than that of the main sequence stars of which they are precursors. The x-ray emission is extremely variable (Montmerle *et al.* 1983), and while it is correlated with stellar optical magnitude it is unrelated with the classic indicators of pre-main-sequence "activity" (such $H_\alpha$ emission). The observed x-ray spectra seem to be intrinsically harder than the spectra of normal main-sequence stars and showed a cut-off at low energies, probably due to absorption by the intervening matter within the star-forming region.

The extensive x-ray surveys of several star formation regions (*cf.* above references) showed that approximately, at least, one third of the classical T Tauri were x-ray sources. Other x-ray sources were discovered to be associated with anonymous K7-M0 stars (previously not suspected to be pre-main-sequence stars) in the

Taurus-Auriga region. Mundt *et al.* (1983) showed that these stars are kinematic members of the Taurus-Auriga star forming complex. Since then many other x-ray sources with similar optical counterparts have been discovered (Walter 1986; Walter *et al.* 1987; Feigelson *et al.* 1987; Walter *et al.* 1988). They are the building blocks of the so-called "Naked T Tauri" class of pre-main-sequence stars (*cf.* Walter, this volume).

### 5.3 The Stellar Evolution and X-ray Emission of Late-type Stars

We conclude by noting a principal lesson we have learned: Stellar evolution determines the mean x-ray luminosity level of late-type stars. In order to summarize our present knowledge we have plotted the x-ray luminosity along an evolutionary track of a 1 $M_\odot$ stars in the H-R diagram (Figure 15). It is evident that hot circumstellar plasma is present essentially through a star's life. pre-main sequence stars are stronger x-ray emitters than main sequence stars. Main-sequence stars show a clear trend of decreasing x-ray emission with increasing age. The available data on giant stars suggest an increase of x-ray emission when, due to the change of interior structure, convection becomes more efficient and a more efficient dynamo mechanism is likely turned on, followed by a sudden drop when, with advancing stellar evolution, further changes of stellar structure makes the dynamo action inefficient. In summary, our present knowledge indicates that evolutionary changes of stellar structure and of the radial distribution of angular momentum are the likely causative agents determining changes in x-ray luminosity, from the pre-main sequence to the giant phase.

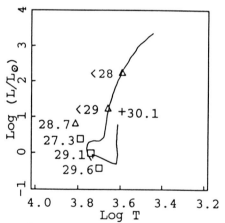

Figure 15. Schematic of the likely temporal evolution of stellar x-ray luminosity along the evolutionary tracks of 1 $M_\odot$ in the H-R diagram. Typical values of log $L_x$ are indicated at several stellar evolutionary stages, by stars at the median of x-ray luminosity function where this has been evaluated (triangles indicate giant stars, squares main-sequence stars, and crosses pre-main sequence stars).

## 6. THE SOFT X-RAY BACKGROUND

While the IPC was conceived as an imaging detector whose primary purpose was that of determining the intensity and spectra of both point-like and diffuse sources, its wide field of view, the substantial sky coverage attained by the end of the mission, and the quality and quantity of information resulting from the standard processing motivated an attempt by us to use the IPC data to study the diffuse soft x-ray background, one of the most intriguing subjects of x-ray astronomy.

The first step was the difficult problem of selecting a subsample of all available fields so that those images where the IPC background was severely influenced by the diffuse solar fluorescent radiation could be removed. In addition, we needed to determine the amount of scatter in the measured background induced both by particle events and the calibration source. To gain insight into the properties of the diffuse background emission, we removed from the analysis many other images where the morphology and the intensity of the x-ray sources made the determination of background uncertain. As a result of this procedure (for more details *cf.* Micela *et al.* 1989b,c), we were left with $\sim$ 1,700 fields, allowing us to study the diffuse soft x-ray background on a spatial scale of 1°, with the capability of characterizing its spectrum.

A brief account of the analysis of the "cleaned" data-set is presented elsewhere in this volume (Micela *et al.* 1989b) where the reader will find relevant figures and a detailed report is in progress (Micela *et al.* 1989c). Here I recall the major results: (i) two large scale structures are clearly present: one in the north hemisphere spatially coincident with the North Polar Spur, the other in the galactic plane region toward the galactic center direction identifiable with a Soft Galactic Ridge; (ii) the intrinsic spectrum of NPS is soft, while that of the Ridge is hard. The median value for the background emission in the featureless region, removing the particle contribution, is $\sim 2 \times 10^{-4} cnt\ s^{-1}\ arcmin^{-2}$ in the [.16 - 3.5] keV band-pass; the background excess in the galactic ridge is of $\sim .5 \times 10^{-4} cnt\ s^{-1}\ arcmin^{-2}$ (Micela *et al.* 1989c).

## 7. THE STELLAR X-RAY CONTENT OF GALAXY AND THE CONTRIBUTION OF STELLAR SOURCES TO THE SOFT X-RAY BACKGROUND

All the above analysis, studies and consideration provides us with the fundamental pieces of information to answer the question of the stellar x-ray make-up of the Galaxy when we reach enough sensitivity to see behind the major contribution of the High Luminosity Compact x-ray sources.

Our present knowledge allow us to identify four major classes of Low Luminosity Stellar Sources, namely: OB stars and pre-main sequence stars, young late-type stars, late-type main sequence, giant stars and RS CVn's. The ready availability of x-ray luminosity functions for essentially all the above classes now makes it possible to answer the question of their contribution to the Galaxy x-ray make-up (including their contribution to the diffuse "Galactic" x-ray background).

Hence the next question is: To what extent can the stellar sources account for the spatial structure seen in the soft x-ray background, and what fraction of its intensity can be accounted for by stellar sources ?

To properly address this problem, we had to be able to predict the number of expected stellar sources as a function of limiting flux. A reasonably accurate solution of this problem essentially requires 3 ingredients: (i) knowledge from complete stellar samples of the x-ray luminosity functions (*cf.* Figure 9); (ii) a numerical model of the Galaxy stellar distribution (for a review *cf.* Bahcall 1986, and Bahcall, Casertano and Ratnacunga 1987 for recent developments ); (iii) a model of the Galactic x-ray absorptionX-ray absorption and an estimate of the typical temperatures of unresolved sources. Our group has developed the necessary tools and have applied this technique to show a statistical excess of yellow stars in the MSS stellar sample (Favata *et al.* 1988), but not in the Deep Survey (Primini *et al.* 1989). These two facts indicate that the excess should be due to x-ray bright stars with relatively small scale height (*i.e.* belonging to the disk population). Very likely these requirements identify a population of RS CVn's emitters (*cf.* also sec. 5.1). Favata *et al.* (1988) have pointed out that although the stellar *logN-logS* number counts are strongly affected by the presence of these stars for surveys with sensitivity levels comparable to the *Einstein* MSS, this is not the case for deeper surveys. Hence at deeper sensitivity the stellar contribution is dominated by other contributors. Rosner *et al.* (1981) using the version of the x-ray luminosity functions available at that time, evaluated the stellar contribution to the x-ray background and especially the contribution at energies below 1 keV. They showed that the major contribution comes from the dM stars, and the stellar contribution is about $\sim$ 20% in the [.28 -1.0 ] keV band-pass. Rosner, Golub and Vaiana (1985) have reported a slightly refined calculation, based on a more refined spatial distribution of M stars than the spherical one adopted by Rosner *et al.* (1981), that confirms the previous finding of Rosner *et al.* (1981).

Adopting the technique of Favata *et al.* (1988) and a slightly refined modeling of x-ray absorptionX-ray absorption (that accounts for source temperature and gas distribution in the Galactic disk) Sciortino *et al.* (1989c) have computed, for various source temperatures and various B-V color index ranges, predicted *logN-logS* relations for stellar sources at various Galactic latitudes. This calculation confirms the previous results of Rosner *et al.* (1981) and indicates that with the typical sensitivity of AXAF or XMM, the dM stars will dominate the realm of stellar sources (*cf.* Figure 16) and that this contribution should account for $\sim$ 15% of the x-ray background in the 0.2 - 4.0 keV energy band. Fleming *et al.* (1989) have proven that the x-ray luminosity level of x-ray selected M stars is in agreement with the x-ray luminosity function estimates from optically selected sample. This EMMS result is a confirmation of all the above predictions.

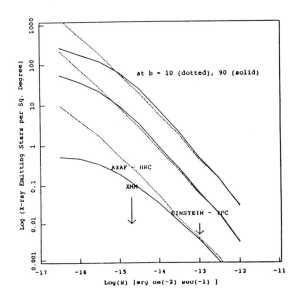

Figure 16. logN-logS for dA (bottom curves), dG (central curves), and dM (dM < dM5; upper curves) stars predicted for two values of Galactic latitude and for the *Einstein* bandpass. Typical limiting sensitivity for *Einstein*, AXAF and XMM are also indicated.

The x-ray background has ben detected over a wide range of energies, from tens of eV to tens of keV. The HEAO-1 data (Worrall *et al.* 1982) have shown in the 2 - 10 keV energy range the presence of a Galactic ridge in the Galactic center direction, and associated with the Galactic disk. These authors suggested that RS CVn-type emitters could account for the ridge emission, but do not consider realistic contributions from other classes of stellar sources. More recently Koyama *et al.* 1986 have reported the detection of the ridge with the TENMA satellite in the 2 - 11 keV band-pass. To account for the prominent 6.7 keV detected iron line they have suggested that the ridge emission can be explained in terms of thin thermal plasma emission from uncatalogued SNR's. However the spectrum of known SNR's is generally softer than that of the ridge emission, hence other contributors have been considered such as the RS CVn's and the CV's (Koyama 1988). The ridge is visible as an emission structure also in the *Einstein* IPC band-pass (Micela *et al.* 1989b,c). A modeling of the stellar contribution is in progress. The narrow extent of the ridge implies small scale height components such as OB stars, pre-main sequence stars, young stars, RS CVn's. OB stars however have a space density too low to be a significant contributors. Among the most likely contributors we are left with the RS CVn's, pre-main sequence stars and young stars, or a suitable combination of them. There are indications that a fraction of the ridge emission can indeed be explained in terms of emission from active young stars (*cf.* Micela *et al.*, 1989b).

We conclude by noting that ROSAT, due a factor of four increase in sensitivity (for the same exposure times), and sky-coverage (in the survey mode a factor $\sim 20$ larger of that of *Einstein*, with typical *Einstein* 2-3 ksec exposure times), is expected to detect approximately a factor 200 more stars than the successful *Einstein Observatory*. A further increase may be expected, under certain circumstances, by the softer x-ray bandpass. AXAF and XMM observatories will probably not be used in a total sky survey mode. In selected regions of the sky most stellar sources will be probed $\sim 100$ times deeper allowing us to detect $\sim 1000$ stars in the entire AXAF-HRC PI program, and to probe the stellar x-ray luminosity functions at significantly deeper sensitivity than any other previous mission.

## ACKNOWLEDGEMENTS

I want to aknowledge the useful discussions with several my colleagues, the contribution of the colleagues on my team, and in particular S. Sciortino. The author acknowledges partial support from the Italian MPI, GNA, ASI.

## REFERENCES

Ayres, T.R., Linsky, J. L., Vaiana, G. S., Golub, L., Rosner, R. 1981, *Ap.J.*, **250**, 293.

Avni, Y., Soltan, A., Tananbaum, H., Zamorani, G. 1980 *Ap.J.*, **238**, 800.

Bahcall, J. N., 1986, *Ann. Rev. Astr. Astrophys.*, **24**, 577.

Bahcall, J. N., and Soneira, R. M. 1980, *Ap.J.Suppl.*, **44**, 73.

Bahcall, J. N., Casertano, S., and Ratnacunga, K.U. 1987, *Ap.J.*, **320**, 515.

Bookbinder, J. 1985, PhD Thesis, Harvard University.

Brinkman, A. C., Gronenschild, E., Mewe, R., Mc Hardy, I., Pyr, J. P. 1985, *Adv. Sp. Res.*, **5**, 61.

Caillault, J.-P. and Helfand, D. J. 1985, *Ap.J.*, **289**, 279.

Caillault, J.-P. and Zoonematkermani, S. 1986, in *Circumstellar Matter - IAU Symp. 122* I. Appenzeller and C. Jordan (eds), (Dordrecht: Reidel), p. 119

Cash, Jr., W. and Snow, Jr., T. P. 1982, *Ap.J.(Letters)*, **263**, L59.

Cassinelli, J. P. 1985, in *The Origin of Nonradiative Heating/Momentum in Hot Stars*, ed. A. B. Underhill and A. G. Michalitsianos, NASA Conference Publ. 2358, p. 2.

Cassinelli, J. P., Olson, G. L. 1979, *Ap.J.*, **229**, 304.

Cassinelli, J. P., Waldron, W. L., Sanders, W. T., Harnden, F. R., Jr., Rosner, R., Vaiana, G. S. 1981, *Ap.J.*, **250**, 677.

Cassinelli, J. P., Swank, J. H. 1983, *Ap.J.*, **271**, 681.

Chleblowski, T. 1989, *Ap.J.*, in press.

Chleblowski, T., Harnden, F. R., Sciortino, S. 1989, *Ap.J.*, **341**, 427.

Collura, A., Sciortino, S., Serio, S., Vaiana, G. S., Harnden, F. R., Jr., Rosner, R. 1989b, *Ap.J.*, **338**, 296.

Favata, F., Rosner, R., Sciortino, S., and Vaiana, G.S. 1988, *Ap.J.*, **324**, 1010.

Feigelson, E. D. and deCampli, W. M. 1981 *Ap.J.(Letters)*, **243**, L89.

Feigelson, E. D. and Kriss G. A. 1981 *Ap.J.(Letters)*, **248**, L35.

Feigelson, E. D. and Nelson, P. I. 1985 *Ap.J.*, **293**, 192.

Feigelson, E. D., Jackson, J. M., Mathieu, R. D., Myers, P. C., and Walter, F. M. 1987, *A.J.*, **94**, 1251.

Feigelson, E. D. and Kriss, G. A. 1989 *Ap.J.*, in press.

Fleming, T. A. 1988, Ph. D. Thesis, University of Arizona.

Fleming, T. A., Liebert, J., Gioia, I. M., Maccacaro, T. 1988, *Ap.J.*, **331**, 958.

Fleming, T. A., Gioia, I. M., Maccacaro, T. 1989, *Ap.J.*, in press.

Gahm, G. F. 1980, *Ap.J.(Letters)*, **242**, L163.

Giacconi, R., *et al.* 1979, *Ap.J.*, **230**, 540.

Gioia, I. M., Maccacaro, T., Schild, R. E., Stocke, J. T., Liebert, J. W., Danziger, I. J., Kunth, D., Lub, J. 1984, *Ap.J.*, **283**, 495.

Gioia, I. M., Maccacaro, T., Wolter, A. 1987 in *Observational Cosmology*, I.A.U. Symp. 124 (Dordrech:Reidel), p. 593

Gorenstein, P., Harnden, Jr., F. R., and Fabricant, D. G. 1981, *I.E.E.E. Nucl.Sc.*, NS-28, 869.

Haisch, B. and Simon T. 1982, *Ap.J.*, **263**, 252.

Harnden, F. R., Jr., *et al.* 1979, *Ap.J.(Letters)*, **243**, L51.

Harnden, Jr., F. R., Fabricant, D. G., Harris, D. E. and Schwarz, J. 1984, *SAO Special Report 393*.

Harnden, F. R., Jr., Vaiana, G. S., Sciortino, S., Micela, G., Maggio, A., Schmitt, J.H.H.M., and Rosner, R., 1988b, in preparation.

Hearn, A. G. 1975, *Astr.Ap.*, **40**, 355.

Hoffleit, D., Jaschek, C. 1984, *The Bright Star Catalog*, Yale University Press.

Isobe, T., Feigelson, E. D., Nelson, P. I. 1986, *Ap.J.*, **306**, 490.

Koyama, K. 1988, review at *The Physics of Neutron Stars and Black Holes*, ISAS Res. Note 391.

Koyama, K., Ikeuchi, S., Tomisaka, K. 1986, *P.A.S.J.*, **38**, 503.

Ku, W. H.-M. and Chanan, G. A. 1979, *Ap.J.(Letters)*, **234**, L59.

Ku, W. H. M., Righini-Cohen, G., and Simon, M. 1982, *Science*, **215**, 61.

Linsky, J. L., Haisch, B. M. 1979, *Ap.J.(Letters)*, **229**, L27.

Long, K., White, R. 1980, *Ap.J.(Letters)*, **239**, L27.

Lucy, L. B. 1982, *Ap.J.*, **255**, 286.

Lucy, L. B., White, R. L. 1980, *Ap.J.*, **241**, 300.

Maccacaro, T., Gioia, I. M., Wolter, A., Zamorani, G., Stocke, J. T. 1988, *Ap.J.*, **326**, 680.

Majer, P., Schmitt, J. H. M. M., Golub, L., Harnden, F. R., Jr., Rosner, R. 1986, *Ap.J.*, **300**, 360.

Maggio, A., Sciortino, S., Vaiana, G. S., Mayer, P., Bookbinder, J., Golub, L., Harnden, F. R. Jr., and Rosner, R. 1987, *Ap.J.*, **315**, 687.

Maggio, A., Vaiana, G. S., Stern, R., Bookbinder, J., Haisch, B., Harnden, F. R., Jr., and Rosner, R. 1989b, *Ap.J.*, in press.

Mangenay, A., Praderie, F. 1984, *Astr.Ap.*, **130**, 143.

Mewe, R., Groneshield, E. H. B. M., Westergard, N. J., Heise, J., Seward, F. D., *et al.* 1982, *Ap.J.*, **260**, 233.

Micela, G., Sciortino, S., and Serio, S. 1984, in *X-ray Astronomy '84*, M. Oda and R. Giacconi, eds., Symposium (Bologna), p. 43.

Micela, G., Sciortino, S., Serio, S., Vaiana, G. S., Bookbinder, J., Golub, L., Harnden, Jr., F. R., and Rosner, R. 1985, *Ap.J.*, **292**, 172.

Micela, G., Sciortino, S., Vaiana, G. S., Schmitt, J. H. M. M., Stern, R. A., Harnden, Jr., F. R., and Rosner, R. 1988, *Ap.J.*, **325**, 798.

Micela, G., Sciortino, S., Vaiana, G. S., Harnden, Jr., F. R., Rosner, R., and Schmitt, J. H. M. M. 1989a, *Ap.J.*, in press.

Micela, G., Harnden, Jr., F. R., Rosner, R. Sciortino, S., Vaiana, G. S., 1989c, *Ap.J.*, to be submitted.

Montmerle, T., Koch-Miramond, L., Falgarone, E. and Grindlay, J. 1983, *Ap.J.*, **269**, 182.

Mundt, R., Walter, F. M., Feigelson, E. D., Finkenzeller, U., Herbig, G., and Odell, A. P. 1983, *Ap.J.*, **269**, 182.

Noyes, R. W., Hartmann, L., Baliunas, S., Duncan, D., Vaugham, A. H. 1984, *Ap.J.*, **279**, 763.

Owocki, S. and Rybicki, G. 1984, *Ap.J.*, **284**, 337.

Owocki, S. and Rybicki, G. 1985, *Ap.J.*, **299**, 265.

Pallavicini, R., Golub, L., Rosner, R., Vaiana, G. S., Ayres, T. and Linsky, J. L. 1981, *Ap.J.*, **248**, 279.

Primini, F., *et al.* 1989, in preparation.

Raymond, J. C. and Smith, B. W. 1977, *Ap.J.Suppl.*, **35**, 419.

Raymond, J. C. 1988 in *Hot Thin Astrophysical Plasmas*, R. Pallavicini (ed.), Dordrecht: Kluwer Publ., p. 3.

Rosner, R. *et al.* 1981, *Ap.J.(Letters)*, **249**, L5.

Rosner, R., Golub, L., Vaiana, G. S. 1985 *Ann.Rev.Astron.Astrophys.*, **23**, 413.

Schmitt, J. H. M. M. 1984, in *Symp. X-ray Astron. 84*, Bologna, Italy, p. 17.

Schmitt, J. H. M. M. 1985, *Ap.J.*, **293**, 178.

Schmitt, J. H. M. M. 1988, in *Hot Thin Plasmas in Astrophysics* ed. R. Pallavicini, 109

Schmitt, J. H. M. M., Golub, L., Harnden, Jr., F. R., Maxson, C. W., Rosner, R., and Vaiana, G. S. 1985a, *Ap.J.*, **290**, 307.

Schmitt, J. H. M. M., Harnden, F. R., Jr., Peres, G., Rosner, R., Serio, S. 1985b, *Ap.J.*, **288**, 751.

Schmitt, J. H. M. M., Pallavicini, R., Monsignori-Fossi, B. C. and Harnden, Jr., F. R. 1987, *Astr.Ap.*, **179**, 193.

Schmitt, J. H. M. M., Micela, G., Sciortino, S., Vaiana, G. S., Harnden, F. R., Jr., Rosner, R. 1989a, *Ap.J.*, submitted.

Schmitt, J. H. M. M., Collura, A., Sciortino, S., Vaiana, G. S., Harnden, F. R. Jr., and Robert, R. 1989b, *Ap.J.*, to be submitted.

Schrijver, C. J., Mewe, R., Walter, F. M. 1984, *Astr.Ap.*, **138**, 258.

Sciortino, Harnden, Jr. F. R., Maggio, A., Micela, G., Vaiana, G. S., Rosner, R., Schmitt, J.H.H.M. 1988, in *Astronomy from Large Database*, ESO Conference and Workshop Proceedings No. 28, p. 483

Sciortino, S., Vaiana, G. S., Harnden, F. R., Jr., Morossi, C., Ramella, M., Rosner, R., Schmitt, J.H.H.M. 1989b, *Ap.J.*, to be submitted.

Sciortino, S., Favata, F., Micela, G., Vaiana, G. S., Harnden, F. R., Jr ., Rosner, R., Schmitt, J.H.H.M. 1989c, *Mem. S.A.It.*, **60**, 125.

Seward, F.D. *et al.* 1979, *Ap.J.(Letters)*, **234**, L55.

Seward, F.D. and Chlebowski, T. 1982, *Ap.J.*, **256**, 530.

Simon, T., Cash, W., and Snow, T. P., Jr. 1985, *Ap.J.*, **293**, 542.

Snow, T. P.,Jr., Cash, W., Grady, C. A. 1981, *Ap.J.(Letters)*, **244**, L19.

Stephien, K. 1988, *Ap.J.*,, **335**, .

Stern, R. A., Zolcinski, M.-C., Antiochos, S. K. and Underwood, J. H. 1981, *Ap.J.*, **249**, 647.

Stern, R. A., Antiochos, S. K., Harnden, F. R., Jr. 1986 *Ap.J.*, **305**, 417.

Swank, J. H. 1984 in *The Origin of Nonradiative Heating/Momentum in Hot Stars*, A.B. Underhill and A.G. Michalitsianos (eds.), NASA Conference Publications 2358.

Swank, J. H., White, N. E., Holt, S. S., Becker, R. H. 1981, *Ap.J.*, **246**, 208.

Vaiana, G. S., *et al.* 1981, *Ap.J.*, **245** , 163.

Vaiana, G. S., Sciortino, S., Serio, S., Golub, L., Maxson, C. W. Maxson, Rosner, R. 1983, *Bull. A.A.S.*, **14**, 945.

Vaiana, G. S. 1983, in *Solar and Stellar Magnetic Fields. Origins and Coronal Effects*, ed. O. Stenflo (Dorderecht: Reidel), 165.

Vaiana, G. S., Harnden, F. R., Jr., Sciortino, S., Micela, G., Maggio, A., Schmitt, J.H.H.M. and R. Rosner 1989, in preparation.

Vihlu, O. 1984 *Astr.Ap.*, **133**, 117.

Waldron, W. L. 1984, *Ap.J.*, **282**, 256.

Walter, F. W., and Kuhi, L. V. 1981, *Ap.J.*, **250**, 254.

Walter, F. M. 1982, *Ap.J.*, **253**, 745.

Walter, F.W. 1983, *Ap.J.*, **274**, 794.

Walter, F. W., Linsky, J. L., Simon, T., Golub, L., and Vaiana, G. S., 1984, *Ap.J.*, **281**, 815.

Walter, F. M. 1986, *Ap.J.*, **306**, 1573.

Walter, F. M., *et al.* 1987, *Ap.J.*, **314**, 297.

Walter, F. M., Brown, A., Mathieu, R. D., Myers, P. C., Vrba. F. J. 1988, *A.J.*, **96**, 297.

Walter, F. M. 1989, *Symp. "From Einstein to AXAF"*, ed. M. Elvis and T. Maccacaro, this volume.

Woolley, R. Epps, E. A., Penston, M. J. and Pocock, S. B. 1970, *Royal Obs. Ann.*, **281**, 815.

Worrall, D.M., Marshall, F.E., Boldt, E.A. and Swank, J.H. 1982, *Ap.J.*, , **255**, 111.

Zoonematkermani, S., and Caillault, J. P. 1987, *B.A.A.S.*, **19**, 717.

# CATACLYSMIC VARIABLES AS X-RAY SOURCES

**Joseph Patterson**
Columbia University

## 1. INTRODUCTION

Cataclysmic variables (hereafter CVs; see Robinson 1976, Cordova and Mason 1983 for general reviews) are close binary stars in which a low-mass, late-type secondary fills its critical Roche surface and transfers matter to a white dwarf primary. Because the gravitational well of the white dwarf is very deep ($\sim 100$ keV/nucleon), it has long been realized that accretion of gas could lead to intense emission of hard ($> 1$ keV) X-rays (Cameron and Mock 1967; Fabian, Pringle and Rees 1976; Katz 1977; Kylafis and Lamb 1979). Because the accretion rates in CVs are in the range $10^{-11}$ to $10^{-7} M_\odot/\text{yr}$, the total expected accretion luminosity ($= GM_{wd}\dot{M}/R_{wd}$) is in the range $10^{32}$ to $10^{36}$ erg s$^{-1}$. At typical distances of 100–300 pc, these stars could be extremely bright X-ray sources, with a flux received at the Earth of $10^{-11}$ to $10^{-7}$erg cm$^{-2}$s$^{-1}$. If this flux appeared predominantly as hard X-rays, then CVs would greatly outnumber all other classes of X-ray emitters in existing catalogues of X-ray sources, with many thousands detectable above the 2–10 keV flux limit of $\sim 2 \times 10^{-11}$erg cm$^{-2}$s$^{-1}$ for the present all-sky X-ray surveys.

However, it is well-known that these catalogues (*e.g.* Forman *et al.* 1978; Warwick *et al.* 1981; McHardy *et al.* 1981; Marshall *et al.* 1979; Bradt, Doxsey and Jernigan 1978; Bradt and McClintock 1983) contain only a few CVs (11 in Bradt and McClintock's comprehensive listing), with X-ray fluxes generally near the threshold for detection. Evidently CVs as a class radiate only a very small fraction of their accretion luminosity in the easily accessible 2–10 keV X-ray band. The launch of the *Einstein* Observatory in 1978 permitted much more sensitive searches, and the results (*e.g.* Cordova, Mason and Nelson 1981; Cordova and Mason 1983) demonstrate that X-ray emission is indeed a general property of CVs, but at typical luminosities of $10^{30}$–$10^{32}$erg s$^{-1}$ – about 2–4 orders of magnitude lower than originally expected. Thus, only a few of the nearest systems have managed to creep into the existing all-sky catalogues of bright sources.

*Why does the efficiency for producing X-rays appear to be so low?* If X-rays are not produced, where does the accretion luminosity emerge? Ten years after the launch of *Einstein*, and six years after its demise, neither theory nor observation has answered these questions, though each has supplied some provocative clues. Let us see how the subject has evolved.

## 2. THE AM HERCULIS STARS

The AM Herculis stars, or "polars", are cataclysmic variables in which white dwarf rotation is synchronous with the binary orbit, and the flow of gas is channeled throughout by the influence of the white dwarf's strong magnetic field. This class was first identified in the mid-1970s, in the following way. Bond and Tifft (1974) included the obscure star AM Her in a list of unclassified and misclassified variable stars, noting that a low-resolution optical spectrum resembled that of a cataclysmic variable. Berg and Duthie (1977) obtained high-speed photometry of AM Her, confirmed the CV identification, and from a coarse positional coincidence suggested its identification with the X-ray source 3U1809+50. From such small beginnings come great revolutions!

Subsequent studies in the 1976 observing season showed a common 3.1 hour period in the radial velocities (Cowley and Crampton 1977), the optical light curve (Szkody and Brownlee 1977), the X-ray light curve (Hearn and Richardson 1977), and – most dramatically – the optical polarization (Tapia 1976, Tapia 1977a, Stockman *et al.* 1977). The large circular polarization strongly suggested an origin in a magnetic field of $10^7 - 10^8$ gauss. And this in turn could explain the hard X-ray emission, since a strong magnetic field would channel the flow of accreting gas radially onto the white dwarf, supplying most of the energy available in the gravitational potential well to a strong shock lying just above the white dwarf surface.

Within that year, surely the *annus mirabilis* of CV research, two other stars were found with the same magnetic field credentials: VV Pup ($P = 1.7$ hours, Tapia 1977b) and AN UMa ($P = 1.9$ hours, Krzeminski and Serkowski 1977). The next year, a fourth member joined the club when the X-ray source 2A 0311-227 was identified with the CV subsequently named EF Eri, which showed a 1.35 hour variation in photometry, radial velocities, and circular/linear polarization. For several years membership stalled at four, but now the ranks are growing at the rate of $\sim 1$/year as a result of new, hard-fought optical identifications of X-ray sources. In Table 1 we present a list of the 15 currently known AM Her stars, along with the references emphasizing the most detailed or the most recent X-ray observations.

By the launch of HEAO-1, it was clear that AM Her represented an important new class of X-ray source, and a beautiful X-ray spectrum was obtained (see Figure 1, from Rothschild *et al.* 1981). The hard X-ray emission was generally consistent with $kT \sim 30$ keV, while at low energies the emission shows a strong soft component with $kT \lesssim 45$ eV. Tuohy *et al.* (1978, 1981) and Fabbiano *et al.* (1981) pointed out that this soft component appears to dominate the total accretion luminosity, by a factor of $\gtrsim 30$ – *depending on what is assumed for the temperature.* If one assumes a blackbody temperature $kT = 45$ eV, then most of the emission is in the HEAO-1 LED bandpass, and the soft and hard X-ray sources have approximately equal luminosities: $L_{sx}/L_{hx} \approx 1$. On the other hand, if $kT$ is substantially lower, then most of the soft source luminosity falls outside the bandpass, and the result is $L_{sx}/L_{hx} \gg 1$.

| Optical Name | X-ray or Other Name | $P_{orb}$ (min) | X-ray References |
|---|---|---|---|
| EF Eri | 2A 0311-227 | 82 | White 1981; Patterson, Williams and Hiltner 1981; Beuermann *et al.* 1987a |
| DP Leo | E1114+182 | 90 | Biermann *et al.* 1985 |
| VV Pup | | 100 | Patterson *et al.* 1984 |
| V834 Cen | E1405-451 | 101 | Jensen *et al.* 1982, Bonnet-Bidaud *et al.* 1985 |
| MR Ser | PG1550+191 | 113 | Watson and King 1988 |
| BL Hyi | H0139-68 | 114 | Agrawal *et al.* 1983, Singh *et al.* 1984, Beuermann *et al.* 1985 |
| ST LMi | CW1103+254 | 114 | |
| | EXO 023432 −5232.3 | 114 | Beuermann *et al.* 1987b |
| | 1E1048.5+5421 | 114 | Morris *et al.* 1987 |
| AN UMa | | 115 | Hearn and Marshall 1979, Osborne 1987 |
| | EXO 033319 −2554.2 | 128 | Osborne *et al.* 1988 |
| AM Her | 3U1809+50 | 185 | Swank *et al.* 1977, Tuohy *et al.* 1981, Rothschild *et al.* 1981, Heise *et al.* 1985, Mazeh *et al.* 1986 |
| | H0538+608 | 186 | Remillard *et al.* 1986 |
| V1500 Cyg | Nova Cyg 1975 | 201 | Stockman, Schmidt and Lamb 1988 |
| QQ Vul | E2003+225 | 222 | Nousek *et al.* 1984, Mukai *et al.* 1985, Osborne *et al.* 1986a, Osborne 1987 |

Table 1 — The AM Herculis Stars

ote: There is a vast literature on these stars; here I only cite key papers reporting X-ray observations. Excellent general review articles, together with a more complete list of references on work at all wavelengths, are given by Liebert and Stockman (1985) and by Bailey (1986).

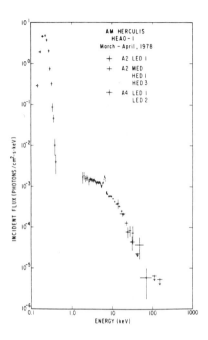

Figure 1. The X-ray spectrum of AM Herculis showing a thermal bremsstrahlung with $kT \sim 30$ keV and a "black body" component with $kT \sim 15 - 45$ eV. (From Rothschild *et al.* 1981.)

This is an issue of great importance, because radial accretion theories produce hard X-rays in a shock standing just above the white dwarf surface, and soft X-rays by reprocessing the downward-going half of the hard X-ray emission in the white dwarf photosphere. (See illustration in Figure 2.) Thus $L_{\mathrm{sx}} \approx L_{\mathrm{hx}}$ is a more or less necessary feature of reprocessing models.

What do the observations have to say about this prediction? Patterson *et al.* (1984) summarized the situation five years ago, when X-ray observations were available for the original four members of the club. We concluded that while the uncertainty in temperature was severe, $L_{\mathrm{sx}}/L_{\mathrm{hx}}$ seemed to be generally in the range 0.5 to 4. The *Einstein* Objective Grating Spectrometer data, discussed by Heise *et al.* (1983; see also Lamb 1985), appeared to settle the question of the temperature of AM Her's soft component in favor of a relatively high temperature ($kT = 40$–$55$ eV), implying a relatively low $L_{\mathrm{sx}}$, so that $L_{\mathrm{sx}} \approx L_{\mathrm{hx}}$ as predicted. Beuermann, Stella and Patterson (1987), analyzing the long *Einstein* IPC observation of EF Eri, also found a sufficient temperature constraint to establish $L_{\mathrm{sx}} \lesssim L_{\mathrm{hx}}$, implying no problem for the standard theory.

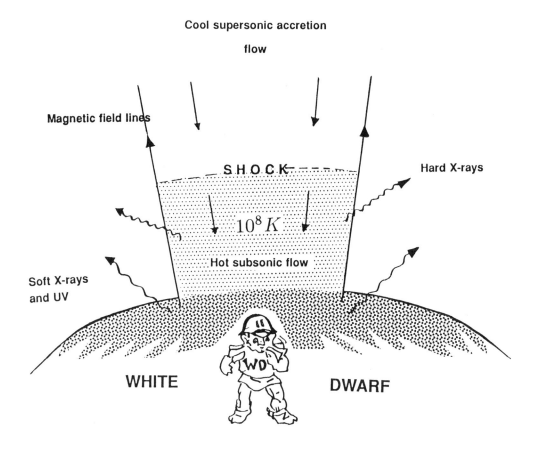

Figure 2. A close-up view of the accretion shock just above the white dwarf. Hard X-rays are produced in the shocked gas, and soft X-rays in the white dwarf photosphere which – according to the simplest theories – reprocesses most of the incident hard X-rays. But are there, perhaps, too many soft X-rays observed?

But in 1983–4, *EXOSAT* observations showed that the AM Her light curve had drastically changed. The soft and hard X-ray fluxes had changed, $L_{sx}$ now exceeded $L_{hx}$ by a factor of $\sim 50$, and the orbital light curves were now 180° out of phase (Heise *et al.* 1985, Mazeh *et al.* 1986). Not surprisingly, this was called the "anomalous" state of AM Her. And the once-banished soft X-ray problem had now returned, in spades.

As of this writing, there is still no solution to the problem. A proposed scheme to conduct heat directly into the star from the post-shock gas has been much discussed, but the major advocates and critics now agree that it is not feasible (Frank, King and Lasota 1988; Imamura *et al.* 1987; Lamb 1985). Kuijpers and Pringle (1982) suggest that accretion energy could be transported directly into the star if the matter falls in the form of "bullets". (This is why the white dwarf in Figure 2 has been equipped with an Army helmet.)

Another test of the standard theory, less commonly discussed, is to look for a correlation in the pulse arrival times of soft and hard X-rays. Here too the results are mixed. Stella *et al.* (1986) found essentially no correlation for AM Her. But Beuermann *et al.* (1987) found for EF Eri a beautiful correlation – both hard and soft X-rays showing 6 minute quasi-periodic oscillations, with 75% cross-correlation at zero lag. This analysis should be carried out on all large data sets when two easily separable components are present.

A great deal of observational effort has been devoted to specifying the system geometry (3-dimensional orientation of magnetic dipole axis) of each individual AM Her star, based on X-rays, polarimetry, and radial velocities. To first order, the results show reasonable consistency with the standard model. An example is seen in Figure 3, which I prepared some years ago when the census of stars was holding steady at four. The soft X-ray light curves show their usual humps, resulting probably from the varying aspect of the heated polar cap. When the polar cap is maximally directed towards us, the soft X-rays should peak, the broad emission lines from the infalling gas should reach maximum recessional velocity, and the "absorption dip" should occur (if ever) as maximum column density passes through the line of sight. And lo! these phenomena occur at about the same time, if you'll permit me some hand-waving about light curve asymmetries and second poles.

Well over a hundred papers have now been written about constraints on system geometry; Mukai (1988) presents the most complete and up-to-date treatment.

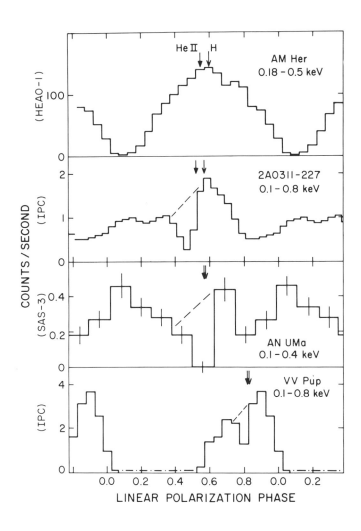

Figure 3. Soft X-ray light curves of AM Her stars in their "high states," with the data smoothed to reduce rapid variability. The dashed lines indicate intervals of assumed X-ray absorption. The dash-dot line for VV Pup indicates an interval not covered by the observations. The phases at which the broad components of the He II λ4686 and Hα emission lines reach maximum recessional velocity are indicated by the heavy and light arrows, respectively.

## 3. THE DQ HERCULIS STARS

While it is still not known exactly how AM Her stars manage to achieve and maintain synchronism, it is clear that synchronism will be broken if the white dwarf's magnetic field is sufficiently reduced, or if the accretion rate is sufficiently raised (Chanmugam and Ray 1984). In such a case an accretion disk forms around the white dwarf, but is disrupted at some inner radius, within which the magnetic field dominates the flow and channels the gas radially onto the white dwarf. The star should then be spun up rapidly by the torques exerted by accreting gas. These stars are "weak sisters" to their more famous and more luminous counterparts, the accretion-powered X-ray pulsars. But do such stars *exist*?

Shortly after the discovery of X-ray pulsars, a gaggle of theorists jumped on this model to explain the very stable 71 second periodicity in the optical light curve of DQ Herculis, an old nova extensively studied by Merle Walker in the 1950s (Lamb 1974; Katz 1975; Bath, Evans, Papaloizou and Pringle 1974; Herbst, Hesser and Ostriker 1974). All predicted that the star should have magnetically channeled radial accretion, and hence should be a strong and pulsating hard X-ray source. To everyone's consternation, the predicted hard X-rays never showed up, and this remains true today.

Meanwhile, observers began to find other stars with stable short periods, suggesting rapid rotation, in the optical light curves. Patterson (1979a, 1979b, 1980) and Patterson and Price (1980) nominated 4 new stars for membership in the class: V533 Her, WZ Sge, AE Aqr, and AO Psc. The last two of these stars promptly displayed X-ray pulses at or near the expected period, confirming the suggestion (AE Aqr: Patterson *et al.* 1980; AO Psc: White and Marshall 1980, Patterson and Garcia 1980). From these results it seemed likely that a new class of cataclysmic variable could be declared, the "DQ Herculis stars".

While HEAO-1 was primarily a *scanning* telescope and hence ill-suited to the discovery or study of periodicities, the pointed observations of *Einstein* and *EX-OSAT* approximately doubled the census of known DQ Her stars, and greatly expanded our knowledge of their X-ray properties. New members were quickly added: EX Hya (Vogt *et al.* 1979, Kruszewski *et al.* 1981), V1223 Sgr (Steiner *et al.* 1981), FO Aqr (Patterson and Steiner 1983), BG CMi (McHardy *et al.* 1984), GK Per (Watson *et al.* 1984), and TV Col (Schrijver et al. 1985). In Table 2 I present a list of the 10 currently known and well-certified DQ Her stars, along with the references reporting the X-ray observations.

| | | | |
|---|---|---|---|
| **Table 2** | | | |
| **The DQ Herculis Stars** | | | |
| Optical Name | X-ray Name | Period (min) | X-ray References |
| AE Aqr | | 0.55 | Patterson, Chincarini, Branch and Robinson 1980 |
| DQ Her | | 1.2 | Chanan *et al.* 1978 |
| GK Per | | 5.9 | Watson, King and Osborne 1984; Norton, Watson and King 1988 |
| V1223 Sgr | 4U1849−31 | 12.4 | Osborne *et al.* 1985 |
| AO Psc | H2252-035 | 13.4 | White and Marshall 1981, Pietsch *et al.* 1987 |
| BG CMi | 3A 0729+103 | 15.2 | McHardy *et al.* 1984, 1987 |
| FO Aqr | H2215-086 | 20.9 | Cook *et al.* 1986 |
| | H0542-407 | 32.0 | Tuohy *et al.* 1986 |
| TV Col | 2A 0526-328 | 32.4 | Schrijver *et al.* 1985 |
| EX Hya | 4U1249−28 | 67.0 | Kruszewski *et al.* 1981 |

In addition to these stars, there are another 5–6 whose credentials are not quite yet in order – because pulsed X-rays have not been found, because there are questions about the stability of the periodicity, or because the period is very close to the orbital period. Candidates in this class include WZ Sge, V533 Her, SW UMa, AM CVn, and TT Ari. Hopefully we can decide soon on these stars' applications for membership.

Some will argue against the inclusion of DQ Her itself in this list, since the star has never been seen to emit pulsed X-rays (or *any* X-rays, for that matter). However, this eclipsing star is observed at a binary inclination $i > 80°$; Petterson (1980) pointed out that a small bloating anywhere on the disk, or even the disk's own curvature, could easily obscure X-rays from the central object. The indirect evidence for a pulsed soft X-ray flux is really quite strong. Chanan, Nelson and Margon (1978) showed that the He II $\lambda4686$ line is pulsed, and that the variations in line profile with the phase of the 71 second periodicity indicate that the line is being excited by a searchlight beam of ionizing radiation rotating around the disk. Since it requires 54 eV to ionize helium twice, I conclude that the beam is rich in photons of energy $> 54$ eV. We would call these "soft X-rays" if we could see them directly.

DQ Her stars are not only weak sisters, they are also plain Janes. Even a devotee of this class, like myself, has to concede that they just lack the glamor of their more magnetic (or more obviously magnetic) relatives. They do not sport strong emission lines which wheel around with huge amplitudes ($K \sim 500$ km/sec). They do not show dramatic, pulsed circular and linear polarization features which are unambiguous signs of a magnetic field (although see West *et al.* 1987). They do not show sharp X-ray eclipses, which unambiguously yield the system geometry. They do not show "low states" in which accretion nearly ceases. They do not show large-amplitude quasi-periodic oscillations at $\sim 1$ Hz, which some have interpreted as a signature of a radial accretion shock (Larsson 1985; Langer, Chanmugam and Shaviv 1982; Middleditch 1982). And, perhaps most significantly, they do not show an observable soft X-ray component, which is quite surprising since such a component is a universal hallmark of the AM Her stars. Possibly this is because the gas accretes along a wider cone of field lines in the DQ Her stars. This would result in a larger area covered at the polar cap, which would lower the temperature of the soft component and hence push it into the EUV. The hard X-ray light curves also tend to suggest a larger polar cap (King and Shaviv 1984a). On the other hand, it is widely believed that $\dot{M}$ is systematically higher in the DQ Her stars, and this would *boost* the soft X-ray emission.

These are unsolved problems. For systems which are supposed to be so fundamentally similar, it is worrying that the AM Her and DQ Her stars resemble each other so little. Now it is true that good theorists, and for that matter bad theorists, can always make excuses for such disappointments. But while not turning our back entirely on suggested excuses, I think the time is ripe for observers to work a little harder here – to explore in greater detail the relationships between these two types of magnetic CV.

## 4. NONMAGNETIC STARS

Probably $\sim 90\%$ of CVs contain white dwarfs insufficiently magnetic to affect the flow of accreting gas. In these stars the accretion disk should extend all the way to the surface of the white dwarf. While the mechanism by which gas is transported inward (and angular momentum outward) through the disk is still not known, the run of temperature in a steady-state disk, which is powered by gravitational energy and radiates locally like a blackbody, is well-known (Lynden-Bell 1969, Shakura and Sunyaev 1973, Bath *et al.* 1974) to be

$$T(R) = T_*(R/R_1)^{-3/4} \left[1 - (R_1/R)^{1/2}\right]^{1/4},$$

where

$$T_* = 4 \times 10^4 K (\dot{M}/10^{16} \text{g/s})^{1/4} (M_1/M_\odot)^{1/4} (R_1/10^9 \text{cm})^{-3/4}.$$

Here $M_1$ is the mass of the white dwarf, $\dot{M}$ is the accretion rate, and $R_1$ is the inner radius of the disk (usually assumed to be the white dwarf radius). When you sum over the disk annuli, extending the integration from $R = R_1$ to $\infty$, you find that for a wide range of frequency (not too close to the highest frequencies radiated in the disk), the flux follows the familiar law $F_\nu \propto \nu^{1/3}$ (Lynden-Bell

1969). Nearly every observer who presents optical/UV fluxes fits his data to some such formula, sometimes including corrections for the disk's outer limit in radius, interstellar reddening, or the contribution from the white dwarf photosphere. Since at typical disk temperatures of $10,000-20,000\ K$ the fluxes are significantly affected by line opacities (especially Lyman $\alpha$), Wade (1984) has published a grid of the emergent fluxes from appropriate stellar atmosphere models, to be used in lieu of the blackbody assumption. This is a substantial improvement; disheveled, well-thumbed copies of Wade's paper are on many desks these days.

But the information contained in the disk light seems to be sparse. After IUE emerged as a flashy new toy in the late 1970s, it became fashionable to derive accretion rates from fits to the optical/UV continuum fluxes (which come from the disk). This approach was more or less doomed to failure, because: (1) the continuum slopes are to first order independent of $\dot{M}$ (that's the point of the "$\nu^{1/3}$ law"); and (2) the effect of deepening Lyman $\alpha$ absorption as the models become cooler is to shuffle more flux into the short wavelengths (say 1400 – 1700Å), cancelling the effect of the cooler blackbodies. Thus there is a cruel conspiracy for optically thick disks to resemble each other quite thoroughly in the presently accessible optical/UV bandpass.

While there are still rear-guard actions being fought, and much quibbling over detail, I think it is fair to say that most experienced observers now believe that for CVs with a sufficiently high $\dot{M}$, the disks *just look optically thick*, and do not contain decipherable information about $\dot{M}$ in their continuum slopes.

At the surface of the white dwarf, the gas must adjust itself from Keplerian rotation to the unknown rotation speed of the white dwarf. A boundary layer forms in this small region, and for a nonrotating or slowly rotating white dwarf the luminosity $L_{\rm BL}$ is about half the total accretion luminosity:

$$L_{\rm BL} \approx (1/2)GM_{\rm wd}\dot{M}/R_{\rm wd}.$$

At what wavelength will this luminosity emerge? Well, in radial accretion there is a total of $\sim 100$ keV/nucleon available, and in a strong shock the characteristic temperature is $\sim 10\%$ of this (Landau and Lifshitz 1959). But half of the infall energy is radiated in the disk, so the expected shock temperature is reduced to $\sim 5$ keV, or perhaps a bit less – excellently placed for detection in the 0.1–4.0 keV window of the *Einstein* IPC, as well as the previous generation of hard X-ray (2–10 keV) experiments.

Are these X-rays observed? To first order, the answer appears to be NO; of the disk accretors, only one – SS Cyg – manages to barely peek over the flux threshold of the *UHURU* catalogue. Cordova, Mason and Nelson (1981) and Cordova and Mason (1983) observed several dozen of the brightest CVs with the IPC, and found nearly all of them to be hard X-ray sources (typically with $kT \sim 10$ keV), but with luminosities several orders of magnitude too low to represent the predicted boundary layer luminosity. Ferland *et al.* (1982) wondered if these observations could be reconciled at all with the very existence of a disk/star boundary layer from which much of the gravitational luminosity is radiated.

## 4.1 The Soft X-ray/EUV Component

Actually, the earlier HEAO-1 observations of SS Cyg and U Gem in outburst contained a very substantial clue to this puzzle. Cordova *et al.* (1980) found an extremely luminous, very soft X-ray source ($kT \sim 30$ eV) in SS Cyg. The estimated luminosity of this source, $2 \times 10^{33}$erg s$^{-1}$, suggests an accretion rate of $10^{-9} M_\odot$/yr, and the estimated temperature of $3 \times 10^5 K$ suggests a radiating area only 0.001 times the surface area of a white dwarf. Similar best-fit estimates for U Gem in eruption were obtained by Cordova *et al.* (1984). The uncertainty in temperature severely affects the derived luminosity and radiating area, but these results do suggest the kind of emission one would expect from an *optically thick boundary layer* (as predicted, *e.g.*, by Pringle 1977).

Patterson and Raymond (1985a) studied all of the *Einstein* data for high-$\dot{M}$ systems, and concluded that the data were consistent with a simple optically thick boundary layer model. A power-law approximation to their results yields an expected soft X-ray flux

$$F_{\mathrm{sx}}\,(0.12 - 1.0\,\mathrm{keV}) = 2.1 \times 10^{-12} \dot{M}_{17.5}^{2.1}\ M_{0.7}^{8.0}\ d_{200}^{-4.5}\ \mathrm{erg\ cm^{-2}s^{-1}}$$

where $M_{0.7}$ is the white dwarf mass in units of $0.7 M_\odot$, $\dot{M}_{17.5}$ is the accretion rate in units of $3 \times 10^{17}$ g/sec, $d_{200}$ is the distance in units of 200 pc, and where an average $N_{\mathrm{H}}$-distance relation is assumed. Although not valid for values of $\dot{M}_{17.5}$, $M_{0.7}$, and $d_{200}$ far from unity, this equation illustrates the great sensitivity to the three critical parameters. Thus the systems most promising for soft X-ray emission should have a massive white dwarf, a high accretion rate, and a small distance (or fairly transparent line of sight). There are only two stars known which satisfy all three criteria: SS Cyg, with $M_1 = 1.0 - 1.3\ M_\odot$, $d = 100$ pc, $\dot{M} = 10^{18}$ g/s in eruption (Stover *et al.* 1980; Cowley, Crampton and Hutchings 1980); and U Gem, with $M_1 = 1.1\ M_\odot$, $d = 80$ pc, $\dot{M} = 10^{18}$ g/s in eruption (Stover 1981). Thus it does not seem mysterious that these two stars are by far the brightest in soft X-rays.

## 4.2 The Hard X-ray Component

Early X-ray observations of SS Cyg (Ricketts, King and Raine 1978; Cordova *et al.* 1980) showed that when the star is in quiescence (*i.e.* when the accretion rate is low), the soft X-rays disappear altogether and the star becomes a weak hard X-ray source, with $kT \sim 30$ keV. Pringle and Savonije (1979) and Tylenda (1981) pointed out that there ought to be a critical $\dot{M}$ below which the boundary layer becomes optically thin, and the less efficient cooling will drive the temperature up to high values, of order $10^8 K$. They roughly estimated that this transition would occur for an accretion rate near $10^{16}$ g/s.

Patterson and Raymond (1985b) collected all of the published hard X-ray fluxes obtained from disk accretors in the IPC, and used estimates of distance, accretion rate, and simultaneous visual magnitude to prepare Figure 4. Here we see the correlation of $L_x$ and $F_x/F_v$ with $\dot{M}$, with the predictions of several models superimposed. Please bear with me as I slog through this quite messy diagram!

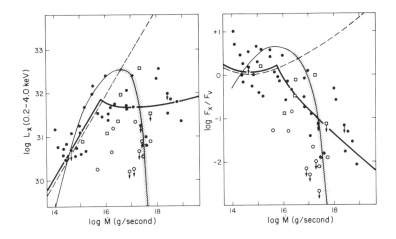

Figure 4. Predicted dependence of $L_x$ and $F_x/F_v$ on accretion rate for three boundary-layer models, with observational data superposed. See text for details.

The dashed line shows the prediction of a maximally naive model, in which the BL is assumed to be optically thin with $kT = 10$ keV. This model works admirably for $\dot{M} < 10^{16}$ g/s, but fails utterly for higher accretion rates. The light solid curve is Tylenda's (1981) model of a turbulent BL, in which $kT$ varies because the cooling is more efficient at higher $\dot{M}$. Again there is approximate agreement for low $\dot{M}$, but failure for higher values. Finally, the heavy solid curve is the attempt by Patterson and Raymond (1985b) to allow for a density gradient in the accretion disk – permitting some hard X-ray emission at the top of the disk where the local density is low. This shows an improved fit to the data, except for the erupting dwarf novae (shown by the open circles) which are seen to be very underluminous.

In Figure 5 we show a picture of this boundary layer structure. The dotted region is optically thin, and radiates an X-ray bremsstrahlung component at $\sim 10^8 K$. The shaded region is optically thick, and radiates nearly half of the accretion energy in a blackbody component at $\sim (1–3) \times 10^5 K$. Sufficiently far from the mid-plane of the disk, the density is too low for efficient cooling, and so even the high-$\dot{M}$ systems radiate a low-luminosity, high-temperature X-ray component.

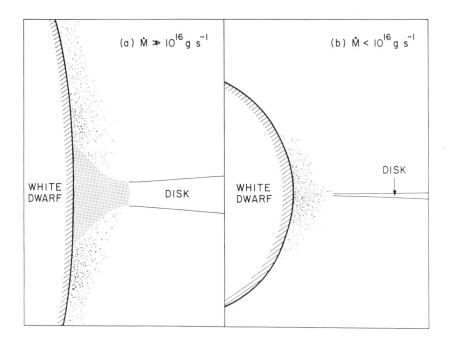

Figure 5. Picture of the boundary layer for (a) high accretion rates and (b) low accretion rates.

## 5. CONCLUSIONS

Let's return to our original question: *why do white dwarfs appear to be such faint X-ray emitters?* There seems to be a conspiracy at work. For low-$\dot{M}$ disk accretors, the efficiency of X-ray production is quite high ($L_x/L_{opt} \sim 1$), but the low accretion rate keeps these sources faint. High-$\dot{M}$ disk accretors develop optically thick boundary layers, and radiate their very substantial luminosity primarily in the EUV. The magnetic accretors have a similar story. The AM Her stars have a low accretion rate, and are kept out of all-sky X-ray catalogues primarily by their low $\dot{M}$, and probably sometimes also by an unexplained fondness for soft X-rays. The DQ Her stars have higher $\dot{M}$, but allot most of their accretion power to other wavelengths, presumably the EUV. The problem of how gas accretes onto these stars is important and unsolved.

Over the next decade, there is plenty of reason to expect great advances in this field. The relatively high-$\dot{M}$ stars are probably spectacular sources at EUV energies, and the flights of ROSAT and EUVE are likely to reveal them – especially nearby dwarf novae in eruption. On the other hand, progress in understanding the accretion processes at work in the low-$\dot{M}$ stars is hampered *principally by the low count rates*, and one would like a low-background, large-area detector sensitive to hard X-rays. For these stars, the show will resume when AXAF flies.

# REFERENCES

Agrawal, P.C., Riegler, G.R. and Rao, A.R. 1983, *Nature*, **301**, 318.

Bailey, J. 1985, in *Recent Results on Cataclysmic Variables, Proc. Bamberg ESA Workshop*

Bath, G.T., Evans, W.D., Papaloizou, J. and Pringle, J.E. 1974, *M.N.R.A.S.*, **169**, 447.

Berg, R.A. and Duthie, J.G. 1977, *Ap. J.*, **211**, 859.

Beuermann, K., Schwope, A., Weissieker, H. and Motch, C. 1985, *Space Sci. Rev.*, **40**, 135.

Beuermann, K., Stella, L. and Patterson, J. 1987a, *Ap. J.*, **316**, 360.

Beuermann, K., Thomas, H.C., Giommi, P. and Tagliaferri, G. 1987b, *Astr. Ap.*, **175**, L9.

Biermann, P., Schmidt, G.D., Liebert, J., Stockman, H.S., Tapia, S., Kuhr, H., Strittmatter, P.A., West, S. and Lamb, D.Q. 1985, *Ap. J.*, **293**, 303.

Bond, H.E. and Tifft, W. 1974 *Pub. A.S.P.*, **86**, 981.

Bonnet-Bidaud, J.M. *et al.* 1985, in *Recent Results on Cataclysmic Variables, Bamberg ESA Workshop*, p. 155

Bradt, H.V. and McClintock, J.E. 1983 *Ann. Rev. Astr. Ap.*, **21**, 13.

Bradt, H.V., Doxsey, R.E. and Jernigan, J.G., 1978, in *X-ray Astronomy*, ed W.A. Baity and L.E. Peterson, (Oxford: Pergamon), p.3

Cameron, A.G.W. and Mock, 1967, *Nature*, **215**, 464.

Chanan, G.A., Nelson, J.E. and Margon, B. 1978 *Ap. J.*, **226**, 963.

Chanmugam, G. and Ray, A. 1984, *Ap. J.*, **285**, 252.

Cook, M.C., Watson, M.G. and McHardy, I.M. 1984, *M.N.R.A.S.*, **210**, 7p.

Cordova, F.A., Chester, T.J., Tuohy, I.R. and Garmire, G.P. 1980, *Ap. J.*, **235**, 163.

Cordova, F.A., Mason, K.O. and Nelson, J.E. 1981, *Ap. J.*, **245**, 609.

Cordova, F.A. and Mason, K.O. 1983, in *Accretion Driven Stellar X-ray Sources*, ed. W.H.G. Lewin and E.P.J. van den Heuvel (Cambridge: Cambridge Univ. Press)

Cordova, F.A., Chester, T.J., Mason, K.O., Kahn, S.M. and Garmire, G.P. 1984, *Ap. J.*, **278**, 739.

Cowley, A.P., and Crampton, D. 1977, *Ap. J. (Letters)* , **212**, L121.

Crampton, D., Cowley, A.P. and Hutchings, J.B. 1980, *Ap. J.*, **241**, 269.

Fabian, A.C., Pringle, J.E. and Rees, M.J. 1976, *M.N.R.A.S.*, **173**, 43.

Fabbiano, G., Hartmann, L., Raymond, J.C., Steiner, J.E., Branduardi-Raymont, G. and Matilsky, T. 1981, *Ap. J.*, **243**, 911.

Ferland, G.J., Langer, S.H., MacDonald, J., Pepper, G.H., Shaviv, G. and Truran, J.W. 1982, *Ap. J. (Letters)*, **262**, L53.

Forman, W. *et al.* 1978, *Ap. J. (Supp.)*, **38**, 357.

Frank, J., King, A.R., and Lasota, J.-P. 1988, *Astr. Ap.*, **193**, 113.

Hearn, D.R. and Marshall, F. 1977, *Ap. J. (Letters)*, **232**, L21.

Hearn, D.R., Richardson, J.A. and Clark, G.W. 1977, *Ap. J. (Letters)*, **210**, L23.

Hearn, D.R. and Richardson, J.A. 1977, *Ap. J.*, **213**, L115.

Heise, J., Kruszewski, A., Chlebowski, T., Mewe, R., Kahn, S. and Seward, F.D. 1984, *Phys. Scr.*, **T7**, 115.

Heise, J., Brinkmann, A.C., Gronenschild, E., Watson, M.G., King, A.R., Stella, L. and Kieboom, K. 1985, *Astr. Ap.*, **148**, L14.

Heise, J., and Verbunt, F. 1988, *Astr. Ap.*, **189**, 112.

Herbst, W., Hesser, J.E. and Ostriker, J.P. 1974, *Ap. J.*, **193**, 679.

Imamura, J.N., Wolff, M.T. and Durisen, R.H. 1984, *Ap. J.*, **276**, 667.

Imamura, J.N., Durisen, R.H., Lamb, D.Q., and Weast, G.J. 1987, *Ap. J.*, **313**, 298.

Jensen, K. *et al.* 1982, *Ap. J.*, **261**, 625.

Katz, J.I. 1975, *Ap. J.*, **200**, 298.

Katz, J.I. 1977, *Ap. J.*, **215**, 265.

King, A.R., Ricketts, M.J. and Warwick, R.S. 1979, *M.N.R.A.S.*, **187**, 77p.

King, A.R. and Shaviv, G. 1984a, *M.N.R.A.S.*, **211**, 883.

King, A.R. and Shavivi, G. 1984b, *Nature*, **308**, 519.

Kruszewski, A., Mewe, R., Heise, J., Chlebowski, T., van Dijk, W. and Ballker, R. 1981, *Space Sci. Rev.*, **30**, 221.

Krzeminski, W. and Serkowski, K. 1977, *Ap. J. (Letters)*, **216**, L45.

Kuijpers, J. and Pringle, J.E. 1982, *Astr. Ap.*, **114**, L4.

Kylafis, N.D. and Lamb, D.Q. 1979, *Ap. J. (Letters)*, **228**, L105.

Kylafis, N.D. and Lamb, D.Q. 1982, *Ap. J. (Supp.)*, **48**, 239.

Lamb, D.Q. 1974, *Ap. J. (Letters)*, **192**, L129.

Lamb, D.Q. 1985, in *Cataclysmic Variables and Low-Mass X-ray Binaries*, ed. D.Q. Lamb and J. Patterson (Dordrecht: Reidel)

Landau, L. and Lifshitz, 1959, *Fluid Mechanics* (Reading: Addison-Wesley)

Langer, S.H., Chanmugam, G. and Shaviv, G. 1982, *Ap. J.*, **258**, 289.

Larsson, S. 1985, *Astr. Ap.*, **145**, L1.

Liebert, J. and Stockman, H.S. 1985, in *Cataclysmic Variables and Low-Mass X-ray Binaries*, ed. D.Q. Lamb and J. Patterson (Dordrecht: Reidel)

Lynden-Bell, D. 1969, *Nature*, **223**, 690.

Marshall, F.E., *et al.*, *Ap. J. (Supp.)*, **40**, 657.

Mason, K.O. 1985, *Space Sci. Rev.*, **40**, 99.

Mason, K.O. *et al.* 1983, *Ap. J.*, **264**, 575..

Mazeh, T. *et al.* 1986, *M.N.R.A.S.*, **221**, 513.

McHardy, I.M. *et al.*, 1981, *M.N.R.A.S.*, **197**, 893.

McHardy, I.M., Pye, J.P., Fairall, A.P., Warner, B., Cropper, M. and Allen, S. 1984, *M.N.R.A.S.*, **210**, 663.

McHardy, I.M., Pye, J.P., Fairall, A.P., and Menzies, J.W. 1987, *M.N.R.A.S.*, **225**, 355.

Middleditch, J. 1982, *Ap. J. (Letters)*, **257**, L71.

Morris, S.L., Schmidt, G.D., Liebert, J., Stocke, J., Gioia, I.M. and Maccacaro, T. 1987, *Ap. J.*, **314**, 641.

Mukai, K. *et al.* 1985, *Space Sci. Rev.*, **40**, 151.

Mukai, K. 1988, *M.N.R.A.S.*, **232**, 175.

Norton, A.J., Watson, M.G. and King, A.R. 1988, *M.N.R.A.S.*, **231**, 738.

Nousek, J.A. *et al.* 1984, *Ap. J.*, **277**, 682.

Osborne, J.P. 1987, in *IAU Colloq. No. 93, Cataclysmic Variables: Recent Multifrequency Observations and Theoretical Developments*, ed. H. Drechsel, Y. Kondo, and J. Rahe (Dordrecht: Reidel)

Osborne, J.P., Rosen, R., Mason, K.O. and Beuermann, K. 1985, *Space Sci. Rev.*, **40**, 143.

Osborne, J.P. *et al.* 1986a, *M.N.R.A.S.*, **221**, 823.

Osborne, J.P. *et al.* 1986b, in *Recent Results on Cataclysmic Variables, Proc. Bamberg ESA Workshop*, p. 161

Osborne, J.P., Giommi, P., Tagliaferri, G. and Stella, L. 1988, *Ap. J. (Letters)*, **328**, L45.

Patterson, J. 1979a, *Ap. J. (Letters)*, **233**, L13.

Patterson, J. 1979b, *Ap. J.*, **234**, 978.

Patterson, J., Branch, D., Chincarini, G. and Robinson, E.L. 1980, *Ap. J. (Letters)*, **240**, L133.

Patterson, J. and Garcia, M. 1980, *IAU Circ. No. 3514*

Patterson, J. and Price, C. 1980, *IAU Circ. No. 3511*

Patterson, J., Williams, G.A. and Hiltner, W. 1981, *Ap. J.*, **245**, 618.

Patterson, J. and Price, C. 1981, *Ap. J. (Letters)*, **243**, L83.

Patterson, J. and Steiner, J.E. 1983, *Ap. J. (Letters)*, **264**, L61.

Patterson, J. and Raymond, J.C. 1985a, *Ap. J.*, **292**, 550.

Patterson, J. and Raymond, J.C. 1985b, *Ap. J.*, **292**, 535.

Patterson, J., Beuermann, K., Lamb, D.Q., Fabbiano, G., Raymond, J.C., Swank, J. and White, N.E. 1984, *Ap. J.*, **279**, 785.

Petterson, J.A. 1980, *Ap. J.*, **241**, 247.

Pietsch, W., Voges, W., Kendziorra, E. and Pakull, M. 1987, in *IAU Colloquium No. 93, Cataclysmic Variables: Recent Multifrequency Observations and Theoretical Developments*, ed. H. Drechsel, Y. Kondo and J. Rahe (Dordrecht: Reidel)

Priedhorsky, W., Marshall, F.J. and Hearn, D.R., 1987 *Astr. Ap.*, **173**, 95.

Pringle, J.E. 1977, *M.N.R.A.S.*, **178**, 195.

Pringle, J.E. and Rees, M.J. 1972, *Astr. Ap.*, **22**, 1.

Pringle, J.E. and Savonije, G.J. 1979, *M.N.R.A.S.*, **187**, 777.

Remillard, R.A., Bradt, H.V., McClintock, J.E., Patterson, J., Roberts, W., Schwartz, D.A. and Tapia, S. 1986, *Ap. J. (Letters)*, **302**, L11.

Ricketts, M.J., King, A.R. and Raine, D.J. 1978, *M.N.R.A.S.*, **186**, 233.

Robinson, E.L. 1976, *Ann. Rev. Astr. Ap.*, **14**, 119.

Rothschild, R.E. *et al.* 1981, *Ap. J.*, **250**, 723.

Schrijver, J., Brinkman, A.C., van der Woerd, H., Watson, M.G., King, A.R., van Paradijs, J. and van der Klis, M. 1985, *Space Sci Rev.*, **40**, 121.

Shakura, N. I. and Sunyaev, R.A. 1973, *Astr. Ap.*, **24**, 337.

Singh, K.P., Agrawal, P.C. and Riegler, G.R. 1984, *M.N.R.A.S.*, **208**, 679.

Steiner, J.E., Schwartz, D.A., Jablonski, F.J., Busko, I.C., Watson, M.G., Pye, J.P. and McHardy, I.M. 1981, *Ap. J. (Letters)*, **249**, L21.

Stella, L., Beuermann, K., and Patterson, J., 1986, *Ap. J.*, **306**, 225.

Stockman, H.S., Schmidt, G.D., Angel, J.R.P., Liebert, J., Tapia, S. and Beaver, E.A. 1977, *Ap. J.*, **217**, 815.

Stockman, H.S., Schmidt, G.D., and Lamb, D.Q., 1988, *Ap. J.*, **332**, 282.

Stover, R.J. 1981, *Ap. J.*, **248**, 684.

Stover, R.J., Robinson, E.L., Nather, R.E. and Montemayor, T.J. 1980, *Ap. J.*, **240**, 597.

Swank, J. *et al.* 1977, *Ap. J. (Letters)*, **216**, L71.

Szkody, P., Schmitt, E., Crosa, L. and Schommer, R. 1981, *Ap. J.*, **246**, 223.

Szkody, P. and Brownlee, D. 1977, *Ap. J. (Letters)*, **212**, L113.

Tapia, S. 1976, *IAU Circ. No. 2987*

Tapia, S. 1977a, *Ap. J. (Letters)*, **212**, L125.

Tapia, S. 1977b, *IAU Circ. No. 3054*

Tuohy, I.R., Lamb, F.K., Garmire, G.P. and Mason, K.O. 1978, *Ap. J. (Letters)*, **226**, L17.

Tuohy, I.R., Mason, K.O., Garmire, G.P. and Lamb, F.K. 1981, *Ap. J.*, **245**, 183.

Tuohy, I.R., Buckley, D.A.H., Remillard, R., Bradt, H.V. and Schwartz, D.A. 1986, *Ap. J.*, **311**, 275.

Tylenda, R. 1981, *Acta Astr.*, **31**, 267.

Vogt, N., Krzeminski, W. and Sterken, C. 1980, *Astr. Ap.*, **85**, 106.

Wade, R. A. 1984, *M.N.R.A.S.*, **208**, 381.

Warner, B. 1983, in *IAU Colloq. No. 72, Cataclysmic Variables and Related Objects*, ed. M. Livio and G. Shaviv (Dordrecht: Reidel)

Warwick, R.S. *et al.*, 1981, *M.N.R.A.S.*, **197**, 865.

Watson, M.G., King, A.R. and Osborne, J.P. 1985, *M.N.R.A.S.*, **212**, 917.

Watson, M.G. and King, A.R. 1988, preprint

West, S.C., Berriman, G., and Schmidt, G.D. 1987, *Ap. J. (Letters)*, **322**, L35.

White, N.E. 1981, *Ap. J. (Letters)*, **244**, L85.

White, N.E. and Marshall, F.E. 1980, *IAU Circ. No. 3514*
White, N.E. and Marshall, F.E. 1981, *Ap. J. (Letters)*, **249**, L25.

# ACCRETING GALACTIC X-RAY SOURCES

**Paul Hertz**

E.O. Hulburt Center for Space Research

Naval Research Laboratory

## 1. INTRODUCTION

The brightest x-ray sources in the sky are Galactic x-ray binaries. These binary star systems consist of a compact object (neutron star, black hole) and a stellar companion. In low mass x-ray binaries (LMXRBs), the companion typically has a mass less than 1 $M_\odot$. The binary system is a compact one and the stellar companion fills its Roche lobe. X-rays are produced by the accretion of material from the companion into an accretion disk, and then onto the compact object.

Although we now understand the basic nature of the $\sim$100 LMXRBs in the "Galactic bulge", Galactic plane, and globular cluster system of the Galaxy, this class of sources was initially identified through a set of common observed properties including (i) high brightness or luminosity, (ii) faint optical counterpart, (iii) position near older Population II regions of the Galaxy, (iv) relatively soft ($\sim$5 keV) spectrum, (v) the absence of pulses or eclipses, and (vi) the presence of x-ray bursts (Lewin and Joss 1983). From these properties, the basic nature of LMXRBs, as described above, was suggested (Joss and Rappaport 1979).

The presence of a low mass, unevolved companion star, their location in older regions of the Galaxy far removed from star forming regions, and the lack of x-ray pulsations (indicating that any magnetic field the neutron star may have had at formation has decayed), all imply that LMXRBs are old systems. However after 20 years of study, and 10 years after we began to understand them, the formation and evolution of LMXRBs is still a matter of question and active research. Detailed studies with the *Einstein Observatory* and other instruments have lead to a detailed understanding of some individual sources; however, broad questions concerning the entire class of LMXRBs remain unanswered. The recent discovery of quasi-periodic oscillations (QPOs) from many of the brightest LMXRBs (*e.g.* Lewin, van Paradijs, and van der Klis 1988) has further stimulated theoretical and observational studies of these sources.

In this review I will use *Einstein Observatory* observations of Galactic accretion sources to highlight recent advances in the understanding of individual LMXRBs. I will treat in some detail advances made in the study of x-ray sources in globular clusters with *Einstein*, including the discovery of a class of low luminosity accreting x-ray sources, with luminosities $10^3$ times less than LMXRBs. I will also discuss the search for low luminosity accretion sources in the Galactic plane. Finally I will indicate some observational tests of current theories which may be possible with future instruments.

## 2. THE EINSTEIN OBSERVATORY AND OBSERVATIONS OF LOW MASS X-RAY BINARIES

*Einstein* carried several instruments which were well suited for the detailed study of individual LMXRBs. It also had capabilities which made possible several surveys of LMXRBs in the Galaxy and in globular clusters that provided important constraints on the formation of LMXRBs as a class. These capabilities may be summarized as imaging, sensitivity, spectroscopy, timing, and archival. I discuss each in turn below.

### 2.1 Imaging

One of *Einstein*'s most widely used capabilities was the arcsecond imaging accuracy of the High Resolution Imager (HRI). This allowed the determination of accurate positions for compact (unresolved) x-ray sources. Since LMXRBs are found in regions with high densities of field stars such as the Galactic plane and bulge, and since the optical counterparts of LMXRBs are typically quite faint (V > 18), the arcsecond error circles obtained with the HRI led to the identification of optical counterparts for a number of LMXRBs including the globular cluster source 4U2129+12 in M15 (Auriere, Le Fevre, and Terzan 1984), the persistent bulge source GX13+1 (Garcia and Grindlay 1987), and the x-ray burster 4U1415-05 (Grindlay *et al.* 1988). A second use of arcsecond x-ray positions was the statistical determination of the mass of globular cluster x-ray sources from the offset of the x-ray source relative to the cluster center (Grindlay *et al.* 1984; see below).

Imaging of diffuse x-ray sources with both the HRI and the Imaging Proportional Counter (IPC) made possible another set of discoveries involving LMXRBs. These included the study of grains in the interstellar medium through the observation of scattering halos about bright x-ray sources (Mauche and Gorenstein 1986), the x-ray morphology of the precessing jets in SS 433 (Watson *et al.* 1983), and the detection of diffuse x-ray emission in several globular clusters (Hartwick, Cowley, and Grindlay 1982).

### 2.2 Sensitivity

*Einstein* was the most sensitive x-ray mission flown to date. This allowed the detection and localization of faint x-ray sources. Several surveys for faint Galactic accretion sources were undertaken. A survey of over 70 Galactic globular clusters led to the discovery of a new class of low luminosity globular cluster x-ray sources (Hertz and Grindlay 1983a,b). A search for faint sources in the Galactic plane was also carried out, although with ambiguous results (Hertz and Grindlay 1984, 1988). These surveys are discussed below.

## 2.3 Spectroscopy

There were several spectrometers on board the *Einstein Observatory*, ranging from the sensitive but low resolution IPC, through the Monitor Proportional Counter (MPC), Solid State Spectrometer (SSS), to the high resolution Focal Plane Crystal Spectrometer (FPCS). Spectroscopy of accreting x-ray sources revealed valuable diagnostics into the physical conditions present in the accretion stream and the accretion disk about the compact object. Spectroscopic results are reviewed elsewhere in this volume (Holt 1989) and will not be discussed in detail here.

## 2.4 Timing

As with any pointed instrument, x-ray sources were monitored for variability with *Einstein*. The sensitivity of *Einstein* allowed the detection of small changes in x-ray flux, or the detection of changes in x-ray flux in short intervals. Timing experiments fell into two categories: studying known variability such as eclipses or pulses, and the search for serendipitous variability in sources observed for other reasons. One example of the former was the study of the partial eclipses in 4U2129+47 (McClintock *et al.* 1982). Serendipitous discoveries included the periodic dips in the light curve of the x-ray burster 4U1415-05 (Walter *et al.* 1982) and x-ray bursts from the "Galactic bulge" source GX17+2 (Kahn and Grindlay 1984).

## 2.5 Archival

*Einstein* observed a large number of known LMXRBs during its operational lifetime. Subsequent analysis of this archival information in a uniform way leads to understanding of the class of LMXRBs in a statistical sense. Studies of the variability of LMXRBs (Maccacaro, Garilli, and Mereghetti 1987) and their spectral characteristics (Garcia 1989) are among the archival studies that have been performed. The *Einstein* archival data bank also supports re-analysis of *Einstein* data in the light of subsequent discoveries. For example the 11 minute binary period of the burster 4U1820-30 in NGC 6624 (Morgan, Remillard, and Garcia 1988) was confirmed in archival *Einstein* MPC data. Also once the importance of millisecond pulsations and QPOs was appreciated, it was possible to search *Einstein* data for evidence of both these phenomena in LMXRBs (Mereghetti and Grindlay 1987).

## 3. LOW MASS X-RAY BINARIES

The study of LMXRBs is a large industry in x-ray astrophysics. I will not review the field here, as that has been done adequately elsewhere (*e.g.*, Lewin and Joss 1983; Mason 1986; Verbunt 1988; White 1989). Many of the individual results from LMXRBs are referred to in Section 2. In this section, I consider two specific aspects of LMXRBs were *Einstein* observations, particularly those for which arcsecond HRI positions identified optical counterparts, have led to better understanding of the nature of the sources.

## 3.1 Evidence for Binary Nature

As mentioned above, the class of LMXRBs was identified through common observational properties, none of which gave evidence of the binary nature of the sources.
In fact, the absence of x-ray eclipses and bright companion stars were two of the
defining characteristics. The success of the model of an accreting compact object in
a close binary when applied to the x-ray pulsars, or high mass x-ray binaries, was
used to infer that a similar model would explain the similarly luminous LMXRBs.
For the x-ray pulsars, Doppler shifted pulsations and x-ray eclipses are common
indicators of binary systems. However pulsations and eclipses are rare in LMXRBs,
and are present in only a few, possibly atypical, systems.

In the last 15 years, binary periods have been confirmed in $\sim$25 LMXRB systems
(Parmar and White 1988). The discovery of so many periods was enabled primarily
in two ways. First, arcsecond positions from *Einstein*, HEAO-1, and EXOSAT led
to secure optical counterparts. The availability of sensitive optical CCD detectors
has made possible the detectability of orbital periods in 17th - 20th magnitude
stars. Second, the long orbital period of EXOSAT made possible continuous observations of LMXRBs, without the interruptions caused by Earth occultation during
observations with low Earth orbit satellites such as the *Einstein*. Short periods ($<$
1 day) with moderate amplitude fluctuation were detectable with EXOSAT.

The periods of LMXRBs fall primarily in the range 1 - 8 hours. This is evidence
for the low mass nature of the companion star. More direct evidence is provided
in those cases where the secondary is directly observable. Although the optical
light from LMXRBs is dominated by x-rays reprocessed in the accretion disk, in
some cases stellar absorption lines from the companion star are visible. Also, some
LMXRBs are transient in nature; when the x-rays are "turned off" the low mass
stellar companion is clearly visible optically (see, *e.g.*, van Paradijs 1983; Bradt and
McClintock 1983).

## 3.2 Evidence for X-ray Binaries in Hierarchical Triples

Although the binary nature of LMXRBs has been firmly established, anomalous
behavior in several LMXRBs has been explained recently as the result of the x-ray
binary being the inner member of a hierarchical triple star system. Sources that lend
themselves to this explanation include: Cyg X$-$3, where the radio and x-ray periods
differ and the phase of the radio period undergoes phase jumps, both of which can
be explained by a third body in the system (Molnar 1986); GX17+2, which has an
anomolously high accretion rate and a short period (Bailyn and Grindlay 1987a);
4U1915-05, which has different optical and x-ray periods (Grindlay *et al.* 1988);
and 4U2129+47, in which an F7 star showing no orbital motion is present when
the x-ray source is turned off (Garcia, Bailyn, and Grindlay 1989). In addition,
long periodic flux variations, such as that observed by Vela 5 in several LMXRBs
(Priedhorsky and Holt 1987), have been attributed to hierarchical triples.

Although the theory can be used to explain flux and period variations, it is difficult to understand how an LMXRB could have evolved in a hierarchical triple: it is likely that the formation of the neutron star would disrupt the system. One suggestion is that the hierarchical triples formed in globular clusters, where this is a more likely scenario (Grindlay 1986). Under this scenario, the LMXRB forms in a globular cluster through tidal capture; the hierarchical triple forms either through tidal capture or exchange scattering. Finally the cluster is disrupted by passage through the Galactic bulge or through encounters with giant molecular clouds.

The lasting value of using hierarchical triples to explain long term periods and period differences in LMXRBs must await more definitive data. This data will come from the Hubble Space Telescope and AXAF.

## 4. GLOBULAR CLUSTER X-RAY SOURCES

Prior to the launch of *Einstein*, seven bright globular cluster x-ray sources (GCXSs) were known, and the identification of several of these with globular clusters was tentative (Grindlay 1977). Although the first GCXSs were identified in the early days of the *UHURU* mission (Giacconi *et al.* 1974), their basic nature remained unclear: were they similar to Galactic plane sources or specific to globular clusters? Theories supporting the latter included massive black holes in the cluster cores accreting gas from the intracluster medium (Grindlay and Gursky 1976). Once LMXRBs were identified as a class, it became clear that GCXSs belonged to the same class (see, *e.g.*, Lewin and Joss 1983). The discovery of x-ray bursts from most GCXSs solidified their identification with accreting neutron stars.

The launch of *Einstein* led to rapid progress in understanding GCXSs, including the determination of the GCXS mass in a statistical manner, the GCXS luminosity function, and the optical counterpart of one GCXS. *Einstein* observations also led to several serendipitous discoveries including a new class of x-ray sources, the low luminosity globular cluster x-ray sources.

### 4.1 The Mass of Luminous Globular Cluster X-ray Sources

It had been noted soon after GCXSs were discovered that the mass of these sources could be determined through statistical techniques (Bahcall and Lightman 1976). Since a globular cluster is a relaxed gravitating system, a class of objects has a spatial distribution in the cluster which is a function of mass only – heavier objects are more centrally concentrated. Thus the mass of the GCXSs could be determined by observing the distribution of the GCXSs relative to the bulk of the stars in the host globular clusters (Lightman, Hertz, and Grindlay 1980). This required the accurate measurement of three parameters: the position of the cluster center, the position of the x-ray source, and the cluster core radius. A pre-*Einstein* determination indicated that GCXSs are more massive than globular cluster field stars, but could not constrain the x-ray source mass (Jernigan and Clark 1979).

In order to determine masses as low as 3-5 times the field star mass, which is expected since GCXSs contain a 1.4 $M_\odot$ neutron star and a low mass companion and the typical field star mass is 0.5 $M_\odot$, the observational parameters needed to be determined with arcsecond accuracy. The optical parameters were determined by digitization of photographic plates (Hertz and Grindlay 1985). HRI observations of 8 GCXSs, and application of the proper statistical test, allowed the GCXS mass to be measured; it is 1.8-3.8 times the field star mass (Grindlay *et al.* 1984). This remains the best measurement of the mass of LMXRBs.

## 4.2 Optical Counterparts of Globular Cluster X-ray Sources

The arcsecond positions of GCXSs determined with the *Einstein Observatory* HRI provided the first real opportunity to detect the optical counterparts of GCXSs. The cores of globular clusters are so crowded that previous x-ray error boxes contained too many optical candidates. Optical counterparts have been searched for in all of the GCXSs by looking for blue excess objects in or near the error box. Unfortunately most searches have been unsuccessful, probably because the fields are too crowded and the optical counterparts are too faint. This situation will be rectified by observations with the Hubble Space Telescope, which will conduct counterpart searches as part of the Guaranteed Time Observations in the early part of its mission.

The one success story has been the globular cluster M15, which contains the x-ray source 4U2129+12. On the basis of positional coincidence and an ultraviolet excess, the star AC211 has been identified as the optical counterpart of 4U2129+12 (Auriere, Le Fevre, and Terzan 1984). AC211 shows several other properties similar to the optical counterparts of LMXRBs (van Paradijs 1983), including photometric variability and emission lines (Auriere *et al.* 1986; Charles, Jones, and Naylor 1986). The identification was secured when an 8.6 hour period was discovered in the optical data (Ilovaisky *et al.* 1987; Naylor *et al.* 1988) and later detected in archival x-ray data (Hertz 1987).

The x-ray source in M15 is unusual for reasons other than having an optical counterpart and a known binary period. These include: (i) 4U2129+12 is the only GCXS which has not been observed to burst, (ii) the x-ray light curve is sinusoidally modulated but varies erratically from cycle to cycle (Hertz 1987), (iii) it has a small mass function (Naylor *et al.* 1988), and (iv) it has a large systemic velocity relative to M15 (Naylor *et al.* 1988; Bailyn, Garcia, and Grindlay 1989). Although the dust has not yet settled, a consistent model for 4U2129+12/AC211 is beginning to emerge. The system has a high inclination angle, and as such the neutron star is obscured by the accretion disk. This effect was predicted a decade ago to account for the apparent paucity of binary modulation in LMXRBs (Milgrom 1978). The x-rays which are observed are scattered off an accretion disk corona into the observer's line of sight. The intrinsic x-ray luminosity is much higher than observed, and in fact the source is too bright to burst (Lewin and Joss 1983). The sinusoidal x-ray modulation is caused by variations in the height of the accretion disk which obscures the accretion disk corona. This effect is observed in other accretion disk corona LMXRBs (*e.g.* White *et al.* 1981; McClintock *et al.* 1982; White and Holt 1982). Finally AC211 may be an evolved star. If so, the large systemic velocity may result from absorption in an outflow from the outer Lagrangian point (Bailyn and Grindlay 1987b; Bailyn, Grindlay, and Garcia 1989).

## 4.3 The Einstein Globular Cluster Survey

One of the questions the *Einstein Observatory* was able to answer was: are there any more GCXSs? Previous surveys (*e.g. UHURU*) were spotty, and it wasn't clear if there was a cutoff in the luminosity function of GCXSs at $10^{36}$ ergs s$^{-1}$, or if there were fainter sources. A survey of over 70 globular clusters with the IPC and HRI (Hertz and Grindlay 1983b) revealed the unexpected result that both answers were correct. There is indeed a cutoff in the luminosity function of GCXSs at $10^{36}$ ergs s$^{-1}$. However, a new class of low luminosity globular cluster x-ray sources (LLGCXs) is detected below $10^{34}$ ergs s$^{-1}$. The gap is extremely significant; when all available data is included, the probability that there are any x-ray sources in the Galaxy with luminosities in the luminosity gap is less than 2% (Hertz and Wood 1985). It seems likely that these sources represent a physically distinct class of objects. In addition, the *Einstein* Galactic Plane Survey (see below) provides evidence for the existence of a class of objects similar to LLGCXs in the Galactic plane.

One possibility (Hertz and Grindlay 1983a) is that the LLGCXs are cataclysmic variable (CVs, see Patterson, this volume), *i.e.*, a white dwarf accreting matter from a Roche lobe filling low mass companion. The evidence for this includes (i) the factor of $\sim 10^3$ difference in luminosity between the GCXSs and the LLGCXs is the difference in the gravitational potential, which powers accretion sources, at the surface of a neutron star and a white dwarf, (ii) the maximum luminosity observed is the theoretical maximum for an accreting white dwarf, (iii) the distribution of positions, though poorly known since most sources were detected with the IPC, indicates that the LLGCXs have lower mass than the GCXSs, and (iv) the inferred luminosity function and number density of LLGCXs is consistent with that for tidally captured white dwarfs.

Alternate identifications have been proposed (*e.g.* quiescent transients; Verbunt, van Paradijs, and Elson 1984) and more detailed calculations indicate there may be some problem forming as many LLGCXs as are observed, if they are tidally captured white dwarfs (Verbunt and Meylan 1988). In order to determine their nature definitively, it will be necessary to detect and study the optical counterparts of the LLGCXs.

### 4.4 The Search for Optical Counterparts to Low Luminosity Globular Cluster X-ray Sources

The CV model predicts quite strongly that the optical counterparts to LLGCXs will have blue excesses and exhibit short timescale variability or flickering. The search for optical counterparts to date has involved searching the relatively large IPC error boxes for stars which show either a blue excess (Grindlay 1986; Margon and Bolte 1987) or short timescale variability (Shara *et al.* 1988). For $\omega$ Cen, smaller EXOSAT CMA error boxes are available (Verbunt *et al.* 1986), but this has not led to an identification.

The failure of these searches to discover the optical counterparts of LLGCXs is a major problem. However, the current error boxes are relatively large and contain hundreds to thousands of resolved stars. Clearly more accurate x-ray positions are needed to unambiguously identify the optical counterparts. These counterparts can then be studied, much as AC211 has been, in order to understand the nature of the LLGCXs. These accurate x-ray positions will require an imaging x-ray telescope, such as the one on AXAF.

## 5. THE EINSTEIN GALACTIC PLANE SURVEY

### 5.1 The X-ray Data

The *Einstein* Galactic Plane Survey (a.k.a. the GPX Survey) is a survey for serendipitous x-ray sources detected at low Galactic latitudes by *Einstein* IPC (Hertz and Grindlay 1984, 1988). It is the low latitude complement to the *Einstein* Medium Sensitivity Survey (Gioia 1989). The GPX Survey was undertaken to study the x-ray content of the Galaxy at moderate flux levels ($10^{-13} - 10^{-11}$ ergs s$^{-1}$ cm$^{-2}$). Questions addressed by such a survey include the number-flux relation, or *logN-logS* curve, at low Galactic latitudes, the number and distribution of low luminosity Galactic x-ray sources, and the search for interesting x-ray selected objects.

The data in the GPX Survey consists of all x-ray sources exceeding a significance threshold in IPC images below 15° Galactic latitude. The original survey consisted of data obtained only by the *Einstein* Consortium members, but an extension utilizing the entire *Einstein* Data Bank is underway (Hertz and Geldzahler 1987). The survey has been conducted in such a way that the sample is well defined. Thus number-flux relations can be determined, and the expected contribution from any known class of x-ray sources can be estimated.

### 5.2 Search for Optical Counterparts

An important part of the GPX Survey is the search for the optical counterparts of the GPX sources. This is a three part process. First, examination of sky survey plates shows which sources are simply nearby bright stars. Second, three color (UBV) CCD images are obtained of the x-ray error boxes, and blue candidates are identified. Third, spectroscopy is obtained on all blue objects in the error boxes, as well as any other optical sources which observing constraints permit.

On the basis of the optical observations, GPX sources can be classified into several categories. These include (i) bright stars, (ii) extragalactic sources, (iii) Galactic sources, (iv) unidentified but possibly coronal emission from a star, and (v) unidentified and non-coronal. In Table 1 the current status of the GPX survey is given.

| Table 1 | | | |
|---|---|---|---|
| Source Classes in the GPX Survey | | | |
| | 5 Sigma Sample | North Sample | South Sample |
| area (sq. deg.) | 276 | 185 | 90 |
| median flux limit (ct s$^{-1}$) | .042 | .032 | .036 |
| total sources | 71 | 67 | 45 |
| coronal sources | 33 | 37 | 19 |
| extragalactic sources | 4 | 4 | 1 |
| Galactic non-coronal sources | 6 | 4 | 4 |
| unidentified sources | 28 | 22 | 21 |
| expected extragalactic sources | 26 | 25 | 10 |

## 5.3 Search for Galactic Accretion Sources

One result of the GPX Survey which is relevant here is the search for low luminosity Galactic accretion sources, similar in nature to the LLGCXs. We do this through a multi-stage process. First, sources with bright stars in the error box are identified. Next a catalog search is performed to identify as many GPX sources as possible with previously cataloged items. Then optical observations are performed, both photometry and spectroscopy, to identify as many blue, variable, or relatively bright objects in error boxes as possible. Finally, using the results from the high latitude *Einstein* Medium Sensitivity Survey (Gioia *et al.* 1984), we estimate the number of extragalactic sources in the GPX sample. Since we have eliminated stars and other Galactic objects in the identification process, the excess sources over the expected number of extragalactic objects (which are predominantly reddened AGNs) are Galactic accretion sources.

The current status of the GPX identification project is presented in Table 2. The GPX Survey can be divided into subsamples by choosing different acceptance criteria. The three subsamples which have been most analyzed are those consisting of 5σ sources (5 Sigma Sample; Hertz and Grindlay 1984), all 4σ sources north of $\delta = -20°$ (North Sample; Hertz and Grindlay 1988), and all 4σ sources south of $\delta = -20°$ (South Sample). From the Table it can be seen that in the North Sample there is no excess of unidentified sources over the expected number of extragalactic sources, while in the South Sample there is a significant excess. We interpret this to be evidence for a class of low luminosity Galactic accretion sources which are concentrated towards the Galactic bulge.

| Table 2 | | | |
|---|---|---|---|
| Estimated Source Content in the GPX Survey | | | |
| | 5 Sigma Sample | North Sample | South Sample |
| coronal sources | 31 (44%) | 35 (52%) | 18 (41%) |
| extragalactic sources | 27 (38%) | 27 (40%) | 11 (24%) |
| Galactic non-coronal sources | 12 (17%) | 5 ( 8%) | 16 (36%) |
| total | 71 | 67 | 45 |

Since the GPX Survey is sensitive to LLGCXs and x-ray bright CVs at distances of 3-10 kpc, these low luminosity accretion sources may represent the first detection of a Galactic bulge population of CVs. The specific nova rate in the bulge of M31 exceeds that in the disk by a factor of 10, and data on the distribution of novae and other CVs in the Galaxy are consistent with this ratio (Ciardullo *et al.* 1987). The recent discovery of three CVs in the southern GPX sample (see below) supports this possibility.

### 5.4 Three X-ray Selected Cataclysmic Variables

The possibility that a class of x-ray bright CVs represent the excess unidentified sources in the South Sample is made all the more likely by the discovery of three new CVs during our identification work at CTIO (Grindlay *et al.* 1987). During our photometry run, these objects were identified as UV-excess objects. Follow-up spectroscopy revealed Balmer lines in emission; one source shows He II in emission and is probably an AM Her star. From the equivalent widths of the Balmer emission lines in all three objects, approximate distances may be derived using the correlations of Patterson (1984). The resulting distances of 1-2 kpc are greater than optically selected CVs, suggesting that the GPX survey samples a much larger volume of the Galaxy for CVs than optical surveys.

### 6. PROSPECTS WITH AXAF

Future instruments will allow as much progress to be made in understanding LMXRBs, GCXSs, LLGCXs, and other accreting Galactic x-ray sources, both individually and as a class, as the **Einstein Observatory** did. Some aspects of these sources can be better studied with increased timing resolution, which will be addressed with timing missions such as the x-ray Timing Explorer (XTE) or the x-ray Large Array (XLA). However AXAF will have a major impact on a broad range of problems in Galactic accretion sources.

What powers the accretion in LMXRBs? What is their evolutionary status? These questions will be better answered when optical counterparts of the most luminous LMXRBs, those in the Galactic bulge, can be studied. This will require high accuracy x-ray positions from AXAF and high resolution optical studies with the Hubble Space Telescope (HST).

What are the LLGCXs? Arcsecond positions from AXAF will allow identification of optical counterparts, either from ground based telescopes in less crowded regions of globular clusters, or with the HST. The increased collecting power of AXAF will yield increased sensitivity to LLGCXs, hence the luminosity function can be probed deeper. The goal would be to study the luminosity function where the difference between LLGCXs and CVs becomes blurred. Can we detect CVs in globular clusters?

What is the geometry of accretion disk corona sources? Can we map out the structure of the accretion stream and the accretion disk? This will require time resolved high resolution spectroscopy, a task only AXAF is well suited for.

Are there additional low luminosity accretion sources in the Galactic plane? Studies of serendipitous AXAF sources, which will come from deeper images and have better determined positions, will yield further clues into the low luminosity content of the Galactic plane.

What new physics is revealed with higher resolution pictures of diffuse sources such as the jet and shell in SS 433? How does the scattering of x-rays off the interstellar medium affect x-ray spectra? What new sources will be discovered serendipitously by AXAF?

It is clear that the future of studying accreting Galactic x-ray sources will shine bright in the AXAF era.

I would like to thank the organizing committee for inviting me to the **Einstein to AXAF Symposium**. This work has been supported in part by the Office of Naval Research.

## 7. REFERENCES

Auriere, M., Le Fevre, O., and Terzan, A. 1984, *Astr.Ap.*, **138**, 415.

Auriere, M., Maucherat, A., Cordoni, J.-P., Fort, B., and Picat, J.P. 1986, *Astr.Ap.*, **158**, 158.

Bahcall, J.N., and Lightman, A.P. 1976, private communication in Bahcall and Wolf (1976).

Bahcall, J.N., and Wolf, R.A. 1976, *Ap.J.*, **209**, 214.

Bailyn, C.D., Garcia, M.R., and Grindlay, J.E. 1989, *Ap.J.*, in press.

Bailyn, C., and Grindlay, J.E. 1987a, *Ap.J.*, **312**, 748.

Bailyn, C.D., and Grindlay, J.E. 1987b, *Ap.J.(Letters)*, **316**, L25.

Bradt, H.V.D., and McClintock, J.E. 1983, *Ann.Rev.Astr.Ap.*, **21**, 13.

Charles, P.A., Jones, D.C., and Naylor, T. 1986, *Nature*, **323**, 417.

Ciardullo, R., Ford, H.C., Neill, J.D., Jacoby, G.H., and Shafter, A.W. 1987, *Ap.J.*, **318**, 520.

Garcia, M.R. 1989, in *Timing Neutron Stars*, ed. H. Ogelman and E.V.D. Heuvel, (Dordrecht: Reidel).

Garcia, M.R., Bailyn, C.D., and Grindlay, J.E. 1989, *Ap.J.(Letters)*, in press.

Garcia, M.R., and Grindlay, J.E., 1987, private communication.

Giacconi, R., *et al.* 1974, *Ap.J.Suppl.*, **27**, 37.

Gioia, I.M. 1989, this volume.

Gioia, I.M., Maccacaro, T., Schild, R.E., Stocke, J.T., Liebert, J.W., Danziger, I.J., Kunth, D., and Lub, J. 1984, *Ap.J.*, **283**, 495.

Grindlay, J.E. 1977, *Highlights of Astronomy*, **4**, 111.

Grindlay, J.E. 1986, in *The Evolution of Galactic X-ray Binaries*, ed. J. Truemper, W.H.G. Lewin, and W. Brinkmann (Dordrecht: Reidel), p.25.

Grindlay, J., Bailyn, C., Cohn, H., Lugger, P., Thorstensen, J., and Wegner, G. 1988, *Ap.J.(Letters)*, **334**, L25.

Grindlay, J., Cohn, H., Lugger, P., and Hertz, P. 1987, IAU Circ., No. 4408.

Grindlay, J., and Gursky, H. 1976, *Ap.J.(Letters)*, **205**, L131.

Grindlay, J.E., Hertz, P., Steiner, J.E., Murray, S.S., and Lightman, A.P. 1984, *Ap.J.(Letters)*, **282**, L13.

Hartwick, L.F., Cowley, A., and Grindlay, J. 1982, *Ap.J.(Letters)*, **254**, L11.

Hertz, P. 1987, *Ap.J.(Letters)*, **315**, L119.

Hertz, P., and Geldzahler, B.J. 1987, *Bull.A.A.S.*, **19**, 680.

Hertz, P., and Grindlay, J.E. 1983a, *Ap.J.(Letters)*, **267**, L83.

Hertz, P., and Grindlay, J.E. 1983b, *Ap.J.*, **275**, 105.

Hertz, P., and Grindlay, J.E. 1984, *Ap.J.*, **278**, 137.

Hertz, P., and Grindlay, J.E. 1985, *Ap.J.*, **298**, 95.

Hertz, P., and Grindlay, J.E. 1988, *A.J.*, **96**, 233.

Hertz, P., and Wood, K.S. 1985, *Ap.J.*, **290**, 171.

Holt, S.S. 1989, this volume.

Ilovaisky, S.A., Auriere, M., Chevalier, C., Koch-Miramond, L., Cordoni, J.P., and Angebault, L.P. 1987, *Astr.Ap.*, **179**, L1.

Jernigan, J.G., and Clark, G.W. 1979, *Ap.J.(Letters)*, **231**, L125.

Joss, P.C., and Rappaport, S. 1979, *Astr.Ap.*, **71**, 217.

Kahn, S.M., and Grindlay, J.E. 1984, *Ap.J.*, **281**, 826.

Lewin, W.H.G., and Joss, P.C. 1983, in *Accretion Driven Stellar X-ray Sources*, ed. W.H.G. Lewin and E.P.J. van den Heuvel (Cambridge: Cambridge University Press), p.41.

Lewin, W.H.G., van Paradijs, J., and van der Klis, M. 1988, *Space Sci. Rev.*, **46**, 273.

Lightman, A.P., Hertz, P., and Grindlay, J.E. 1980, *Ap.J.*, **241**, 367.

Maccacaro, T., Garilli, B., and Mereghetti, S. 1987, *A.J.*, **93**, 1484.

Margon, B., and Bolte, M. 1987, *Ap.J.(Letters)*, **321**, L61.

Mason, K.O. 1986, in *The Physics of Accretion onto Compact Objects*, ed. K.O. Mason, M.G. Watson, and N.E. White, (Berlin: Springer), p.29.

Mauche, C., and Gorenstein, P. 1986, *Ap.J.*, **302**, 371.

McClintock, J.E., London, R.A., Bond, H.E., and Grauer, A.D. 1982, *Ap.J.*, **258**, 245.

Mereghetti, S., and Grindlay, J.E. 1987, *Ap.J.*, **312**, 727.

Milgrom, M. 1978, *Astr.Ap.*, **67**, L25.

Molnar, L.A. 1986, in *The Physics of Accretion onto Compact Objects*, ed. K.O. Mason, M.G. Watson, and N.E. White, (Berlin: Springer), p.313.

Morgan, E.H., Remillard, R.A., and Garcia, M.R. 1988, *Ap.J.*, **324**, 851.

Naylor, T., Charles, P.A., Drew, J.E., and Hassall, B.J.M. 1988, *M.N.R.A.S.*, **233**, 285.

Parmar, A.N., and White, N.E. 1988, *Mem.Soc.Astr.It.*, **59**, 7.

Patterson, J. 1984, *Ap.J.Suppl.*, **54**, 443.

Priedhorsky, W.C., and Holt, S.S. 1987, *Space Sci.Rev.*, **45**, 291.

Shara, M.M., Kaluzny, J., Potter, M., and Moffat, A.F.J. 1988, *Ap.J.*, **328**, 594.

van Paradijs, J. 1983, in *Accretion Driven Stellar X-ray Sources*, ed. W.H.G. Lewin and E.P.J. van den Heuvel (Cambridge: Cambridge University Press), p.189.

Verbunt, F. 1988, in *Physics of Neutron Stars and Black Holes*, ed. Y. Tanaka, (Tokyo: Universal Academy Press), p.159.

Verbunt, F., and Meylan, G. 1988, *Astr.Ap.*, **203**, 297.

Verbunt, F., Shafer, R.A., Jansen, F., Arnud, K., and van Paradijs, J. 1986, *Astr.Ap*, **168**, 169.

Verbunt, F., van Paradijs, J., and Elson, R. 1984, *M.N.R.A.S.*, **210**, 899.

Walter, F., Bowyer, S., Mason, K., Clarke, J., Henry, J., Halpern, J., and Grindlay, J. 1982, *Ap.J.(Letters)*, **253**, L57.

Watson, M., Willingale, R., Grindlay, J., and Seward, F. 1983, *Ap.J.*, **273**, 688.

White, N.E. 1989, *Astr.Ap.Rev.*, **1**, in press.

White, N.E., Becker, R.H., Boldt, E.A., Holt, S.S., Serlemitsos, P.J., and Swank, J.H. 1981, *Ap.J.*, **247**, 994.

White, N.E., and Holt, S.S. 1982, *Ap.J.*, **257**, 318.

# HIGH RESOLUTION X-RAY SPECTROSCOPY OF THERMAL PLASMAS

**Claude R. Canizares**

Department of Physics and Center for Space Research
Massachusetts Institute of Technology

## 1. INTRODUCTION

Information in astronomy lives in a four-dimensional hyperspace. As with the familiar Minkowski space of special relativity, one dimension is temporal, but only two are spatial, corresponding to the two dimensional sky. The fourth dimension is spectral (I am ignoring the two additional dimensions representing polarization, which are likely to remain "compacted" for the foreseeable future). In most areas of astronomy, the spectral dimension is the most laden with quantitative information about the emitting object. But this is also the dimension in which it is most difficult to achieve the kind of resolution and sensitivity required to extract all this information. Through most of the history of x-ray astronomy the spectral dimension was virtually inaccessible. It was *Einstein's* great advance in spatial resolution that permitted us, for the first time, to realize legitimately high resolution and sensitivity in the spectral dimension. *Einstein* inaugurated a new era of quantitative x-ray astrophysics which AXAF will have the power to exploit.

Steve Holt (this volume) relates some of the results from the Solid State Spectrometer (SSS) on *Einstein* . I will concentrate on reviewing highlights of the Focal Plane Crystal Spectrometer (FPCS; see Giacconi *et al.* 1979) results on thermal plasmas, particularly supernova remnants (SNRs) and clusters of galaxies. During *Einstein's* short but happy life, we made over 400 observations with the FPCS of 40 different objects. Three quarters of these were objects in which the emission was primarily from optically thin thermal plasma, primarily supernova remnants (SNRs) and clusters of galaxies. Thermal plasmas provide an excellent illustration of how spectral data, particularly high resolution spectral data, can be an important tool for probing the physical properties of astrophysical objects.

The spectra of thermal plasmas bristle with tens to hundreds of emission lines. This means that the number of potential measurements greatly exceeds the number of physical parameters one might plausibly hope to constrain, at least to get global properties of sources. This central characteristic of high resolution spectroscopy in astronomy holds throughout the electromagnetic spectrum and explains why in every band spectroscopy is the tool for probing the physics of celestial sources. X-ray spectroscopy of thermal plasmas has the additional feature that it allows us to make contact with laboratory physics, with studies of controlled nuclear fusion, and with studies of the solar corona. This is a luxury which is lacking, for example, in spectroscopy of most relativistic and non-thermal plasmas.

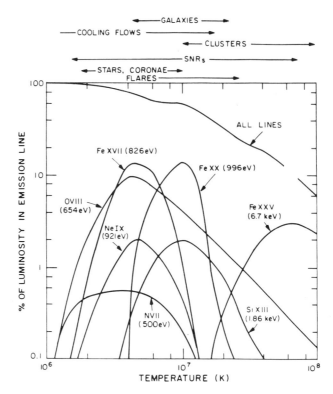

Figure 1. The fraction of the luminosity emerging in selected emission lines for an optically thin, thermal plasma with solar abundances. The appropriate temperature ranges for various astronomical objects are indicated.

Figure 1 shows the fraction of power emitted by a thermal plasma with solar abundances in various spectral lines and in all lines. One can see that a large fraction of the luminosity of a thermal source may emerge in only a few lines. Thus one can often learn a great deal by observing carefully chosen, narrow slices of the spectrum and measuring individual line strengths. This means that even a scanning spectrometer like the FPCS can be extremely productive. The classes of x-ray sources that are primarily thermal emitters are indicated along the top of the figure. The plasma in these systems covers 15 decades of density, from $\sim 10^{-2}$ to $10^{-3}$ cm$^{-3}$ in clusters of galaxies to $\sim 10^{12}$ cm$^{-3}$ in stellar coronae. Nevertheless, they share a common micro-physics, and it is this fact that allows us to decode their spectra.

To inject a historical perspective, prior to *Einstein* there was one marginal observation of an x-ray emission line from a celestial source. In the year of the *Einstein* launch, the OVIII Lα line from Puppis A had just been detected at three-sigma confidence (Zarnecki *et al.* 1978). The advance provided by the FPCS is illustrated by our observation of the same line (shown in Figure 2 together with several other oxygen lines), which we measured in various portions of the SNR.

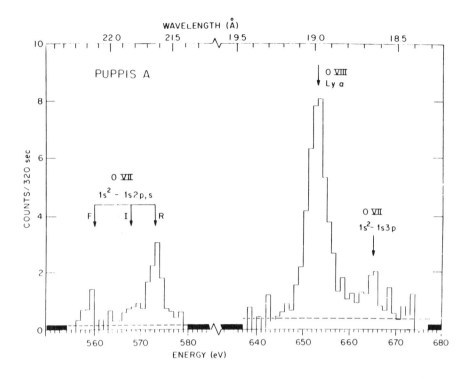

Figure 2. Emission lines of He-like and H-like oxygen from the interior of Puppis A as measured by the FPCS (from Winkler *et al.* 1981). The background level is indicated by the dashed line. The bold bars indicate regions of the spectrum that were not scanned by the instrument.

The main requirement of high resolution spectroscopy is the ability to isolate individual lines, measure their strengths, and then apply the powerful diagnostics that have been developed over the past several decades in studies of solar and laboratory plasmas. The details of our application of plasma diagnostics to thermal plasmas are beyond the scope of this paper (*e.g.* see Canizares 1989). Here I will simply give a few illustrative examples from our studies of SNRs and clusters.

## 2. DIAGNOSTICS OF PHYSICAL PARAMETERS

### 2.1 Puppis A

Figure 2 shows several cleanly resolved oxygen lines from the interior of Puppis A. The relative strengths of these lines can be used to find the physical properties of the source independent of abundance, since they all come from the same element. For example, the O VIII Lyman $\alpha$ line has a neighbor, only 10 eV away, which comes from the 3p-1s transition of O VII. The relative strengths of these two lines is a nearly model-independent measure of the relative population of the H-like and He-like ionization stages of oxygen. Another example is ratio of intensities of the

O VIII Lyman$\alpha$ and $\beta$ lines. This is independent of the population of ionization stages, because both are H-like lines, is a weak function of temperature and a reasonably strong function of absorbing column density along the line-of-sight.

A third illustration is given by the line ratios of the triplet of 2p-1s lines from He-like oxygen. In high density plasmas, like stellar coronae, one particular ratio is an indicator of particle density. In more diffuse coronae, in SNRs and in clusters, the appropriate ratio is a sensitive diagnostic of the relative ionization state of the material. In particular, it indicates whether the plasma is at equilibrium, is rapidly ionizing or is cooling and recombining. For example, the rate of cascades following radiative recombination of H-like ions in an ionizing plasma is suppressed relative to that in a plasma at equilibrium; this in turn suppresses two of the three lines in the triplet (see Canizares 1989 and references therein). We saw this signature in every SNR we could study, including the 20,000 year-old Cygnus Loop (Vedder *et al.* 1986), indicating that none of them is in ionization equilibrium.

Because there are so many lines from a thermal plasma, it is not necessary to hang ones conclusions on a single line ratio. In the interior of Puppis A, for example, we have four independent line ratios (see Figure 3; Fischbach *et al.* 1989), each of

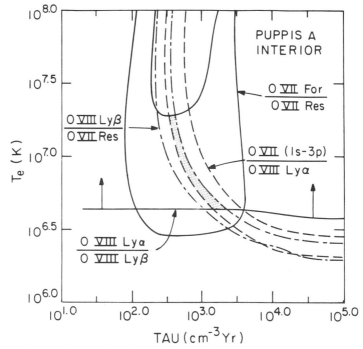

Figure 3. Constraints on the allowed values of electron temperature and ionization time for the interior cf Puppis A derived from four line ratios as indicated. The curves are 90% confidence intervals of allowed parameter space for each ratio; the shaded region is the intersection of all the allowed regions (from Fischbach *et al.* 1989).

which provides a constraint on two physical parameters: electron temperature and ionization time (the ionization time is the product of the age of the plasma and the electron density — it is proportional to the number of electron-ion collisions since the plasma was heated and thus provides an "ionization age"). In addition to defining the allowed ranges for these parameters, in several cases we have over-constrained the problem and can thus check for internal consistency. This helps validate our assumptions (*e.g.* the approximate uniformity of the plasma in the region) and increases our confidence in the conclusions.

To continue the example of Puppis A, having derived physical parameters allows us to make contact with certain astrophysical properties of the object. Our studies of three different regions, the interior, the bright eastern knot, and the eastern shock front, yielded values of electron temperature and ionization time. The latter can be combined with kinematic ages to give measures of the post-shock electron densities, which range from $\sim 0.4$ cm$^{-3}$ in the interior to $\sim 10$ cm$^{-3}$ in the bright eastern knot. In each case, our values derived solely from emission line strengths are in good agreement with values derived independently from x-ray imaging studies (*e.g.* see Fischbach *et al.* 1989, Fischbach 1989, Canizares 1989). This gives us additional confidence in the power and accuracy of our methods.

## 2.2 Casseopeia A

Cas A illustrates another power of high resolution spectroscopy. This SNR also has distinct emission regions — namely the forward shock in the circum-stellar medium and the reverse shock in the stellar ejecta. The latter is the cooler and denser of the two. As Cas A is much smaller than Puppis A (4' vs. 30') we could not resolve these regions spatially with the FPCS. But we did resolve them spectrally, in much the same way that an optical spectroscopist can easily resolve the lower-density narrow-line region from the denser broad-line region in a quasar.

Figure 4 shows a region of allowed temperature/ionization-time parameter space defined by the ratio of H-like to He-like lines of Si and S (Markert *et al.* 1989). The box indicates the parameters favored by a model fit to the low-resolution, wide-band spectrum from *TENMA* (Tsunemi *et al.* 1988). The agreement is excellent. However, the allowed values of the parameters as constrained by the comparable line ratio of Ne is completely incompatible. It is reasonable to identify the bulk of the Si and S line emission with the hotter primary shock, while assigning the bulk of the Ne line emission to the cooler reverse shock. If we make the plausible assumption of approximate pressure equilibrium between the two regions, we can deduce their densities. The derived values are again in good agreement with those from imaging studies (see Canizares 1989, Markert *et al.* 1989).

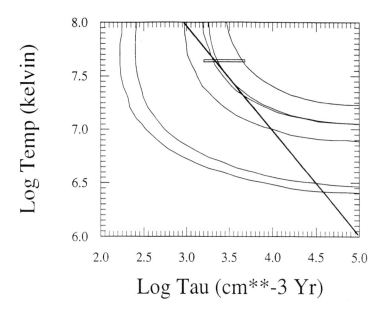

Figure 4. Same as figure 3 but for Cas A. The box shows the best fit parameters from Tsunemi *et al.* (1988). The overlapping contours in the upper right are for Si and S, while the one lower and to the left is for Ne. The straight line is a locus of constant pressure (see text; from Markert *et al.* 1989).

## 2.3 Clusters of Galaxies

The most important results of our FPCS studies of clusters of galaxies involved the spectral separation of thermal components in a manner analogous to that just described for Cas A. Although the densities are several orders of magnitude lower and the size scales are vastly larger, the microphysics in these clusters is very similar to that in SNRs (one important parametric difference is age — the clusters are almost surely in ionization equilibrium; Canizares, Markert and Donahue 1988). In several cases we used the FPCS data to spectrally isolate that small fraction of the cluster emission coming from a cooling flow. Andy Fabian's paper (this volume) describes this in much greater detail.

Figure 5 shows a slice of the spectrum of the central region of the Perseus cluster. Although the bulk of the plasma in Perseus is at a temperature of $\sim 10^8$ K, our observation revealed a strong Fe XVII line. Fe XVII is fully ionized at $10^8$ K (see Figure 1), so the presence of this line requires the presence of a significant quantity of cool gas in the central portions of the cluster. Similar studies of several clusters with the FPCS (Canizares, Markert and Donahue 1988) and SSS (Mushotzky and Szymkowiak 1988) not only give evidence for cool gas, but also provide a quantitative measure of the rate at which matter is cooling. These results, based entirely on spectral methods, agree very well with comparable measures derived by completely independent imaging analyses.

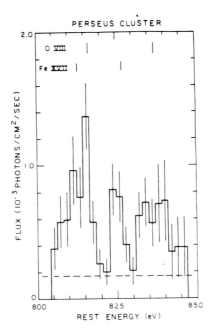

Figure 5. FPCS observations of line emission from the central region of the Perseus Cluster (from Canizares *et al.* 1988).

## 3. ELEMENTAL ABUNDANCES

One of the most important goals of astronomical spectroscopy is the determination of elemental abundances. For a handful of cases, we had sufficient data to compare the abundances of oxygen and iron, taking account of physical conditions like ionization disequilibrium. We found several cases of oxygen enrichment, the best being Puppis A. Figure 6 shows that the oxygen (also neon) lines dominate the spectrum in the region 800-1000 eV, whereas in a solar spectrum of the same region it is the Fe XVII line at 824 eV that dominates. A careful analysis, which accounts for differences in temperature and ionization state, confirms that oxygen is several times more abundant in Puppis A than in the Sun (Canizares *et al.* 1983, Canizares 1989).

In the SNR N132D (in the Large Magellanic Cloud), the overabundance of oxygen is even more striking. Figure 7 shows slices of the spectrum observed with the FPCS. (To refer back to a point I made in the Introduction, although we observed only a tiny fraction of the x-ray spectrum of this source, these lines account for 30-40% of the total luminosity.) The O VIII Lyman $\alpha$, $\gamma$, and $\delta$ lines are all strong, while the Fe XVII line at 824 eV, which dominates the solar spectrum and is even present in Puppis A, appears extremely weak if not entirely absent. Again more detailed models confirm the overabundance of oxygen.

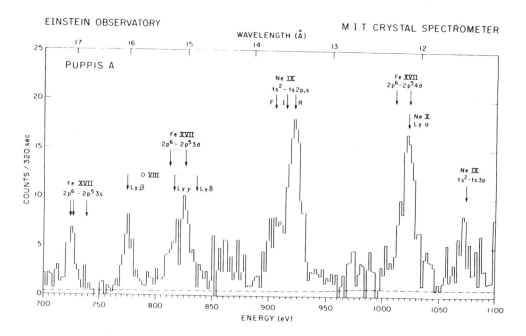

Figure 6. FPCS Observations of a portion of the spectrum of the interior of Puppis A. The background level is indicated by the dashed line.

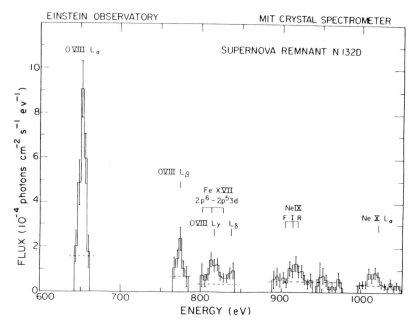

Figure 7. FPCS observations of a portion of the spectrum of N132D. The background level is indicated by the dashed lines.

We were initially surprised to see a similar overabundance of oxygen in our studies of two cluster sources, Perseus and M87 in Virgo. Figure 5 shows unusually strong O VIII Lyman $\gamma$ and $\delta$ lines compared to the Fe XVII line in Perseus. In the context of a plausible cooling flow model, this implies an oxygen overabundance by a factor of $\sim 3$ relative to Fe. A similar result was obtained for M87 (Canizares *et al.* 1982). It is likely that all the excess oxygen we detect has a similar origin: the intracluster gas that is now cooling in Perseus and M87 was probably enriched by a short-lived population of high mass stars like the progenitors of Puppis A and N132D.

Oxygen is not always overabundant, however. In one very different set of FPCS observations, we studied the interstellar oxygen K absorption edge in the spectrum of the Crab nebula (see Figure 8; Schattenburg and Canizares 1986). We were able to determine that the interstellar oxygen abundance is within $\sim$30% of the solar value. One advantage of making this measurement in the x-ray is that we are comparatively insensitive to depletion of oxygen in grains, a problem that complicates similar studies in the ultraviolet. As long as the grains are less than a few microns across, they are transparent to the x-rays.

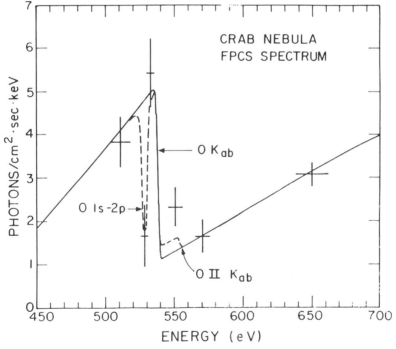

Figure 8. FPCS observation of the interstellar oxygen absorption edge in the spectrum of the Crab. The solid curve is a best-fit model, which gives an oxygen abundance of $1.3 \pm 0.3$ times solar. The dashed lines are the result of model predictions for the O 1s-2p absorption line and the O II K edge (from Schattenburg and Canizares 1986).

## 4. DOPPLER VELOCITIES

One of the powers of high resolution spectroscopy is the ability to measure Doppler velocities. We were able to do that with the FPCS for Cas A (Markert *et al.* 1983). The FPCS has one-dimensional imaging capability, and our observations of both the Si and S lines showed that the lines from the NW portions of the remnant are Doppler shifted relative to those from the SE. The mean velocity difference is $\sim 5000$ km s$^{-1}$. This is the first and, so far, the only kinematic measurement of the hot gas in an SNR. Combined with estimates of the total mass of the remnant, it implies a total kinetic energy of $\sim 4 \times 10^{51}$ erg s$^{-1}$.

## 5. HIGH RESOLUTION SPECTROSCOPY ON AXAF

I have summarized the kinds of studies that we were able to do with even the limited sensitivity of the FPCS. Now I will look ahead to the power that AXAF will provide for high resolution spectroscopy. Figure 9 shows the resolving powers

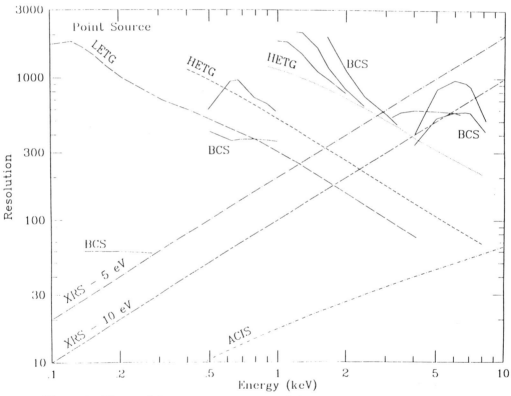

Figure 9. The resolving power vs. energy of the spectrometers on AXAF for the study of point sources. For extended sources, the LETG and HETG are not effective and the BCS has slightly lower resolution. The two XRS curves correspond to two different performance parameters as indicated.

of all the spectrometers selected for Phase B study. These are the x-ray Spectrometer/Calorimeter (XRS), the Bragg Crystal Spectrometer (BCS), the Low and High Energy Transmission Gratings (LETG and HETG), and the AXAF CCD Imaging Spectrometer (ACIS). Not all are equally effective on all types of sources. For example, the transmission gratings are most effective for point sources, although they can be used for slitless spectroscopy of bright features in extended sources as well. Furthermore, the instruments differ in effective area vs. energy, field of view, imaging capability, and instantaneous spectral coverage (for further detail see Weisskopf 1989, Canizares 1989). It is clear that multiple instruments are required to achieve the desired resolving powers of ~1000 over the full two decades of the AXAF bandwidth, and so these instruments complement each other in several ways. The AXAF instruments will detect lines with fluxes as small as $\sim 10^{-6}$ photons cm$^{-2}$ s$^{-1}$, which is nearly three orders of magnitude below the weakest lines we detected with the FPCS. Steve Holt's paper (this volume) shows the power of the XRS. I will give a few examples of the other high resolution spectrometers.

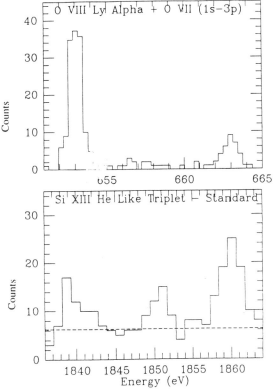

Figure 10. (Top) A simulated BCS observation of the O VIII Lyman $\alpha$ (left) and O VII 1s-3p (right) as might be observed for a thermal plasma. The resolution in this setting is 0.9 eV. The flux corresponds to a source with one third the luminosity of N132D. Observation of the two lines would require $\sim$ 5 ksec. (Bottom) A simulated BCS observation of the Si XIII He-like triplet. The resolution is 3.4 eV. The full scan would require 20 ksec for the an assumed line flux 35 times smaller than that of Cas A.

The Bragg Crystal Spectrometer (BCS) on AXAF is a scaled up version of the FPCS, achieving better sensitivity and higher spectral resolution over a wider energy range. For example, Figure 10a shows a simulated spectrum of the O VIII Lyman $\alpha$ and O VII 3p-1s lines which I used to illustrate the plasma diagnostics of Puppis A. It requires a spectral resolution of several eV to cleanly separate these lines, which are only 10 eV apart. The BCS will have 0.7 eV resolution at this energy in the "Hi-Resolution" mode for point sources, and 0.9 eV resolution in the standard setting for extended sources. The high resolution is even important for higher energy lines, like the triplet of He-like Si XIII (see Figure 10b). Here the standard setting of the BCS gives a resolution of 3.4 eV, and the Hi-Resolution setting gives 1.9 eV.

AXAF will be able to exploit the measurement of Doppler velocities in a wide variety of sources, such as stellar coronae, SNRs, and active galactic nuclei. Figure 11 shows that even velocities of $\sim 600$ km s$^{-1}$ will be easily resolvable in a spectrum obtained with the $\sim$1 eV resolutions attained near 1 keV with the High Energy Transmission Grating Spectrometer. The BCS, which achieves still higher spectral resolution, can resolve velocities as small as a few hundred km s$^{-1}$.

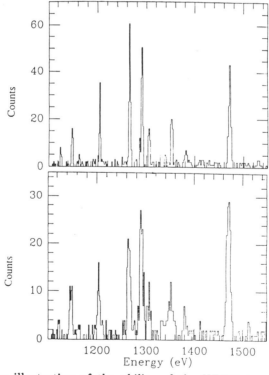

Figure 11. An illustration of the ability of the HETG to resolve Doppler broadening. The same spectrum is shown for zero broadening (top) and for broadening by a line-of-sight velocity dispersion of 600 km/s. The simulation assumes a thermal source with temperature $4 \times 10^7$ K and flux of $2.6 \times 10^{-12}$ erg cm$^{-2}$s$^{-1}$ observed for 20 ksec.

The richness of thermal spectra is convincingly illustrated in Figure 12, which shows a simulation of an observation of Capella with the Low Energy Transmission Grating Spectrometer. The inset shows that the O VII He-like triplet is well resolved. Such data will be a gold mine of plasma diagnostics for the study of stellar coronae.

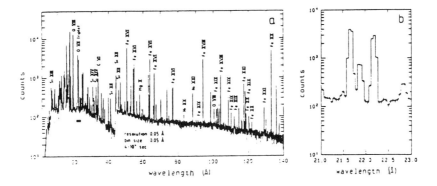

Figure 12. A simulated 10 ksec observation of Capella with the LETG. Panel a) shows the entire spectrum in 0.05 A bins, and panel b) shows the region around the O VII He-like triplet expanded for clarity (from Brinkman *et al.* 1987).

## 6. CONCLUSION

The present state of high resolution x-ray spectroscopy bears some resemblance to the state of x-ray studies of extragalactic sources after *UHURU*. In 1972, we had detected x-rays from only a few clusters, one quasar, and one Seyfert galaxy (see Kellogg 1973). *UHURU* had just managed to reach the tips of the luminosity functions of these classes, whetting the appetites of all x-ray astronomers with the possibilities of future studies. The numbers of sources increased somewhat with *Ariel-V* and SAS-3, but it was *Einstein* that really opened up the field of extragalactic x-ray astronomy.

I believe that *Einstein* did for high resolution x-ray spectroscopy what *UHURU* did for extragalactic x-ray astronomy, giving us tantalizing results on a few sources in each category. AXAF will do for high resolution x-ray spectroscopy what *Einstein* did for extragalactic x-ray astronomy, allowing us to apply these qualitatively more powerful tools to every class of x-ray source, putting the physics into X-ray Astronomy to make it X-ray Astrophysics. This will be the hallmark of the AXAF era.

Numerous individuals have made great contributions to the FPCS program. I would like to acknowledge in particular the contributions of Thomas Markert and P. Frank Winkler to the studies of thermal plasmas which I have reviewed here.

# REFERENCES

Brinkman, A. C., van Rooijen, J. J., Bleeker, J. A. M., Dijkstra, J. H., Heise, J., de Korte, P. A. J., Mewe, R. and Paerels, F. 1987, *Astron. Lett. Comm.*, 26, 73.

Canizares, C. R. 1989, in Gorenstein, P. and Zombeck, M. V. (eds.), *IAU Colloquium No. 115, High Resolution X-ray Spectroscopy of Cosmic Plasmas*, (in press).

Canizares, C. R. 1989, in *Advances in Space Research* (in press).

Canizares, C. R., Clark, G. W., Jernigan, J. G., and Markert, T. H. 1982, *Ap. J.*, 262, 33.

Canizares, C., Markert, T. H., and Donahue, M. 1988, in Fabian, A. C. (ed.), *Cooling flows in Clusters and Galaxies*, (Kluwer), 63.

Canizares, C., Winkler, P. F., Markert, T. H., and Berg, C. 1983, in Danziger, J. and Gorenstein, P. (eds), *Supernova Remnants and Their X-ray Emission*, (Reidel), 205.

Fischbach, K. F. 1989, M.I.T. PhD Thesis.

Fischbach, K. F., Bateman, L. M., Canizares, C. R., Markert, T. H., and Saez, P. J. 1989, in Gorenstein, P. and Zombeck, M. V. (eds.), *IAU Colloquium No. 115, High Resolution X-ray Spectroscopy of Cosmic Plasmas*, (in press).

Giacconi, R., *et al.* 1979, *Ap. J.*, 230, 540.

Kellogg, E. 1973 in Bradt, H. and Giacconi, R. (eds.), *X- and Gamma-Ray Astronomy*, (Reidel), 171.

Markert T., Blizzard, P., Canizares, C., and Hughes, J. 1988, in Roger, R. S. and Landecker, T. L. (eds) *Supernova Remnants and the Interstellar Medium*, (Cambridge Univ. Press), 129.

Markert T., Blizzard, P., Canizares, C., and Hughes, J. 1989, in preparation.

Markert, T.H., Canizares, C.R., Clark, G.W., and Winkler, P.F. 1983, *Ap. J.*, 268, 134.

Mushotzky, R. and Szymkowiak, A. 1988 in Fabian, A. C. (ed.), *Cooling flows in Clusters and Galaxies*, (Kluwer), 53.

Schattenburg, M. L. and Canizares, C. R. 1986, *Ap. J.*, 301, 759.

Tsunemi, H., Manabe, M., Yamashita, K., and Koyama, K. 1988, *Publ. Astron. Soc. Japan*, 40, 449.

Vedder, P.W., Canizares, C. R., Markert, T. H., and Pradhan, A. K. 1986, *Ap. J.*, 307, 269.

Weisskopf, M. C. 1989, *Space Sci. Rev.* (in press).

Winkler, P.F. Canizares, C.R., Clark, G.W., Markert, T.H., and Petre, R. 1981, *Ap. J.*, 245, 574.

Zarnecki, J. C., Culhane, J. J., Toor, A., Seward, F. D., and Charles, P. A. 1978, *Ap. J. (Letters)*, 219, L17.

# SPECTRA OF NON-THERMAL X-RAY SOURCES

S. S. Holt

Laboratory for High Energy Astrophysics

NASA/Goddard Space Flight Center

## 1. INTRODUCTION

In anticipating how I could contribute to this meeting, I assumed that Claude Canizares would be asked to summarize results obtained from the Focal Plane Crystal Spectrometer (FPCS), and that I would similarly review results obtained from the Solid State Spectrometer (SSS). There exists a decade of tradition associated with this specific demography in the presentation of spectroscopic results from *Einstein*, and I had no reason to suspect that such a precedent was not so well-established as to be virtually inviolate. Besides, I have a complete set of the required SSS viewgraphs, and by now I can prepare and deliver my talk without having to take the trouble to learn anything.

I was therefore lulled by a false sense of security when the co-chairs of this meeting suggested that Claude and I use titles for our talks that reflected differences in scientific rather than instrumental content. Since the wonderful FPCS results on emission lines from thermal plasmas demanded that Claude discuss thermal sources, I absent-mindedly accepted a title containing the modifier "non-thermal" for its spectral content. Of course, the program was printed well before I got around to shuffling the viewgraphs. I realized too late, therefore, that my very best examples of SSS capabilities were associated with the thermal sources that I had so casually let slip away.

Faced with this dilemma, there were several choices open to me. I could avoid the situation completely by staying away from the meeting with some lame excuse. I could ignore the title and give a talk with the standard SSS examples (in fact there seems to be a tradition in astronomy of speakers that go out of their way to disdain the advertised titles of their talks). Or I could "bite the bullet" and actually spend a little time developing the presentation that I promised the organizers that I could be counted on to deliver. For better or worse, I chose the latter.

Table 1 summarizes, in a rather subjective way, the extent to which I would generally categorize the X-ray spectra of astrophysical systems as "non-thermal" or "thermal." I have used the common practice of associating "non-thermal" with "continuum" and "thermal" with "line-dominated" in the table, but I have not applied these equivalences rigorously where consistency would override reason. In particular, there are two cases of spectra which are well-fit to bremsstrahlung continua that are treated quite differently in the table. In the case of the Cosmic X-ray background, I chose to use the lack of emission lines to define the source as "continuum" because the bremsstrahlung fit might be fortuitous. On the other hand, I have placed the spectra of clusters of galaxies solidly (and incorrectly, in a literal sense) in the "line-dominated" column because of the clear physical association with bremsstrahlung; here the gas is so hot that lines are a trivial fraction of the total energy emitted, but those lines provide important information about the intracluster gas.

In accordance with the title of my talk, the spectroscopic results discussed here will be culled from the three leftmost entries in Table 1: binary X-ray sources with degenerate members, active galactic nuclei, and the Cosmic X-ray background.

| Table 1 |
|---|
| Characterization of the Spectra of X-ray Sources |

| Continuum | Line-dominated |
|---|---|
| | Stars |
| | Non-degenerate binaries |
| | CVs |
| Neutron-star binaries | |
| | Supernova remnants |
| | Galaxy haloes and ridges |
| Active Galactic Nuclei | |
| | Clusters of galaxies |
| Cosmic X-ray Background | |

## 2. CONTINUUM SPECTROMETERS

The simplest characterization of a continuum is an index corresponding to a power law approximation. It is straight-forward to determine the relative importance of various instrument parameters in establishing the precision with which a power law index can be measured. Interestingly, the relative importance of resolving power, efficiency and bandpass are "inverted" in comparison with their importance for line detection. In either case the dependence on average efficiency is the same, *i.e.* proportional to its square root. For lines, the resolving power is important and the bandpass irrelevant (provided that it contains the line). In contrast, the determination of a best power law index is independent of the resolving power if the bandpass is as large as a few resolution elements, but its dependence on bandpass is linear.

For identical bandpasses, therefore, the precision with which a power law index can be measured will scale like the square root of the number of detected counts. In the case of the *Einstein* Observatory, the SSS and the Imaging Proportional Counter (IPC) both have near-unit efficiency over most of the telescope bandpass, and collect orders of magnitude more photons than do the dispersive *Einstein* Observatory spectrometers for the same sources over similar accumulation times. Their chief drawbacks are that the IPC has so poor a spectral resolving power that the bandpass is only a few resolution elements wide, and the SSS has its bandpass artificially restricted on the low energy side by the effective detector window. Table 2 lists approximate values for the energy ranges over which some high throughput "continuum spectrometers" operate, and their resolving power at 6 keV (note that the *Einstein* bandpass did not actually extend must beyond 4 keV). The last three table entries are "noise-limited" with energy resolution that is approximately independent of energy (*i.e.* resolving power R proportional to E), while the others have R proportional to the inverse square root of E.

Most of the experimental results quoted in the remainder of this talk will have been obtained (or simulated) with instruments listed in Table 2.

| Table 2 High Efficiency X-ray "Spectrometers" | | |
| --- | --- | --- |
| Type | Energy Range (keV) | R(6 keV) |
| Mechanically Collimated Proportional Counter | | |
| HEAO-1 A2 | 0.2 − 60 | |
| EXOSAT ME | 1.5 − 10 | 6 |
| *GINGA* | 2 − 30 | |
| Gas Scintillation PC | | |
| EXOSAT | 2 − 15 | 12 |
| *TENMA* | 1.5 − 35 | |
| X-ray Telescope Imaging PC | | |
| *Einstein* IPC | 0.1 − 4 | 3 |
| Si (Li) | | |
| *Einstein* SSS | 0.6 − 4 | 35 |
| CCD | | |
| AXAF ACIS | 0.3 − 8 | 50 |
| Calorimeter | | |
| AXAF XRS | 0.3 − 10 | 1000 |

## 3. SIMPLE CONTINUUM PERTURBATIONS

### 3.1 Absorption

The simplest perturbation to a power law continuum involves a parameter corresponding to photons being removed from the source spectrum below a "low energy cutoff" arising from material in the line-of-sight to the X-ray source. It is usual to assume cosmic abundances in this material and to express this parameter as an equivalent column density of hydrogen atoms (H-atoms per square centimeter). The most recent reports of the X-ray continua of quasars and Seyferts measured with the *Einstein* IPC and *EXOSAT* ME still fit to three simple parameters: normalization (source strength), power law index, and hydrogen column density. The geometry corresponding to this model is an absorber along the line-of-sight that is distant from the source.

With sufficient resolving power, the constituent atomic (or even ionization) species in this absorbing medium can be detected from characteristic absorption edges. The approximate energies of the K-complexes are given by $10Z^2$ eV, so that the energy resolution required to separate the abundant even-Z elements is $(dE/dZ) \times (dZ=2) = 40Z$. Comparing this with resolving powers available with the continuum spectrometers of Table 2, note that those which are not noise-limited have energy resolution proportional to the square root of the energy. If iron were not super-abundant relative to its nearest even-Z neighbors, even proportional counters would be capable of measuring relative abundances by separating the absorption edges from a cold medium (*i.e.* 800 eV required, compared to 1 keV achievable near Fe), but such counters would be incapable of separating those edges for lower-Z elements.

### 3.2 Scattering

Suppose, instead, that there is no material whatsoever along the line-of-sight. The observed spectrum might still differ from that emitted directly from the source because there might be a component of scattered and/or fluoresced radiation from nearby material (even from material behind the source!); in this case, the scattered/fluoresced photons would simply be added to the source spectrum. The contribution to the observed intensity from Compton scattering will have a shape similar to the unscattered continuum, and will have an intensity that depends strongly on the distance from the source to the absorber (actually, the solid angle subtended by the scatterer at the source), and more weakly on its thickness once the optical depth for scattering exceeds unity.

### 3.3 Fluorescence

Fluorescence, in this context, is the production of characteristic line radiation from the recombination of material in the scatterer or absorber that is ionized by the primary continuum. For X-rays, the dominant fluorescent line species (the only one detectable at current or near-term expected sensitivities) is from Fe-K, *i.e.* arising from continuum photons with energies in excess of the Fe-K-shell ionization energy. The fluorescence yield for Fe-K is approximately 1/3, more than an order of magnitude higher than that for other common species, so that a continuum irradiating a medium subtending a large solid angle can produce measurable Fe-K fluorescence as its only detectable line feature.

Clearly, an absorbing medium in the line of sight will also produce scattered continuum and fluorescent Fe-K. If the absorbing medium is distant from the source, however (actually, the relevant condition in this context is that it subtends a small solid angle at the source), there will be a trivial amount of fluorescent output at the observer.

## 4.  EXAMPLES OF CONTINUUM PERTURBATION BY EXTRA-SOURCE MATERIAL

Figure 1 (from Makishima 1986) summarizes the effects of various geometries in terms of the equivalent continuum width (EW) of Fe-K fluorescence, *i.e.* the width (in eV) of the observed continuum near the fluorescent feature that contains the same number of photons as does that feature. Each of the traces in the figure is calculated with the aid of a Monte Carlo code for the interaction of photons injected into a medium of uniform density, with geometries as indicated in the accompanying cartoons. Most of the data in the figure were obtained from the *TENMA* satellite, and virtually all are consistent with the Fe-K emission arising in cold material.

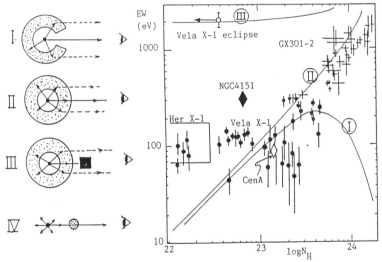

Figure 1.  The equivalent width (EW) of Fe-K fluorescence, defined as the ratio of the observed 6.4 keV line flux to the observed 6.4 keV continuum, as a function of column density. The solid traces are those expected from the models depicted in the corresponding cartoons on the left, where the column densities are in those in the scattering/fluorescing medium. For the data plotted, the column densities are along the line of sight. (From Makashima 1986)

## 4.1 Case IV – Simple Absorption

Because all sources have some amount of interstellar medium between the source and observer, there will be some contribution from Case IV absorption whether or not there is also a contribution from more complicated photon transport near the source. Note that there is no Case IV model trace in Figure 1 because there is no measurable fluorescence into the line of sight.

Essentially all of the pre-*Einstein* work on X-ray binaries was performed with the simple assumption of Case IV. In the case of the Be-star/neutron-star binary system GX 301-2, for example, observations with proportional counters from OSO-8 measured low energy absorption that was interpreted as $> 10^{23}$ H-atoms/cm$^2$ of cold material along the line-of-sight, along with its attendant Fe-K edge (Swank *et al.* 1976). Such Be-star/neutron- star systems have since been associated with "hard transient" sources that appear at quasi-regular intervals; the conventional wisdom assigns the regularity in their appearance to the passage of the neutron star in its eccentric orbit through the equatorial effluence of the Be star, as sketched in Figure 2.

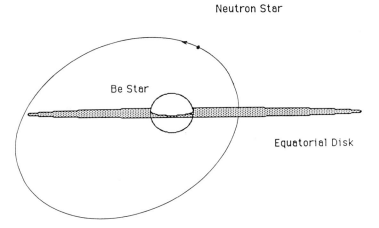

Figure 2. A sketch of the typical geometry of a "hard transient" binary system consisting of a massive Be star and a neutron star in an eccentric orbit about the center of mass. The Be star is presumed to be expelling material in its equatorial plane, and the transient X-ray source appears when the neutron star dips into this plane of accretion fuel.

Newer data have borne out this interpretation of the "hard transient" phenomenon. The continuum spectroscopic data of Figure 3 (from Cook and Warwick 1987) demonstrates the rise in both absorption and intensity as the neutron star enters the equatorial plane of the Be star that provides the accretion fuel. Data with better spectral resolution from a similar episode in a different source (GX 304-1 measured with the *Einstein* SSS) is capable of directly determining the abundance of silicon in the equatorial plane of the Be star from the depth of its K-absorption edge (Holt and Kelley 1989), and future measurements with the AXAF calorimeter will allow such direct abundance measurements of virtually all the abundant even-Z elements from oxygen through iron.

The Case IV approximation for most Be-neutron star systems is not bad because the absorption along the line of sight dominates the scattered contribution to the observed spectrum by a very wide margin. The approximation is even better for any instance in which the absorption is dominated by the interstellar medium rather than material local to the source.

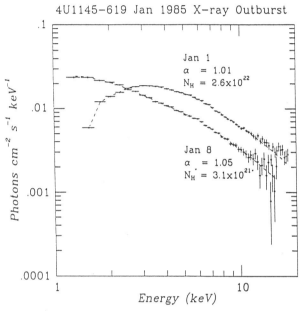

Figure 3. The derived pulse-averaged photon spectra from the EXOSAT ME detectors for the 1 January 1985 and 8 January 1985 observations of 4U1145−619. The simple interpretation in terms of Figure 2 is that on 1 January the neutron star was closer to the ecliptic plane of the Be star, so that both the hard intensity and absorption were higher than on 8 January. (From Cook and Warwick 1987)

A particularly interesting application of the Case IV approximation is to data obtained from the *Einstein* Objective Grating Spectrometer (OGS) in the direction of the BL Lac object PKS 2155-304. Canizares and Kruper (1984) measured an apparent absence of continuum photons from the source in a band at least 50 eV wide close to where absorption lines from ionized oxygen are expected (the red shift of the source is not known directly, and so must be inferred from indirect evidence). This spectral anomaly has been interpreted as an O VIII absorption trough (*i.e.* arising from an absorption line in material with a wide range of velocities). Although the cross sections are high, narrow X-ray absorption lines are not directly measurable because the EWs cannot be larger than the actual line widths. If due to true intergalactic absorption, this result could have cosmological significance. It is more likely due to a local "wind" however, in which case the geometry may allow the next generation of observations to measure corresponding fluorescence features and other absorption/fluorescence combinations.

## 4.2 Case II — Immersion in a Uniform Medium

Case II is the idealization of a source that is embedded in a medium of uniform density which can absorb, scatter and fluoresce. The immersion of an X-ray source into such a medium will result in a contribution from scattering to the observed continuum of about 5% for a medium with radius equivalent to a hydrogen column density of $10^{23}$ H-atoms/cm$^2$, but will be more like 1/3 of the observed intensity (*i.e.* half of the absorbed unscattered source intensity) for a medium with column density an order of magnitude larger.

As the thickness of the scattering material increases, the EW keeps increasing while the fraction of radiation that escapes the absorber keeps decreasing, as evidenced by the GX 301-2 data depicted in Figure 1. GX 301-2 is a Be-star/neutron-star system; in the previous section such sources were discussed as being potentially well-described by Case IV rather than Case II (and, further, this is the same source for which the Swank *et al.* (1976) data were used as exemplary of Case IV). The resolution of these apparent inconsistencies is that the GX 301-2 system is not a particularly good example of the common "hard transient" situation sketched in Figure 2; it is better described by a neutron star passing through a dense isotropic wind rather than through a thin layer of material.

The Case II Fe-K fluorescence EW scales approximately linearly with the log of the line-of-sight column density until the material becomes thick to the fluorescence itself. When the medium is so thick that no radiation can escape in the forward direction, none will reach the observer from anywhere else in the medium.

These models can be applied to AGN as well as X-ray binaries. Case II may be a reasonable lowest-order approximation to a Seyfert I galaxy, where a central ionizing source is surrounded by an absorbing "broad line region" (BLR). Here SSS data were useful in determining that all the X-ray data from NGC 4151 were consistent with an approximately spherical distribution of cold BLR clouds that covered approximately 90% of the source (Holt *et al.* 1980). Subsequent *EXOSAT* data demonstrated that this very simple model could not explain everything (Pounds, et al. 1986), but this geometry is probably still a good first approximation for other Seyferts and quasars.

Since the EW magnitude should be linear with covering factor for the BLR clouds, it is not disconcerting that the Cen A data point in Figure 1 is below the trace for complete coverage. It is astonishing that the NGC 4151 data seem to fall above the trace, however. Source anisotropies may be invoked to resolve this inconsistency. Another possible explanation is that the primary NGC 4151 intensity has pronounced variability on timescales shorter than the size of the BLR (about one light month). Since most of the fluorescence EW is not from the material in the forward direction, the fluorescence should lag the continuum from which it arose by about one month, and *EXOSAT* data taken near the time of the *TENMA* data point plotted for NGC 4151 suggest that its continuum was, indeed, brighter a month earlier.

The AXAF spectrometers may be capable of extracting an extraordinary wealth of information from a large sample of AGN. Figure 4 is a simulation of idealized Fe-K features that might be measured in an observation of NGC 4151 with the AXAF calorimeter. For simplicity, the features in the simulated spectrum are broadened only by the 10 eV resolution of the instrument (if these features are created in the BLR, the several thousand kilometers per second of the optical emission lines will produce an actual broadening of approximately 100 eV). The Fe edge is typical of values measured for NGC 4151, and the strong emission feature at 6.4 keV is the corresponding fluorescence line. Just for fun, the line at 6.7 keV is arbitrarily included at 10% of the fluorescence line; if there is a hot intercloud medium in pressure equilibrium with the broad line clouds, just such a line may well be present in the spectrum. Since the line centroid energies can be determined to approximately 1 eV, the X-ray redshifts can be determined to $10^{-4}$.

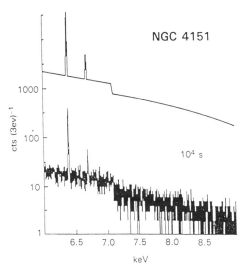

Figure 4. Simulated data from NGC 4151 that might be obtained with AXAF calorimeter in an exposure of $10^4$ seconds (lower trace) or an infinite exposure (upper trace). The simulated Fe lines are not broadened here, as they would be if they were produced in the BLR. The bright feature at 6.4 keV is that expected from the observed absorption edge if there is spherical symmetry. The 6.7 keV feature is arbitrarily included at one-tenth the magnitude of the 6.4 keV feature. (From Holt 1987)

## 4.3 Case I — Scattering and Fluorescence Without Absorption

The geometry of Case I provides for the contribution of scattering and fluorescence to the observed spectrum where there is little absorption in the line of sight. The *Einstein* SSS measurement of the Her X−1 1.24-s pulse period has sufficient energy resolution to completely separate radiation components more than about 200 eV apart, and is therefore able to demonstrate that the < 1 keV pulse is out of phase with that at higher energies. This is precisely what would be expected from a pulsar beam that illuminates material accumulated at the Alfven surface that is sufficiently thick to provide a measurable scattered component; note that in this case the radiation from the illuminating source is not isotropic. Figure 5 exhibits Her X−1 phase-dependent spectra synthesized from data obtained with *Einstein* SSS, OSO-8 and HEAO-1 A2. Note that the spectrum is hardest in phase with the direct pulse, while the "off-pulse" spectrum exhibits a softer scattered component and a prominent Fe-K fluorescence feature.

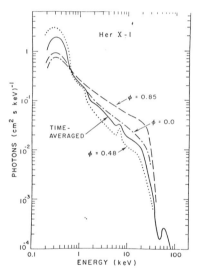

Figure 5. Composite X-ray spectrum of Her X−1 as a function of pulse phase. Solid curve: time-averaged spectrum, showing an excess of soft (< 0.5 keV) photons, an Fe-K emission line at 6.4 keV, a near-power-law (index 0.5) for $2 < E < 20$ keV that breaks sharply > 20 keV, and structure near 60 keV that is suggestive of electron cyclotron resonance. Dashed and dotted curves: spectra near the maximum and minimum, respectively, of the hard X-ray pulse, showing that the peak intensitites in both the soft X-rays and the Fe-K fluorescence are out of phase with this hard pulse. (From McCray et al. 1982)

## 4.4 Case III — X-Ray Eclipses

Case III represents an asymmetric medium that is the complement of that for Case I; Case I is for a relative deficiency of material in the line of sight, while in Case III the medium is much thicker in the line-of-sight than in other directions. This situation might be expected during the eclipse of a neutron star by its non-degenerate companion in an X-ray binary system, where the Fe-K EW should be very large from the material fluorescing near the line-of-sight to the eclipse boundary; this is exactly what is observed from eclipses of Vela X−1, for example, as demonstrated in Figure 1.

Another striking example is for NGC 1068, the prototypical Seyfert 2 galaxy. Optical polarization measurements by Antonucci and Miller (1985) were responsible for the realization that Seyfert 2s may be obscured Seyfert 1s if the generic Seyfert geometry is more cylindrical than spherical, with a thick torus of scattering/absorbing material. Such a a system may be identified as a Seyfert 1 if it is observed along a line-of-sight near its system axis of symmetry, such that obscuration by the thick torus is minimized, or as a Seyfert 2 when it is "eclipsed" by the torus. In the latter case, the only observed X-rays will be those that are scattered and fluoresced from the thinner material above and below the torus. A great triumph for this model was the X-radiation subsequently measured from NGC 1068 with the *GINGA* satellite (Makino *et al.* 1988). At virtually the same distance as NGC 4151, its continuum is two orders of magnitude of lower, but its Fe-K EW is an order of magnitude higher.

## 5. AGN CONTINUA

The first opportunity to systematically study a large sample of the continuum spectra of AGN was provided by the large area proportional counters of the HEAO-1 mission. With a typical bandpass of 2-40 keV (limited by interstellar absorption at low energies and by photon statistics at high energies), measurements from the HEAO-1 A2 counters demonstrated remarkable uniformity in the best-fit slopes of AGN (Mushotzky 1984). For a total sample of 30 Seyfert 1s observed with HEAO-1 A2, the best fit power index alpha is $0.65 \pm 0.17$. With the exception of possible 100-eV EW Fe fluorescent features and their corresponding absorption edges (the detection of these features was not possible with the typical statistics of HEAO-1 A2 measurements), the spectra appear to be featureless and inflectionless over the whole bandpass (dynamic range of 10, with typical resolving power of 6). The HEAO-1 A4 experiment extended the spectral measurements of a handful of individual AGN (as well as the summed contribution from others) to > 100 keV (Rothschild *et al.* 1983). Quasars appear to have spectra similar to Seyfert 1s and will be included here under the generalized heading "AGN" but the pathological spectra of BL Lac-type sources (*e.g.* Urry, Mushotzky and Holt 1986) will not.

New measurements over the same bandpass as that of HEAO-1 A2 from *EXOSAT* and *GINGA* have verified this remarkably consistent continuum index. Approximately four dozen *EXOSAT* AGN (mostly Seyfert 1s) have $\alpha = 0.70 \pm 0.16$ (Turner 1988) and a dozen deep exposures of quasars with *GINGA* (Makino *et al.*

1988) are also consistent both with the average HEAO-1 A2 index and the dispersion in that index. Although the spectra measured with the *EXOSAT* ME detectors are virtually identical with those measured from HEAO-1 A2, approximately 30% of these sources exhibit "soft components" in the LE (low energy) detectors (Turner, 1988). When there is no such "soft component" present, the indices determined from the combination of the spectra measured in the LE and ME detectors agree with those from HEAO-1 A2.

New analyses of sources measured with the *Einstein* IPC are also yielding new information about the continuum spectra of AGN. Here the bandpass is restricted (the effective dynamic range for most measurements is $\sim 10$) and the spectral resolution is almost non-existent ($R < 2$ at 1 keV), but the new analyses are yielding very interesting correlations.

Wilkes and Elvis (1987) performed a detailed study of 33 low redshift quasars with signal-to-noise $> 10$. Unless the typical spectra are inverted (*i.e.* index $< 0$), this threshold-limited condition biases in favor of spectra with low-energy excesses (*i.e.* either in the source spectra themselves or owing to low line-of-sight absorption). On a statistical basis, it also biases in favor of radio-strong rather than optically-strong quasars. Wilkes and Elvis discovered that there is a distinct difference between the best-fit X-ray indices of those that are radio-quiet and those that are radio-loud (where the separation of these two classes is defined as unity in the ratio of 5 GHz core flux density to B-band flux density). Figure 6 demonstrates this effect. The radio-loud quasars have X-ray spectra that generally have best-fit indices near 0.5, while the radio-quiet sources have X-ray spectra are distinctly softer, with best-fit indices near 1.0.

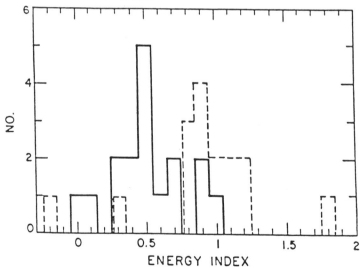

Figure 6. Double histogram showing the distribution of indices for $> 4$ keV X-ray continua for radio-loud (solid line) and radio-quiet (dotted line) quasars. (From Wilkes and Elvis 1987).

Digging deeper into the *Einstein* IPC data bank, Urry and Kruper (1988) find an average spectral index of about 0.9 for a sample of almost 100 Seyfert galaxies, which would be classed as radio-quiet if considered to be an extension of the Wilkes and Elvis sample. Canizares and White (1989) have literally extended the quasar sample to higher redshifts, and have confirmed the correlation of low energy X-ray spectral index with radio behavior. Figure 7 demonstrates that radio-quiet quasars have spectral indices $> 1$ for $z > .5$, while radio-loud quasars have an average spectral index approximately 0.5. In addition, they demonstrate that if the latter group is further bifurcated into flat and steep spectrum radio sources, that flatter spectrum radio sources yield flatter soft X-ray spectra.

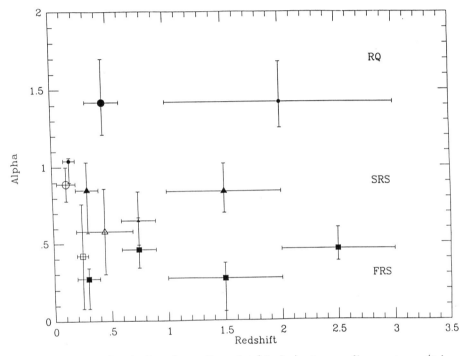

Figure 7. power-law indices for radio-quiet (circles), steep-radio-spectrum (triangles) and flat-radio-spectrum (squares) quasars, as a function of redshift. The open symbols are the data of Wilkes and Elvis that are plotted in Figure 6. (From Canizares and White 1989)

Returning to the steepening at low energies that seems to characterize at least some AGN, there may be a natural (at least in a qualitative sense) explanation. The "big blue bump" in the UV continuum of quasars is currently thought to be associated with an accretion disk that feeds a central black hole. It is conceivable that the shortest wavelength portion of this bump adds to the underlying "canonical" spectrum characteristic of AGN to produce the steepening at low energies (*e.g.* see Figure 8). This model does not however attempt to explain the flatter than "canonical" slopes of the radio-loud quasars.

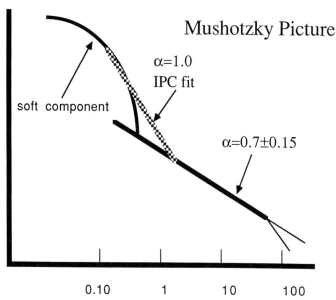

Figure 8. A simple qualitative explanation for the steepening of the X-ray spectra of AGN at low energies, including the fortuitous average value of 1.0 (From Mushotzky 1988).

Wilkes and Elvis agree that the accretion disk continuum may contribute to the low energy excess, but further suggest that the flatter spectra of the radio-loud sources may hint at a synchrotron-self-Compton origin for the component with index about 0.5. They argue that all AGN may possess a component with index approximately unity (*i.e.* distinctly steeper than "canonical") which connects directly to the infrared continuum. Figure 9 describes this general situation, where the positions of the breaks will obviously differ from source to source. The efficacy of the Wilkes and Elvis suggestion will be directly testable with the AXAF calorimeter, where literally thousands of AGN can yield continuum spectra with $> 1000$ photons in observing times $< 10,000$ seconds, with $R > 100$ over the whole 0.3 - 10 keV bandpass.

## 6. THE COSMIC X-RAY BACKGROUND

The Cosmic X-Ray Background (CXB), formerly called the Diffuse X-Ray Background, has been known since the discovery of the first X-ray source more than 25 years ago. Measured carefully with the HEAO-1 A2 experiment (Marshall *et al.* 1980), the CXB spectrum above 3 keV is remarkably well-fit with a 40 keV bremsstrahlung continuum. Above 100 keV, it is entirely conceivable that the contribution from Seyfert 1 galaxies can completely explain the currently unresolved CXB. Below 3 keV, it has been difficult to measure the CXB owing to both emission and absorption components in the galaxy; lacking direct measurements, many investigators simply extrapolate the $> 3$ keV spectrum to lower energies.

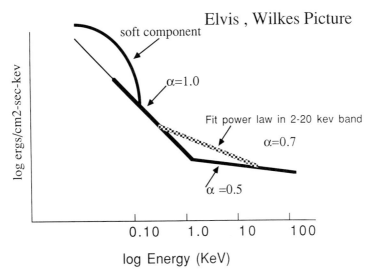

Figure 9. A qualitative description of the Wilkes and Elvis (1987) explanation for the variety of indices and low energy steepening observed in the spectra of AGN, including the fortuitous average value of 0.7. (From Mushotzky 1988)

Taken at face value, a 40 keV bremsstrahlung component that uniformly permeates the universe with a magnitude necessary to explain the CXB has approximately 1% of the energy density of the cosmic microwave background.

There are lots of reasons why an explanation of the CXB in terms of a hot uniform intergalactic plasma is untenable (*e.g.* Field and Perrenod 1977). Most conventional explanations of the CXB are therefore associated with the superposition of sources that are currently below threshold for individual detection.

Even before the *Einstein* results, it was fashionable to argue that quasars like 3C 273 could easily sum to explain the CXB; in fact, the value of the CXB was reason enough to speculate that the quasar source counts as a function of magnitude had to flatten so as not to produce enough X-ray emission to far exceed the CXB (Setti and Woltjer 1979). Even the earliest *Einstein* results seemed to bear out this prediction; quasars were found with remarkable efficiency in X-rays (Tananbaum *et al.* 1979), and approximately one-quarter to one-third of the CXB could apparently be explained from the first *Einstein* Deep Survey (Giacconi *et al.* 1979).

A lingering problem facing the resolution of the CXB with conventional AGN is the spectral inconsistency. The CXB spectrum is much flatter below 10 keV than the 0.7 that is characteristic of the AGN sample. The invocation of a harder spectrum at higher energies that is redshifted into the $<$ 10 keV range is not confirmed by the sample of nearby AGN for which there currently exist higher energy data. Rather than referring to this as a "problem," perhaps it might be more appropriately called a "constraint" on any potential explanation. Interestingly, the larger the fraction of the CXB that is directly attributable to foreground AGN, the more restrictive this constraint becomes, as demonstrated in Figure 10.

A second constraint arises from the fluctuation data. If the CXB is truly diffuse, then the fluctuations in the number of photons detected in source-free IPC pixels will be just those expected from Poisson statistics on the average value. If, on the other hand, the preponderance of the CXB arises in sources just below the threshold for detection, the RMS variations will be higher owing to domination of

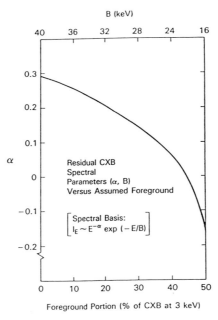

Figure 10. Spectral parameters (power law index $\alpha$ and exponential cutoff B) for the residual CXB as a function of the assumed percentage of AGN foreground at 3 keV. (From Boldt 1988)

the fluctuations by the number of sources rather than by the number of photons. Hamilton and Helfand (1987) have found that the fluctuations in source-free IPC pixels constrain the number of sources that are responsible for the CXB to no less than one per arc minute. One possible explanation is that the luminosity distribution of the contributing sources is considerably flattened; Schmidt and Green (1986) have independently determined that the quasar luminosity distribution that might be required to replicate the CXB is, indeed, considerably flattened — so much so that, in my opinion, it virtually dooms the old idea that ordinary quasars can be responsible for the CXB. The CXB must therefore originate in either a truly diffuse manner, or in a population of "dinosaur" AGN that are no longer present at low redshift. There are many good reviews of contemporary theoretical ideas about the nature of these beasties (*e.g.* Boldt 1988) so that I shall not belabor such issues here.

I must confess to some personal satisfaction about the probable demise of quasars as the primary contributor to the CXB. At the first of these *Einstein* celebrations (High Energy Astrophysics Division Meeting of the American Astronomical Society, held here in Cambridge in January of 1980) there were not many who could resist the appeal of the early *Einstein* quasar and deep survey counts as an explanation of the CXB. Since astronomy is an observational rather than an experimental science, we must usually content ourselves with consistency in lieu of proof, so that the attraction of joining the intuitively sensible quasar "bandwagon" for the CXB is completely understandable.

Nevertheless, there were some disquieting problems with this apparent consistency even then. The disagreement between AGN and the CXB spectra at energies beyond the *Einstein* bandpass is one such problem. The fact that most of the energy density in the CXB is, in fact, at these higher energies rather than in the *Einstein* band is another. A third is that the *Einstein* measurements that discredited the bremsstrahlung spectrum (by explaining a large fraction of its low energy intensity in terms of deep survey sources) are made at energies lower than where the CXB is actually measured. In an attempt to demonstrate that a resolution of the CXB in terms of a quasar (or an even more exotic dinosaur AGN) population was premature at that time, I constructed a remarkably prophetic scenario (Holt 1981) that I cannot resist resurrecting here. I freely admit that it was based more on "orneryness" than on any real physical intuition.

I made one simple assumption about the X-ray sky at $\sim 1$ keV: that "quasars" had steeper spectra than did "canonical" AGN at low energies, so that the $\sim 1$ keV CXB was therefore in excess of the extrapolation from higher energies because of the contribution from unresolved steep-spectrum quasars. It is interesting that the results of the new spectral surveys discussed above have "confirmed" that there are such steep spectrum sources to which the *Einstein* HRI is sensitive. Similarly, the new determination of the $\sim 1$ keV CXB from the *Einstein* IPC (Helfand, this meeting) is in excess of the extrapolation from higher energies by approximately an amount equal to the previously inferred direct contribution to the CXB from the *Einstein* deep survey sources. Therefore, there is no current experimental restriction to the true CXB being completely attributable to a 40 keV bremsstrahlung spectrum!

Perhaps AXAF will provide the evidence to resolve the CXB puzzle that we had hoped would come from *Einstein*. Even AXAF will suffer from the unavoidable fact that the large fraction of the CXB energy density lies beyond its high-energy cutoff, but AXAF will possess three attributes that will make it much better suited than *Einstein* for CXB studies. Most obvious is its well-advertised increase in sensitivity: both its larger aperture and its better spatial resolution contribute here.

Second is its increased bandpass: although the ~10 keV high-energy AXAF cutoff is still below the CXB energy density peak, there will exist at least some overlap with the measured CXB > 3 keV. Third, the AXAF continuum spectrometers have sufficient resolving power to study the > 3 keV continua of sources measured in deep exposures that will be uncontaminated by possible lower energy steepening. This combination may provide the archæological tools with which to unEarth direct evidence for the existence of the flat-spectrum dinosaurs that are required to produce the CXB.

## REFERENCES

Antonucci, R. R. J. and Miller, J. S. 1985, *Ap.J.*, **297**, 621.
Boldt, E. A. 1988, in *Physics of Neutron Stars and Black Holes*, ed. Y. Tanaka, Universal Academy Press, Inc, Tokyo, p. 333.
Canizares, C. R. and Kruper J. 1984, *Ap.J.(Letters)*,**278**, L99.
Canizares, C. R. and White, J. L. 1989, *Ap.J.*, (in press).
Cook, M. C. and Warwick, R. S. 1987, *MNRAS*, **227**, 661.
Field, G. and Perrenod, S. 1977, *Ap.J.*, **215**, 717.
Giacconi, R., *et al.* 1979, Âp.J.(Letters), **234**, L1.
Hamilton, D. and Helfand, D. 1987, *Ap.J.*, **318**, 93.
Holt, S. S. 1981, in *X-Ray Astronomy with the Einstein Satellite*, ed. R. Giacconi, Reidel, Dordrecht, p.173.
Holt, S. S. 1987, *Astrophys. Lett. and Comm.*, **26**, 61.
Holt, S. S. *et al.* 1980, *Ap.J.(Letters)*, **214**, L13.
Holt, S. S. and Kelley, R. L. 1989 (preprint).
Makino, F. *et al.* 1988, in *Physics of Neutron Stars and Black Holes*, ed. Y. Tanaka, Universal Academy Press, Inc, Tokyo, p. 357.
Makishima, K. 1986, in *The Physics of Accretion onto Compact Objects*, eds. K. O. Mason, M. K. Watson and N. E. White, Springer Verlag, Berlin, p.249.
McCray, R. A. *et al.* 1982, *Ap.J.*, **262**, 301.
Marshall, F. E. *et al.* 1980, *Ap.J.*, **235**, 4.
Mushotzky, R. F. 1984, *Adv. Space Res.*, **3**, 157.
Mushotzky, R. F. 1988, in *Active Galactic Nuclei*, eds. H. R. Miller and P. J. Wiita, Springer-Verlag, Berlin, p.239.
Pounds, K. *et al.* 1986, M̂NRAS, **218**, 685.
Rothschild, R. E. *et al.* 1983, *Ap.J.*, **269**, 423.
Schmidt, M. and Green, R. F. 1986, *Ap.J.*, **305**, 68.
Setti, G. and Woltjer, L. 1979, *Astron. Ap.*, **76**, L1.
Swank, J. H. *et al.* 1976, *Ap.J.(Letters)*, **209**, L57.
Tananbaum, H. *et al.* 1979, *Ap.J.(Letters)*, **234**, L9.
Turner, T. J. 1988, Ph.D. thesis, Univeristy of Leicester.
Urry, C. M. and Kruper, J. 1988 (preprint).
Urry, C. M., Mushotzky, R. F. and Holt, S. S. 1986, *Ap.J.*, **305**, 369.
Wilkes, B. J. and Elvis, M. 1987, *Ap.J.*, **323**, 243.

# X-RAYS FROM SPIRAL AND STARBURST GALAXIES

## G. Fabbiano

Harvard-Smithsonian Center for Astrophysics

## 1. INTRODUCTION

The study of the X-ray properties of normal galaxies as a class was made possible by the launch of the *Einstein Observatory* in November 1978 (Giacconi *et al.* 1979). Before then, with the exclusion of the bright X-ray sources associated with Seyfert nuclei (Elvis *et al.* 1978; Tananbaum *et al.* 1978), only four galaxies had been detected in X-rays: the Milky Way, M31, and the Magellanic Clouds (see Helfand 1984a and references therein). The *Einstein* X-ray observations of well over 100 galaxies have been reported in the literature to date, and data on a similar number can still be found in the *Einstein* data bank. Some galaxies were detected with enough detail to allow a study of their X-ray morphology, spectra, and individual sources, and to make comparisons with optical, infrared and radio data. For all the galaxies, values of the X-ray flux, or even upper limits to this flux in the case of non-detections, could be used to explore average sample properties. These observations have shown that normal galaxies of all morphological types are spatially extended sources of X-ray emission with luminosities in the range of $\sim 10^{38}$ erg s$^{-1}$ to $10^{42}$ erg s$^{-1}$. Although this is only a small fraction of the total energy output of a normal galaxy, X-ray observations are uniquely suited to study phenomena that are otherwise elusive. These include the end products of stellar evolution (supernova remnants and compact remnants), and a hot phase of the interstellar medium.

To mention some of the unexpected results, these observations have led to the discovery of plumes of hot gas ejected by starburst nuclei (*e.g.* M82, NGC 253, NGC 3628), and to the study of small active nuclei, which are important for understanding the full range of the AGN phenomenon. Hot X-ray halos have been discovered in early-type galaxies, and provide a *potentially* very powerful means for measuring their mass. The implications of these results range from new insights on the composition and evolution of X-ray emitting sources in spiral galaxies, and their relationship with star formation activity and cosmic ray production, to the formation of the intracluster medium and the origin of the x-ray background. An up to date summary of the state of this subject can be found in a paper that I have recently written for *Annual Reviews of Astronomy and Astrophysics* (Fabbiano 1989). In this talk, which is closely based on this review, I will concentrate on the results of the *Einstein* observations of spiral and starburst galaxies.

## 2.  THE X-RAY EMISSION OF SPIRAL GALAXIES

### 2.1  Sources of X-rays

Spiral galaxies are extended and complex X-ray sources with total luminosities in the *Einstein* band ($\sim 0.2 - 3.5$ keV) of $\sim 10^{38}$ erg s$^{-1}$ to a few $10^{41}$ erg s$^{-1}$ (Fabian 1981; Long and Van Speybroeck 1983; Fabbiano 1984, 1986a). Observations of the Milky Way and of the Local Group galaxies (*e.g.* see Fabian 1981; Helfand 1984) suggest that a good fraction of this X-ray emission is due to a collection of individual bright sources, such as close accreting binaries with a compact companion and supernova remnants, with luminosities ranging from $\sim 10^{35}$ erg s$^{-1}$ up to a few $10^{38}$ erg s$^{-1}$ (*e.g.* M31, Figure 1).

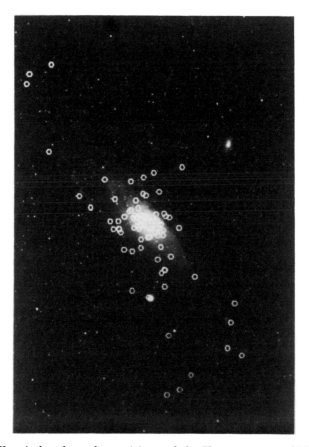

Figure 1. The circles show the positions of the X-ray sources of M31, superimposed onto an optical photograph (courtesy of L. Van Speybroeck). Notice the clustering of sources on the bulge.

Only a few very bright X-ray sources can be detected in the *Einstein* images of more distant galaxies, which typically appear as extended X-ray emission regions (*e.g.* Fabbiano and Trinchieri 1987). However, there is reason to believe that most of the X-ray emission of these galaxies is due to sources akin to those detected in the Local Group. A comparison of the fraction of the X-ray emission resolved in individual bright sources, versus that which appears diffuse, is consistent with the bulk of the emission originating from individual sources below threshold: more distant galaxies, with higher point source thresholds, have a relatively larger 'diffuse' emission than less distant galaxies (Fabbiano 1988a). Moreover, the X-ray spectra of these galaxies, although ill-defined, are consistent with the hard spectra expected from binary X-ray sources (Fabbiano and Trinchieri 1987; Trinchieri *et al.* 1988; Fabbiano 1988b). Finally, the X-ray luminosities are linearly correlated with the emission in the optical B band (see Figure 2). This correlation is similarly tight for early-type bulge-dominated spirals, and for late-type disk/arm-dominated galaxies: for all of them the ratios of monochromatic (2 keV) X-ray to optical (B) flux densities cluster around $10^{-7}$. This results suggests that the X-ray emission is mostly due to sources constituting a constant fraction of the stellar population, in

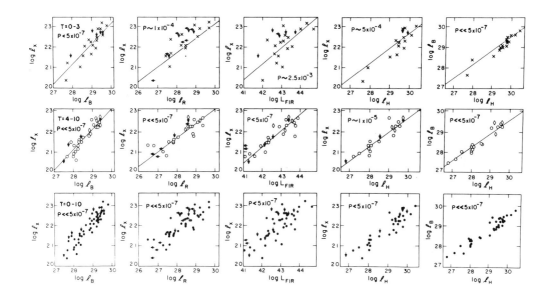

Figure 2. Scatter diagrams of the X-ray, B, radio continuum (R), near-infrared H, and far-infrared (FIR) luminosities for the X-ray sample of spiral galaxies. Crosses identify the early-type spirals (T=0-3); open circles the late-type galaxies (T=5-10); diamonds the four galaxies with T=4; and dots the entire range of morphological types (T=0-10). The non-parametric Spearman Rank probabilities of chance correlation are indicated in each case. The diagonal lines represent the average functional relationships derived for the late-type spirals (Fabbiano *et al.* 1988).

agreement with the conclusion of the detailed X-ray observations, that show that the X-ray emitting population is likely to be dominated by binary X-ray sources (Long and Van Speybroeck 1983; Fabbiano *et al.* 1984b; Fabbiano 1984; Fabbiano and Trinchieri 1985).

Stars emit coronal X-rays with luminosities of $10^{28}$ erg s$^{-1}$ to $10^{33}$erg s$^{-1}$ (Vaiana *et al.* 1981). However, except perhaps at the lowest energies and in some starburst regions, stars do not contribute significantly to the total X-ray emission of galaxies, since the X-ray to optical ratios measured in spiral galaxies (Long and Van Speybroeck 1983; Fabbiano and Trinchieri 1985) are larger than those expected from a normal stellar population (Topka *et al.* 1982; Helfand and Caillault 1982), and the average X-ray spectrum of spiral galaxies appears harder than that of the stellar emission (kT $> 2$ keV in galaxies, Fabbiano and Trinchieri 1987; kT $\sim 0.5 - 1$ keV in stars, Helfand and Caillault 1982). Nuclear sources, either connected with star formation activity, or with nonthermal Seyfert-like activity, can also be present and contribute various amounts to the X-ray emission (see Fabbiano 1989).

Diffuse X-ray emission from inverse Compton scattering of the radio electrons off the optical-infrared photons or synchrotron emission are not likely to contribute significantly to the total X-ray emission (Fabbiano et al 1982). Diffuse thermal emission from a hot phase of the interstellar medium, heated by supernovae, can instead be present. Supernovae release $\sim 10^{42}$ erg s$^{-1}$ in a galaxy, and it has been suggested that hot gaseous coronae, or galactic fountains could be produced, and should be visible in soft X-rays in the *Einstein* range (*e.g.* Spitzer 1956; Cox and Smith 1974; Bregman 1980a, b; Corbelli and Salpeter 1988). There is evidence of soft thermal diffuse emission both in the Galactic plane and in the LMC (*e.g.*, McCammon *et al.* 1983; Marshall and Clark 1984; Singh *et al.* 1987), and perhaps in M33 (Trinchieri et al 1988). An unsuccessful search for this type of emission in more distant galaxies has been made by Bregman and Glassgold (1982) and McCammon and Sanders (1984). The lack of intense diffuse soft X-ray emission could imply that most of the supernova energy is radiated in the unobservable far UV (Cox 1983). This type of soft X-ray emission could instead be present in the edge-on spiral galaxy NGC 4631 where this component could have an X-ray luminosity of $5 \times 10^{39}$ erg s$^{-1}$, which represents $\sim 13\%$ of the total emission in the *Einstein* band (Fabbiano and Trinchieri 1987).

## 2.2 Types of 'Evolved' X-ray Sources

In the *pre-Einstein* era, the X-ray sources of the Milky Way were classified as young Population I 'spiral arm' sources with massive early-type stellar counterparts and Population II or 'bulge' sources, with low mass stellar counterparts (*e.g.* van den Heuvel 1980). The *Einstein* imaging observations of spiral galaxies have led us to modify this classification and to gain new insight into the evolution of binary X-ray sources. We can now identify a 'spiral arm', a 'bulge', and a 'disk' component of the X-ray emitting population.

### 2.2.1 Spiral-arm sources

The presence of often very bright X-ray sources associated with the spiral arms and HII regions is immediately demonstrated by the X-ray images of the Local Group and of other relatively nearby galaxies. Bright point sources are detected in the spiral arms of M31 (Van Speybroeck *et al.* 1979; see Figure 1). The point-like X-ray sources detected in M33 appear to be associated with young Population I indicators, with the exception of a strong nuclear source (Long *et al.* 1981b; Markert and Rallis 1983; Trinchieri *et al.* 1988). One of these sources (M33 X-7) has a variable light curve, consistent with a 1.8-day eclipsing binary period, similar to those of some massive galactic X-ray binaries (Figure 3; Peres *et al.* 1988).

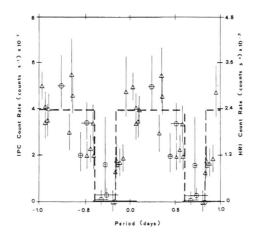

Figure 3. The light curve of the X-ray source M33 X-7 (Peres *et al.* 1989).

### 2.2.2 Bulge sources

A dominant bulge component of the X-ray emitting population and a number of globular cluster sources are evident in M31 (see Figure 1; Van Speybroeck *et al.* 1979; Van Speybroeck and Bechtold 1981; Long and Van Speybroeck 1983; see Helfand 1984a). The average X-ray spectrum of the bulge sources is similar to those of low-mass X-ray binaries in the Galaxy (Figure 4; Fabbiano et al 1987; Ohashi *et al.* 1988). Their luminosity is in the range of that of Galactic low-mass binaries, and some variability has been reported (Van Speybroeck and Bechtold 1981; McKechnie *et al.* 1984), consistent with the hypothesis of their being powered by accretion onto a compact object. The widespread presence of a bulge component of the X-ray luminosity in bulge-dominated early-type spirals is suggested by statistical analyses of the sample of spirals observed with *Einstein* . Correlations are found between the X-ray emission and other variables, including the radio continuum, the near-infrared H band, and the far-infrared *IRAS* emission (Fabbiano and Trinchieri

1985; Fabbiano *et al.* 1988). These correlations are all very tight in late-type galaxies where the bulge contribution is smaller or absent. In the subsample of bulge-dominated galaxies, instead, the correlations between radio continuum and/or far-infrared luminosities with any of the other emission bands show a considerable amount of scatter and sometimes also a shift in zero point (Figure 2). In particular, for a given X-ray luminosity, there is a clear deficiency of radio continuum emission, when bulge-dominated spirals are compared to disk/arm-dominated galaxies, while no differences are seen in the X-ray and optical (B) correlations. Since most of the optical and near-infrared emission of early-type spirals is dominated by the emission of the bulge (Kent 1985), these differences suggest that the radio continuum and the far-infrared are mainly related to the stellar population of the disk, while the X-rays originate in both the disk and the bulge components.

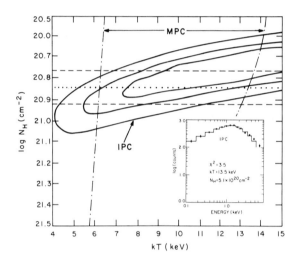

Figure 4. 99%, 90%, and 68% (from outside in) confidence contours for the spectral fitting of the *Einstein* IPC data of the bulge of M31 to a thermal bremsstrahlung spectrum with low-energy absorption. The horizontal dotted line indicates the line of sight extinction ($N_H$) and the two dashed lines its 90% confidence values. The IPC data and the best fit spectrum are shown in the insert. The dot-dashed lines represent the 90% MPC confidence contours (Fabbiano *et al.* 1987).

### 2.2.3 Disk sources

The close resemblance of the radial profile of the X-ray surface brightness of a few face-on spirals with that of the optical light of their exponential disk suggests the presence of a third component of the X-ray emission, associated with the stellar population of the disks (see Fabbiano 1986a). This effect was first seen in M83 (Figures 5 and 6; Trinchieri *et al.* 1985), and then in M51 (Palumbo *et al.* 1985), and possibly NGC 6946 (Fabbiano and Trinchieri 1987) and M81 (Fabbiano 1988a; see also M33, Trinchieri *et al.* 1988). In particular, in M51, the X-ray profile

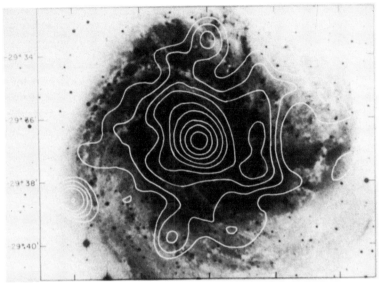

Figure 5. The *Einstein* HRI X-ray map of M83 (Trinchieri *et al.* 1985). X-ray emission can be seen from most of the optical disk. A bright source is visible on a spiral arm.

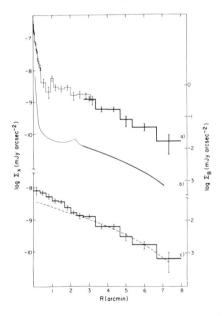

Figure 6. Radial profile of the X-ray surface brightness of M83 (Trinchieri *et al.* 1985; curves with error bars; the solid dark curve is from the IPC, the thin curve from the HRI), compared with the optical surface brightness profile. (curves without error bars; the dotted is the optical profile convolved with the IPC point response function).

is significantly different from the Hα profile, where the arms are very prominent, and follows the exponential disk distribution, as also do the radio continuum and the CO profiles (Figure 7). These observations open interesting possibilities for our understanding of the origin of low-mass X-ray binaries. The nature of these sources, which constitute ∼ 60% of the Galactic sources, and which are also called 'galactic bulge sources' and 'Population II' sources in the X-ray literature (*e.g.* van den Heuvel 1980; Helfand 1984), is one of the still open problems of 'classical' X-ray astronomy. Suggestions on their origin have included capture of neutron stars in the Galactic bulge (van den Heuvel 1980), remnants of disrupted globular clusters (Grindlay 1984, 1985), or the evolved remnants of low-mass binary systems in the Galactic disk (*e.g.* Gursky 1976; Rappaport *et al.* 1982; Nomoto 1984; van den Heuvel 1984 and references therein). The observations of external galaxies, morphologically similar to the Milky Way, suggest that a good fraction at least of these sources may originate from the evolution of binary systems belonging to the disk stellar population, rather than originating from dynamical evolution (Fabbiano 1985a, 1986a; Trinchieri *et al.* 1985). However (Fabbiano 1985a, 1986a) there is a relative excess of X-ray emission over the disk emission in the innermost disk region seen in both M83 and M51 (Trinchieri *et al.* 1985; Palumbo *et al.* 1985), which, scaling by distance and dimensions, roughly coincides with what has been called the X-ray Galactic Bulge. This excess emission could either indicate an intrinsically brighter population of X-ray sources, analogous to the bright 'bulge' Galactic sources; or it could point to an enhanced past episode of star formation in the inner disk, in contrast perhaps with steady star formation in the disk as a

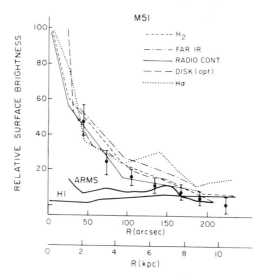

Figure 7. Radial profile of the X-ray surface brightness of M51 (the points; Palumbo *et al.* 1985), together with radial profiles at other wavelengths from Scoville and Young (1983) and Klein *et al.* (1984).

whole (*e.g.* Vader *et al.* 1982); or finally, it could suggest the presence of an additional component of the X-ray binary source population, which could be related to the disruption of globular clusters (Grindlay 1984, 1985).

## 2.3 'Super-Eddington' Sources

Only a very few of the sources detected in spiral galaxies can be chance superpositions of background or foreground objects, given the statistics of the serendipitous *Einstein* source detections (Gioia *et al.* 1984; see Fabbiano and Trinchieri 1987). Most of these sources are therefore in the galaxies, often in the spiral arm, and their X-ray luminosities can be very high indeed. They are typically well above the Eddington limit for accretion onto a 1 $M_\odot$ compact object, which is $\sim 1.3 \times 10^{38}$ erg s$^{-1}$, and can be as bright as a few $\times 10^{39}$ erg s$^{-1}$. These sources have been detected in a number of spirals, including M83, M51, NGC 253, NGC 4631, NGC 6946, M101, and M81 (Long and Van Speybroeck 1983; Fabbiano and Trinchieri 1984; Trinchieri *et al.* 1985; Palumbo *et al.* 1985; Fabbiano and Trinchieri 1987; Fabbiano 1988a), and 8 bright sources have also been reported in the central starburst region of M82 (Watson *et al.* 1984). We exclude bright nuclear regions from the present discussion, and we concentrate instead on the more puzzling galactic sources. About 36 sources more luminous than $\sim 2 \times 10^{38}$ erg s$^{-1}$ have been reported to date, 16 of them more luminous than $10^{39}$ erg s$^{-1}$.

What are these sources? In one case the answer is simple: one of these sources is SN1980k detected in NGC 6946, $\sim 35$ days after maximum light (Canizares *et al.* 1982b). The variability reported for some bright sources in M101 suggests point-like objects, possibly bright accretion binaries (Long and Van Speybroeck 1983). If these sources are mostly complex emission regions, we would be faced with several bright sources (*e.g.* $10^{37}$ erg s$^{-1}$) in volumes with typical dimensions of a few hundred parsecs to a kiloparsec (Fabbiano and Trinchieri 1987). These sources are not typically in bulges, were such crowding could be expected, *e.g.* M31 (see earlier discussion). If these sources are truly single objects, they could indicate the presence of massive black holes in these galaxies. It is however possible that the distances of some galaxies might have been overestimated, making these sources to appear more luminous that they are in reality. For instance, estimates of the distance of NGC 4631 range from 12 Mpc (Sandage and Tamman 1981) to 3 Mpc (Duric *et al.* 1982); using the lower estimate, the luminosity of the source reported by Fabbiano and Trinchieri (1987) as $\sim 1.4 \times 10^{40}$ erg s$^{-1}$ would become $\sim 9 \times 10^{38}$ erg s$^{-1}$. However, this still exceeds the Eddington luminosity of an accreting neutron star.

## 3. STARBURST GALAXIES AND NUCLEAR OUTFLOWS

### 3.1 Widespread Starburst Activity in Peculiar Galaxies

Bluer 'starburst', often interacting, galaxies tend to have enhanced X-ray emission, when compared with galaxies with redder, more normal, colors (see Figure 8; Fabbiano *et al.* 1982; Stewart *et al.* 1982; Fabbiano *et al.* 1984b). The X-ray emission of these galaxies tends to originate from spatially extended regions, excluding a purely nonthermal nuclear origin, and their X-ray spectra exclude on average very soft emission, suggesting that the X-ray emission is not dominated by the thermal emission of a gaseous halo (Fabbiano *et al.* 1982).

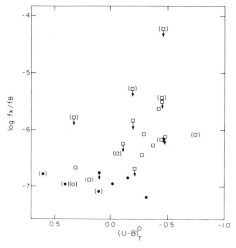

Figure 8. The log of X-ray to blue flux ratios ($f_x/f_B$) plotted versus the (U-B)$^o_T$ color. The squares identify the peculiar galaxies of Fabbiano *et al.* (1982); from Fabbiano *et al.* 1984.

The bulk of the X-ray emission of these galaxies can be understood in terms of a number of young supernova remnants and massive X-ray binaries, with X-ray luminosity possibly enhanced by the low metallicity of the accreting gas, similar to those observed in the Magellanic Clouds (Fabbiano *et al.* 1982; Stewart *et al.* 1982). Although this explanation is not applicable in general (Fabbiano *et al.* 1982), the integrated coronal emission of the young stellar population (see Vaiana *et al.* 1981 for typical values) may dominate in very young starbursts, where the ultraviolet *IUE* spectra suggest the presence of a very large number of OB stars (Moorwood and Glass 1982; Fabbiano and Panagia 1983). Recently, Ward (1988) has reported a correlation between the Brackett $\gamma$ line emission from a sample of starburst nuclei and their X-ray emission, which is interpreted in terms of a relationship between the number of ionizing photons produced by the OB stars and the associated X-ray binary population, as estimated by Fabbiano et al (1982).

## 3.2 Starburst Nuclei

There are galaxies in which the starburst activity is confined to the nuclear regions. The first reported instance of X-ray emission from this type of nucleus is that of NGC 7714 by Weedman *et al.* (1981), who associate this emission with the type of activity discussed above, as opposed to Seyfert-type nonthermal activity. A number of starburst nuclei, often embedded in an otherwise normal spiral galaxy, have been studied in X-rays. They include the Milky Way Galactic Center region (Watson *et al.* 1981), M82 (Van Speybroeck and Bechtold 1981; Watson *et al.* 1984; Biermann 1984; Kronberg *et al.* 1985; Schaaf *et al.* 1988; Fabbiano 1988b), NGC 253 (Van Speybroeck and Bechtold 1981; Fabbiano and Trinchieri 1984; Fabbiano 1988b), M83 (Trinchieri *et al.* 1985), M51 (this nucleus also contains a Seyfert component, which does not dominate the X-ray emission; Palumbo *et al.* 1985), NGC 6946 and IC 342 (Fabbiano and Trinchieri 1987), and NGC 3628 (Fabbiano *et al.* 1989, in preparation). A common characteristic of the emission spectrum of these nuclei is their intense far-infrared emission (see Figure 9; Fabbiano and Trinchieri 1987 and references therein), indicative of dusty nuclear regions heated by newly formed early type stars. The X-ray emitting regions are extended, whenever observed with high enough spatial resolution, and in M82 there is evidence of a population of bright individual sources (Watson et al 1984).

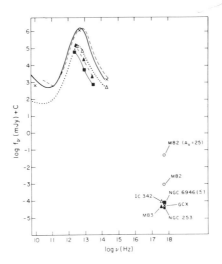

Figure 9. Radio through X-ray emission spectra of starburst nuclei (Fabbiano and Trinchieri 1987).

Typical X-ray luminosities of these nuclear regions are in the $10^{39}$ erg s$^{-1}$ range, except for the Galactic Center region, which is ~1000 times less luminous. To explain this emission requires, in different cases, different amounts of evolved sources (supernova remnants and X-ray binaries), superimposed on the integrated stellar emission from a young stellar population. The X-ray spectra of two of these nuclei,

NGC 253 and M82 (Fabbiano 1988b; Schaaf *et al.* 1988), are intrinsically absorbed, consistent with the presence of a dusty emission regions. In M82, in particular, it is possible that two different spectral components are present: a softer one, possibly dominated by the newly formed stars and by the interstellar medium shock-heated by the frequent supernovae; and a harder one, which could be due to binary X-ray sources with large intrinsic absoption cut-off, or else to inverse Compton emission, resulting from the interaction of the infrared photons in the nucleus with the relativistic electrons responsible for the radio emission.

### 3.3 Nuclear Outflows

Perhaps the most unexpected result of the *Einstein* observations of these nuclei has been the discovery of extended emission components, suggestive of gaseous outflows from the nuclear regions, in the edge-on galaxies M82, NGC 253 (Figure 10) and, more recently, NGC 3628. In M82, a correspondence between the region of extended X-ray emission ($\sim$90$''$ radially from the nucleus) seen in the *Einstein* High resolution imager (HRI) with the H$\alpha$ filaments was first noticed by Van Speybroeck and Bechtold (1981). Watson *et al.* (1984) then suggested that this X-ray 'halo' is likely to be thermal emission of shock-heated gas escaping the nuclear region. They point out that only 2% of the energy released by supernovae in the nucleus, exploding at a rate of 0.2 yr$^{-1}$ over a time scale of 10$^7$ years, is needed to heat the gas to X-ray temperatures. In NGC 253 the presence of an extended source, positionally coincident with a region of noncircular motions (Demoulin and Burbidge 1970),

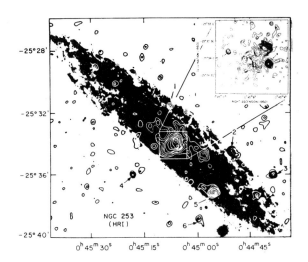

Figure 10. The *Einstein* HRI contour map of NGC 253, smoothed with a 7$''$ gaussian. The insert shows a higher resolution map of the nuclear region. Here the shaded oval represents the nuclear starburst region seen in the radio and infrared. A plume of X-ray emission can be seen extending along the southern minor axis. This was interpreted as evidence of hot outflowing gas (see Fabbiano and Trinchieri 1984 and references therein).

was also first reported by Van Speybroeck and Bechtold (1981). Fabbiano and Trinchieri (1984) studied this nuclear region, and identified an emission region associated with the starburst nucleus proper, and a 'jet-like' feature, or 'plume' of emission extending for ∼60" along the southern minor axis, which they suggested could be due to a bipolar nuclear outflow, similar to the one seen in M82, collimated by the galaxy disk and seen in projection along the minor axis. They suggested that the northern side of the outflow would not be visible because the soft X-ray photons would be absorbed by the interstellar medium in the disk of NGC 253. The subsequent report of a OH line emission plume from the dusty northern side (Turner 1985) confirms this picture. Optical work on the emission line gas velocity fields in these two galaxies is also in agreement with the picture of gaseous outflows (McCarthy *et al.* 1987; Bland and Tully 1988).Theoretical models of this phenomenon have been proposed by Chevalier and Clegg (1985) and Tomisaka and Ikeuchi (1988).

Analyzing the lower resolution, but more sensitive, *Einstein* IPC images of these two galaxies, Fabbiano (1988b) found evidence of diffuse X-ray emission at large radii in the northern side of NGC 253, which could be related to the nuclear outflow; in M82 instead there is clear evidence of an X-ray halo, elongated along the minor axis, and extending as far as ∼ 9 kpc from the nucleus (Figure 11; see also Kronberg *et al.* 1985). This halo is not likely to be bound to the galaxy, and the hot gas may be leaving the system at a rate that could be as high as 0.7 $M_\odot yr^{-1}$. These estimates are at the moment rather uncertain, since neither the gas volume filling factor, nor its emission temperature are really known. However, taken at face value, they would imply a maximum lifetime of $7 \times 10^8$ years for the starburst, since the mass present in the nuclear region is ∼ $5 \times 10^8 M_\odot$ (Rieke *et al.* 1980), unless the outflowing gas, cooling at large radii, would flow in again in a galactic fountain, or

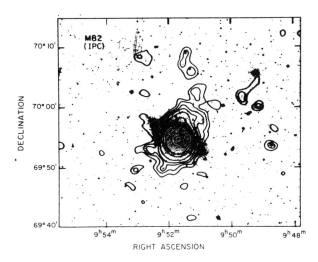

Figure 11. The *Einstein* IPC contour map of M82 (Fabbiano 1988b).

fresh gas inflow from the intergalactic medium would take place to fuel the nucleus. Relatively short lifetimes ($< 2 \times 10^9$ years) are also suggested by the OH data for the starburst at the nucleus of NGC 253 (Turner 1985).

Fabbiano (1984) remarked that these gaseous outflows should be visible in X-rays in many galaxies, since starburst or low-activity nuclei are quite common (Keel 1983). With the present data it is impossible to distinguish them from the underlying disk emission in face-on galaxies, such as M83 and M51 (Trinchieri *et al.* 1985; Palumbo *et al.* 1985); they should, however, be obvious in edge-on galaxies. Very recently, the reanalysis of the *Einstein* data of the edge-on NGC 3628 in different energy bands has shown an elongated soft emission region associated with the nucleus, suggestive of this phenomenon. The presence of a gaseous plume has been confirmed by subsequent optical observations (Fabbiano *et al.* 1989, in preparation).

Different authors have pointed out that these outflows, if generally associated with violent star formation activity, could be responsible for the formation and enrichment of a large part of the gaseous intracluster medium (Heckman *et al.* 1987; Fabbiano 1988b). In particular, if a relatively small galaxy like M82 can expell of the order of 1 $M_\odot \mathrm{yr}^{-1}$, a primordial large elliptical system could expell 1000 times this amount. Therefore some 1000 such systems in a cluster, undergoing violent star formation over a period of $10^8$ years, could produce the $\sim 10^{14} M_\odot$ of gas that are now found in clusters of galaxies (Jones and Forman 1984). This type of scenario has been modeled by Mathews (1988a).

## 4. GALAXIES AND THE X-RAY BACKGROUND

### 4.1 Before the *Einstein* Observatory

The extragalactic X-ray background was discovered in 1962 in the same rocket flight that led to the discovery of the first extrasolar source of X-rays, Sco X−1 (Giacconi *et al.* 1962). Since then a great deal of effort has been spent to understand if this radiation is due to the integrated contributions of different classes of discrete sources, or if diffuse processes are responsible for it (see Boldt 1987; Giacconi and Zamorani 1987). Based on the four galaxies then known to emit X-rays, The Milky Way, M31, and the Magellanic Clouds, Silk (1973) estimated that normal galaxies would contribute $\sim 10\%$ to the x-ray background. This estimate was then revised downward to less than $\sim 1\%$ (Rowan-Robinson and Fabian 1975; van Paradijs 1978; Worrall *et al.* 1979; see Fabian 1981) for a variety of reasons. One is the assumption that the X-ray luminosity of normal galaxies, being dominated by the low-mass binary sources, would not be correlated with the optical luminosity, but with some function of the galaxy mass (Rowan-Robinson and Fabian 1975); however, we now know that this is not so, as discussed earlier in this review. Other reasons are the effect of the redshift on the spectrum of the galactic sources (van Paradijs 1979); and the failure to detect a volume limited sample of galaxies with the HEAO-1 satellite (Worrall *et al.* 1979). However, this survey observed galaxies which are all of very low optical luminosity and as such cannot be considered representative of normal galaxies as a whole (Fabian 1981; Elvis *et al.* 1984).

## 4.2 After the *Einstein* launch

Based on the X-ray to optical ratios of normal galaxies observed with the *Einstein* Observatory (Fabbiano and Trinchieri 1985; Trinchieri and Fabbiano 1985), Giacconi and Zamorani (1987) extimate instead that the integrated emission of normal galaxies contributes $\sim$ 13% of the 2 keV extragalactic x-ray background (see also Setti 1985). If one includes in this estimate the contribution of low-activity nuclei present in a fraction of these galaxies, and the effect of starburst activity, this contribution could be significantly larger: Elvis *et al.* (1984) estimate that the former could contribute $\sim$ 20% of the X-ray background in absence of evolution; and Weedman (1987), using X-ray to optical ratios measured in starburst galaxies (Fabbiano *et al.* 1982), and the $60\mu m$ luminosity function derived from the *IRAS* survey, finds that in absence of evolution these galaxies could account for $\sim$ 13% of the 2 keV background (see also Giacconi and Zamorani 1987). If one assumes that starburst activity was much more pronounced in the past (Bookbinder *et al.* 1980; Stewart *et al.* 1982), starburst galaxies could be responsible for the bulk of the background (Giacconi and Zamorani 1987; Weedman 1987). Although this possibility is rejected by Giacconi and Zamorani, because it would predict a surface density of 21-23 mag. galaxies inconsistent with optical searches, the presence of large amount of dust in these systems, suggested by the *IRAS* data, leaves this a still viable option (Weedman 1987).

Even if galaxies contribute substantially to the 2 keV X-ray background, their contribution to the x-ray background in a harder spectral range is rather uncertain, and rests upon the spectral characteristics of their X-ray emission (van Paradijs 1978; Giacconi and Zamorani 1987; Weedman 1987). The spectra of spiral galaxies are consistent with a relatively hard X-ray emission (kT > 2 keV; Fabbiano and Trinchieri 1987), and X-ray binaries can have hard X-ray spectra (with kT $\sim$ 20 keV). Moreover, a hard spectral component may be present in the starburst galaxy M82 (Fabbiano 1988b; Schaaf *et al.* 1988). However, it is unlikely that their contribution would be substantial above 10 keV, expecially considering that the spectra of far-away galaxies would be red-shifted. The soft X-ray spectrum of the nucleus of M81 (Fabbiano 1988a) introduces an additional source of uncertainty in the estimate of Elvis *et al.* (1984) of the contribution of low-activity nuclei to the 2-10 keV X-ray background, since most of the X-ray luminosity of these sources could be emitted in a softer energy range. On the other hand, we do not know how common and how bright the optically quiet X-ray active nuclei, such as those of M33 and NGC 1313 are (see Fabbiano 1988a and refs. therein). Their inclusion could raise the estimate of the contribution to the x-ray background, although their X-ray spectra (Trinchieri *et al.* 1988; Fabbiano and Trinchieri 1987) suggest that even this type of source should contribute mainly in the soft energy range. Future X-ray observations will help to constrain the spectral range in which galaxies may contribute to the extragalactic X-ray background, and will give us a better estimate of the contribution of low-activity nuclei.

## 5. THE FUTURE

The *Einstein* Observatory has opened up the study of normal galaxies in X-rays. However, we must not forget that what we have learned so far about the X-ray properties of normal galaxies has been the result of limited exploratory observations. Future X-ray satellites with increased sensitivity, and higher spatial and spectral resolution, will be essential for answering the many open questions resulting from the present work and for expanding and deepening our knowledge of these systems.

The German X-ray satellite *ROSAT*, which is scheduled to be launched in 1990, will increase the number of galaxies mapped in X-rays with a good sensitivity to low surface brightness features, and the Japanese *ASTRO-D* (to be launched in 1993) will allow the study of galaxies' spectral properties with a ten-fold increased spectral resolution (but only 2 arcminute spatial resolution). The next major US X-ray astronomy endeavour, *AXAF*, with its sub-arcsecond spatial resolution and good spectral resolution, will allow the study of the luminosity function of X-ray sources in nearby spiral galaxies, down to to limiting luminosities at least 100 times smaller than the present ones, and will be able to detect single early-type stars in the Magellanic Clouds. Spectral parameters or X-ray colors will be measured for these sources, helping in establishing their nature (*e.g.* black hole candidates versus X-ray pulsars, see White and Marshall 1984). With *AXAF* and the European *XMM*, with its larger collective area and sensitivity to low surface brightness features, the astronomical community will be able to address some of the outstanding questions on the X-ray properties of spiral and starburst galaxies. These include the study of a hot phase of the interstellar medium, the study of nuclear outflows and long-term variability studies of bright binary sources and low-activity nuclei. It will be also possible to establish the luminosity functions and the spectral characteristics of individual X-ray sources in different spiral galaxies, and thus investigate the nature and evolution of these sources in different environments.

## ACKNOWLEDGEMENTS

This work was supported by NASA contract NAS8-30751.

## REFERENCES

Biermann, P. 1984, *Seventh European Regional Astronomy Meeting, Frontiers of Astronomy and Astrophysics*, ed. R. Pallavicini, 191. Florence: Italian Astronomical Society

Bland, J., Tully, R. B. 1988, *Nature*

Boldt, E. 1987, *Phys.Rep.* 146:215

Bookbinder, J., Cowie, L. L., Krolik, J. H., Ostriker, J. P., Rees, M. 1980, *Ap.J.* 237:647

Bregman, J. N. 1980a, *Ap.J.* 236:577

Bregman, J. N., Glassgold, A. E. 1982, *Ap.J.* 263:564

Canizares, C. R., Kriss, G. A., Feigelson, E. D. 1982b, *Ap.J.(Letters)* 253:L17

Chevalier, R. A., Clegg, A. W. 1985, *Nature* 317:44

Corbelli, E., Salpeter, E. E. 1988, *Ap.J.* 326:551

Cox, D. P. 1983, *IAU Symp. No. 101, Supernova Remnants and their X-ray Emission*, ed J. Danziger and P. Gorenstein (Dordrecht: Reidel) 385

Cox, D. P., Smith, B. W. 1974, *Ap.J.(Letters)* 189:L105

Demoulin, M.-H., Burbidge, E. M. 1970, *Ap.J.* 159:799

Duric, N., Crane, P. C., Seaquist, E. R. 1982, *A.J.* 87:1671

Elvis, M., Maccacaro, T., Wilson, A. S., Ward, M. J., Penston, M. V., Fosbury, R. A. E., Perola, G. C. 1978, *MNRAS* 183:129

Elvis, M., Soltan, A., Keel, W. C. 1984, *Ap.J.* 283:479

Fabbiano, G. 1984, *X-ray Astronomy '84*, ed. M. Oda, R. Giacconi, 333. Tokyo: Institute of Space Astronautical Science

Fabbiano, G. 1985a, *Japanese-U.S. Seminar on Galactic and Extragalactic Compact X-ray Sources*, ed. Y. Tanaka, W. H. G. Lewin, 233. Tokyo: ISAS

Fabbiano, G. 1986a, *Pu.A.S.P.*, 98:525

Fabbiano, G. 1988a, *Ap.J.* 325:544

Fabbiano, G. 1988b, *Ap.J.* 330:672

Fabbiano, G. 1989, *Ann.Rev.Astron.Astrophys.*, Vol. 27, in press

Fabbiano, G., Feigelson, E., Zamorani, G. 1982, *Ap.J.* 256:397

Fabbiano, G., Gioia, I. M., Trinchieri, G. 1988, *Ap.J.* 324:749

Fabbiano, G., Gioia, I. M., Trinchieri, G. 1989, preprint

Fabbiano, G., Klein, U., Trinchieri, G., Wielebinski, R. 1987, *Ap.J.* 312:111

Fabbiano, G., Panagia, N. 1983, *Ap.J.* 266:568

Fabbiano, G., Trinchieri, G. 1984, *Ap.J.* 286:491

Fabbiano, G., Trinchieri, G. 1985, *Ap.J.* 296:430

Fabbiano, G., Trinchieri, G. 1987, *Ap.J.* 315:46

Fabbiano, G., Trinchieri, G., Macdonald, A. 1984b, *Ap.J.* 284:65

Fabbiano, G., Trinchieri, G., Van Speybroeck, L. S. 1987, *Ap.J.* 316:127

Fabian, A. C. 1981, *The Structure and Evolution of Normal Galaxies*, ed. S. M. Fall, D. Lynden-Bell, 181. Cambridge: University Press

Giacconi, R., Gursky, H., Paolini, F., Rossi, B. 1962, *Phys.Rev.Lett.* 9:439

Giacconi, R., et al. 1979, *Ap.J.* 230:540

Giacconi, R., Zamorani, G. 1987, *Ap.J.* 313:20

Gioia, I. M., et al. 1984, *Ap.J.* 283:495

Grindlay, J. E. 1984, *Adv.Space Res.* 3:19-27

Grindlay, J. E. 1985, *Japanese-U.S. Seminar on Galactic Extragalactic Compact X-ray Sources*, ed. Y. Tanaka, W. H. G. Lewin, 215. Tokyo: ISAS

Gursky, H. 1976, *IAU Symp. 73, Structure and Evolution of Close Binary Systems*, ed. P. Eggleton, S. Mitton, J. Whelan, 19. Dordrecht: Reidel

Heckman, T. M., Armus, L., Miley, G. 1987, *A.J.* 93:276

Helfand, D. J. 1984a, *Pu.A.S.P.* 96:913

Helfand, D. J., Caillault, J.-P. 1982, *Ap.J.* 253:760

Jones, C., Forman, W. 1984, *Ap.J.* 276:38

Keel, W. C. 1983, *Ap.J.* 269:466

Kent, S. M. 1985, *Ap.J.Suppl.* 59:115

Klein, U., Wielebinski, R., Beck, R. 1984, *Astr.Ap.* 135:213

Kronberg, P. P., Biermann, P., Schwab, F. R. 1985, *Ap.J.* 246:751

Long, K. S., D'Odorico, S., Charles, P. A., Dopita, M. A. 1981b, *Ap.J.(Letters)* 246:L61

Long, K. S., Van Speybroeck, L. P. 1983, *Accretion Driven X-ray Sources*, ed. W. Lewin, E. P. J. van den Heuvel, 117. Cambridge University Press

Markert, T. H., Rallis, A. D. 1983, *Ap.J.* 275:571

Marshall, F. J., Clark, G. W. 1984, *Ap.J.* 287:633

Mathews, W. G. 1988a, preprint.

McCammon, D., Burrows, D. N., Sanders, W. T., Kraushaar, W. L. 1983, *Ap.J.* 269:107

McCammon, D., Sanders, W. T. 1984, *Ap.J.* 287:167

McCarthy, P. J., Heckman, T., van Breugel, W. 1987, *A.J.* 93:264.

McKechnie, S. P., Jansen, F. A., deKorte, P. A. J., Hulscher, F. W. H., van der Klis, M., Bleeker, J. A. M., Mason, K. O. 1984, *X-ray Astronomy '84*, ed. M. Oda, R. Giacconi, 373.

Moorwood, A. F. M., Glass, I. S. 1982, *Astr.Ap.* 115:84

Nomoto, K. 1984, *Problems of Collapse and Numerical Relativity*, ed. D. Bancel, M. Signore, 89. Dordrecht: Reidel

Ohashi, T., *et al.* 1988, preprint

Palumbo, G. G. C., Fabbiano, G., Fransson, C., Trinchieri, G. 1985, *Ap.J.* 298:259

Peres, G., Reale, F., Collura, A., Fabbiano, G. 1988, *Ap.J.*, in press

Rappaport, S., Joss, P. C., Webbink, R. F. 1982, *Ap.J.* 254:616

Rieke, G. H., Lebofsky, M. J., Thompson, R. I., Low, F. J., Tokunaga, A. T. 1980, *Ap.J.* 238:24

Rowan-Robinson, M., Fabian, A. C. 1975, *MNRAS* 170:199

Sandage, A., Tamman, G. A. 1981, *A Revised Shapley-Ames Catalog of Bright Galaxies*, Pub. No. 635. Washington: Carnegie Institution

Schaaf, R., Pietsch, W., Biermann, P. L., Kronberg, P. P., Schmutzler, T. 1988, preprint

Scoville, N., Young, J. S. 1983, *Ap.J.* 265:148

Setti, G. 1985, *Nonthermal and Very High Temperature Phenomena in X-ray Astronomy*, ed. G. C. Perola, M. Salvati, 159. Rome: Istituto Astronomico,

Universita "La Sapienza"

Silk, J. 1973, *Ann.Rev.Astr.Ap.* 11:269

Singh, K. P., Nousek, J. A. Burrows, D. N., Garmire, G. P. 1987, *Ap. J.* 313:185

Spitzer, L. 1956, *Ap.J.* 124:20

Stewart, G. C., Fabian, A. C., Terlevich, R. J., Hazard, C. 1982, *MNRAS* 200:61P

Tananbaum, H., Peters, G., Forman, W., Giacconi, R., Jones, C., Avni, Y. 1978, *Ap.J.* 223:74

Tomisaka, K., Ikeuchi, S. 1988, *Ap.J.* 330:695

Topka, K., Avni, Y., Golub, L., Gorenstein, P., Harnden, F. R. Jr., Rosner, R., Vaiana, G. S. 1982, *Ap.J.* 259:677

Trinchieri, G., Fabbiano, G. 1985, *Ap.J.* 296:447

Trinchieri, G., Fabbiano, G., Palumbo, G. G. C. 1985, *Ap.J.* 290:96

Trinchieri, G., Fabbiano, G., Peres, G. 1988, *Ap.J.* 325:531

Turner, B. E. 1985, *Ap.J.* 299:312.

Vader, J. P., van den Heuvel, E. P. J., Lewin, W. H. G., Takens, R. J. 1982, *Astr.Ap.* 113:328

Vaiana, G., *et al.* 1981, *Ap.J.* 245:163

van den Heuvel, E. P. J. 1980, *X-ray Astronomy*, ed. R. Giacconi, G. Setti, 119. Dordrecht: Reidel

van den Heuvel, E. P. J. 1984, *Seventh European Regional Astronomy Meeting, Frontiers of Astronomy and Astrophysics*, ed. R. Pallavicini, 167. Florence: Italian Astronomical Society

van Paradijs, J. 1978, *Ap.J.* 226:586

Van Speybroeck, L., Bechtold, S. 1981, *X-ray Astronomy with the Einstein Satellite*, ed. R. Giacconi, 153. Dordrecht: Reidel

Van Speybroeck, L., Epstein, A., Forman, W., Giacconi, R., Jones, C., Liller, W., Smarr, L. 1979, *Ap.J.(Letters)* 234, L45

Ward, M. J. 1988, *MNRAS* 231:1P

Watson, M. G., Stanger, V., Griffiths, R. E. 1984, *Ap.J.* 286:144

Weedman, D. W. 1987, *Star Formation in Galaxies*, ed. C. J. Lonsdale Persson, 351. NASA

Weedman, D. W., Feldman, F. R., Balzano, V. A., Ramsey, L. W., Sramek, R. A., Wu Chi-Chao 1981, *Ap.J.* 248:105

White, N. E., Marshall, F. E. 1984, *Ap.J.(Letters)* 281:354

Worrall, D. M., Marshall, F. E., Boldt, E. A. 1979, *Nature* 281:127

# COOLING FLOWS IN CLUSTERS OF GALAXIES

**A. C. Fabian**
Institute of Astronomy
Cambridge, ENGLAND

## 1. INTRODUCTION

Cooling flows appear to be relatively common in the centers of the hot gaseous atmospheres of clusters and groups of galaxies. They also occur in many early-type galaxies. The gas is densest in the core of a cluster and its cooling time due to the emission of X-rays such as those observed, $t_{cool}$, is shortest there. A cooling flow is formed when $t_{cool}$ is less than the age of the system, $t_a(\sim H^{-1})$.

In the cases considered here, $t_{cool}$ exceeds the gravitational free-fall time, $t_{grav}$, within the cluster (except perhaps in some very small region at the center), so, for a cooling flow,

$$t_a > t_{cool} > t_{grav}.$$

The flow takes place because the gas density has to rise to support the weight of the overlying gas.

If that is not immediately clear, consider the gaseous atmosphere trapped in the gravitational potential well of the cluster or galaxy to be divided into two parts at the radius, $r_{cool}$, where $t_{cool} = t_a$. The gas pressure at $r_{cool}$ is determined by the weight of the overlying gas, in which cooling is not important. Within $r_{cool}$, cooling is tending to reduce the gas temperature and so the gas density must rise in order to maintain the pressure at $r_{cool}$. The only way for the density to rise (ignoring matter sources within $r_{cool}$, which is a safe assumption in a cluster of galaxies) is for the gas to flow inward. This is the cooling flow.

If the initial gas temperature exceeds the virial temperature of the central galaxy (which is generally the case for rich clusters but not for poor ones or individual galaxies) then the gas continues to cool as it flows in. However, when the temperature has dropped to the virial temperature of the central galaxy, the gas heats up as it flows further in due to the release of gravitational energy. The gas temperature eventually drops catastrophically in the core of the galaxy if its gravitational potential flattens there. The net result is that the gas within $r_{cool}$ radiates its thermal energy plus the $PdV$ work and gravitational energy released in the flow.

This is how an idealized, homogeneous cooling flow, in which the gas has a unique temperature and density at each radius, will behave. Observations of real cooling flows shows that they are inhomogeneous and must consist of a mixture of temperatures and densities at each radius. The homogeneous flow is, however, still a fair approximation of the mean flow.

In this paper, which is an expanded version of my contribution to IAU Colloquium 115 (Fabian 1989), I give a personal view of the current status of the observational and theory of cooling flows. More general reviews have been made by Fabian, Nulsen and Canizares (1984) and Sarazin (1986, 1988) and some other points of view may be found in the Proceedings of a NATO Workshop (Fabian 1988a). As explained above, the cooling flow mechanism is very simple, although the details of its operation are not. The primary evidence for them is in the X-ray observations. There is no evidence at other wavelengths for the large mass deposition inferred from the X-ray data. I discuss this point more fully later, but it should be stressed that large amounts of distributed low-mass star formation at other wavelengths need not be detectable if the gas is initially at X-ray emitting temperatures. This is, perhaps, the crux of the controversial aspect of cooling flows. They are difficult to prove or disprove in wavebands other than the X-ray. The X-ray evidence is, for me, sufficiently compelling that the existence of large cooling flows is a reasonable and straightforward conclusion.

It was *UHURU* observations of clusters that first showed the mean cooling time of the gas in the cores of clusters to be close to a Hubble time (Lea *et al.* 1973). X-ray measurements from the Copernicus satellite showed that the core emission in the Perseus and Centaurus clusters was highly peaked (Fabian *et al.* 1974; Mitchell *et al.* 1975). These, and theoretical considerations, led Cowie and Binney (1977), Fabian and Nulsen (1977) and Mathews and Bregman (1978) to independently consider the effects of significant cooling of the central gas, *i.e.* cooling flows. The process was noted by Silk (1976) as a mechanism for the formation of central cluster galaxies from intracluster gas at early epochs. It has been pointed out to me by T. Gold that Gold and Hoyle (1959) proposed the process for galaxy formation thirty years ago.

## 2. X-RAY EVIDENCE FOR COOLING FLOWS

### 2.1 X-ray Images

A sharply-peaked X-ray surface brightness distribution is indicative of a cooling flow. It shows that the gas density is rising steeply towards the center of the cluster or group since the emissivity depends upon the square of the gas density and only weakly on the temperature[1].

Most of the images have been obtained with the *Einstein Observatory* and with EXOSAT, although the peaks were anticipated with data from the Copernicus satellite (Fabian *et al.* 1974; Mitchell *et al.* 1975), from rocket-borne telescopes

---

[1]The spectroscopic data discussed in Section 2.2 rules out models in which the increased X-ray emission is due to populations of point sources around the central galaxy or to inverse Compton emission. Other points against such interpretations are the lack of peaks around the dominant ellipticals in the Coma cluster and the lack of any detailed spatial correlation with radio emission.

(Gorenstein *et al.* 1977) and with the modulation collimators on SAS 3 (Helmken *et al.* 1978).

Deprojection, or modelling, of the X-ray images shows that $t_{cool} < H_0^{-1}$ within the central 100kpc or so of more than 30 to 50 per cent of the clusters well-detected with the *Einstein Observatory* (Stewart *et al.* 1984b; Arnaud 1988) [2]. Whether $H_0^{-1}$ should be used for $t_a$ is debatable (but see §2.2), and it is not obvious how to extrapolate from 'well-detected' clusters to all clusters. Inspection of the results shows that reducing $t_a$ by 2, say, does not much change the fraction of clusters which contain cooling flows. The spatial resolution of the commonly-used IPC was not sufficient to resolve the central regions of the fainter or more distant clusters. The measured $t_{cool}$ is then an upper limit. The overall picture is that the prime criterion for a cooling flow, $t_{cool} < 10^{10}$yr, is satisfied in a large fraction of clusters. It is also satisfied in a number of poor clusters and groups (Schwartz, Schwarz and Tucker 1980; Canizares, Stewart and Fabian 1983; Singh, Westergaard and Schnopper 1986). Cooling flows must be both common and long-lived, in order that such a high fraction of peaked clusters is observed.

The mass deposition rate, $\dot{M}$, due to cooling (*i.e.* the accretion rate, although this is a poor term since most of the gas does not much change its radius) can be estimated from the X-ray images by using the luminosity associated with the cooling region (*i.e.* $L_{cool}$ within $r_{cool}$) and assuming that it is all due to the radiation of the thermal energy of the gas, plus the *PdV* work done.

$$L_{cool} = \frac{5}{2} \frac{\dot{M}}{\mu m} kT,$$

where $T$ is the temperature of the gas at $r_{cool}$. Values of $\dot{M} = 50 - 100$ M$_\odot$Yr$^{-1}$ are fairly typical for cluster cooling flows. ($L_{cool}$ is similar to the excess luminosity measured by Jones and Forman 1984.) Some clusters show $\dot{M} \sim 500$ M$_\odot$Yr$^{-1}$ (*e.g.* PKS 0745−191, A1795, A2597 and Hydra A). The main uncertainties in the determination of $\dot{M}$ are the gravitational potential within the cluster core and $t_a$. Assuming $t_a \sim 10^{10}$yr, the estimates of $\dot{M}$ are probably accurate to within a factor of 2 (Arnaud 1988).

Since we often measure a surface brightness profile for the cluster core (where The X-ray emission is well-resolved), we have $L_{cool}(r)$ which can be turned into $\dot{M}(r)$, the mass deposition rate within radius $r$. Generally,

$$\dot{M}(r) \propto r.$$

---

[2]More than two-thirds of the 50 X-ray brightest clusters in the Sky (see list in Lahav *et al.* 1989) have cooling flows, according to unpublished work of A. Edge and K.Arnaud. Since the luminosity associated with the flow does not dominate the total X-ray emission, this high fraction is not a simple consequence of the clusters being X-ray bright. It is due to the data on them generally being of the best quality (*i.e.* many X-ray counts detected and the core well-resolved).

This means that the surface brightness profiles are less peaked than they would be if all the gas were to flow to the center. This means that the gas must be inhomogeneous, so that some of the gas cools out of the flow at large radii and some continues to flow in. The actual computation of $\dot{M}(r)$ is in detail complicated, since we need to take into account how the gas cools and any gravitational work done, but since plain cooling dominates in clusters, a simple analysis gives a fair approximation to the profile (see Fabian, Arnaud and Thomas 1986; Thomas, Fabian and Nulsen 1987; White and Sarazin 1987abc).

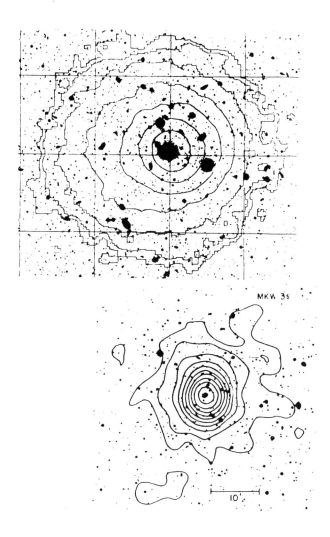

Figure 1. IPC X-ray surface brightness contours of the Perseus cluster (upper) and MKW3s (lower, from Kriss, Cioffi and Canizares 1983) superimposed on an optical image of the cluster. Note that the contours peak onto the central cluster galaxy. The mass deposition rates are about 200 and 100 $M_\odot Yr^{-1}$ respectively.

## 2.2 X-ray Spectra

Key evidence that the gas does actually cool is given by moderate to high resolution spectra of the cluster cores. Canizares *et al.* (1979, 1982), Canizares (1981); Mushotzky *et al.* (1981) and Lea *et al.* (1982) used the Focal Plane Crystal Spectrometer (FPCS) and the Solid State Spectrometer (SSS) on the *Einstein Observatory* to show that there are low temperature components in the Perseus and Virgo clusters, consistent with the existence of cooling flows. Detailed examination of the line fluxes and of the emission measures of the cooler gas by Canizares, Markert and Donahue (1988) and Mushotzky and Szymkowiak (1988) shows that, in the case of the Perseus cluster, the gas loses at least 90 per cent of its thermal energy and that the mass deposition rates are in agreement with those obtained from the images. Good agreement is obtained also in several other clusters. The SSS results show that the emission measures vary with temperature in the manner expected from a cooling gas. *The importance of these data cannot be overemphasized since they show that the gas does cool.* Any 'alternative interpretation' of the images must confront this spectroscopic evidence successfully.

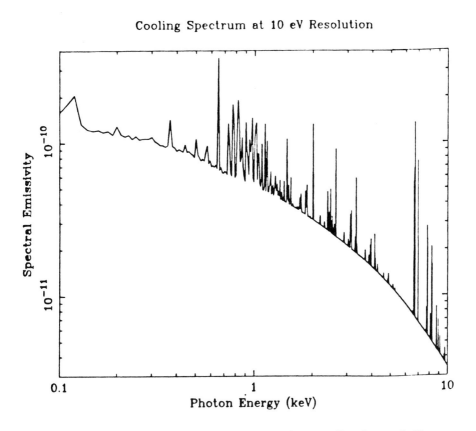

Figure 2. X-ray spectrum at 10 eV resolution of gas cooling from 6 keV.

The cooling time of the gas in the Perseus cluster which emits the FeXVII line ($T < 5 \times 10^6$K) is less than $3 \times 10^7$yr. Since the emission measure of this gas agrees with that inferred from the gas cooling at the higher temperatures which dominate the images and the SSS result, we must conclude that the flow is steady (Nulsen 1988). The shape of the continuum and line spectrum observed with the SSS is consistent with the same mass deposition rate at all X-ray temperatures (as expected) so we must again conclude that the flow is long-lived. It cannot be some intermittent or transient phenomenon only a billion years old.

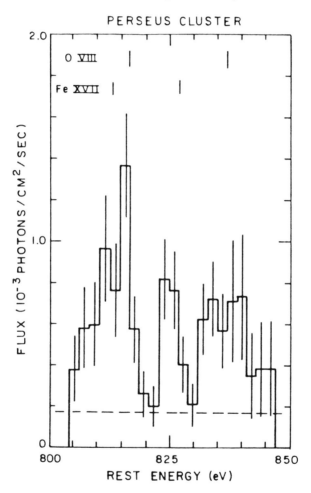

Figure 3. Part of FPCS spectrum of the Perseus cluster. Note the prominent emission lines of OVIII and Fe XVII (from Canizares, Markert and Donahue 1988). The emission measure of the gas producing the Fe XVII lines is consistent with about 200 $M_\odot$Yr$^{-1}$ of gas cooling through $\sim 5.10^6 - 10^6$K. This is the $\dot{M}$ found from imaging and SSS studies, which are most sensitive to higher temperature gas (typically $5.10^6 - 3.10^7$K).

## 2.3 Summary

The overwhelming evidence of the images and spectra shows that cooling does occur at a steady rate over long times (at least several billion years). Since mass is then cooling out of the hot phase at rates of hundreds of solar masses per year an inflow must occur. We do not expect yet to have direct evidence of any inward flow since the velocity is highly subsonic at $\sim 10$ km s$^{-1}$.

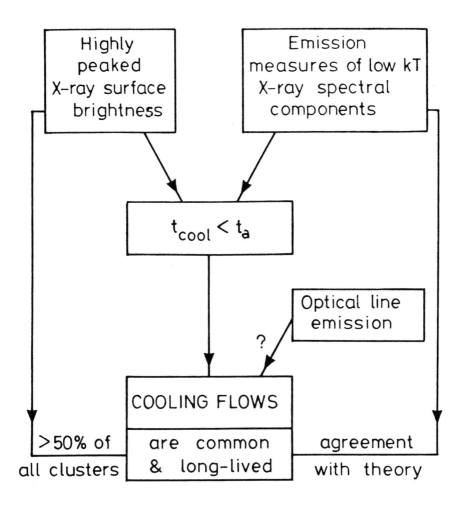

Figure 4. The evidence for Cooling Flows.

Cooling flows are common and all of the nearest clusters (Virgo, Centaurus, Hydra, Fornax, Perseus, Ophiuchus...) contain one apart from the Coma cluster. Clusters such as the Coma cluster and CA 0340-54 with 2 large central galaxies orbiting each other are the main class of clusters that does not show strongly peaked emission. Even they could contain disrupted flows. The motion of the central galaxies means that there is no focus for the flow (Fabian, Nulsen and Canizares 1984). Many flows are observed out to a redshift of 0.1, with more distant ones being in 3C 295 (Henry and Henriksen 1987) and 1E 0839+2938 (Wolter *et al.* 1988). I suspect that *all* clusters of galaxies will be found to have gas cooling out at rates exceeding a few solar masses per year when we have the improved spectral and spatial response of future missions such as AXAF.

The current values of $\dot{M}$ are probably good to a factor of 2 (Arnaud 1988) and could be higher if there are denser blobs beyond $r_{cool}$ (Thomas *et al.* 1987). I am not aware of any alternative interpretation of the X-ray data which explains both the peaked images and the X-ray spectra.

## 3. THE FATE OF THE COOLED GAS

The accumulated mass of cooled gas can be considerable;

$$\dot{M}t_a = 10^{12} \left( \frac{\dot{M}}{100 M_\odot \mathrm{Yr}^{-1}} \right) \left( \frac{t_a}{10^{10} \mathrm{yr}} \right) M_\odot.$$

This is a significant fraction of the mass of the central galaxy. It suggests that we are witnessing the continued formation of that galaxy, which is typically one of the largest galaxies known.

If the gas forms stars, then cooling flows are some of the largest and strongest regions of star formation in our part of the Universe. Even a casual comparison of a central cluster galaxy and a spiral galaxy such as our own, which is thought to be forming stars at a rate of $3 - 10$ $M_\odot \mathrm{Yr}^{-1}$, shows that cooling flows must form low-mass stars (Fabian, Nulsen and Canizares 1982; Sarazin and O'Connell 1983). Massive stars would make central cluster galaxies much bluer than they are. The absence of massive blue stars means that star formation with a spiral galaxy initial-mass-function is almost non-existent in the central galaxies of cooling flows.

It should be stressed that the cooling gas is not directly detected once it has cooled below about $3 \times 10^6 \mathrm{K}$. If it recombines and forms low-mass stars ($\langle M_* \rangle \ll 0.5 M_\odot$) in a distributed manner ($M(r) \propto r$) then there is no reason for it to have been seen.

There is, however, plenty of evidence for dark matter in clusters and low-mass stars are one plausible form of dark matter. The manner of the mass deposition with radius, $\dot{M}(r) \propto r$, leads to an isothermal halo which is consistent with the dark matter distribution around large galaxies (*e.g.* M87, Stewart *et al.* 1984a; Mould *et al.* 1987). Cooling flows are a source of baryonic dark matter.

## 3.1 Heating

Since the implied star formation rates are so large and there is little sign of it optically, there have been a number of studies suggesting that the rates have been grossly over-estimated. Some heat source that balances the cooling is the obvious solution. Cosmic rays (Tucker and Rosner 1982), conduction (Bertschinger and Meiksin 1986), supernovae (Silk *et al.* 1986) and galaxy motions (Miller 1986) have all been invoked as heat sources. Unfortunately for these models, the X-ray spectra indicate cooling without heating. None of the models proposed so far is able (or even attempts!) to account for the X-ray line emission. There are other problems with these heat sources as well (see Fabian 1988b; Bregman and David 1988).

It has been suggested by Hu (1988) that cooling flows began only recently. This reduces that total accumulated mass but does not explain the lack of blue stars. The initial-mass-function must be different from our local initial-mass-function even if the flow is recent.

The total level of heating necessary to balance the cooling is very large, $\sim 10^{62}$ erg for a large flow over $t_a$ and so if some heat source is found that can accommodate the X-ray spectral measurements successfully, it must be one of the major (unseen!) energy flows in the Universe! Whilst the luminosity of a cooling flow may be only 10 per cent of the total cluster X-ray luminosity and the mass lost through cooling a negligible drain on the enormous outer atmosphere, the cooling luminosity is a major loss of energy from the cluster core. Whatever is eventually decided about cooling flows, they cannot be an insignificant process.

## 3.2 Star Formation in Cooling Flows

As already mentioned, the average mass of a star formed in a cooling flow must be considerably smaller than in our Galaxy. In particular, the fraction of the mass turned into massive OB stars must be very small, since there is little ultraviolet light seen with the IUE (*e.g.* Fabian, Nulsen and Arnaud 1984). (The shortest cooling times for the X-ray emitting gas ($\sim 3 \times 10^7$yr) are comparable to the lifetime of B stars, so intermittency of the flow cannot be important here.) In understanding why there are these differences, it would be helpful to have a predictive theory of star formation for our Galaxy. As we do not, we must look for differences. Some are a) the lack of dust[3] in gas that has cooled from $T > 10^7$K (Draine and Salpeter 1979), which presumably means that there are no molecular clouds such as give birth to massive stars in our Galaxy; b) the thermal pressure of the gas is 100 – 1000 times higher than in our interstellar medium; c) differential motions, cloud masses, angular momentum and magnetic field strengths may be different.

---

[3]Hintzen and Romanishin (1988) find that the central parts of the cooling flow galaxy 3A 0335+096 are red. Whether this is due to red stars or to dust is not clear. NGC 4696 in the Centaurus cluster, which has a cooling flow, also has a dust lane, so some dust is found in the middle of cooling flows. It may just be due to stellar mass loss from the galaxy itself.

The general statement about the necessity for low mass stars applies to the bulk of the cooled gas. In the center, there are often seen optical emission line blobs or filaments, which may be atypical of most of the flow. These blobs may give rise to higher mass stars. There is some excess blue light observed at the centers of many cooling flows and it does correlate in strength with the mass deposition rate (Johnstone, Fabian and Nulsen 1987). Spectral fits of the blue light together with upper limits from IUE spectra show that the upper mass limit for stars must be around $1.5 - 2M_\odot$ there. Some F and early G stars are seen. Of course, the best place to look for the bulk of the cooled gas is at large radii where the underlying stellar light of the galaxy is least, so the contrast is highest.

Small cooling flows occur continuously in most elliptical galaxies which are not in rich clusters. Ram-pressure stripping removes the gas from galaxies within clusters. This is observed in the Virgo cluster where M86 has a plume of X-ray emission to the NW (Forman *et al.* 1979). A faint diffuse patch of optical light is observed coincident with this plume and may be due to stars formed from the cooling gas (Nulsen and Carter 1986).

An exciting possibility is that the Giant, Red Envelope Galaxy (GREG; Maccagni *et al.* 1988) found from the *Einstein* Medium Sensitivity Survey is showing the low-mass stars formed in a cooling flow (Johnstone and Fabian 1989). This galaxy lies in the center of an X-ray luminous poor group of galaxies and has a normal de Vaucouleurs profile in the V-band. A large $r^{-1}$ envelope appears in the i-band, which can plausibly be explained as due to $0.5M_\odot$ stars. If due to starlight, then the envelope requires an exceptional imf of the kind required by cooling flows. Its profile is consistent with mass deposition in a cooling flow, which must have been more massive in the past.

## 3.3 The Behavior of Gas Blobs

The distributed manner of the mass deposition shows that the cooling flow is in-homogeneous. This means that it contains blobs of gas that are denser than the surrounding gas. How these blobs behave is ill-understood. Malagoli *et al.* (1987) and Balbus (1988) have shown that cooling flows are not expected to be thermally unstable and cannot generate sizable blobs from initially infinitesimal perturbations. A region that is slightly overdense with respect to its surroundings will fall ahead of the flow under gravity and join a region of similar properties to itself. Computations of the oscillations of overdense blobs are discussed by Loewenstein (1989).

Nulsen (1986) has pointed out that the gravitationally-induced motions of a blob relative to its surroundings will cause it to break up and so increase its area-to-mass ratio and have a lower terminal velocity. Magnetic fields then 'pin' blobs to the flow so that they do become unstable.

Another property of a flow is turbulence, or at least chaotic motions (Loewenstein and Fabian 1988; Pringle 1988). This stirs the gas and reduces the tendency of dense blobs to flow inward. The motion of cluster galaxies, of subcluster infall and of the flow itself promotes chaotic motions. Turbulence of the hot gas can also explain the large velocity spread seen in the optical line-emitting filaments and blobs common at the centers of flows and can help to heat the cold blobs. Stars formed in a turbulent flow will have non-radial orbits. Future high-resolution X-ray spectroscopy can test this idea since X-ray line widths should exceed thermal values.

## 4. OPTICAL EVIDENCE FOR COOLING FLOWS

The presence of optical emission-line filaments and blobs in many cooling flows (note that some, *e.g.* A2029; Hu, Cowie and Wang 1985; Johnstone, Fabian and Nulsen 1987, do not have any detectable emission) can be used as a diagnostic of the conditions in those flows. We can also use them to identify candidate cooling flows where X-ray observations are not available or at higher redshifts if there is some resemblance to nearby flows.

One property that can be obtained from the optical spectra is the gas pressure, in particular from the [SII] lines. In the Perseus cluster, these lines change their ratio indicating high density and thus high pressure within 5kpc of the nucleus (Johnstone and Fabian 1988). Since the pressure has risen above the X-ray inferred pressure (from the X-ray surface brightness) at 20 kpc, the mean gas temperature must be down to the virial temperature of the central galaxy, NGC1275, of about $10^7$K. (Hydrostatic equilibrium requires that the pressure increases inward once the gas has cooled to the local virial temperature.) This is further confirmation that the gas has cooled there below the outer temperature of $\sim 8 \times 10^7$K. Since the gas pressure is so high there, the magnetic pressure cannot be more than about twice the gas pressure (and is probably less than that).

The velocity spread of the optical lines does also indicate large chaotic motions in the hot gas, as discussed earlier. The origin of the optical line emission remains a problem. As already mentioned in §3, the cooling gas by itself cannot produce much detectable optical emission; some of the gas must be held at $\sim 10^4$K in order that we do see the optical lines. This means that there is a distributed, weak heat source at the centers of many cooling flows (see *e.g.* Johnstone and Fabian 1988). The optical emission is typically less than 1% of $L_{cool}$ (and always less than 10%) and so this heat source cannot significantly affect the cooling flow itself. It is likely that the turbulent energy of the cooling gas is the heat source (see also Heckman *et al.* 1989).

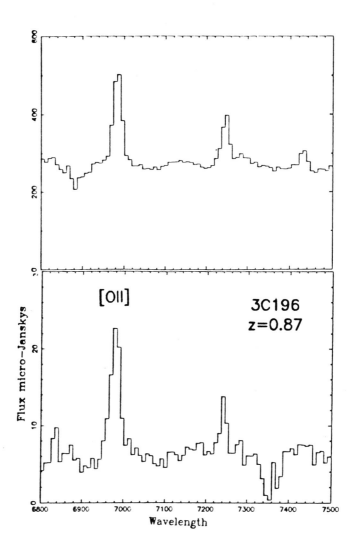

**Figure 5.** Optical spectra of the radio-loud quasar 3C196 on-nucleus (top panel) and 3 arcsec off-nucleus (bottom panel) showing the increased equivalent width of the [OII] line off-nucleus. Comparison of the [OIII] and [OII] emission shows that the surroundings of 3C196 are at high gas pressure, consistent with a $\sim 100$ $M_\odot Yr^{-1}$ cooling flow (Crawford 1988).

The optical line emission of nearby flows shows some resemblances to the neb-ulosities surrounding many distant (3CR) radio galaxies and radio-loud quasars (Fabian *et al.* 1986, Hintzen and Romanishin 1986). One way to check whether they are indeed embedded in hot cooling gas is to measure the gas pressure. If the emission-line gas is very extended and has a pressure, $nT > 10^5$ $cm^{-3}K$, then it is likely to be confined, or it would rapidly disperse. The confining hot gas is then part of a cooling flow unless its temperature exceeds $5 \times 10^7$K. The arguments against either unconfined or gravitationally confined gas are discussed in detail in Fabian *et al.* (1987).

The pressure of gas around quasars cannot easily be measured from the [SII] lines Owing to the large distances of the quasars. We estimate the pressure by an indirect photoionization method which assumes that the ionization state of, say, oxygen is due to the competition between photoionization by the UV radiation from the quasar nucleus and recombination. The higher the gas pressure and so density, the more rapid the recombination and the lower the ionization state. Basically, if we use the forbidden oxygen lines to measure the ionization state of the gas,

$$\frac{[\text{OIII}]}{[\text{OII}]} \propto f\left(\frac{L_{ion}}{nR^2}\right),$$

where the function $f$ is obtained by computation (using G. Ferland's code, CLOUDY), $L_{ion}$ is the ionizing luminosity of the quasar nucleus, obtained from an interpolation between the observed IUE and X-ray emission, $R$ is the radial distance of the nebulosity where the intensity ratio of [OIII] to [OII] line is measured and $n$ is the gas density. The gas pressure is then $\sim 10^4 n$ cm$^{-3}$K. We have found that 3C 48 at redshift $z = 0.37$ and several other radio-loud quasars out to $z = 0.87$ (Fabian *et al.* 1988; Crawford and Fabian 1989) have pressures (and pressure profiles in some cases) indicating surrounding cooling flows with $\dot{M} \sim 100$ $M_{\odot}$Yr$^{-1}$. There is other evidence that radio-loud quasars lie in clusters and groups from galaxy counts around the quasars (Yee and Green 1984).

This work suggests that cooling flows can have been more common in the past and possibly more massive, if the high line luminosities of the distant 3CR galaxies (*e.g.* Spinrad and Djorgovski 1984) may be scaled up from NGC1275. We have suggested that cooling flows surrounded the dominant galaxies in groups and poor clusters at earlier epochs and these have since merged with other groups to form the present day clusters (Fabian *et al.* 1986; Fabian 1988a). Only one cooling flow survives around the largest (or least disturbed) galaxy. Some of the evolution of radio-loud quasars and of radio sources may be due to the evolution of their surrounding cooling flows.

## 5. WHAT NEXT?

There is much left to do. On the theoretical side there are many unsolved problems on the behavior of blobs and on how the flow became, or becomes, inhomogeneous. What are the length scales involved? How does the star formation take place? How is conduction suppressed? On the observational side we clearly need to find the cooled gas in some form, hopefully as stars. This can be achieved with sensitive searches for an $r^{-1}$ surface brightness profile at red wavelengths. Is GREG unique?

Of greatest need are more X-ray images and more X-ray spectra. ROSAT, BBXRT, ASTRO-D and AXAF will supply them. Spatially-resolved, high-resolution spectroscopy is the ultimate goal. Then we can tell whether cooling flows really do have something to do with galaxy formation in general. We need AXAF with the 4 major instruments.

## ACKNOWLEDGEMENTS

I thank the Royal Society for supporting my work.

## REFERENCES

Arnaud, K.A., 1988. In *Cooling Flows in Clusters and Galaxies*, ed. A.C.Fabian, Reidel, 31.

Balbus, S., 1988. *Ap.J.*, **328**, 395.

Bertschinger, E. and Meiksin, A., 1986. *Astrophys.J*, **306**, L1.

Bregman, J.D. and David L.P., 1988. *Ap.J.*, **326**, 639.

Canizares, C.R., Clark, G.W., Markert, T.H., Berg, C., Smedira, M., Bardas, D., Schnopper, H. and Kalata, K., 1979, *Ap.J.*, **234**, L33.

Canizares, C.R., 1981. In *X-ray Astronomy with the Einstein Satellite* ed. R. Giacconi, Reidel, 215.

Canizares, C.R., Clark, G.W., Jernigan, J,G. and Markert, T.H., 1982. *Ap.J.*, **262**, L33.

Canizares, C.R., Stewart, G.C. and Fabian A.C., 1983. *Astrophys. J.*, **272**, 449.

Canizares, C.R., Markert, T.H. and Donahue, M.E., 1988. In *Cooling Flows in Clusters and Galaxies*, ed. A.C.Fabian, Reidel, 63.

Cowie, L.L. and Binney, J., 1977. *Ap.J.*, **215**, 723.

Crawford, C.S., 1988. PhD Thesis, University of Cambridge.

Crawford, C.S., Arnaud, K.A., Fabian, A.C. and Johnstone, R.M., 1989. *M.N.R.A.S.*, **236**, 277.

Crawford, C.S. and Fabian, A.C., 1989. Preprint

Draine, B.T. and Salpeter, E.E., 1979. *Ap.J.*, **231**, 77.

Fabian, A.C. *et al.* , 1974. *Ap.J.*, **189**, L59.

Fabian, A.C. and Nulsen, P.E.J., 1977. *M.N.R.A.S.*, **180**, 479.

Fabian, A.C., Hu, E.M., Cowie, L.L and Grindlay, J.,1981. *Ap.J.*, **248**, 47.

Fabian, A.C., Nulsen, P.E.J. and Canizares, C.R., 1982. *M.N.R.A.S.*, **201**, 933.

Fabian, A.C., Nulsen, P.E.J. and Canizares, C.R., 1984. *Nature*, **311**, 733.

Fabian, A.C., Nulsen, P.E.J. and Arnaud, K.A., 1984. *M.N.R.A.S.*, **208**, 179.

Fabian, A.C., Arnaud, K.A., Nulsen, P.E.J. and Mushotzky, R.F., 1986. *Ap.J.*, **305**, 9.

Fabian, A.C., Arnaud, K.A. and Thomas, P.A., In *Dark Matter in the Universe*, eds. J. Kormendy and G.R. Knapp, Reidel, 201.

Fabian, A.C., Crawford, C.S., Johnstone, R.M. and Thomas, P.A., 1987. *M.N.R.A.S.*, **228**, 963.

Fabian, A.C., 1988a. In *Cooling Flows in Clusters and Galaxies*, ed. A.C.Fabian, Reidel, 315.

Fabian, A.C., 1988b. In *Hot Thin Plasmas in Astrophysics*, ed. R. Pallavicini, Reidel, 293.

Fabian, A.C., Crawford, C.S., Johnstone, R.M., Allington-Smith, J.R. and Hewett, P.C., 1988c. *M.N.R.A.S.*, **235**, 13P.

Fabian, A.C., 1989. In *Hot Astrophysical Plasmas*, ed. P.Gorenstein and M. Zombeck, in Press.

Forman, W., Schwarz, J., Jones, C., Liller, W. and Fabian, A.C., 1979. *Ap.J.*, **234**, L27.

Gold, T. and Hoyle, F., 1958. In *Paris Symposium on Radio Astronomy*, ed. RN Bracewell, Stanford Univ. Press, 574.

Gorenstein, P., Fabricant, D., Topka, K., Tucker, W. and Harnden, F.R., 1977. *Ap.J.*, **216**, L95.

Heckman, T. *et al.* 1989. Preprint

Helmken, H., Delvaille, J.P., Epstein, A., Geller, M.J., Schnopper, H.W. and Jernigan, J.G., 1978. *Ap.J.*, **221**, L43.

Hintzen, P. and Romanishin, W., 1986. *Ap.J.*, **311**, L11.

Hintzen, P. and Romanishin, W., 1988. *Ap.J.*, **327**, L17.

Henry, J.P and Henriksen, M.J., 1986. *Ap.J.*, **301**, 689.

Hu, E.M., 1988. In *Cooling Flows in Clusters and Galaxies*, ed. A.C.Fabian, Reidel, 73.

Hu, E.M., Cowie, L.L. and Wang, 1985. *Ap.J.(Supplement)*, **59**, 447.

Johnstone, R.M., Fabian, A.C. and Nulsen, P.E.J., 1987. *M.N.R.A.S.*, **224**, 75.

Johnstone, R.M. and Fabian, A.C., 1988. *M.N.R.A.S.*, **233**, 581.

Johnstone, R.M. and Fabian, A.C., 1989. *M.N.R.A.S.*, **237**, 27P.

Jones, C. and Forman, W., 1984. *Ap.J.*, **276**, 38.

Kriss, G.A., Cioffi, D.F. and Canizares, C.R., 1983. *Ap.J.*, **272**, 439.

Lahav, O., Edge, A.C., Fabian, A.C. and Putney, A., 1989. *M.N.R.A.S.*, in press.

Lea, S.M., Silk, J., Kellogg, E. and Murray, S., 1973. *Ap.J.*, **184**, L105.

Lea, S.M., Mushotzky, R.F. and Holt, S.S., 1982. *Ap.J.*, **262**, 24.

Loewenstein, M., 1989. *M.N.R.A.S.*, in press.

Loewenstein, M. and Fabian, A.C., 1988. Preprint

Maccagni, D., Garilli, B., Gioia, I.M., Maccacaro, T., Vettolani, G. and Wolter, A., 1988. *Ap.J.*, **334**, L1.

Mathews, W., G. and Bregman, J.N., 1978. *Ap.J.*, **244**, 308.

Malagoli, A., Rosner, R. and Bodo, G., 1987. *Ap.J.*, **319**, 632.

Miller, L., 1986. *M.N.R.A.S.*, **220**, 713.

Mitchell, R.J., Charles, P.A., Culhane, J.L., Davison, P.J.N. and Fabian, A.C., 1975. *Ap.J.*, **200**, L5.

Mould, J.R., Oke, J.B. and Nemec, J.M., 1987. *Astr.J.*, **92**, 53.

Mushotzky, R.F., Holt, S.S, Smith, B.W., Boldt, E.A. and Serlemitsos, P.J., 1981. *Ap.J.*, **244**, L47.

Mushotzky, R.F. and Szymkowiak, A.E. , 1987b. In *Cooling Flows in Clusters and Galaxies*, ed. A.C.Fabian, Reidel, 47.

Nulsen, P.E.J., 1986. *M.N.R.A.S.*, **221**, 377.

Nulsen, P.E.J. and Carter, D., 1987. *M.N.R.A.S.*, **225**, 935.

Nulsen, P.E.J., 1988.In *Cooling Flows in Clusters and Galaxies*, ed. A.C.Fabian, Reidel, 378.

Pringle, J.E., 1989. *M.N.R.A.S.*, in press.

Sarazin, C.L., 1986. *Rev. Mod. Phys.*, **58**, 1.

Sarazin, C.L., 1988. *X-ray Emission from Clusters of Galaxies*, C.U.P.

Sarazin, C.L. and O'Connell, R.W., 1983. *Ap.J.*, **258**, 552.

Schwartz, D.A., Schwarz, J. and Tucker, W.H., 1980. *Ap.J.*, **238**, L59.

Silk, J., 1976. *Ap.J.*, **208**, 646.

Silk, J., Djorgovski, G., Wyse, R.F.G. and Bruzual, G.A., 1986. *Ap.J.*, **307**, 415.

Singh, K.P., Westergaard, N.J. and Schnopper, H.W., 1986. *Ap.J.*, **308**, L51.

Spinrad, H. and Djorgovski, G., 1984. *Ap.J.*, **280**, L9.

Stewart, G.C., Canizares, C.R., Fabian, A.C. and Nulsen, P.E.J., 1984a. *Ap.J.*, **278**, 536.

Stewart, G.C., Fabian, A.C., Jones, C. and Forman, W., 1984b. *Ap.J.*, **285**, 1.

Thomas, P.A., Fabian, A.C. and Nulsen, P.E.J., 1987. *M.N.R.A.S.*, **228**, 973.

Tucker, W.H. and Rosner, R., 1982. *Ap.J.*, **267**, 547.

White, R.E. and Sarazin, C.L., 1987. *Ap.J.*, **318**, 612.

White, R.E. and Sarazin, C.L., 1987. *Ap.J.*, **318**, 621.

White, R.E. and Sarazin, C.L., 1987. *Ap.J.*, **318**, 629.

Wolter, A. *et al.* 1988. Preprint

Yee, H.K.C. and Green, R.F., 1984. *Ap.J.*, **280**, 79.

# SKY SURVEYS WITH EINSTEIN

**Isabella M. Gioia**

Harvard-Smithsonian Center for Astrophysics, Cambridge, MA

Istituto di Radioastronomia del CNR, Bologna, Italy

## 1. INTRODUCTION

I was asked to talk at this symposium, celebrating the 10th Anniversary of the *Einstein* launch, on surveys performed with the *Einstein Observatory*. Unlike the surveys performed with the detectors on board *UHURU*, Ariel V or HEAO-1 satellites, which discovered a few hundred sources above a limiting sensitivity of $10^{-11}$ erg cm$^{-2}$ s$^{-1}$, the imaging instruments on board the *Einstein Observatory* were not designed to conduct surveys of large areas of sky. Nonetheless *Einstein* provided a large amount of data which was used to search for new X-ray sources. The extragalactic source-counts were extended to fluxes of the order of a few times $10^{-14}$ erg cm$^{-2}$ s$^{-1}$, a factor of 1000 deeper than detectable by previous instruments, and the nature and type of objects contributing to the diffuse x-ray background were studied.

Since the early times after the launch, systematic studies of serendipitous *Einstein* sources have been carried out by several observers with interests in both galactic and extragalactic astronomy. I list some of these authors' work in Table 1.

| Table 1 Studies of serendipitous *Einstein* Sources | |
|---|---|
| Grindlay *et al.* (1980) | |
| Chanan, Margon and Downes (1981) | |
| Kriss and Canizares (1982) | |
| Reichert *et al.* (1982) | Serendipitous searches for AGN |
| Katgert, Thuan and Windhorst (1983) | |
| Margon, Downes and Chanan (1985) | |
| Pravdo and Marshall (1984) | |
| Kriss and Canizares (1985) | |
| Vaiana *et al.* (1981) | The *Einstein* Stellar survey |
| Helfand and Caillault (1982) | Field star surveys |
| Caillault *et al.* (1986) | |
| Hertz and Grindlay (1984) | The *Einstein* Galactic Plane survey |
| Hertz and Grindaly (1988) | |
| Maccacaro *et al.* (1982) | |
| Stocke *et al.* (1983) | The *Einstein* Medium Sensitivity Survey |
| Gioia *et al.* (1984) | |

The majority of these studies were not surveys in the strict sense of the word: in several cases no analyses requiring flux completeness were performed. However, these systematic searches for sources added much to our knowledge of the behavior in the X-ray domain of the different classes of astronomical objects and, in many istances, led to the study of their properties at different wavebands.

## 2. THE DEEP SURVEYS

The first X-ray survey undertaken with the *Einstein Observatory* which appeared in the literature was reported by Giacconi *et al.* in 1979. They covered about 1 sq. deg of sky in the Draco and Eridanus region down to sensitivities of $1.3 \times 10^{-14}$ erg cm$^{-2}$ s$^{-1}$ in the 1-3 keV band, with the primary goal of studying the number-counts relationship and investigating the nature of the extragalactic x-ray background. Giacconi *et al.* extended the observed number-intensity distribution of extragalactic sources by 3 decades in flux down to $S = 2.6 \times 10^{-14}$ erg cm$^{-2}$ s$^{-1}$ in 1-3 keV. Murray (1981) added a new point to the source counts at $1.3 \times 10^{-14}$ using additional data from the Pavo and Ursa Minor regions. About the same sensitivity was achieved by Griffiths *et al.* (1983), using the deep survey data of the Pavo region. A Super Deep Survey was also carried out with the HRI in the Ursa Minor region. This was the most sensitive X-ray survey carried out by *Einstein* having a total of 330,000 seconds and a limiting flux of $1.3 \times 10^{-14}$ erg cm$^{-2}$ s$^{-1}$ (Murray *et al.* 1989, Primini *et al.* 1989). The main result from these surveys was that about 30-35% of the total x-ray background in the 1-3 keV band is resolved into discrete sources. Unlike the less sensitive *UHURU* and Ariel V surveys which found approximately an equal number of clusters and Seyfert galaxies, at these low flux levels the predominant contributors to the X-ray background were found to be quasars.

Although the deep surveys provide a very faint limit in the X-ray flux, a completely identified sample is not yet available given the extreme faintness of the large majority of the optical counterparts. The value for the source density found by Giacconi *et al.* in 1979 of N $(>S)= (6.3 \pm 2.6) \times 10^4$ extragalactic sources sterad$^{-1}$ at $S = 2.6 \times 10^{-14}$ erg cm$^{-2}$ s$^{-1}$ in 1-3 keV was in close agreement with the number expected by extrapolating, with the Euclidean slope, the *logN-logS* curve derived at higher fluxes by the previous X-ray missions. This agreement was not expected and could have been fortuitous. The deep surveys are in fact sampling regions of the universe where the Euclidean approximations are no longer valid, and regions where evolutionary effects have been detected already at other wavelengths. Moreover no information was available on the behavior of the number-flux relationship at fluxes intermediate between the deep survey limit and the *UHURU*/Ariel V/HEAO-1 limit, nor was the nature of the sources detected in the deep survey fields clearly established. A project to study in detail the shape of the number-count relationship at fluxes intermediate between the deep surveys and the "bright" surveys was then undertaken at the Center for Astrophysics. This project, the *Einstein Observatory* Medium Sensitivity Survey (MSS), consists of a systematic search for serendipitous X-ray sources detected in a large number of high galactic latitude IPC images taken by different observers with the purpose of studying preselected objects (see Section

4).

## 3. THE EINSTEIN SLEW SURVEY

Recently Martin Elvis and collaborators have started the analysis of the IPC data gathered while the *Einstein* satellite was slewing from one target to the other (see Elvis, Plummer and Fabbiano, this volume). They will produce an all-sky survey for bright sources, about $10^{-12}$ erg cm$^{-2}$ s$^{-1}$, which will allow a direct comparison, using the same instrument, of the source densities with the Medium and Deep survey source densities and will make easier the extrapolation to the high energy surveys.

## 4. THE EINSTEIN MEDIUM SURVEY

From now on I will concentrate on the *Einstein* Medium Survey because it is the major of the many efforts made to study the serendipitous sources detected by *Einstein*. A large number of people have been involved in this project during the years, and many still are. At present the Medium Survey "team" is composed by T. Maccacaro, A. Wolter, R. Schild, J. Stocke, S. Morris and myself. I will mention first some of the results obtained from the original Medium Sensitivity Survey sample which was 100% optically identified and then I will present new preliminary results from the Extended Medium Sensitivity Survey, focusing on the extragalactic population. Interesting results have also been obtained for the stellar sample of the survey. They are presented and discussed in Fleming (1988).

We have analyzed 1435 IPC fields covering 780 deg$^2$ of the high galactic latitude sky ($| b | > 20°$) and we have detected 835 sources at a SNR $\geq 4$ in the flux range $6 \times 10^{-14}$ to $10^{-11}$ erg cm$^{-2}$ s$^{-1}$ in (0.3-3.5) keV energy band. The criteria for inclusion in the survey of the IPC images and IPC sources are described in detail by Gioia *et al.* (this volume). The published Medium Survey samples (Maccacaro *et al.* 1982; Stocke *et al.* 1983; Gioia *et al.* 1984) consisted of 112 sources, all of them spectroscopically identified. Given the completeness of the sample and of the optical identification, we used these 112 sources to investigate in detail the behavior of the different classes of X-ray selected objects. We were able to derive the X-ray *logN-logS* curves for the extragalactic population as a whole, and for the AGN and clusters of galaxies separately (shown in Figure 1). The *logN-logS* relationship for the extragalactic sample was found to be consistent with the Euclidean model. We have shown that this result is the sum of two different contributions given by the AGN and clusters of galaxies, a sum which mimics the Euclidean slope in the number-flux space. In particular, the AGN have a steeper slope, reflecting the fact that they are an evolving population of objects, and the clusters of galaxies have a slope flatter than the Euclidean value (Gioia *et al.* 1984). No flattening is evident at the low flux end of the AGN *logN-logS* curve, implying the existence of a large number of X-ray sources too faint to be detected with the *Einstein Observatory*. While we have been able to determine the X-ray luminosity function and its evolution with cosmic time for AGN (Maccacaro, Gioia and Stocke, 1984), the most numerous class of objects found in X-ray surveys at these flux limits, for the other two classes of objects, clusters of galaxies and BL Lacs, we were limited by the available statistics and were only able to obtain preliminary results on their possible evolution (Maccacaro *et al.* 1984). I will discuss the BL Lac sample in more detail later on.

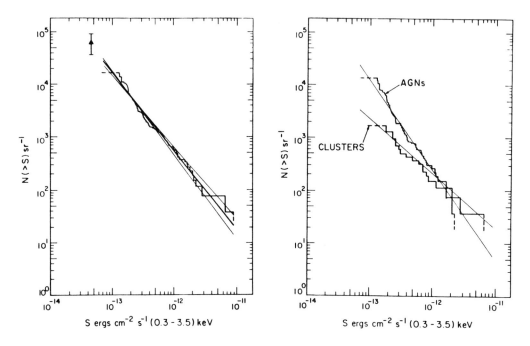

Figure 1. The *logN-logS* curves for the entire extragalactic population and for AGN and clusters of galaxies (from Gioia *et al.* 1984).

For the AGN X-ray selection gives us the opportunity to discover large numbers of relatively low luminosity, low redshift objects, a subset of AGN not easily found by other means. This is illustrated in Figure 2 which presents the Hubble diagram for the UVX quasars (from Boyle *et al.* 1987) and the Medium Survey AGN (filled circles). It is evident that X-ray selected AGN occupy an otherwise rather unpopulated region of the diagram. At any given redshift they are in fact characterized, on the average, by a fainter magnitude and thus luminosity than those of the UVX quasars. Therefore X-ray surveys complement optical surveys and have the potential to sample a different region of the underlying quasar population. It is important to study these X-ray selected AGN which come from the boundary region between "classical" quasars and Seyfert galaxies in order to fill in important gaps in our understanding of the quasar phenomenon. For clusters of galaxies X-ray selection offers an efficient method for finding high redshift objects. For BL Lacs X-ray selection may be our best tool to define complete samples of large sizes.

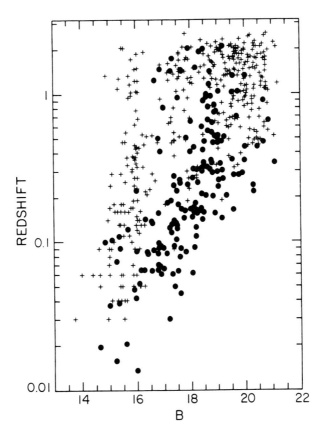

Figure 2. Hubble diagram for UVX quasars (crosses, from Boyle *et al.* 1987) and X-ray selected AGN (filled circles) (from Maccacaro *et al.* 1989).

## 5. IDENTIFICATION AND CLASSIFICATION

The Extended Medium Sensitivity Survey (EMSS), Gioia *et al.* (1989), consists of about eight times as many sources as were in the original MSS. As with the previous MSS sample, we are committed to achieve a 100% identification rate. Optical identification of X-ray sources, especially at these low fluxes, is very energy and time consuming and requires allocation of large amounts of large telescope time. In fact, the IPC error circle of about 50″ radius generally contains 1-6 optical candidates.

### 5.1 Search in the literature

A preliminary work of classification and identification has been done to reduce the amount of optical telescope time required to observe every single object inside the 835 source error boxes. Since we have reanalyzed, for reasons of homogeneity, IPC images used by other observers, and re-detected sources which had already been discovered and identified, we have performed an extensive search of the literature to use available spectroscopic data obtained by others.

## 5.2  Cross correlation with other catalogs

We have examined the source error circles for coincidences with optical, radio and infrared objects from the other catalogs. Each positional match between a serendipitous source and a cataloged object has been evaluated to ascertain the correctness of the coincidence.

## 5.3  X-ray to visual flux ratio

We have then optimized the search for the correct identification using the $f_x/f_o$ ratio for each source (log $f_x/f_o = \log f_x + m_v/2.5 + 5.37$, where $f_x$ is the X-ray flux in erg cm$^{-2}$ s$^{-1}$ in 0.3-3.5 keV). The major advantage of the method is that only the visual magnitude of the optical counterpart, as estimated from the POSS prints, and the X-ray flux of the source are needed. It is a reliable method using $f_x/f_o$ versus $f_o$ to separate galactic from extragalactic sources prior to spectroscopy. When an optical counterpart is found in the error circle with a $f_x/f_o$ appropriate for its class, we do not always continue to observe objects as we did in the first MSS sample, unless other obvious candidates are present (*e.g.* a faint BSO, a radio source).

## 5.4  Spectroscopy and CCD photometry

Finally, spectroscopic observations are performed using the MMT, (and Kitt Peak 90″ in the past) in the Northern hemisphere, the ESO 1.5 and 3.6 m and Las Campanas 2.5 m in the Southern hemisphere. We have also initiated collaborations with several astronomers to follow-up and study in more detail selected classes of objects. A parallel program of photometry with the CfA CCD camera on the 61 cm telescope of the Whipple Observatory is being carried out by R. Schild for the extragalactic population of the survey north of declination −20°. As of today, 704 sources have been identified corresponding to a rate of 84%. Among the identifications the most numerous class of extragalactic objects is represented by AGN (358), followed by clusters of galaxies (87), BL Lac objects (31) and "normal" galaxies (15). 213 stars have also been identified.

The brightest sources, and those already known and catalogued, have been identified first. We are now left with 131 sources which are the among the faintest in the survey and thus the toughest to identify. The fact that the unidentified sources cluster at the faint end of the flux distribution allows us to define flux limited subsamples virtually fully identified and to begin the scientific analysis and interpretation of the data. Also, we have classified the sources still unidentified using the $f_x/f_o$ method so as to be able to obtain some preliminary results on studies for which completeness of the identification is not critical. In particular we have been able to study the the X-ray energy distribution of X-ray selected AGN and clusters of galaxies (Maccacaro *et al.* 1988). We have found that the AGN are characterized, in the soft X-ray band by a variety of spectral indices. Their average energy slope is different from the slope that we obtain for clusters of galaxies and galaxies ($\alpha_{AGN} = 1$ against $\alpha_{CL} = 0.5$), which gives us confidence of the correctness of our analysis. Particularly interesting though is the derivation of the number-counts relationship for BL Lac objects (Maccacaro *et al.* 1989). This has been computed using a subset of the EMSS by applying a "blind cut" to the survey, that is using only the sky North of $\delta = -20°$ since we have had so far more telescope time in the Northern hemisphere. The resulting sample is completely identified above a flux of $10^{-12}$ and contains only 4 unidentified sources in the 5 to $10 \times 10^{-13}$ region and 18 more in the flux region region 3 to $5 \times 10^{-13}$. The flattening of the counts and the redshift information available, has allowed us to set constraints on the BL Lac luminosity function and evolution. In order to reproduce the observed count distribution, we find that the BL Lacs have to be characterized by a relatively flat luminosity function (see Figure 3).

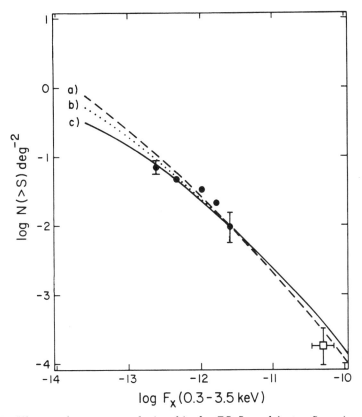

Figure 3. The number-counts relationship for BL Lac objects. Superimposed is the *logN-logS* resulting from the integration of a single power law luminosity function of differential slope $\gamma = 2.0$ and a) no redshift cutoff (*i.e.* $z_{max} = 4$), b) $z_{max} = 1.0$, and c) $z_{max} = 0.5$ (from Maccacaro *et al.* 1989).

Within the assumption of a single power law, the data suggest a luminosity function slope of $\gamma \sim 2$, and no evidence for a cosmological evolution of the kind needed to explain the quasar data.

A number of studies, like those outlined above, require a fully, or almost fully identified sample of sources. At present we are close to concluding the gathering of the data, and we are looking forward, upon completion of the optical identification program, to beginning the most exciting part of the work, that is the scientific data analysis. I would like to end this talk by noting that we still sit in front of computer consoles, look at X-ray images of the sky and make new discoveries using data from a telescope which has not existed for over 7 years! And this says something about how important *Einstein* has been.

I am indebted to Tommaso Maccacaro for many discussions and to the other members of the EMSS team for their continuous contribution to the success of this project. This work has received partial financial support from NASA contract NAS8-30751, and from the Scholarly Studies Program of the Smithsonian Institution through grants SS48-8-84 and SS88-03-87.

## REFERENCES

Boyle, B.J., Fong, R., Shanks, T., Peterson, B.A., *M.N.R.A.S.*, 227, 717, 1987.

Chanan, G.A., Margon, B., and Downes, R.A., 1981, *Ap.J. (Letters)*, 243, L5.

Caillault, J., Helfand, D.J., Nousek, J.A., and Takalo, L.O., 1986, *Ap.J.*, 304, 318.

Elvis, M., Plummer, D., and Fabbiano, G., this volume.

Fleming, T.A., 1988, Ph. D. Thesis, University of Arizona.

Giacconi, R. *et al.*, 1979, *Ap.J. (Letters)*, 234, L1.

Gioia, I.M., Maccacaro, T., Schild, R.E., Stocke, J.T., Liebert, J.W., Danziger, I.J., Kunth, D., and Lub, J., 1984, *Ap.J.*, 283, 495.

Gioia, I.M., Maccacaro, T., Morris, S.L., Schild, R.E., Stocke, J.T., A. Wolter, 1989, *Ap.J.*, submitted.

Gioia, I.M., Maccacaro, T., Morris, S.L., Schild, R.E., Stocke, J.T., A. Wolter, this volume.

Griffiths, R.E., *et al.*, 1983, *Ap.J.*, 269, 375.

Grindlay, J.E., Steiner, J.E., Forman, W.R., Canizares, C.R., and McClintock, J.E., 1980, *Ap.J. (Letters)*, 239, L43.

Helfand, D.J., and Caillault, J.-P., 1982, *Ap.J.*, 253, 760.

Hertz, P. and Grindlay, J.E., 1984, *Ap.J.*, 278, 137.

Hertz, P. and Grindlay, J.E., 1988, *A.J.*, 96, 233.

Katgert, P., Thuan, T.X., and Windhorst, R.A., 1983, *Ap.J.*, 275, 1.

Kriss, G.A., and Canizares, C.R., 1982, *Ap.J.*, 261, 51.

Kriss, G.A., and Canizares, C.R., 1985, *Ap.J.*, 297, 177.

Maccacaro, T., *et al.*, 1982, *Ap.J.*, 253, 504.

Maccacaro, T., Gioia, I.M., and Stocke, J.T., 1984, *Ap.J.*, 283, 486.

Maccacaro, T., Gioia, I.M., Maccagni, D., and Stocke, J.T., 1984, *Ap.J. (Letters)*, 284, L23.

Maccacaro, T., Gioia, I.M., Wolter, A., Zamorani, G., and Stocke, J., 1988, *Ap.J.*, 326, 680.

Maccacaro, T., Gioia, I.M., Schild, R.E., Wolter, A., Morris, S.L., and Stocke, J.T, 1989, Proceedings of the meeting on *BL Lac Objects: 10 years after*, Villa Olmo, Como, Italy, in press.

Margon, B., Downes, R.A., and Chanan, G.A., 1985, *Ap.J.Suppl.*, 59, 23.

Murray, S.S., 1981, in *X-ray Astronomy with the Einstein Satellite*, page 281.

Murray, S.S., Jones, C. and Forman, W., 1989, *Proceeding of the Liller Symposium*, in press.

Pravdo, S.H., and Marshall, F.E., 1984, *Ap.J.*, 281, 570.

Primini, F.A., Murray, S.S., Burg, R., Huchra, J., Schild, R., and Giacconi, R., 1989, in preparation.

Reichert, G.A., Mason, K.O., Thorensten, J.R., and Bowyer, S., 1982, *Ap.J.*, 260, 437.

Stocke, J.T., Liebert, J., Gioia, I.M., Griffiths, R.E., Maccacaro, T., Danziger, I.J., Kunth, D., and Lub, J., 1983, *Ap.J.*, 273, 458.

Vaiana, G.S., *et al.*, 1981, *Ap.J.*, 245, 163.

# INTERPRETATION OF X-RAY SOURCE COUNTS AND A PROGNOSIS FOR AXAF

**Maarten Schmidt**

Palomar Observatory, California Institute of Technology

## 1. INTRODUCTION

The organizers of the conference asked me originally to talk about quasars and AGNs. For many years, much of my work on quasars in the optical regime has been of statistical nature, involving the luminosity function of quasars and its evolution with redshift. Extension of this work to x-ray energies is of particular interest, since quasars and AGNs are important contributors to x-ray source counts. Since BL Lacs objects, galaxies, and clusters of galaxies also play a role in statistics of x-ray sources, I decided to address the general issue of the interpretation of extragalactic x-ray source counts.

In this review, we will be primarily interested in the interpretation of x-ray source counts based on complete x-ray samples. The only optical evidence that we will use consists of the classification of objects based on the optical identification, and the redshift. Since we will limit ourselves to x-ray counts and x-ray luminosity functions, I will not discuss the relation between x-ray and optical or radio properties.

In sections 2 and 3, we discuss the derivation of the luminosity function and of the source counts. We analyze in section 4 the two complete x-ray samples available at present. This will involve consideration of instrumental effects in the detection of clusters of galaxies, of evolution for quasars, and of absorption in low-luminosity AGNs. On the basis of the luminosity functions developed, we present in section 5 a source count prognosis for AXAF. Finally, we discuss in section 6 briefly the advantages of x-ray astronomy for statistical studies of extragalactic objects.

## 2. DERIVATION OF LUMINOSITY FUNCTION

Consider an x-ray sample that is complete over a given area of sky to a limiting flux $S_{lim}$ in a given energy band. Since the luminosity function that is derived from such a sample is a linear combination of the contributions from the sample sources, we start by considering just one source only.

Hypothetically, we move this source radially away from us, maintaining all its absolute properties. As the redshift $z$ increases, the flux $S$ declines: let $S(z)$ be the flux-redshift relation for this source. We can invert this and derive $z(S)$, which is the redshift at which the given source would have an observed flux $S$. Now let $V(< z)$

be the co-moving observable volume out to redshift $z$, for a given cosmological model. Substitution of $z(S)$ in $V(< z)$ produces $V(> S)$, which is the volume over which the source will be observed to have a flux of $S$ or greater.

Since our sample was complete to flux $S_{lim}$, $V(> S_{lim})$ is the volume over which the given source will be included in the sample. Since the luminosity function is the space density of sources, as a function of their luminosity, the one source provides a contribution of $1/V(> S_{lim})$ to the luminosity function at the source's luminosity $L$. The total luminosity function is the sum of the contribution of the $n$ sources that comprise the complete sample,

$$\Phi(L) = \sum_{i=1}^{n} \frac{\delta(L - L_i)}{V_i(> S_{lim})}$$

## 3. DERIVATION OF SOURCE COUNTS

We now derive the source counts corresponding to the luminosity function just determined. Let us return to the one source in the original complete sample, which yielded a space density of $1/V(> S_{lim})$. At a flux larger than $S$, this source can be seen over a volume $V(> S)$ and therefore we should observe $V(> S)/V(> S_{lim})$ of such sources brighter than $S$. The total source counts are again the sum of the $n$ individual contributions,

$$N(> S) = \sum_{i=1}^{n} \frac{V_i(> S)}{V_i(> S_{lim})}$$

We illustrate this procedure by considering the counts produced by just one source in the HEAO-1 A-2 survey (Piccinotti *et al.* 1982). We assume that it has a power law spectrum with energy spectral index $-0.7$. We employ a cosmological model with Hubble constant $H_0 = 50$ km s$^{-1}$Mpc$^{-1}$, and $q_0 = 0.5$. We also assume that there is a redshift cutoff at $z_{max} = 2$, similar to that found for quasars. Figure 1 shows the counts generated by the source, depending on its x-ray luminosity $HX = L_x(2\text{-}10 \text{ keV})$.

At the lowest luminosity illustrated, $\log HX = 41$, the counts follow the 3/2 law expected in Euclidean space for fluxes larger than $\log S(2 - 10 \text{ keV}) \approx -13$. For sources of higher luminosity, the slope at a given flux becomes progressively smaller. Clearly, for most extragalactic x-ray sources, the $-3/2$ law is a very poor approximation.

The striking differences in the predicted source counts for different luminosities are also reflected in the contribution to the x-ray background. In our example, the source counts corresponding to a single source of luminosity $\log HX = 41$ contribute 13% to the observed background at 2 keV. For $\log HX = 43$ the contribution is 1% and for $\log HX = 45$ it is only 0.1%.

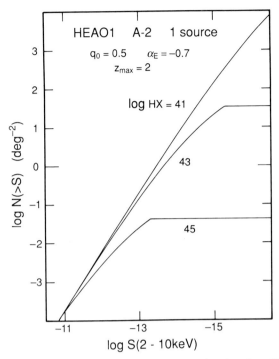

Figure 1. Source counts based on one single source luminosity $HX$, hypothetically observed in the HEAO-1 A-2 survey.

The non-parametric procedure for the derivation of luminosity function and source counts, described above, is transparent and accurate. It takes into account cosmology, in terms of $S(z)$ and $V(< z)$, and the spectral energy distribution, through $S(z)$. No assumptions are needed about the shape of the luminosity function. If there is a need to account for luminosity or density evolution, then the luminosity or number density can be varied as a function of redshift for each of the $n$ sources contributing to the luminosity function. The r.m.s. error of the predicted source counts (due to the sampling error in the sample used to generate the source counts) can be estimated by assigning an r.m.s. error of $\pm 100\%$ to each of the $n$ contributions. For further details about the non-parametric derivation of the luminosity function, the reader is referred to Schmidt and Green (1986).

## 4. TWO COMPLETE X-RAY SAMPLES

There are two x-ray samples for which optical identifications and redshifts are essentially complete. These are the HEAO 1 A-2 survey (Piccinotti *et al.* 1982) and the *Einstein* Medium Sensitivity Survey (MSS) (Maccacaro *et al.* 1982, Gioia *et al.* 1983, Stocke *et al.* 1984). Sky coverage, energy bands, flux limits and numbers detected in the two surveys are given in Table 1. The MSS has a distribution of flux limits versus sky coverage. The flux limit given in Table 1 is an effective limit;

this limit would produce the observed number of sources if it applied uniformly to the entire area of the survey.

| Table 1 X-ray Surveys with Complete Identifications | | |
|---|---|---|
| | **HEAO1** A-2 | **Einstein** MSS |
| Area | 27,000 deg$^{-2}$ | 89.1 deg$^{-2}$ |
| Energy | 2–10 keV | 0.3–3.5 keV |
| Limit | $3 \times 10^{-11}$ cgs | $3 \times 10^{-13}$ cgs |
| BL Lacs | 4 | 4 |
| Galaxies | 1 | 3 |
| Clusters | 30 | 20 |
| Quasars | 1 | 23 |
| AGNs | 20 | 32 |

Since the two surveys differ in flux limit by a factor of around 100, a detailed comparison of their content is of interest. In making the comparison, we generally use the sample with the larger number of objects of a given class to predict the expected number in the other sample, following the procedure described in the preceding sections.

*BL Lacs*: The two samples, both very small, are consistent with each other for a uniform space distribution.

*Galaxies*: Based on the three galaxies in the MSS, we expect one galaxy in the HEAO 1 A-2 for a uniform space distribution. Numbers are very small so there is large uncertainty.

*Quasars*: We define as quasars those active galactic nuclei with optical absolute magnitude $M_B < -23$. Besides the one quasar (3C 273) in the A-2 sample, there is the BQX sample, a small subsample of the Bright Quasar Survey (*cf.* Schmidt and Green 1986). Both of these samples contain fewer objects than are predicted from the 23 quasars in the MSS for a uniform space distribution. This constitutes pure x-ray evidence for the evolution of quasars, independent of optical evidence. We invoke luminosity-dependent density evolution to fit both the x-ray counts, as well as the total surface density of quasars with $z < 2$ of around 70 deg$^{-2}$. This evolution is somewhat different from that used by Schmidt and Green (1986).

*Clusters of galaxies*: Schmidt and Green (1986) found that there was a large difference between the luminosity distributions of clusters in the two samples. Further study shows that there is only a discrepancy at the bright end: based on the A-2 sample, we expect 25 clusters with $\log HX > 44.4$ in the MSS, but none are observed. Part of the explanation is probably that the MSS detection efficiency for clusters is low. If part of the discrepancy is due to cluster evolution, it would have to be very steep. For clusters of lower luminosity, $\log HX < 44.4$, the two samples are consistent with a uniform space distribution.

*AGNs*: We define as AGNs (or Seyfert galaxies) those active galactic nuclei with $M_B > -23$. Both samples contain substantial numbers of AGNs. There is a discrepancy opposite in sign from that found for clusters: based on the A-2 sample, we expect 25 AGNs with $\log HX > 43.5$ in the MSS, but only 9 are observed. Following Reichert *et al.* (1985), we explain this as a consequence of absorption by clouds of $6 \times 10^{22}$ H atom cm$^{-2}$ with a coverage of 70%.

## 5. PROGNOSIS FOR AXAF

In preparation for the AXAF mission, it is of interest to have a preview of expected x-ray source counts. We have based these on the evaluation of the HEAO-1 A-2 and *Einstein* MSS samples, discussed in the preceding section. The x-ray counts for the different classes of objects are shown in Figure 2. At this point, I would like to stress the uncertainty associated with this prognosis. As we saw in the preceding section, the reconciliation of the contents of the HEAO-1 A-2 sample and the *Einstein* MSS sample require invoking evolution (for quasars), instrumental effects (for clusters), and absorption (for AGNs of lower x-ray luminosity). There is considerable uncertainty associated with each of these interpretations. In addition, we have assumed a uniform space distribution for galaxies, BL Lacs, clusters of galaxies and AGNs. Each of these objects probably exhibits some cosmological evolution.

The predicted number of x-ray sources with $\log X(0.5 - 2.0 \text{ keV}) > -15$ is 800 deg$^{-2}$, of which AGNs contribute 500 deg$^{-2}$. At this flux, total counts vary approximately as $S^{-0.9}$. The total x-ray background produced by discrete sources in this prognosis is 52% of the observed background at 2 keV, and 30% at 10 keV.

I have also considered in alternative scenario, in which the entire background is accounted for by discrete sources. We achieve this by postulating that the AGNs evolve in number and spectrum such that the background at 2 keV and at 10 keV is entirely due to discrete sources. This is achieved if the AGNs show density evolution $e^{2.15\tau}$ and have an energy spectral index of $-0.7 + 0.6\tau$, where $\tau$ is the light-travel time in terms of the age of the universe. In this scenario, the number of AGNs with $\log S(0.5 - 2.0 \text{ keV}) > -15$ increases to 2500 deg$^{-2}$. The median redshift of these AGNs would be around 1.0. I want to stress, that there is no physical basis for this AGN evolution scenario. It does illustrate, though, that AXAF may provide important clues to the composition and the nature of the x-ray background.

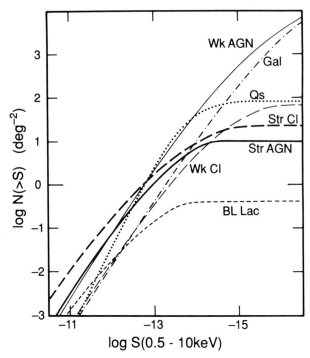

Figure 2. Source count prognosis for AXAF. See text for a discussion of the uncertainties associated with this prognosis.

## 6. DISCUSSION

There are two reasons why the statistics of x-ray sources can make a major contribution to studies of the luminosity function and evolution of extragalactic objects. First, among the objects found in x-rays are BL Lacs, clusters of galaxies, and Seyfert galaxies. These are all objects which are relatively difficult to find optically. For BL Lacs, no complete optical sample exists. Clusters of galaxies at large redshift are difficult to find optically, due to their low contrast versus foreground field galaxies. Only one well defined complete optical sample of Seyfert galaxies, based on the CfA redshift survey, is available. For none of these objects do we have any optical evidence about their cosmological evolution. X-ray surveys, such as the all-sky survey planned for ROSAT, will make important contributions in this area.

The second reason why x-ray surveys provide interesting statistical information is that the observed x-ray background is so relatively low. Already, a substantial part can be explained by a conservative extrapolation of observed x-ray source counts. The evolution of the major contributors to the x-ray background, in particular that of AGNs, is strongly constrained by the total observed background. A similar situation does not prevail at optical, infrared and radio wavelengths.

## REFERENCES

Gioia, I.M., Maccacaro, T., Schild, R.E., Stocke, J.T., Liebert, J.W., Danziger, I.J., Kunth, D., and Lub, J. 1983, *Astrophys.J.*, **283**, 495.

Maccacaro, T., Feigelson, E.D., Fener, M., Giacconi, R., Gioia, I.M., Griffiths, R.E., Murray, S.S., Zamorani, G., Stocke, J., and Liebert, J. 1982, *Astrophys.J.*, **253**, 504.

Piccinotti, G., Mushotzky, R.F., Boldt, E.A., Holt, S.S., Marshall, F.E., Serlemitsos, P.J., and Shafer, R.A. 1982, *Astrophys.J.*, **253**, 485.

Reichert, G.A., Mushotzky, R.F., Petre, R., and Holt, S.S. 1985, *Astrophys.J.*, **296**, 69.

Schmidt, M. and Green, R.F. 1986, *Astrophys.J.*, **305**, 68.

Stocke, J.T., Liebert, J., Gioia, I.M., Griffiths, R.E., Maccacaro, T., Danziger, I.J., Kunth, D., and Lub, J. 1983, *Astrophys.J.*, **273**, 458.

# X-RAY ASTRONOMY BEYOND AXAF AND XMM

Riccardo Giacconi

Space Telescope Science Institute

## 1. INTRODUCTION

To discuss the future of x-ray astronomy can be a trivial or an extremely difficult exercise. Trivial in that experience shows that nothing or very little will happen in the development of x-ray observatories in the next twenty years, if it is not underway now. Extremely difficult because the next group of missions already on the horizon, promises orders of magnitude improvements in the measurements of almost all the relevant parameters of cosmic x-ray sources. Improvements of this magnitude have, in the past, led to the discovery of unsuspected phenomena in known sources or of entirely new classes of sources. These discoveries in turn tend to open entirely new fields of research and determine the direction of future activities. Thus the observation of compact x-ray sources in binary systems in 1970 has led to entirely new research programs which have been and will continue to be fruitfully pursued. Similarly the discovery of high temperature X-ray emitting gas in clusters of galaxies, was not only of great intrinsic significance, but has also led to powerful new investigations in cluster formation and evolution which will come to full bloom in the next decade.

I have chosen as my task to discuss what might be interesting to do after the ROSAT, ASTRO-D, SPECTRUM X, XTE, AXAF and XMM missions have been carried out, in whatever time interval that might happen. My discussion should therefore be considered an intellectual rather than a temporal extrapolation.

## 2. TECHNICAL TRENDS

In a paper presented at the International Symposium on The Physics of Neutron Stars and Black Holes in Tokyo, I tried to summarize the past and expected improvements in sensitivity, angular and spectral resolution. From figures 1, 2 and 3 we see the rapid improvements brought about by technological advances such as the use of orbiting spacecrafts rather than rockets in the early 70's, and the use of focusing optical and imaging detectors rather than collimated counters in the late 70's.

In the early 90's, the All Sky Survey from ROSAT will yield an enormous wealth of synoptic observations to a sensitivity exceeding those attained in medium *Einstein* surveys. The deep surveys of ROSAT will be deeper than those of *Einstein* by almost an order of magnitude. In the mid 90's we can expect a further order of magnitude improvement in sensitivity to point sources through the use of high resolution telescopes and imaging detectors on AXAF. Orders of magnitude

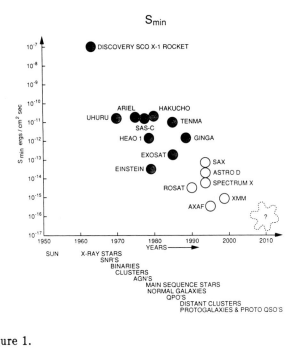

Figure 1.

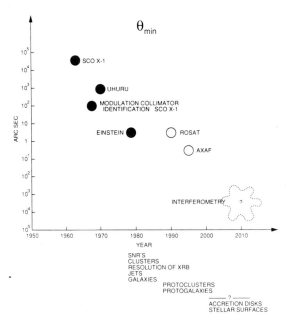

Figure 2.

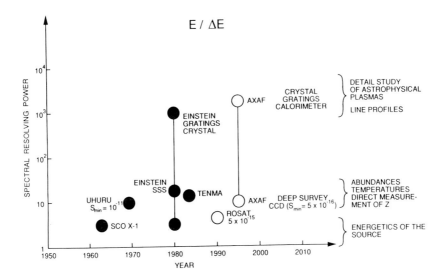

Figure 3.

improvements in the spectroscopy of faint sources due to the use of high spectral resolution non-dispersive spectrometers and Objective grating spectrometers (AXAF).
Major improvements in high energy spectroscopy for moderately weak point and
extended sources through the use of high throughput optics (XMM).

However if we try to anticipate future developments, we find it hard to imagine
easy ways to improve observational capabilities through the introduction of new
technology. Improving angular resolution will certainly be an extremely taxing
effort. Use of x-ray interferometry, if it can be implemented at all, will be limited
to very specialized problems. Sensitivity can only be improved at this point, by
the achievement of larger and larger telescope areas since the detectors are close
to theoretical performance. If one considers the very substantial engineering and
financial resources required by AXAF and XMM, this will be no easy task.

In spectroscopy, detectors with near unity quantum efficiency and a few eV
resolution will already operate in signal limited conditions. Again higher resolution
will require larger telescope collecting area.

From the above I concluded that what is ahead of us will be a much harder
road to follow than in the past, much more similar to the conditions prevailing in
a mature discipline than in one in rapid development. As was the case in optical
astronomy in the 50's through the 70's, the post AXAF-XMM period could be a
time in x-ray astronomy denoted by slow technical improvements, yet extremely
productive scientifically. Careful planning of the observational programs and stren-
uous efforts in data reduction and interpretation will be the typical approach. And

yet, reasonable as the above considerations may be, I found them quite unsatisfactory and a sign of closure rather than opening of the field. I attempted to find a way out by reflecting on a previous experience of mine.

## 3. PAST PREDICTIONS AND PROPOSALS

In September 1963, Herbert Gursky and I offered (Giacconi 1963) to NASA a program of x-ray astronomy which encompassed a number of rocket and satellite payloads to be flown over the 1963-1968 period. It is useful to recollect that at the time only three celestial sources of x-rays were known (beside the Sun): Sco X−1, the Crab Nebula and Cyg X−1.

The proposals were contained in a document entitled "An Experimental Program of extra-Solar X-ray Astronomy" (ASE-449, September 1963). The first figure (figure 4) from that document shows a time table for the execution of a number of programs beginning with rocket flights, a scanning satellite, which became "*UHURU*", a 30 cm x-ray telescope on OAO (never approved) and culminating in a 1.2 meter diameter x-ray telescope which we proposed should be launched in the "very distant future", 1968 (figure 5). While the true embodiment of the 1.2 meter telescope will only be realized in the AXAF mission in 1996, it is clear that important steps have been taken along the way, for instance with the launch of "*Einstein*", although with only 60 cm aperture and 10 years late.

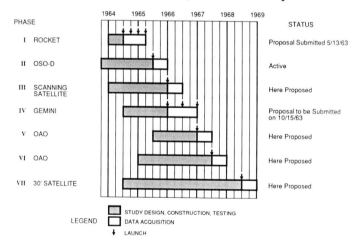

PROPOSED TIME SCHEDULE

Figure 4.

Of course, these proposals were hopelessly romantic in their timing projections. However, it is interesting that they held so well over more than two decades. I believe they did so because they were not simply fashionable or responding to political expediency or opportunity, but rather because they were based on solid scientific objectives (figure 6) and on reasonable projections of technological developments.

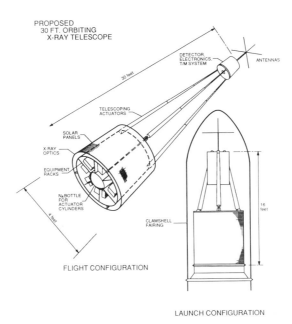

PROPOSED
30 FT. ORBITING
X-RAY TELESCOPE

Figure 5.

One could object that the 1963 predictions became true because they were some-what in the nature of self-fulfilling prophecies, since we, the authors, were ourselves planning to carry out these programs. That may be partly true; yet I thought I would attempt today to follow the same approach when it is obvious that I will not be the one to carry out, most if any of the new predictions. It will be interesting to see what will come of it.

The approach then is to consider some of the scientific problems that may still remain unsolved, after all the currently planned missions are over, to find opportunities for useful developments. As we shall see this line of reasoning leads, in the examples I have chosen, to investigations which are outside the scope of current missions and of great scientific interest in themselves.

## 4. SCIENTIFIC QUESTIONS

### 4.1 The Study of X-ray Stellar Coronae

High spectral resolution ($R > 500$) study of stellar coronae can be carried out with AXAF by use of high angular resolution imaging detectors and high dispersion objective gratings. A typical observation requires of order of $10^4$ seconds. Since many of the phenomena under study are time variable with time constants of order months (*e.g.* Sun) or shorter, several observations will be required to elucidate the physical processes occurring at each source. This may require $10^5$ sec total obser-vation for object. It is clear that a synoptic program of stellar coronae observations cannot effectively be carried out with less than tens of objects for stellar class, (see

## II.  REQUIREMENTS FOR FUTURE X-RAY OBSERVATIONS

THE DISCOVERY OF GALACTIC X-RAYS, TOGETHER WITH THE VARIOUS HYPOTHESIS THAT HAVE BEEN PUT FORWARD TO EXPLAIN THEM RAISE TWO OBVIOUS QUESTIONS WHICH MUST BE ANSWERED IN FUTURE OBSERVATIONS:

1.  WHAT ARE THE PRECISE POSITION, DISTANCES AND DIMENSIONS OF THE DISCRETE SOURCES?

2.  DO ALL X-RAYS COME FROM DISCRETE SOURCES OR IS THERE A GENERAL DIFFUSE BACKGROUND?

A GREAT ELABORATION OF THE OBSERVATIONAL TECHNIQUES WILL CLEARLY BE REQUIRED IN ORDER TO ANSWER THESE QUESTIONS AND THE NEW ONES WHICH WILL ARISE IN THE COURSE OF THE DEVELOPMENT OF X-RAY ASTRONOMY.  IN COMMON WITH TECHNICAL DEVELOPMENTS FOR ASTRONOMICAL OBSERVATIONS IN OTHER REGIONS OF THE SPECTRUM, THOSE NEEDED FOR X-RAY ASTRONOMY WILL BE DIRECTED TOWARD:

1.  ALL-SKY SURVEYS WITH INCREASED ANGULAR RESOLUTION AND INCREASED SENSITIVITY TO DISTINGUISH DISCRETE SOURCES AND THE DIFFUSE BACKGROUND;

2.  HIGHER RESOLUTION STUDIES OF THE STRUCTURE OF INDIVIDUAL SOURCES;

3.  INCREASED SPECTRAL RESOLUTION BOTH FOR DISCRETE AND DIFFUSE SOURCES;

4.  STUDY OF THE DETAILED PROPERTIES OF X-RAY EMISSIONS SUCH AS SECULAR CHANGES AND POLARIZATION.

Figure 6.

for instance the complex relationship between magnetic field intensity, rotational speed and x-ray luminosity for M dwarf stars). Thus a minimum and modest program of stellar x-ray coronal studies may well use in excess of $10^7 - 10^8$ seconds and be limited to very bright stars, a commitment of time to a single discipline well beyond what may be possible with AXAF.

I believe a dedicated explorer class missions may well be the answer to the problem we posed. Such a mission consisting of an x-ray telescope, objective grating and imaging detectors requires no technological improvement. It was proposed as early as 1979 for the U.S. Space Program by Giuseppe Vaiana and his collaborators. Given the extremely tight resource constraints of the U.S. Explorer program at the time, this "Stellar Coronal Observatory" could not be implemented, notwithstanding the favorable review. The same concept proposed for the Italian Space Program in 1981 (OOXA) was not accepted. It may finally be brought to fruition as the proposed ROSAT II.

## 4.2   The Formation and Evolution of Large Scale Structure

Study of the formation, chemical and dynamical evolution of clusters of galaxies at very great distances ($z \gg 1$) may well hold the key to the solution of some of the most exciting problems in cosmology. Cluster formation and evolution is directly linked to the original spectrum of density fluctuations. The chemical evolution of the gas contained in clusters furnishes an integral quantitative measure of galactic evolution. The morphology and temperature of the gas in the clusters can be used to establish the prevalence of cooling flows at early epochs and their role in galaxy formation. The distribution of distant clusters in the sky can be used to trace the large scale structure of the Universe. Yet the prospects to carry out this work in the forthcoming missions are not completely assured. The very considerable observational capabilities of AXAF in this field (figure 7) could be used to effect if we already had an unbiased sample of extremely distant clusters ($1 < z < 3$ for instance) in a contiguous region of, let us say, 1000 square degrees which we could systematically study. It is not obvious how such a sample could be obtained, but x-ray observations may furnish the most powerful available approach. Richard Burg and I have been engaged in a collaborative effort over the last several months to elucidate this problem.

The ROSAT all sky survey will be limited by sensitivity and angular resolution ($\sim 20''$) to the certain x-ray identification of relatively nearby clusters ($z \lesssim 0.5$). For a few candidate distant clusters ($\sim 100$) detailed morphological studies could be obtained with the ROSAT HRI detector ($\sim 5''$ angular resolution). But measurement of redshifts and detailed kinematic studies can only be carried out through the extremely laborious process of optical observation and characterization of tens of thousands of sources. The optical magnitude of the galaxies for the more distant sources ($z > 1$) will be fainter than 23 $m_v$.

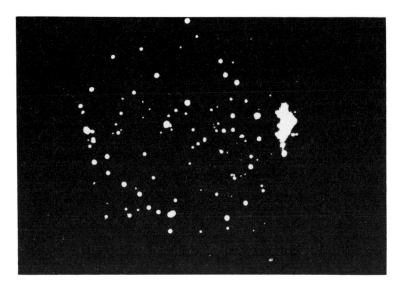

Figure 7.

It would be extremely useful if we could recognize the existence of a cluster by directly measuring the angular extent of the x-ray emission at any z. The problem can be separated into two components.

First the characteristic dimensions of structures collapsing at a given z differ for different cosmological models. We can for instance estimate the linear diameter by using the scaling relations given by Kaiser (Kaiser 1986) and further discussed by Cavaliere and Colafrancesco (Cavaliere and Colafrancesco 1988) and Shaeffer and Silk (Shaeffer and Silk 1988). The characteristic radius of a structure which has collapsed at redshift z is given by

$$R \propto (1+z)-\frac{(5+n)}{(n+3)} \tag{1}$$

where n is the spectral index of the power law describing the mass scale of the initial fluctuations in density $\delta$, $|\delta|^2 \propto k^n$. In the hierarchical clustering scenario $-3 < n < 1$, the preferred index $n_{eff} \simeq -1$ for "adiabatic" fluctuations or $n_{eff} \simeq 1.5$ for the isothermal case. For $n = -1$

$$R \propto (1+z)^{-2} \tag{2}$$

since the characteristic core radius of clusters for $z \leq 0.2$ is of order 250 kpc (Jones and Forman 1984), characteristic core radii for clusters collapsing at z of 1 and 2 would be of order 62 and 28 kpc. Shaeffer and Silk derive somewhat different scaling reactions, which include non-linear effects to comply with the virialization conditions. Within our range of values we find that the predicted radii for the core x-ray emitting regions are practically identical.

Second, the apparent angular size of a cluster is set by cosmology

$$\theta = \frac{2R(1+z)^2}{D_L} \tag{3}$$

where $R$ is the core radius and $D_L$ is the luminosity distance. For a Friedman cosmology with $q_0 = 1/2$ and $H_0 = 50$ km s$^{-1}$Mpc$^{-1}$ the above linear dimensions computed above would correspond to angular diameters of 15 to 7 arc seconds.

AXAF possesses the angular resolution and spectroscopic capability to directly determine the extension of a cluster as well as the ability to directly measure its redshift (if the x-ray emitting gas in enriched). Therefore, in principle, AXAF could carry out a direct survey for very distant clusters. If we assume, however, that the required resolution is finer than 5 arc sec we find that the field of view of AXAF within which that resolution is achieved is of order of 8 arc min radius. At the *Einstein* Deep Survey sensitivity limit, AXAF could carry out the survey with a $10^3$ second exposure in each field. A 1000 square degrees area could then be searched in $1.6 \times 10^7$ seconds. Even this large amount of investment of AXAF observational time however would fall short of our desire to observe *very* distant clusters ($z > 2$) since we would reach a limiting sensitivity of only $5 \times 10^{-14}$ erg cm$^{-2}$ sec$^{-1}$ arc min$^{-2}$ quite short of that required to detect clusters at that redshift. XMM, notwithstanding its throughput, is severely limited at a resolution of $\sim 20''$ arc sec for this kind of problem.

## 5. A NEW X-RAY SURVEY WITH AN EXPLORER CLASS MISSION

While it would be very difficult to envisage multi-year programs on AXAF to carry out the deep surveys for clusters of galaxies that I discussed, it would be perfectly suitable to do so with an Explorer class mission which I call the SXT (for small x-ray telescope). But even if we had all the time in the world could we achieve the required sensitivity with a small telescope? I believe the answer may be yes. The thought is that in signal limited conditions collecting area can be traded for time of observation. Thus to achieve the AXAF limit at $10^3$ sec one could use a telescope with 1/10 the area at $10^4$ sec. But if the field of view remained at 8 arc min radius it would take us 10 years to conduct the survey. Here is where a new technical development could save us. *If we could enlarge the region in the focal plane in which we could achieve the require 5 arc sec resolution to, a 30 arc minutes radius, the time required for the survey would be shortened by a factor of about 14, to less than a year.* A specialized survey for clusters could then be conducted with a telescope $\sim 1/3$ AXAF in length, $\sim 1/27$ AXAF in weight provided only that we could obtain a field of view (with $5''$ arc sec resolution) of 30 arc min. CCD detectors could yield sufficient angular resolution (if the effective focal length is $> 1$ meter) to positively identify clusters (at any z) and sufficient energy resolution to directly determine z for the brightest by measuring the redshift of characteristic lines like iron. In a few years this survey would go much deeper than the ROSAT sky survey limit and provide an unbiased flux limited catalogue for AXAF study. The technology of building x-ray optics with relatively wide fields of view at reasonable resolution does not appear beyond reach. *The very high reflection efficiencies currently achieved permits the design of optical systems with > 2 reflections. Such systems could, at least in principle, permit the required correction of aberrations over a larger field.*

## 6.  LXT (A 2.4m TELESCOPE, THE SUCCESSOR OF AXAF AND XMM)

The technology study to implement the survey Explorer class mission described above should be initiated now if we wish to fly the mission in time to affect even the latter part of the AXAF operational life. Apart from its scientific interest, a successful technological development effort would lead logically to the next step in the design of the Observatory class missions. The scientific justification for a successor to AXAF is obvious if we consider that even with the combined efforts of the small x-ray Explorer and AXAF we will be restricted to clusters at z< 1.5. To go really to z of 2 or 3 we need to achieve much higher sensitivity.

Taking a leaf from the preceding discussion I can imagine an improved observatory for the post AXAF-XMM era. In my opinion it should have the characteristics shown in table 1, where it is compared with SXT, AXAF and XMM. The asterisked quantities under the SXT and LXT heading denote performance that can be achieved only through technological developments.

| Table 1 | | | | |
|---|---|---|---|---|
| | AXAF | XXM | SXT | LXT |
| D | 1.2 m | 3×.7 | .24 m | 2.4 m |
| f | 10m | 7.5 m | 1 m | 20 m |
| $\epsilon_A$ | <10% | 40% | 10% | 40% * |
| A | 1500 cm$^2$ | 6000 cm$^2$ | 200 cm$^2$ | 17600 cm$^2$ |
| $\delta$ | < 1″ | ~ 20″ | < 5″ | < 1″ |
| $\theta_{FOV}(1″)$ | 4.5′ | — | — | 30′ * |
| $\theta_{FOV}(5″)$ | 8′ | — | 30′* | 30′ * |
| $\theta_{FOV}(20″)$ | 18′ | 18′ | 30′* | 30′ * |
| $D_{det}$ | 14 cm | 12.5 cm | 1.5 cm | 42 cm * |

The most important are:

1. The efficiency of utilization of the telescope aperture which is $\sim 10\%$ in AXAF and of 40% in XMM. What is required is to develop the XMM thin mirrors telescope technology to achieve the AXAF optical tolerances. Difficult? Yes but perhaps not impossible.

2. The field of view in which a resolution of $\sim 1''$ arc second can be maintained. While for shallower surveys (AXAF limit in $10^3$ sec is $5 \times 10^{-14}$ erg cm$^2$ sec$^{-1}$ arc min$^{-2}$) $5''$ resolution may be sufficient to exclude confusion from point sources, at $6 \times 10^{-17}$ erg cm$^{-2}$ sec$^{-1}$ arc min$^{-2}$ a resolution of order of 1 arc sec is required. This development seems to me much more difficult to achieve even with corrective optics but perhaps not hopeless.

3. Finally the development of large (42 cm diameter) arrays of CCD is required.

If we could carry out such a program of technological advances, the LXT would offer truly remarkable performance improvements. While the area for spectroscopy and single source sensitivity is improved by a factor of 16, the field of view at high resolution is improved by more than 40. The speed with which surveys can be carried out would therefore improve by factors of $16 \times 40$, a stupendous 640!!

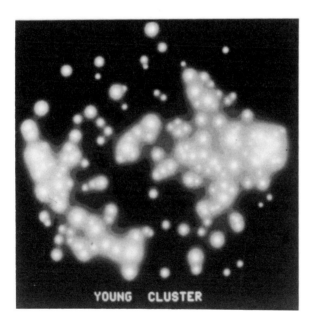

Figure 8.

## 7. SUMMARY AND CONCLUSIONS

In discussing with you my predictions for the future of x-ray astronomy I have stressed *the continued and important role that explorer class missions could carry out*. I have also attempted to imagine a successor to the Observatory class missions currently under construction. My belief is that we have gone about as far as one can with the development of detectors (still a high angular resolution, wide field imaging calorimeter for high resolution, non-dispersive spectroscopy would be welcomed) and that we must go back and *redesign the x-ray optics for corrected wide fields*. Also that *we can gain as much by improving the filling factors as by increasing the size of the telescopes*. This can be achieved by *adapting the XMM technology in high resolution optics*. I am convinced that LXT would provide enormous observational improvements in many disciplines of astronomy. Still even if its only purpose was to elucidate the large scale structure of the Universe it may well be worth it.

In this meeting in which we celebrate our collective efforts and scientific success in a mission now ten years old, it is good to look beyond the AXAF-XMM era ten years from now. I am profoundly convinced that the greatest scientific harvest in x-ray astronomy is yet to come.

## REFERENCES

Cavaliere, A. and Colafrancesco, S. 1988, *Ap.J.*, **331**, 660.

Giacconi, R. 1963, *An Experimental Program of Extra Solar X-ray Astronomy* (ASE-**449**, September, 1963) Smithsonian Archives.

Jones, C. and Forman, W. 1984, *Ap.J.*, **276**, 38.

Kaiser, N. 1988, *M.N.R.A.S.*, **222**, 373.

Shaeffer, R. and Silk, J. 1988, *Ap.J.*, **333**, 509.

# PART 2
# NEW RESEARCH WITH EINSTEIN

# THE IMPACT OF EINSTEIN OBSERVATIONS ON OUR UNDERSTANDING OF LOW MASS STAR FORMATION

**Frederick M. Walter**

CASA, University of Colorado

## 1. INTRODUCTION

Prior to 1980, the world of pre-main sequence stars, if not well understood, was at least well defined. The Herbig and Rao (1972) catalog listed 69 pre-main sequence stars in Tau-Aur, with the vast majority clearly being T Tauri stars. The characteristics of the classical T Tauri stars include strong H$\alpha$ emission, with $W_\lambda(H\alpha)>5$-10Å; forbidden line emission; continuum ultraviolet and IR excesses; veiling of the absorption line spectrum; significant stellar variability; Li I $\lambda$6707Å absorption; and association with dark clouds and/or emission nebulosities (Strom *et al.* 1975).

Star forming regions were observed extensively with the *Einstein* Observatory. Ku and Chanan (1979) and Pravdo and Marshall (1980) showed the abundance of stellar X-ray sources in the Orion Nebula. Montmerle *et al.* (1983) studied 47 X-ray sources in the $\rho$ Oph complex. Feigelson and DeCampli (1981), Gahm (1981), and Walter and Kuhi (1981, 1984) observed selected classical T Tauri stars. About $\frac{1}{3}$ of the known T Tauri stars were detected as X-ray sources, yet the vast majority of the X-ray sources detected were coincident with anonymous stars *not* suspected to be pre-main sequence stars.

In the grand tradition of X-ray astronomy, X-ray astronomers trouped to telescopes to identify the optical counterparts. Feigelson and Kriss (1981) and Walter and Kuhi (1981) showed that 5 of the counterparts were K7-M0 stars, above the main sequence, with strong Li I absorption. Mundt *et al.* (1983) showed that these stars were kinematic members of the Tau-Aur star formation complex. Since then, Walter (1986) and Walter *et al.* (1987) have discussed additional members of this class of *naked T Tauri Stars* (*NTTS*). Feigelson *et al.* (1988) provided finding charts for X-ray selected pre-main sequence star candidates in the general vicinity of Tau-Aur, and Walter *et al.* (1988) discussed 35 X-ray selected and optically confirmed *NTTS* in Tau-Aur.

## 2. WHY THE NAKED T TAURI STARS HAD TO BE X-RAY SELECTED

The *NTTS* are low mass pre-main sequence stars which lie above the Zero Age Main Sequence, exhibit strong Li I absorption, and are found in regions of star formation. Yet while the classical T Tauri stars are readily discovered through their H$\alpha$ emission in objective prism surveys, or through their IR excesses, the *NTTS* lack both characteristics. The only optical *survey* which has succeeded in selecting for *NTTS* was the Lk Ca II objective prism survey (Herbig, Vrba, and Rydgren 1986). This technique is sensitive to Ca II H and K emission *only* in stars cooler than about spectral type K5 (see Figure 15 of Walter *et al.* 1988). As young active stars the *NTTS* possess extreme solar-like coronae and are X-ray sources with luminosities $L_X \approx 10^{30-31}$ erg s$^{-1}$, they were easily detectable by the *Einstein* Observatory.

## 3. HOW MANY NAKED T TAURI STARS ARE THERE?

I and my colleagues (A. Brown (JILA), R.D. Mathieu (Wisconsin), and F.J. Vrba (USNO)) are continuing to study the *NTTS* identified in the *Einstein* images. We have made the following progress:

- **Tau-Aur**. The basic results have been published (Walter *et al.* 1988). There are 44 *NTTS*. In Tau-Aur the *NTTS* outnumber the classical T Tauri stars by about 10:1.

- **Sco/Oph**. We have identified some 30 *NTTS* within the II Sco OB association. This is consistent with a Salpeter (1954) initial mass function from 0.5 through 10 M$_\odot$. A large fraction of the X-ray sources near the $\rho$ Oph cloud are also *NTTS*.

- **CrA**. We have identified 7 *NTTS* within one degree of the R CrA cloud.

- **Ori**. We have identified 76 *NTTS* in the belt of Orion. As in II Sco, it had been though that low mass stars did not form in the same regions where the high mass stars formed, but the X-ray observations have thrown this assertion into question. Elsewhere in Orion, we have identified two dozen *NTTS* in the region of $\lambda$ Ori and the B35 molecular cloud, and around the periphery of M42.

## 4. WHAT THE NAKED T TAURI STARS TELL US ABOUT PRE-MAIN SEQUENCE *STARS*

We now know that the *NTTS* form the dominant population of low mass stars in regions of star formation. Because the *NTTS* are unencumbered by the effects of disks and boundary layers, we can study the properties of the *stars* at ages of $10^5$ to $10^7$ years. For example:

- Walter *et al.* (1988) discussed the decay of coronal surface fluxes with time.

- We have identified 9 spectroscopic binaries (one SB2) in addition to the SB2 V826 Tau among the *NTTS*. Binaries can yield stellar masses and facilitate observational tests of the theoretical isochrones and evolutionary tracks. Only one classical T Tauri star (V4046 Sgr) is a confirmed SB.

- We have constructed H-R diagrams for Tau-Aur, II Sco, and Ori OBIb. With these we can probe the history of the low mass star formation in these regions.

- Star formation appears *not* to be bimodal, at least in the sense that both low and high mass stars form in the same region in numbers consistent with a Salpeter initial mass function.

- There appear to be differences in the stellar populations in the different associations. There is evidence for either two episodes of star formation in Tau-Aur, or a shift of the locus of star formation.

## 5. WHAT CAN WE EXPECT FROM ROSAT AND AXAF?

The 1988 update of the HRC (Herbig and Bell 1988) lists 170 pre-main sequence stars in Tau-Aur. *Of the 91 additions, 35 were discovered through their X-ray emission.* Many of the others are faint visual companions to previously known classical T Tauri stars. Most of our increased understanding of the pre-main sequence evolution of the low mass stars is a direct consequence of follow-up studies of the *NTTS*, which have only been discovered efficiently through imaging X-ray observations.

Although Herbig (1978) anticipated the discovery of relatively inactive low mass pre-main sequence stars, he predicted an older, post-T Tauri population. The discovery of the *NTTS* population was unexpected. Follow-up observations of regions of star formation with ROSAT (to study spatial distributions) and AXAF (to study individual stars in detail and more distant OB associations), in conjunction with multispectral ground- and space-based observations, promise to bring the studies of the *NTTS* to maturity.

The important lesson to be gained from this investigation is that the exploration of new domains, either in space or in limiting magnitude, is likely to have unanticipated consequences. We must be ready for, and be able to follow up on, surprises. Without the X-ray observations, the *NTTS* would likely still be anonymous stars. *But without the ability to identify the X-ray counterparts optically and follow up with a comprehensive program of ground-based spectroscopy and photometry, we would likely still be utterly ignorant of the nature of the naked T Tauri stars.*

This research at the University of Colorado has been supported by grants from the NASA ADP program, the NSF, and a Fullam award.

## REFERENCES

Feigelson, E.D. and DeCampli, W.M. (1981), *Ap.J.(Letters)*, **243**, L89.
Feigelson, E.D. and Kriss, G.A. (1981), *Ap.J.(Letter)*, **248**, L35.
Feigelson, E.D., *et al.* (1987), *Astron. J.*, **94**, 1251.

Gahm, G. (1981), *Ap.J.(Letter)*, **242**, L163.

Herbig, G.H. (1978), in *Problems of Physics and Evolution of the Universe* (Yervan:Academy of Sciences of the Armenian SSR), p.171.

Herbig, G.H., and Bell, K.R. (1988), *Lick Obs. Bull.* 1111.

Herbig, G.H., and Rao, N.K. (1972), *Ap.J.*, **174**, 401.

Herbig, G.H., Vrba, F.J., and Rydgren, A.E. (1986), *Astron. J.*, **91**, 575.

Ku, W.H.-M. and Chanan, G.A. (1979), *Ap.J.*, Lett. **234**, L59.

Montmerle, T., *et al.* (1983), *Ap.J.*, **269**, 182.

Mundt, R., *et al.* (1983), *Ap.J.*, **269**, 229.

Pravdo, S.H. and Marshall, F.E. (1981), *Ap.J.*, **248**, 591.

Salpeter, E.E. (1954), *Ap.J.*, **121**, 161.

Strom, S., Strom, K., and Grasdalen, G. (1975), *Ann.Rev.Astr.Ap.*, **13**, 187.

Walter, F.M. (1986), *Ap.J.*, **306**, 573.

Walter, F.M., *et al.* (1987), *Ap.J.*, **314**, 297.

Walter, F.M. and Kuhi, L.V. (1981), *Ap.J.*, **250**, 254.

Walter, F.M. and Kuhi, L.V. (1984), *Ap.J.*, **284**, 194.

Walter, F.M., *et al.* (1988), *Astron. J.*, **297**, 279.

# THE EINSTEIN SURVEY OF O STARS

**S. Sciortino, G. S. Vaiana**
Osservatorio Astronomico di Palermo

**F. R. Harnden, Jr.**
Harvard-Smithsonian Center for Astrophysics

**M. Ramella, C. Morossi**
Osservatorio Astronomico di Trieste

**R. Rosner**
Department of Astronomy and Astrophysics and
Enrico Fermi Institute, The University of Chicago

**J.H.M.M. Schmitt**
Max Planck Institut für Extraterrestrische Physik

## 1. INTRODUCTION

Chlebowski, Harnden and Sciortino (1989) have recently performed a complete x-ray survey of Galactic O stars observed with the *Einstein Observatory* deriving x-ray luminosities or upper limits for 288 stars. The x-ray luminosities have been computed merging all the available images, taking into account instrumental response, interstellar absorption and assuming a source temperature of 0.5 keV. As final result both values of $L_x/L_{bol}$ (almost distance independent) and of $L_x$ have been evaluated and catalogued. We report in this paper a brief account of some of the main results of a detailed analysis of this - presently the largest - sample of x-ray surveyed O stars (Sciortino *et al.*. 1989).

## 2. SAMPLE CHARACTERISTICS

The $m_v$ distribution of the x-ray surveyed sample is statistically indistinguishable from that of a volume limited optical-based sample of Galactic O stars (Garmany, Conti and Chiosi 1981) complete up to 2.5 kpc. The x-ray sample is therefore statistically representative of the O star population in the Galaxy. The majority of surveyed O stars are inside a circle of 4 kpc radius around the Sun and are members of associations (Humphreys and Mc Elroy, 1984; Chlebowski, Harnden and Sciortino 1989). A comparative formal analysis of heliocentric distance distributions of both the x-ray sample and the volume-limited optical sample shows that these two samples can be considered statistically complete only up to 2.5 kpc. Hence we decided to perform all further analysis of the x-ray data on the volume limited sample.

A formal two-sample test shows that the $m_v$ and $m_{bol}$ distributions of entire x-ray sample and of the volume-limited x-ray sample cannot be drawn from the same parent population (confidence level 99.8% for $m_v$ and 99.99% for $m_{bol}$), thereby shows the occurrence of a selection effect in the sampling of more distant stars. This strengthens the need to limit all further analysis to the volume-limited x-ray sample.

## 3. ANALYSIS AND RESULTS

For the volume limited sample we have constructed, using both detections and upper bounds, the maximum likelihood integral x-ray luminosity function shown in Figure 1.

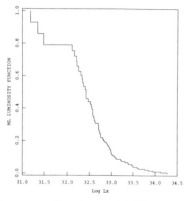 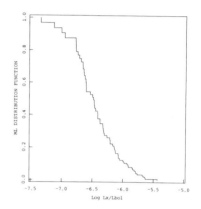

Figure 1. Maximum likelihood integral x-ray luminosity function.

Figure 2. Maximum likelihood integral $L_x/L_{bol}$ distribution function

The median of $\log L_x$ is 32.4 (erg s$^{-1}$) with $3\sigma$ most probable value in the range (32.2-32.6). Three characteristics of this distribution should be noted: (a) the majority $\approx 70\%$ of O stars have an x-ray luminosity between $\approx 3 \times 10^{31}$ erg s$^{-1}$ and $1.5 \times 10^{33}$ erg s$^{-1}$, (b) a tail of high luminosity objects ($L_x > 1.5 \times 10^{33}$ erg s$^{-1}$) extending up to $L_x \approx 2 \times 10^{34}$ represents $\approx 10\%$ of the entire population; (c) about 20% of O stars have $L_x \leq 3 \times 10^{31}$ erg s$^{-1}$.

The low $L_x$ tail can be explained by selection effects due to a combination of stellar distances and achievable sensitivities that do not allow a complete sampling of x-ray faint O stars at large distances.

We have verified that the high $L_x$ tail cannot be accounted for by stellar multiplicity. However the fraction of detected stars (19%) with $L_x$ (detection and upper bounds) above $1.5 \times 10^{33}$ erg s$^{-1}$ is smaller in the volume limited sample than in the entire sample (33%). This lead us to adopt the working hypothesis that the high-$L_x$ tail is associated with the sampling of stars at larger mean distances (*i.e.* belonging to a larger space volume) in a few particularly high-sensitive observations. To confirm this hypothesis we have verified that: (a) the distance distributions of O stars with $L_x$ above $1.5 \times 10^{33}$ erg s$^{-1}$ and that of O stars with $L_x$ between $3 \times 10^{31}$ erg s$^{-1}$ and $1.5 \times 10^{33}$ erg s$^{-1}$ are statistically significantly different, stars having higher

$L_x$ being more distant; (b) the distribution of x-ray flux limiting sensitivities is independent of stellar distance; (c) the distributions of flux limiting sensitivity for the stars, respectively, in the bulk and in the tail of $L_x$ distribution are statistically distinguishable (confidence level greater than 99.999%). Hence we conclude that our working hypothesis is consistent with all experimental evidence.

When proper corrections for absorption are applied, the x-ray luminosity function of the volume limited sample cannot be distinguished from that of the entire sample and, more important, from that of those stars at distances greater than 2.5 kpc. Hence the present data suggests that heliocentric distance is not a relevant parameter in determining the typical x-ray luminosity of O stars in the Galaxy. A similar result holds also if we consider Galactocentric distance as independent parameter. The Galactic longitude distribution of stars in the tail of the x-ray luminosity function is statistically indistinguishable from that of stars of the entire x-ray sample. Present data do not show (with the only statistically-marginal exception of Cyg OB2 O stars) evidence of clustering of highly x-ray luminous stars in given association(s).

The $\log(L_x/L_{bol})$ maximum likelihood integral distribution function for the volume-limited sample is shown in Figure 2. The mean value of $\log(L_x/L_{bol})$ is $-6.46$, the $3\,\sigma$ range is $[-6.59, -6.32]$. We note that this mean value is slightly less negative than the "canonical" value of $-7$ previously derived from smaller size sample (*cf.*, Harnden *et al.* 1979, Seward *et al.* 1979, Long and White 1980, Pallavicini *et al.* 1981).

The comparison of $\log(L_x/L_{bol})$ distribution of the volume limited sample with that of the entire x-ray sample and with that of farther sample (distance $> 2.5$ kpc) does not allow us to statistically distinguish these three distributions. We have not found compelling evidence for dependence of $\log(L_x/L_{bol})$ on either heliocentric or Galactocentric distances, or on Galactic longitude, just as for $L_x$. The $(L_x/L_{bol})$ ratio of Cyg OB2 stars are indistinguishable from that of the other OB association members.

In proceeding beyond this analysis we have considered the relationship between the x-ray properties and the other physical characteristics of the sample stars. Such an analysis must take proper account of the upper limits. In the following, we have adopted the technique for correlation analysis in the presence of censored data described by Schmitt (1985).

We have used our volume-limited statistically complete sample to test for the existence of a proportionality relationship between $L_x$ and $L_{bol}$. Using a non-parametric formulation, we find that:

$$\log L_x = 1.0(+0.14/-0.15)\log L_{bol} - 6.2(+5.6/-5.6), \ (\text{erg s}^{-1}) \qquad (1)$$

with a correlation coefficient of $r \approx 0.72$ [0.66,0.78]. An analogous analysis extended to the entire sample yields essentially the same result. Hence our analysis confirms the previous finding of a linear relationship between $L_x$ and $L_{bol}$ based on smaller size samples (Pallavicini *et al.* 1981).

In order to study other possible stellar parameters which might influence the x-ray luminosity of O stars, we have correlated $L_x$ with the following additional stellar parameters: $T_{eff}$, $R/R_\odot$, $M/M_\odot$, g, $vsini$ (the projected equatorial rotation speed), $vsini/V_{crit}$ (with $V_{crit}$ the critical breakup equatorial rotation speed), $\dot{M}$, $V_\infty$ (wind terminal velocity), $F_m = \dot{M} \, V_\infty$ (total wind momentum flux) and $L_w = \frac{1}{2}\dot{M} \cdot V_\infty^2$ (wind kinetic luminosity).

Using logarithmic quantities for each of the parameters considered, we determined (including upper bound information) the linear correlation coefficient $r$, the slope $B$, the intercept $A$, and the 68% confidence intervals which were derived via bootstrap simulations. Many of the parameters correlate to some degree with $\log L_x$, while $\log(vsini)$ and $\log(vsini/V_{crit})$ definitively do not correlate. There is evidence for strong correlation with $\log(M/M_\odot)$, $\log L_{bol}$, $log(R/R_\odot)$, and $\log F_m$ (or alternatively $\log L_w$).

However, many of the correlations we have found can be generally explained in terms of the $L_x - L_{bol}$ correlation. In fact the dependencies on mass and radius (*i.e.* from "size" of star) and $T_{eff}$ can be explained in terms of functional relations predicted by stellar structure models relating $L_{bol}$ with $T_{eff}$, R, and M. The relation between $\dot{M}$ and $L_{bol}$ (*cf.* Lamers 1981, Garmany and Conti 1984, Wilson and Dopita 1985) accounts for the (rather weak) dependence we have found between $L_x$ and $\dot{M}$.

It is noteworthy, the correlation between $L_x$ and $F_m$ (or $L_w$) is better than that with respect to the correlation of $L_x$ with $\dot{M}$ or $V_\infty$ alone. The formal relation between $L_x$ and $F_m$ is:

$$\log L_x = 0.50(+0.12/-0.10)\log F_m + 34.0(+0.3/-0.2), \text{ (erg s}^{-1}) \qquad (2)$$

with a correlation coefficient $r = 0.69[-0.08, +0.07]$. This relation suggests a dependence of $L_x$ as the square root of $F_m$.

We conclude noting that none of the investigated parameters alone accounts for much of the observed variance in $\log L_x$. A possible explanation could be furnished by the evidence of long-term variability in the x-ray emission of O stars (Snow, Cash and Grady 1981; Collura *et al.* 1989).

This work has been supported by NASA grants NAG 8-445 and NAS8-30751, and by the Ministero per la Pubblica Istruzione and Piano Spaziale Nazionale-Consiglio Nazionale delle Ricerche (Italy), and by the Smithsonian Institution (RR).

# REFERENCES

Chlebowski, T., Harnden, F. R., Jr. and Sciortino S. 1989, *Ap.J.*, **June 1** .

Collura, C., Sciortino, S., Serio, S., Vaiana, G. S., Harnden, F. R., Jr., and Rosner, R. 1989, *Ap.J.*, **338**, 296.

Garmany, C., Conti, P. S. and Chiosi, C. 1981, *Ap.J.*, **263**, 777.

Garmany, C. and Conti, P. S. 1984, *Ap.J.*, **284**, 705.

Harnden, F. R. Jr., *et al.* 1979, *Ap.J.(Letters)*, **234**, L51.

Humphreys, R. and Mc Elroy, D. B. 1984 *Ap.J.*, **284**, 560.

Lamers, H.J.G.L.M. 1981, *Ap.J.*, **245**, 593.

Long, K. S., and White, R. L. 1980, *Ap.J.(Letters)*, **239**, L65.

Pallavicini, R., Golub, L., Rosner, R., Vaiana, G. S., Ayres, T., and Linsky, J. L. 1981, *Ap.J.*, **248**, 279.

Sciortino, S., Vaiana, G. S., Harnden, F. R., Jr., Ramella, R., Morossi, C., Rosner, R., and Schmitt, J. H. M. M. 1989, *in preparation.*

Schmitt, J.H.H.M. 1985, *Ap.J.*, **293**, 176.

Seward, F. D., and Chlebowski, T. 1982, *Ap.J.*, **256**, 530.

Snow, T. P., Cash, W., and Grady, C. A., 1981, *Ap.J.(Letters)*, **244**, L55.

Wilson, I.R.G., and Dopita, M.A. 1985, *Astr.Ap.*, **149**, 295.

# THE X-RAY EMISSION OF LATE-TYPE EVOLVED STARS

**A. Maggio, G. S. Vaiana**[1]
Osservatorio Astronomico di Palermo

**B. M. Haisch, R. A. Stern**
Lookheed Palo Alto Research Laboratory

**F. R. Harnden, Jr.**
Harvard-Smithsonian Center for Astrophysics

**R. Rosner**
Department of Astronomy and Astrophysics,
and Enrico Fermi Institute, Chicago University

## 1. INTRODUCTION

We report on the main results of an extensive x-ray survey carried out on a magnitude-limited sample of late-type giant and supergiant stars (Maggio *et al.* 1988).

We have selected all the optical candidates in the Bright Star Catalogue, with luminosity classes from III to I and spectral types in the range F to M. The observed sample comprises 380 stars which fell into one of the *Einstein* Observatory Imaging Proportional Counter (IPC) fields of view, either as targets or serendipitously. This sample is about ten times larger than any previously x-ray studied sample of evolved stars. The whole optical population is limited to apparent visual magnitude $m_v < 7.1$, but it is complete only to $m_v \sim 6.3$ (Bahcall, Casertano and Ratnatunga 1987). We have indication that the subsample of stars with reliable trigonometric parallaxes, $\pi > 0.01$, is complete only up to $m_v < 5$. On the basis of accurate statistical tests on the distributions of B-V colors, distances, and absolute visual magnitudes, we have also established that our observed sample is indistinguishable from the optical parent population.

For this survey we have employed the final REV-1 x-ray data processing (Harnden *et al.* 1984).

## 2. RESULTS

Considering the sample limited to $m_v < 6.3$ (278 stars), more than 40% of the F and G giants have been detected, both in the total sample and in the single star sample. Among the K giants, 14% have been detected (all of them early K), but most of the x-ray sources have been identified with RS CVn - like or multiple

---

[1] Also Harvard-Smithsonian Center for Astrophysics

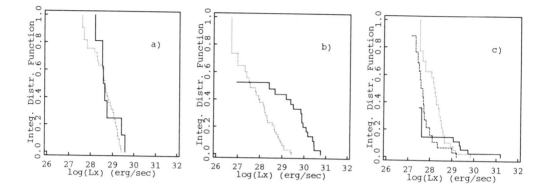

Figure 1. Comparison of x-ray luminosity functions for giants and main sequence stars: (a) Giant F stars (solid), vs. F dwarfs, with $0.3 \leq$ B-V $\leq 0.5$
(dotted), adapted from Schmitt et al. (1985); (b) Giant G stars (solid), *vs.*
F7-G9 dwarfs (dotted), adapted from Maggio *et al.* (1987); (c) Giant K stars
(solid), vs. old disk K dwarfs (dashed), and young disk K dwarfs (dotted),
adapted from Bookbinder (1985). The x-ray luminosity functions for the giant
stars include all the objects with $m_v < 5$ and D $< 100$ pc.

systems, and only 2 detections (4%) refer to single stars. None of the 29 M giants
have been detected, except for one spectroscopic binary. Finally, we report 7 detections out of 46 bright giants or supergiants observed, only one of the sources being
a single star ($\alpha$ Car, F0II).

Using both detections and upper limits, we have computed maximum-likelihood
x-ray luminosity functions (0.2-4 keV) for giant stars of various spectral types, with
distances D $< 100$ pc and $m_v < 5$, and compared these with luminosity functions
previously obtained for main-sequence stars (Figure 1). A complementary analysis
has been performed also on the sample of stars with no indication of binarity. The
main results can be summarized as follows:

(a) The observed F giants (most of them slightly evolved beyond the main sequence)
show x-ray emission in the narrow range of luminosity $10^{38} < L_x \lesssim 4 \times 10^{39}$erg s$^{-1}$,
similar to the range spanned by dwarf stars of similar spectral type. (b) The G
giants show a range of emission more than three orders of magnitude wide: some
presumed single G giants exist with x-ray luminosities comparable to RS CVn -
like systems ($L_x \gtrsim 10^{30}$ erg s$^{-1}$), while some late G giants have upper limits on the
x-ray emission below typical solar values ($L_x \lesssim few \times 10^{27}$ erg s$^{-1}$). The mean x-ray
luminosity for the single G giants ($< \log(L_x) >= 28.7 \pm 0.4$) is about a factor

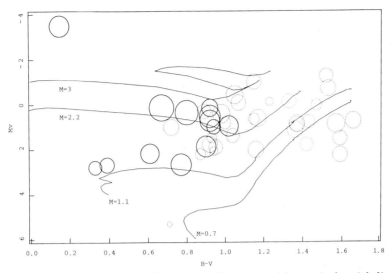

Figure 2. H-R diagram in which each star is indicated by a circle with linear size proportional to $\log(L_x)$ (detected sources: solid lines) or to the upper limit on $\log(L_x)$ (upper limits: dotted lines). Only single stars within 100 pc are represented. Evolutionary tracks for stars with the indicated masses are also shown as solid lines.

of ten higher than for G main-sequence stars. (c) The K giants as a whole have an observed x-ray emission level significatively lower than F and G giants (median $L_x \lesssim 10^{28}$ erg s$^{-1}$), but few high luminosity sources also exist ($L_x \gtrsim 10^{29}$ erg s$^{-1}$). Although most of the latter are identified with binary systems, in some well studied cases (*e.g.* the Hyades giants) the substantial x-ray emission is probably due to a K0III star, the primary component of the system, and not to a companion. These K0 giants appear to be similar to the contiguous G giants and to the young-disk K dwarfs, with respect to the x-ray luminosity. (d) M giants and F to M supergiants have been sampled with low number statistics and sensitivities in this survey because of their low space density and large distances: we can only state that their median x-ray luminosity must be lower than $\sim 10^{29}$ erg s$^{-1}$.

We have confirmed that the drop in the number of detections at spectral type K (B-V $\sim$ 1.1) is statistically significant. This drop, already known as the x-ray boundary line (Ayres *et al.* 1981), is now established on a firm statistical ground, and not simply as a suggestion based on limited samples. We are in no position to explore with equivalent statistical certainty a luminosity class dependence of such a drop.

In Figure 2 we show an H-R diagram in which x-ray detections (solid line circles) and upper limits (dotted circles) are plotted as bubbles whose size is proportional to $\log(L_x)$. Only single stars within 100 pc are represented. Evolutionary tracks (Sweigart and Gross 1978; Mengel *et al.* 1979, Becker 1981) have been also superimposed. We note the existence of single G giants with x-ray luminosities $10^{30} - 10^{31}$ erg s$^{-1}$ located near the track for stars with $\sim 2$ M$_\odot$, at B-V $\sim$ 0.6-0.8. Their main-sequence progenitors are probably A stars which do not show evidence of being strong x-ray emitters (Schmitt *et al.* 1985). We believe, on the basis of the solar analogy, that the heightened level of x-ray activity we observe is the result of the onset of an efficient magnetic dynamo which is likely triggered by the development of a convection zone.

Finally, we have found that the surface rotational velocity is poorly correlated with the x-ray luminosity for the giant stars as a whole, quite in contrast with the situation for main sequence stars. Most likely, this is the result of the influence on the x-ray magnetic activity of other parameters linked to the properties of the stellar internal structure. This conclusion is corroborated by the prediction made by Gilliland (1985) of a substantial increase of the convective turnover time, $\tau_c$, during the stellar evolution, and a corresponding decrease of the Rossby number (an indicator of the efficacy of the dynamo $\alpha$ effect).

## ACKNOWLEDGEMENTS

This work has been supported by NASA contract NAS8-30751 and H-89765B, the Lockheed Independent Research Program, and the Italian CNR, MPI, and CR-RNSM.

## REFERENCES

Ayres, T. R., Linsky, J. L., Golub, L., and Rosner, R. 1981, *Ap.J.*, **250**, 293.

Bahcall, J. N., Casertano, S., and Ratnatunga, K. U. 1987, *Ap.J.*, **320**, 515.

Becker, S. A. 1981, *Ap.J.Suppl.*, **45**, 475.

Bookbinder, J. 1985, Ph. D. Thesis, Harvard University.

Gilliland, R. L. 1985, *Ap.J.*, **299**, 286.

Harnden, F. R., Jr., Fabricant, D. G., Harris, D. E., and Schwartz, J. 1984, *SAO Rept.*, No. 393.

Harnden, F. R., Jr., Sciortino, S., Maggio, G., Micela, M., Vaiana, G. S., and Schmitt, J. 1989, this volume.

Maggio, A., Sciortino, S., Bookbinder, J., Harnden, F. R., Jr., Golub, L., Majer, P., Rosner, R. and Vaiana, G. S. 1987, *Ap.J.*, **315**, 687.

Maggio, A., Vaiana, G. S., Haisch, B. M., Stern, R. A., Bookbinder, J., Harnden, F. R., Jr., and Rosner, R. 1988, *Ap.J.*, submitted.

Mengel, J. G., Sweigart, A. V., Demarque, P., and Gross, P. G. 1979, *Ap.J.Suppl.*, **40**, 733.

Schmitt, J. H. M. M., Golub, L., Harnden, F. R., Jr., Maxon, C. W., Rosner, R., and Vaiana, G. S. 1985, *Ap.J.*, **290**, 307.

Sweigart, A. V., and Gross, P. G. 1978, *Ap.J.Suppl.*, **36**, 405.

# X-RAY OBSERVATIONS OF PLANETARY NEBULAE

**K. M. V. Apparao and S. P. Tarafdar**
Tata Institute of Fundamental Research
Homi Bhaba Road, Bombay 400 005, India

## SUMMARY

The *Einstein* satellite was used to observe 19 planetary nebulae and X-ray emission was detected from four planetary nebulae. The *EXOSAT* satellite observed 12 planetary nebulae and five new sources were detected. An *Einstein* HRI observation shows that *NGC 246 is a point source, implying that the X-rays are from the central star.* Most of the detected planetary nebulae are old and the X-rays are observed during the later stage of planetary nebulae/central star evolution, when the nebula has dispersed sufficiently and/or when the central star gets old and the heavy elements in the atmosphere settle down due to gravitation. However in two cases where the central star is sufficiently luminous X-rays were observed, even though they were young nebulae; the X-radiation ionizes the nebula to a degree, to allow negligible absorption in the nebula. Temperature $T_x$ is obtained using X-ray flux and optical magnitude and assuming the spectrum is blackbody. $T_x$ *agrees with Zanstra temperature obtained from optical Helium lines.*

## 1. INTRODUCTION

The study of planetary nebulae holds interest, since they occur near the end point of evolution of some stars. Observation of X-ray emission from the remnant and/or the central star will give a temperature and clues to physical processes occurring. We have used the archival data from observation of planetary nebulae by the X-ray detecting instruments in the *Einstein* and *EXOSAT* satellites to detect X-ray emission from planetary nebulae.

## 2. OBSERVATIONAL DETAILS

Nineteen planetary nebulae were observed with the IPC instrument of the *Einstein* satellite, of which 4 were detected. The band pass of the IPC is 0.2 - 3.5 keV; most of the flux in the detected planetary nebulae is however in the low energy 0.2 - 0.8 keV range. The fluxes corrected for interstellar extinction are given in Table 1, together with distances, sizes and the extinction parameter C = 1.45 E (B-V). Twelve planetary nebulae were observed with the *EXOSAT* soft X-ray CMA detector with various filters. Eight planetary nebulae were detected, out of which three were detected earlier by *Einstein*. The energy range of the detector with the filters is between 0.05 - 2 keV. The fluxes are corrected for interstellar extinction and the luminosities for the detected planetary nebulae are given in Table 1. The *Einstein* HRI was used to observe NGC 246. This observation shows that the source is a point source indicating that the X-ray flux comes from the central star.

| Planetary Nebula | Distance (pc) | C[a] | Size " | $f_x$ $(10^{-13}$ erg cm$^{-2}$s$^{-1})$ Einstein | EXOSAT | log L$_x$[c] (ergs s$^{-1}$) |
|---|---|---|---|---|---|---|
| | | | | **Table 1** X-ray Observations of Planetary Nebulae | | |
| NGC 246 | 370 | 0.02 | 230 | 15+1.6 | 31.4± 3 | 34.62 |
| NGC 1360 | 260 | 0.02 | 393 | 1.0 | 38 ± 1.8 | 35.04 |
| NGC 1514 | 590 | 0.74 | 114 | < 1.1 | | |
| NGC 2392 | 1089 | 0.24 | 15 | < 2.8 | | |
| NGC 3242 | 787 | 0.21 | 16 | < 2.1 | | |
| NGC 4361 | 939 | 0.19 | 45 | < 2.4 | 3.9 ± 0.6 | 36.04 |
| NGC 6818 | 1459 | 0.18 | 17 | < 2.9 | | |
| NGC 6853 | 400 | 0.02 | 343 | 7.5 ± 1.5 | | |
| NGC 6905 | 1844 | 0.11 | 35 | < 1.5 | | |
| NGC 7293 | 116 | 0.0 | 769 | 2.7 ± 0.7 | 6.9 ± 1.9[b] | 30.04 |
| IC 4642 | 2905 | 0.12 | 17 | < 3.0 | | |
| A21 (YM 29) | 290 | 0.50 | 615 | < 1.8 | | |
| A30 | 1480 | 0.42 | 111 | < 2.8 | | |
| A31 | 240 | 0.1±0.4 | 972 | < 2.5 | | |
| A33 | 580 | 0.05 | 268 | 1.3 ± 0.4 | | |
| A63(UU SGE) | 2154 | 1.63 | 34 | < 2.1 | | |
| A78 | 1830 | 0.20 | 102 | < 2.8 | | |
| He 1–5(FG SGE) | 2300 | 0.81 | 30 | < 1.2 | | |
| VV 1–7 | 603 | 0.43 | 249 | < 1.1 | | |
| | | | | | | |
| NGC 1535 | 1700 | 0.09 | 18 | | 3.2 ± 0.4 | 35.90 |
| A36 (PK 318+41.1) | 312 | 0.05 | 196 | | 1.5 ± 0.4 | 35.11 |
| NGC 3587 | 572 | 0.01 | 196 | | 2.7 ± 0.4 | 32.80 |
| NGC 6853 | 400 | 0.02 | 348 | | 17 ± 1[b] | 34.15 |
| IC 4997 | 1210 | 0.52 | 2 | | < 1.3 | |
| NGC 3242 (PK261+32.1) | 730 | 0.21 | 16 | | < 1.1 | |
| A35 (PK303+40.1) | 200 | 0.0 − 0.76(?) | 800 | | < 1.4 | |
| A31 (PK219+31.1) | 240 | 0.0 ± 0.4 | 972 | | < 1.8 | |

[a] extinction parameter C=1.45 E(B-V).
[b] L3000 filter, except NGC 7293 which is L4000.
[c] corrected for extinction.

## 3. DISCUSSION

It is seen from Table that most of the planetary nebulae except NGC 1535 and NGC 4361 have a large size and/or small extinction. It seems that the X-rays are detected when the nebula has dispersed sufficiently for nebular absorption to be small. In the case of two planetary nebulae the nebular extinction is large.

Correction for the nebular extinction gives implausibly large X-ray emission from the central star. If the nebula itself is to give this flux, then an unreasonably large flux of high velocity wind or hard photons would be needed to heat the nebula. We have examined the possibility that the high X-ray luminosity of these planetary nebulae results in complete ionization of the carbon nuclei responsible for absorption of the soft X-rays. We studied the "Strömgren Sphere" and found that all carbon is in the CV, CVI, CVII ionization states. We also could reproduce the observed CIV λ 1550, HeII λ 1640 obtained from IUE observation and Hβ fluxes as emanating from the ionized nebula. Thus it seems the X-ray flux "burns" its way through the nebula in the case of these young planetary nebulae.

The locations of the planetary nebulae whose X-rays are detected are plotted in an HR diagram in Figure 1.

The results from the *Einstein* observations and detailed conclusions are given in S. P. Tarafdar, K. M. V. Apparao (1988, *Ap. J.*, **327**, 342).

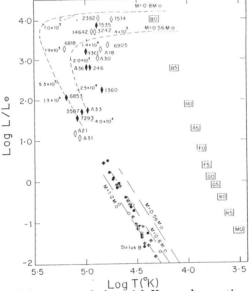

Figure 1. Position of planetary nebulae with X-ray observations (diamonds) are shown in the HR diagram. Symbols are filled for planetaries with positive observations and are open for those with upper limits. NGC or Abell numbers are indicated for each planetary nebula. For reference, positions of main-sequence stars are marked with squares, with spectral types inside the squares. White dwarfs with X-ray observations are marked with closed circles or open circles depending on positive or negative X-ray observations. A cross has been superposed on circles to indicate white dwarfs other than DA type. The evolutionary tracks of $0.6M_\odot$ and $0.8M_\odot$ degenerate carbon-oxygen cores with hydrogen and helium shell burning are presented by dashed curves. Evolutionary ages are marked at a number of points on each track. The long-dashed curves through the positions of white dwarfs are cooling tracks of $0.6M_\odot$, $0.8M_\odot$ stars starting at the end of evolution.

# A NEW BOUNDARY FOR
# THE VELA SUPERNOVA REMNANT

**F.D. Seward**

Harvard-Smithsonian Center for Astrophysics

The Vela supernova remnant, at a distance of 0.5 kpc, is one of the closest SNR to Earth. It was first detected as a conglomerate of radio sources, Vela X, Y, and Z (Milne and Manchester 1986).

This radio-bright region contains prominent optical filaments. The western boundary of the radio source, in particular, coincides with filaments forming part of a "D" shaped configuration (Miller 1973, Elliot, Goudis, and Meaburn 1976). The remnant in 1978 seemed well defined by radio and optical emission and was confined within an approximately circular area with diameter of 5.5°. The Vela Pulsar is interior to this region and located 1.4° from the center. Early rocket observations (Seward *et al.* 1971; Gorenstein, Harnden, and Tucker 1974; and Moore and Garmire 1975) found a bright, diffuse, soft x-ray source filling this circular region in a very lumpy manner.

The remnant was surveyed with the *Einstein* IPC in 1979-80. 33 fields covered the Vela XYZ radio source. The resultant merged x-ray image (Figures 1 and 2) shows the rim of the remnant in the north and east, fades below threshold in the west, and in the south, the boundary extends beyond the observed region.

Just northwest of the Vela SNR is the x-ray bright SNR Puppis A, located 1-2 kpc distant. The superposition with the Vela remnant is accidental. There is a faint x-ray emitting arc extending from the northern boundary of the Vela remnant through the image of Puppis A. The spectrum of this arc is soft, like that of the Vela remnant, not like that of Puppis A. Therefore, this arc is not a ghost image (a problem in the vicinity of bright sources) of Puppis A but the northwest rim of the Vela Remnant.

X-ray coverage in the south is, unfortunately, almost absent, but there are two *Einstein* fields of interest. One is centered on the Wolf-Rayet star, $\gamma$ Velorum, which appears as a faint x-ray source in the center of the field. This field shows no soft diffuse x-rays which might be associated with the Vela remnant. The boundary is probably northeast of this field. The second isolated field, to the south, however, does show diffuse x-ray emission along the northern edge.

In 1977-78 the remnant was photographed through narrow band filters by Parker, Gull and Kirshner (1979) as part of an emission line survey of the galactic plane. Four 5007 Å [OIII] images from this survey are combined here to cover the region

**X-ray      0.6-4.5 keV**

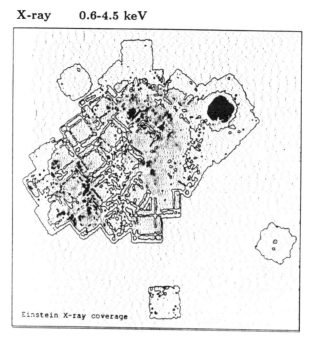

Figure 1. *Einstein* IPC map of hard x-rays from the Vela SNR and from Puppis A. The region surveyed is outlined.

**X-ray      0.1-0.6 keV**

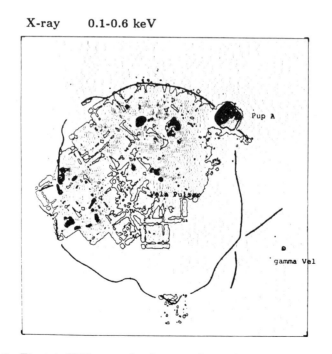

Figure 2. *Einstein* IPC map of soft x-rays from the Vela SNR and from Puppis A. Heavy lines shows the proposed boundary.

around the Vela remnant (Figure 3). Bright filaments are clearly seen along the northern and eastern boundaries, and particularly in the D-shaped central region. This is the extent of the generally-accepted SNR. There are also bright filaments in the south and west. The southernmost filaments coincide with the diffuse x-ray emission in the isolated southern x-ray field suggesting that this is the southern boundary of the remnant. There is no bright [OIII] emission from the x-ray arc extending in front of Puppis A.

[OIII]

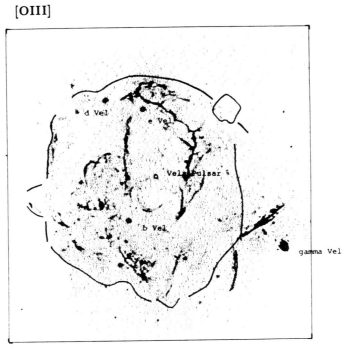

Figure 3. [OIII] map showing the brightest filaments. Heavy line shows the proposed boundary. Four bright early stars and the pulsar are indicated.

There are two bright filaments about 1° from γ Vel. These may indicate the western boundary and are of particular interest since they might result from an interaction between the expanding Vela remnant and the wind from the star γ Vel. Wolf-Rayet stars have strong winds and the distance to γ Vel is about the same as that to the Vela remnant.

A deep Hα Schmidt picture of the western half of the remnant, showing the "D", the filaments by γ Vel, and also diffuse emission was obtained by Elliot, Goudis, and Meaburn (1976). Smith (1978), considering *Ariel* V observations of the Vela pulsar, suggested a 7° diameter boundary of the supernova remnant extending to the filaments east of γ Vel. The soft x-ray imager on *EXOSAT* was used to search for x-rays from those filaments but the result (Smith and Zimmerman, 1985) was null. No x-rays were detected from the proposed western edge. *EXOSAT*, however, was not sensitive enough to see the faint northwest arc in front of Puppis A which is clear in the *Einstein* observation. It is possible that x-ray emission from the proposed western boundary was also too faint for *EXOSAT* detection.

The infrared data (Figure 4) are complex. Bright regions in the galactic plane obscure information from the eastern and northern rim of the Vela remnant. Puppis A, however is bright in both infrared and x-ray wavebands and the morphology in both bands is quite similar. This is not the case for the Vela remnant. There is no infrared emission, for example, from the x-ray bright regions in the east.

**Infrared Ratio     (60 microns)/(100 microns)**

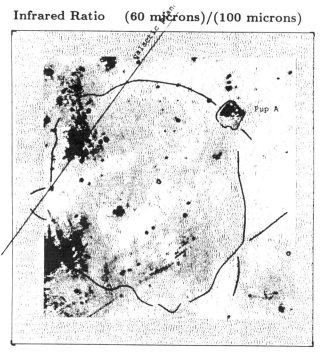

Figure 4. Infrared emission from the region. The heavy line shows the proposed boundary.

There is some infrared structure which seems to follow the northern boundary of the remnant and the [OIII] filamentary structure extending north-south in the center of the remnant. Figure 4 shows not the surface brightness, but the ratio of 60 $\mu$m to 100 $\mu$m surface brightness. This is an indication of warm dust and has been found to trace the morphology of some other remnants. Although the assignment of features to the SNR shell is not straightforward, we have identified an infrared feature as a possible western boundary. There are also the two bright [OIII] filaments outside this feature.

Bignami and Caraveo (1988) have speculated recently that the Vela Pulsar may not be connected with the Vela Remnant because of its location $\sim 1.4°$ from the center of the old boundary. The new boundary is extended to the south and west giving the pulsar a more central location and invalidating part of their argument. The x-ray morphology of the Vela remnant is similar to that of Puppis A, brightest close to the Galactic plane and fading as distance from the plane increases.

The Vela Remnant appears to be the result of a shock propagating into a non-uniform circumstellar medium. The x-ray brightest regions are not on the rim but in the interior. These are probably "warm" gas evaporated from denser clouds associated with the optical filaments. Most filaments surround or are coincident with the brighter x-ray regions. A spectral analysis by Kahn *et al.* (1985) shows that the x-rays can be attributed to thermal radiation from hot gas which, although varying in density and temperature within the remnant, is in approximate pressure equilibrium.

The remnant is, like Puppis A, brightest where material is densest. The northeast boundary is well-determined and the bright clumps are in this half of the remnant, which is in the galactic plane. Away from the plane, towards the southwest, the x-ray emission fades to invisibility or to a very faint shock front. An early rocket flight (Seward *et al.* 1971) surveyed the southwest region and showed evidence for some very soft x-ray emission. However, nothing above background was detected by *EXOSAT*. Knowledge of the southwest x-ray emission must await the next satellite-borne x-ray telescope.

We thank Ted Gull for the generous loan of pictures from the narrow-band optical survey and for discussions concerning the significance of the optical filaments. We thank Jack Hughes for the computer program used to make the illustrations. This work was supported by NASA contract NAS8-30751 and by contract JPL957636, funding obtained from the IRAS data-bank program.

## REFERENCES

Bignami, G.F., and Caraveo, P.A., 1988, *Ap. J.(Letters)*, **325**, L5.

Elliot, K.H., Goudis, C., and Meaburn, J., 1976 *M.N.R.A.S.* **175**, 605.

Gorenstein, P., Harnden, F.R. Jr., and Tucker, W.H., 1974 *Ap.J.* **192**, 661.

Kahn, S.M., Gorenstein, P., Harnden, F.R.,Jr., and Seward, F.D., 1985, *Astrophys. J.* **299**, 821.

Milne, D.K., and Manchester, R.N., 1986 *Astron. Astrophys.* **167**, 117.

Moore, W.E., and Garmire, G.P., 1975, *Ap.J.* **199**, 680.

Parker, R.A.R., Gull, T. R., and Kirshner, R.P., 1979, *NASA SP-434* U.S. Government Printing Office.

Seward, F.D., Burginyon, G.A., Grader, R.J., Hill, R.W., Palmieri, T.M., Stoering, J.P., 1971, *Ap.J.*, **169** 515.

Smith, A., 1978 *M.N.R.A.S.* **182**, 39P.

Smith, A., and Zimmerman, H.U. 1985, *Space Sci. Rev.* **40**, 487.

# LARGE-SCALE STRUCTURES OF THE SOFT X-RAY BACKGROUND

G. Micela, S. Sciortino, G.S. Vaiana

Osservatorio Astronomico di Palermo

F.R. Harnden, Jr.

Harvard-Smithsonian Center for Astrophysics

R. Rosner

Department of Astronomy and Astrophysics
and E. Fermi Institute, The University of Chicago

The Imaging Proportional Counter (IPC) on board the *Einstein* Observatory has measured the soft (0.15-3.5) keV x-ray background in ~4000 directions unevenly spaced on the sky. In the present study we have rejected those exposures with a) possible contamination due to solar x-ray emission observed by the telescope as fluorescent Earth atmospheric emission and b) with exposure time less than 500 sec to reduce the effects of poor count statistics. The number of IPC exposures which we have used in the present study is ~2000. Further work is in progress to improve the screening procedure to remove those exposures whose background measurements could be contaminated by strong (diffuse) x-ray sources. However we expect to have already removed a major fraction of significantly contaminated exposures from the present analysis.

In figure 1a we show the background density (BD) distribution (BD is in units of $10^4$ sec$^{-1}$ arcmin$^{-2}$), and the corresponding quartile subdivision. Data treated as outliers with respect to the observed distribution are also indicated. In figure 1b we show the distribution obtained from Monte Carlo simulations. These simulations take into account the Poisson statistics of counts and the observation time of each exposure. They also use the median background density to compute expected number of background counts in each field. A comparison of the observed and predicted distributions shows that the observed spread cannot be explained only by count statistics. We have found that spread in residual fluorescent emission and in particle flux contributes no more than 25% of the observed variance.

In figure 2 we show the sky distributions of the exposures with BD less than the median value (figure 2a) and of those with BD greater than the median value (figure 2b). Two large-scale structures are clearly present:

a) A structure in the north hemisphere (approximately between 0° and 40° in longitude and between 30° and 60° in latitude), spatially coincident with the North Polar Spur (NPS). A similar structure has been observed at radio wavelength.

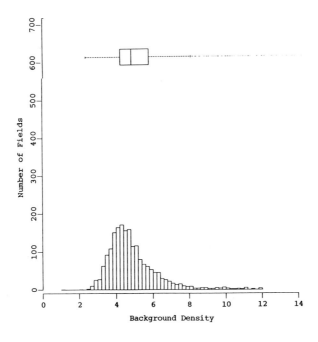

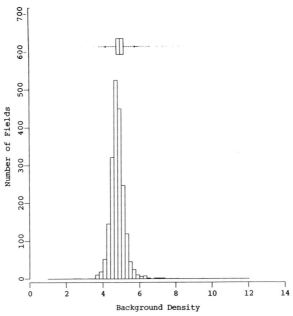

Figure 1. (a) Observed background density distributions; (b) Simulated background density (BD) distribution based on hypothesis of Poisson statistic of counts. Corresponding quartile subdivisions are also indicated.

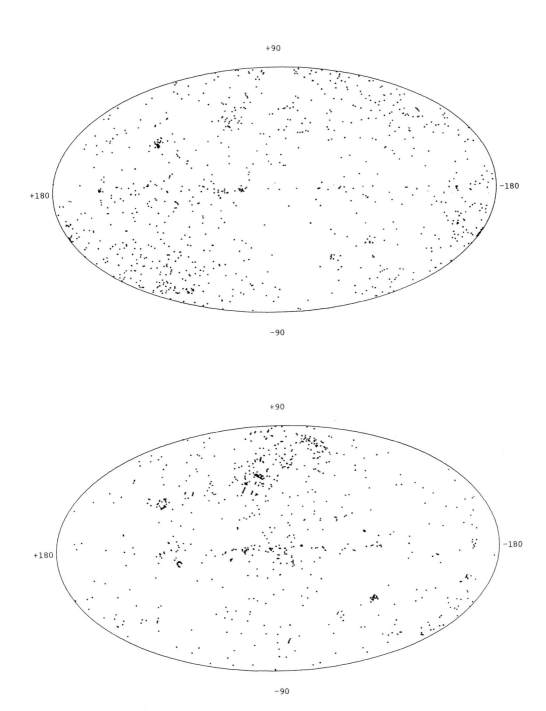

Figure 2. (a-*top panel*) Sky Distributions of IPC pointings with BD below the median value, and (b-*bottom panel*) above the median value.

b) A structure on the Galactic plane toward the Galactic Center identifiable with
a Soft Galactic Ridge.

Previous evidence of these structures have been reported at somewhat different
x-ray energies: the Galactic ridge has been seen with HEAO-1 (Worrall *et al.*, 1982)
in the (2-10) keV energy band, and with *TENMA* (Koyama *et al.*, 1986) in the (2-
10) keV energy band. The "NPS" structure has been reported by McCammon *et
al.* (1983) in the (0.13-0.284) keV energy band. The "NPS" is the softest structure
we observe in the IPC data, with a substantial fraction of energy in the (0.15-0.8)
keV energy band.

We note that the Ridge structure is completely absent in the soft band (0.15-0.8)
keV, while the NPS is present both in the soft and in the hard (0.8-3.5) keV band.

The moderate energy resolution of the IPC ($\Delta E/E \approx 1$ at 1 keV) allows us to
study the gross features of the background spectrum. The (0.15-3.5) keV band is
divided in 9 Pulse Independent (PI) channels.

We have subdivided the sky in four regions, where the x-ray background emission
shows different median intensities and different shapes of median spectra. The
selected regions are indicated in figure 3. The emission has the lowest intensity in
the region 4c. We have assumed the median x-ray background measured in this
region as the minimal background emission level present in the entire sky and have
evaluated the spectrum of excess in the other regions as the difference between the
median spectrum of each region and the template median spectrum of region 4c.

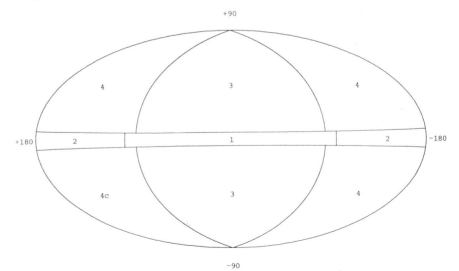

Figure 3. Adopted sky subdivision. Region 4c is the region with the lowest
background level.

Figure 4 shows the median spectrum of the region 4c with errors estimated via bootstrap simulations. The plotted spectrum includes substantial contribution of particles, mainly in PI channels 2-5, and of calibration source, mainly in PI channels 9-10. A detailed analysis of these contributions is in progress, a preliminary result indicate a flattening of the PI spectrum of background when these contributions are subtracted. Figures 5a-c show the excess spectra respectively for region 1 (which includes the Galactic Ridge), region 2 (which includes the anticenter region) and region 3 (which includes the NPS). It is evident that the excess in the ridge region is only in the hard part of spectrum, while the excess in the NPS is much softer, indicating a different origin of the spectral excesses for these two regions.

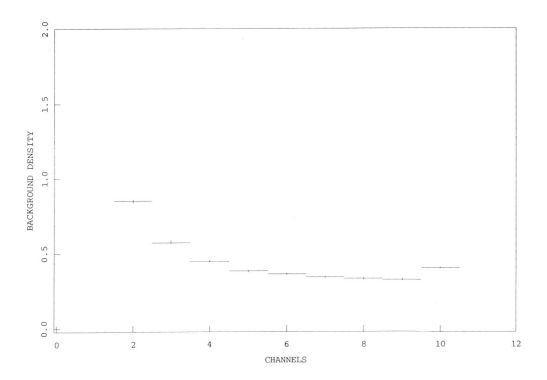

Figure 4. Median template spectrum obtained from data of region 4c.

## ACKNOWLEDGEMENTS

We would like to thank the CNR/Italy and MPI/Italy (GM, SS, GSV), the Smithsonian Institution (RR), and NASA (FRH, GSV) for partial support of this work.

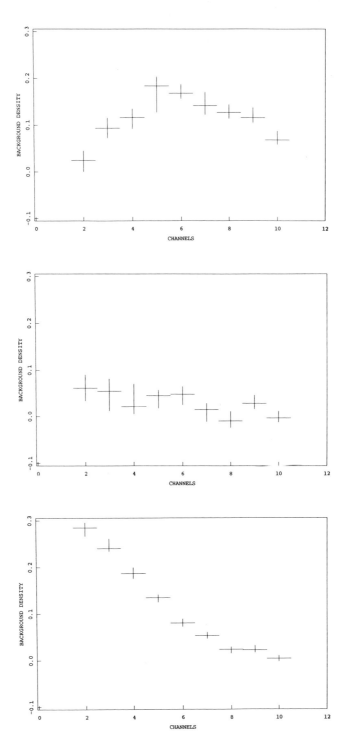

Figure 5. Excess spectra with respect template median spectrum of region 4c for: (5a) region 1 (which includes the Galactic Ridge); (5b) region 2 (which includes the anticenter region); (5c) region 3 (which includes the NPS).

# REFERENCES

Koyama, K., Makishima, K., Tanaka, Y., Tsunemi, H., 1986, *Publ. Astron. Soc. Japan*, **338**, 121.

McCammon, D., Burrows, D.N., Sanders, W.T., and Kraushaar, W.L., 1983, *Ap.J.*, **269**, 107.

Worrall, D.M., Marshall, F.E., Boldt, E.A., Swank, J.H., 1982, *Ap. J.*, **255**, 111.

# A REVIEW OF RESULTS OBTAINED WITH THE MPC TIME INTERVAL PROCESSOR

R.F. Elsner and M. C. Weisskopf

NASA Marshall Space Flight Center

## 1. INTRODUCTION

The Time Interval Processor (*TIP*) was part of the Monitor Proportional Counter (*MPC*) of the *Einstein Observatory*. The *MPC* was an *UHURU* class argon-gas-filled proportional counter. The *TIP* provided high time resolution data, but no photon energy information, over the entire 1.1–22 keV bandwidth by measuring the time between events. The time intervals were accurate to one microsecond or 1.6 percent, whichever was greater, and so this depended on the counting rate. The *TIP* was also severely telemetry limited and utilized a buffer memory to attempt to cope with high counting rates. When the memory filled, no data were recorded until all the stored information was read out. This resulted in 2.56 second gaps whenever the counting rate exceeded approximately 30 counts per second, unless the crystal spectrometer was at the focus of the telescope when the corresponding numbers were 0.852 seconds and approximately 100 counts per second, respectively. The frequency of these gaps increased as the counting rate increased. The *MPC* and the *TIP* are described in detail in a number of publications including Gaillardetz *et al.* (1978), Grindlay *et al.* (1980), and Weisskopf *et al.* (1981).

Although the *TIP* suffered from the drawbacks described above, and the *MPC* received the lowest priority relative to the other instruments, many interesting results were obtained. These are highlighted in what follows.

## 2. A0538−66

A0538-66 is a recurrent transient x-ray source in the Large Magellanic Cloud. Both the x-ray source and its massive Be star optical counterpart undergo outbursts at intervals which are multiples of 16.7 days (White and Carpenter 1978; Johnston *et al.* 1979; Skinner 1980; Skinner *et al.* 1980; Charles *et al.* 1983). The outbursts are typically brief, normally lasting less than 1 day, although some may last a significant fraction of the 16.7 day cycle. The source also exhibits extended quiescent periods, lasting months to years, during which no outbursts are seen. The peak x-ray luminosity during outburst approaches $10^{39}$ erg s$^{-1}$. Between outbursts and during quiescent periods, only upper limits to the x-ray luminosity, $\leq 2 \times 10^4$ the peak luminosity during outbursts (see, *e.g.*, Stella *et al.* 1986), have been obtained. A0538-66 has been in an extended quiescent period since 1983, although recent optical observations (Thompson 1987) indicate that a period of activity may have begun near the end of 1987.

Skinner *et al.* (1982) discovered in *TIP* data a persistent periodicity at a mean heliocentric period of 69.2126 ms in one of two *Einstein* observations of A0538-66 carried out as part of the *Einstein Guest Observer Program*. Application of standard pulse arrival time techniques clearly showed that the period was changing throughout the approximately 4000 s observation at a rate of $(5.01 \pm 0.86) \times 10^{-10}$ s s$^{-1}$. No pulsations were detected in the second *Einstein* observation, although they would have been if the pulse fraction and shape had been the same as during the more intense first observation.

Skinner *et al.* concluded that a changing Doppler shift due to orbital motion of the x-ray source about its companion star accounted for most of the observed period changes. From the mass and radius of the companion deduced from optical data taken during quiescent periods (Charles *et al.* 1983) and the laws of orbital mechanics, assuming an orbital period of 16.7 days, Skinner *et al.* also concluded that the neutron star and its companion were essentially in contact during the outburst showing pulsations. In addition, they found that the measured change in pulse period required an eccentricity $e \geq 0.4$. If one assumed that the two objects were indeed in contact throughout most of the outburst, then even higher values of $e \geq 0.6$ were required. This picture implied that the outbursts were due to mass transfer near periastron. Finally, using the theory of torques exerted on an accreting rotating magnetized neutron star (see, *e.g.*, Ghosh and Lamb 1979), Skinner *et al.* established an upper limit to the surface magnetic field of the neutron star given by $B \leq 10^{11}$ gauss.

These conclusions were based on only 4000 s of data. As remarked by Skinner *et al.*, observations of pulsations over longer periods of time would yield a great deal of information about this unique and scientifically important system. The existence of the 69 ms pulsations, although strongly present in the *TIP* data, are not yet confirmed due to the extended quiescent period begun in 1983 and lasting throughout the lifetime of the *EXOSAT* satellite. If an active period has recently begun (Thompson 1987), and if 69 ms pulsations are detected, then perhaps the *GINGA* satellite or other future missions will have the opportunity to study the dynamics of the A0538-66 system in great detail.

## 3. CYGNUS X–3

Cygnus X–3 is an unusual object that has been observed at virtually all wavelengths of the electromagnetic spectrum (see Cordova 1986 and Bonnet-Bidaud and Chardin 1988 for reviews of its many unique properties). Its x-ray flux is modulated at a period of 4.8 hours, which is assumed to be an orbital period in most of the many models for this system. The evolution of this 4.8 hour periodicity may hold the key to understanding many of the unique properties of this object.

In order to investigate earlier reports (Manzo *et al.* 1978, Lamb *et al.* 1979) that the 4.8 hour periodicity was increasing, Elsner *et al.* (1980) used *TIP* data to extend the history of measurements of the 4.8 hour period. They then were able to confirm the existence of an increase in period and to provide a more accurate measurement of the rate of change of the period. Other investigators (see, *e.g.*, Molnar 1988, van der Klis and Bonnet-Bidaud 1988 and references therein) have continued to extend the time history of the 4.8 hour periodicity.

Elsner *et al.* also suggested that the asymmetric average 4.8 hour x-ray light curve exhibited by Cygnus X−3 might arise as a result of an eccentric binary orbit. They pointed out that if this was so, then it was possible that the observed increase in period might be due to apsidal motion rather than a real increase in orbital period. If this were true, then eventually the rate of increase of the 4.8 hour period should slow down and the sign of the change in period should reverse. Elsner *et al.* examined the constraints on the nature of the companion star imposed by an apsidal motion model. They also discussed models in which such apsidal motion would lead to changes in the average shape of the x-ray light curve. These models were extended by Ghosh *et al.* (1981), who investigated how and to what degree the light curve would change due to apsidal motion.

These results, obtained as part of the program of *TIP* data analysis, have recently become topical again. The observed lack of change in the average x-ray light curve of Cygnus X−3 over a period of 15 years has recently been used to argue against apsidal motion as a significant cause of the change in 4.8 period (Molnar 1988). However, the recently updated time history of the 4.8 hour period provides evidence that the rate of period increase may have decreased (van der Klis and Bonnet-Bidaud 1988). If interpreted as due to apsidal motion, the change in the period derivative is consistent with an eccentricity of $e \simeq 0.25 \pm 0.01$ and an apsidal period of roughly 27 years (van der Klis and Bonnet-Bidaud 1988). While the specific models discussed by Ghosh *et al.* appear to be ruled out by the lack of changes in the x-ray light curve, apsidal motion remains a viable candidate for the cause of changes in the 4.8 hour period of Cygnus X−3. Of course, it should not be forgotten that real changes in the orbital period may also occur due to the transfer and perhaps loss of mass and angular momentum in the system (van der Klis and Bonnet-Bidaud 1988). Based on the recent results given in van der Klis and Bonnet-Bidaud (1988), the apsidal motion model first proposed by Elsner *et al.* should be either confirmed or rejected by the end of this century. If confirmed, then significant constraints will be placed on the nature of the unseen low mass companion to the x-ray emitting compact object.

## 4. LMC X−4 AND THE LONG TERM VARIABILITY OF FAST ROTATORS

LMC X−4 is an x-ray binary in the Large Magellanic Cloud with a 1.4 day orbital period (Chevalier and Ilovaisky 1977, Li *et al.* 1978), a massive OB companion to the neutron star (Sanduleak and Philip 1977), and a 30.5 day variation in its x-ray intensity (Lang *et al.* 1981). *SAS-3* observations led to the discovery of 13.5 s pulsations during flaring episodes (Kelly *et al.* 1983) and to the determination of the orbital elements for the binary system.

Naranan *et al.* (1985) analyzed *TIP* data taken when LMC X−4 was in the *MPC* field of view. They confirmed the existence of the 13.5 s pulsations and updated the spin history of the neutron star, obtaining a marginal determination of the period derivative. Using either this value for the period derivative or the upper limit established from the *SAS-3* data alone, they showed that LMC X−4 is most likely a fast rotator (Ghosh and Lamb 1979; Elsner and Lamb 1977) in the sense that the Keplerian angular velocity at the inner edge of the disk at the magnetospheric boundary is roughly comparable to the rotational angular velocity of the neutron star. This conclusion in turn led to an estimate B $\simeq 10^{13}$ gauss for the strength of the magnetic field at the surface of the neutron star.

Naranan *et al.* went on to examine the long term variability of the fast rotators LMC X−4 and Her X−1, and the intermediate to fast rotator SMC X−1. Using methods previously applied to the long term variability of Cygnus X−1 (Gies and Bolton 1984) and SS 433 (Hut and van den Heuvel 1981), they showed that the expected precession periods of the companions to these x-ray sources were significantly longer than the observed timescales (30–60 days) of the long term variability for these three binary systems. This conclusion was especially firm for those systems, LMC X−4 and SMC X−1, with massive companions. Thus, one model (Roberts 1974) for the long term variability in these systems, which invoked the precession of a tilted accretion disk slaved to the precession of the companion star, could be ruled out.

Finally, Naranan *et al.* briefly speculated that there might be a relationship between the 1–2 month periodic or quasi-periodic intensity variations exhibited by LMC X−4, Her X−1, and SMC X−1 and the fact that all three x-ray sources appear to be intermediate to fast rotators. If real, this relationship suggests that physical processes near the inner edge of the disk at the magnetospheric boundary might be responsible for the observed long term intensity variations. A review of the evidence that this is so in the case of the 35 day cycle of Her X−1 appears in Priedhorsky and Holt (1987). Under these conditions, misalignment of the neutron star magnetic and rotation axes, expected since x-ray pulses are observed, coupled to neutron star precession (Lamb *et al.* 1975; Baym *et al.* 1976) might provide the clock underlying the observed long term variability. We note that Trümper *et al.* (1986) has reported evidence for such precession in Her X−1 based on large changes in pulse shape between the main-on and short-on states. However, Soong, Gruber, and Rothschild (1987) report a significant change in pulse shape during a brief 20 hour period, which they claim cannot be explained by precession with a 35 day period. Thus, the presence of neutron star precession in these x-ray sources remains an open question.

## 5. GX 5−1 AND QUASI-PERIODIC OSCILLATIONS

In 1985 quasi-periodic oscillations (QPO) at frequencies in the range 20–50 Hz were discovered in *EXOSAT* data taken with the bright 'Galactic bulge' x-ray source GX 5−1 in the field of view of the on-board proportional counters (van der Klis *et al.* 1985). This discovery generated great interest and excitement because of the lack of previous diagnostic tools available for the study of 'Galactic bulge' and other low mass x-ray binaries, the relationship between these systems and x-ray burst sources, and the existence of evolutionary scenarios suggesting that such systems are the progenitors of millisecond pulsars. Since the original discovery, QPO have been found in a number of low mass systems exhibiting a wide variety of temporal and spectral behaviors. Recent reviews of the observations and of present theoretical understanding appear in Lewin *et al.* (1988) and Lamb (1988), respectively.

Immediately after the announcement of the discovery of QPO in the x-ray flux from GX 5−1, Elsner *et al.* (1986) looked at previously unanalyzed *TIP* data taken with GX 5−1 in the field of view of the MPC. Analysis of the average power spectrum using the techniques discussed by Leahy *et al.* (1983a) revealed the unambiguous presence of a broad peak at ∼ 16 Hz as well as a feature rising at lower frequencies. This work confirmed the presence of QPO and the so-called low frequency noise in the average power spectrum from GX 5−1. QPO had also been discovered in *EXOSAT* data taken with Cygnus X−2 in the field of view of the on-board proportional counters (Hasinger *et al.* 1986). Analysis of the average power spectrum using *TIP* data from Cygnus X−2 revealed the presence of low frequency (∼ 2–10 Hz) structure (Elsner *et al.* 1986). Because of reduced sensitivity and the probable mixing of different intensity and spectral states, it was not possible to unambiguously determine whether this structure resulted from low frequency noise, QPO, or both. Although QPO were soon discovered in *EXOSAT* data from a number of sources, confirmation of their existence in data taken from GX 5−1 with a different instrument inspired confidence that the original discovery (van der Klis *et al.* 1985) represented a real effect rather than an instrumental phenomenon.

Lamb (1986) and Lamb *et al.* (1985) pointed out that a number of physical models for QPO behavior could be represented mathematically by shot noise models consisting of a steady component plus a sum of randomly occurring oscillating shots. Elsner *et al.* (1986) recalculated the power spectrum in order to include the effects of the binning of the data. The analysis of *TIP* data from GX 5−1 then led naturally into an exploration of various extensions of the simplest shot noise model for QPO, including the effects of coherence (Elsner *et al.* 1987) and of decay of the amplitude of oscillation (Shibazaki *et al.* 1987). In addition, analytical forms were calculated for the cross-correlation functions and cross-spectra (Shibazaki *et al.* 1988) and for the third moments (Elsner *et al.* 1988) expected for QPO shot noise models. The time lags expected as a function of energy under conditions of interest for QPO sources were also studied using Monte-Carlo simulations (Bussard *et al.* 1988; Bussard *et al.* 1989). Finally, studies of the second order auto-correlation functions and corresponding bi-spectra expected for QPO shot noise models are presently underway in order to provide a new tool for determining and studying the shapes of the shots (Elsner and Bussard 1989).

## 5. PULSAR SPIN HISTORIES AND PULSE PROFILES

Analysis of *TIP* data from accretion and rotation powered pulsars has contributed to our knowledge of the spin histories of these objects as well as provided accurate measurements of their 1–20 keV pulse profiles. Pulse arrival time analysis of *TIP* data has also contributed to improved knowledge of the orbital elements for certain accretion powered pulsars. Determinations of the spin histories of accretion powered pulsing x-ray sources are important for tests of accretion torque theories (see Ghosh and Lamb 1979, Lamb 1985; Henrichs 1982; Priedhosky 1986; White and Stella 1988; and references therein). In addition, studies of the response of the neutron star spin to external torques for both accretion and rotation powered pulsars provides important clues to the internal structure of neutron stars (see, *e.g.*, Lamb 1985). Determination of x-ray pulse profiles are important as in principle they can constrain models for x-ray emission mechanisms (see, *e.g.*, White *et al.* 1983; Harding *et al.* 1984; Mészáros 1984). In this last regard, the usefulness of the *TIP* data is limited because, even though it provided high time resolution, the *TIP* provided no information on the energy dependence of the pulse profile over its 1–20 keV energy band. However, the high time resolution did prove useful for studies of pulse shape variations. The studies described above depend on accurate knowledge of the orbital elements for binary sources, which can only be obtained from pulse arrival time analysis of high time resolution data.

As part of the *Einstein Guest Observer Program*, Boynton, Deeter and collaborators included *TIP* data in their timing studies of Her X−1. Their work provided an improved ephemeris for the HZ Her/Her X−1 binary system (Deeter *et al.* 1981). In addition, their study of pulse period fluctuations in Her X−1, together with a similar study for the Crab pulsar not including *TIP* data, showed that these neutron stars behaved essentially as rigid bodies, thus providing important new constraints on the neutron star structure for these sources (Boynton and Deeter 1979; Boynton 1981; Deeter 1981).

In addition to the timing studies of A0538−66, LMC X−4, and Her X−1 discussed above, the *TIP* data provided useful updates to the spin histories and precise measurments of the 1–20 keV pulse profiles for the accretion powered x-ray pulsars 4U1626-67 (Elsner *et al.* 1983), GX 1+4 (Elsner *et al.* 1985), SMC X−1 (Darbro *et al.* 1981), and X−Per (Weisskopf *et al.* 1984). Although the previous spin history of 4U1626-67 had been consistent with spin-up at a constant rate, Elsner *et al.* (1983) found evidence for a spin-down episode in the *TIP* data taken from this source. They also found that the 1–20 keV 7.7 s pulse shape varied significantly on a time scale of minutes. These shape variations correlated in a complex way with the large intensity variations observed from this source.

Elsner *et al.* found variations in the high frequency structure of the 2 min 1–20 keV pulse profile of the slow rotator GX 1+4 (Elsner *et al.* 1985). They speculated that these variations might be due to plasma sheets, filaments, and blobs penetrating deep into the magnetosphere away from the magnetic poles (see Elsner and Lamb (1976) for a discussion of why this may happen in the magnetospheres of slow rotators but not fast ones). Thus future sensitive x-ray observations of the pulse to pulse variability of GX 1+4 may provide a valuable probe of the magnetosphere of this neutron star. Although period fluctuations were present, the spin history of GX 1+4 through 1980, including the *TIP* data, was consistent with secular spin-up on a time scale of approximately 40 years, the shortest such time scale for accretion powered pulsars. However, recent *GINGA* results (Makishima *et al.* 1988) indicate that GX 1+4 has been spinning down during an extended low first noticed in 1983 (Hall and Davelaar 1983).

Weisskopf *et al.* (1984) examined the spin history of X−Per including period measurements obtained with *TIP* data. While the period history of X−Per exhibited significant scatter, they found no evidence for a secular spin-up. They established a lower limit to the period derivative consistent with the compact object being either a neutron star or a white dwarf. In view of this result, the low x-ray luminosity, $\sim 5 \times 10^{33}$ erg s$^{-1}$, and long pulse period, $\sim 835$ s, they concluded that the nature of the compact object in this system remained unknown.

The imaging instruments on board the *Einstein Observatory* discovered two rotation powered fast pulsars (Seward and Harnden 1982; Seward *et al.* 1984). For the 150 ms pulsar in the supernova remnant MSH 15-52 Seward and Harnden 1982), *TIP* data were used to obtain more precise period measurements and to refine its spin history (Weisskopf *et al.* 1983a). For both the 150 ms pulsar and the 50 ms pulsar discovered in SNR 0540-693 in the Large Magellanic Cloud (Seward *et al.* 1984), *TIP* data were used to determine the pulse profile and to measure the strength of the pulsed component at energies above those accessible to the imaging instruments (Seward *et al.* 1985).

## 6. MARCH 5 $\gamma$–RAY BURST

The *TIP* was in operation on 5, March 1979 when an intense and unusual $\gamma$-ray burst took place. The unusual intensity and spectrum of this now famous event permitted its chance detection through secondaries. The high time resolution facilitated an accurate measurement of the onset time which first appeared in Weisskopf *et al.* (1980). This unique data set was also searched, in vain, for evidence of neutron star vibration (Weisskopf *et al.* 1981). However, Cline *et al.* (1982) combined the

*TIP* measurement of the onset time with data from the interplanetary gamma-ray burst network to produce a 0.1 square arc-minute error box identifying the event with the supernova remnant N49 in the Large Magellanic Cloud.

## 6. ANALYSIS TECHNIQUES

The high time resolution provided by the *TIP* and *MPC*, together with the relatively large data base including essentially all the brighter galactic x-ray sources, meant that a serious effort could be expended on analyzing these data for evidence of periodic and aperiodic time variability. This in turn led to the development and refinement of a number of analysis techniques, including accounting for statistical fluctuations, finite length data sets, binning of data, *etc.* Leahy *et al.* (1983a) gave the first complete exposition on both power spectrum analysis and epoch folding for data of this type, together with a discussion of search strategies and a comparison of the realm of applicability for both techniques. Leahy *et al.* applied their work to a search for periodic pulsations in the x-ray emission from four globular cluster sources, but found only upper limits. This was not too surprising since those neutron stars that undergo nuclear flashes are expected to have magnetic fields too weak to funnel accreting matter. Despite this uncontroversial result, this paper is one of the most often cited of the *TIP* papers, testifying to the utility of the section on analysis techniques.

In a subsequent paper, Leahy *et al.* (1983b) studied the Rayleigh test and compared it to epoch folding. They concluded that the former is more sensitive for searches for the broad, relatively smooth pulses characteristic of accreting pulsing x-ray sources, but that it is not as useful for searches for the narrow pulses characteristic of radio pulsars. Both the Rayleigh test and power spectrum analysis were used by Weisskopf *et al.* (1983b) in a study of the time variability of the x-ray emission from LMC X−3 as observed with the *TIP*. These techniques were also utilized by Grindlay *et al.* (1984) for setting low upper limits to the presence of pulsations in *TIP* data from SS 433.

Finally, a number of techniques for analyzing the characteristics of aperiodic time variability have also been developed and extended. A summary of these studies appears at the end of the section on GX 5−1 and QPO sources.

## REFERENCES

*Three asterisks (\*\*\*) at the end of a citation indicate that, at least in part, involved or was derived from analysis of TIP data.*

Baym, G., Lamb, D. Q., and Lamb, F. K. 1976, *Ap. J.*, **208**, 829.
Bonnet-Bidaud, J. M., and Chardin, G. 1988, *Phys. Reports*, in press.
Boynton, P. E. 1981, in *IAU Symposium 95, Pulsars*, ed. W. Sieber and R. Wielebinski (Dordrecht: Reidel), pp. 279–290.

Boynton, P. E., and Deeter, J. E. 1979, in *Compact Galactic X-ray Sources*, ed. F. K. Lamb and D. Pines (Urbana: UIUC Physics Department), pp. 168–180.

Bussard, R. W., O'Dell, S. L., and Elsner, R. F. 1989, in preparation, ***.

Bussard, R. W., Weisskopf, M. C., Elsner, R. F., and Shibazaki, N. 1988, *Ap. J.*, **327**, 284, ***.

Charles, P. A. *et al.* 1983, *M.N.R.A.S.*, **202**, 657.

Chevalier, C., and Ilovaisky, S. A. 1977, *Astr. Ap.*, **59**, L9.

Cline, T. L., *et al.* 1982, *Ap. J. Letters*, **255**, L45, ***.

Cordova, F. A. 1986, *Los Alamos Science*, Spring issue.

Darbro, W., Ghosh, P., Elsner, R. F., Weisskopf, M. C., Sutherland, P. G., and Grindlay, J. E. 1981, *Ap. J.*, **246**, 231, ***.

Deeter, J. E. 1981, *Ph. D. thesis*, University of Washington, ***.

Deeter, J. E., Boynton, P. E., and Pravdo, S. H. 1981, *Ap. J.*, **247**, 1003, ***.

Elsner, R. F., and Lamb, F. K. 1976, *Nature*, **262**, 356.

Elsner, R. F., and Lamb, F. K. 1977, *Ap. J.*, **215**, 897.

Elsner, R. F., Ghosh, P., Darbro, W., Weisskopf, M. C., Sutherland, P. G., and Grindlay, J. E. 1980, *Ap. J.*, **239**, 335, ***.

Elsner, R. F., Darbro, W., Leahy, D., Weisskopf, M. C., Sutherland, P. G., Kahn, S. M., and Grindlay, J. E. 1983, *Ap. J.*, **266**, 769, ***.

Elsner, R. F., Weisskopf, M. C. Apparao, K. M. V., Darbro, W., Ramsey, B. D., Williaims, A. C., Grindlay, J. E., and Sutherland, P. G. 1985, *Ap. J.*, **297**, 288, ***.

Elsner, R. F., Weisskopf, M. C., Darbro, W., Ramsey, B. D., Williams, A. C., Sutherland, P. G., and Grindlay, J. E. 1986, *Ap. J.*, **308**, 655, ***.

Elsner, R. F., Shibazaki, N., and Weisskopf, M. C. 1987, *Ap. J.*, **320**, 527, ***.

Elsner, R. F., Shibazaki, N., and Weisskopf, M. C. 1988, *Ap. J.*, **327**, 742, ***.

Elsner, R. F., and Bussard, R. W. 1989, in preparation, ***.

Gaillardetz, R., Bjorkholm, P., Mastronardi, R., Vanderhill, M., and Howland, D. 1978, *IEEE Trans. Nucl. Sci.*, **NS-25**, 437, ***.

Ghosh, P., and Lamb, F. K. 1979, *Ap. J.*, **234**, 296.

Ghosh, P., Elsner, R. F., Weisskopf, M. C., and Sutherland, P. G. 1981, *Ap.J.*, **251**, 230, ***.

Gies, D. R., and Bolton, C. T. 1984, *Ap. J. Letters*, **276**, L17.

Grindlay, J. E., *et al.* 1980, *Ap. J. Letters*, **240**, L121, ***.

Grindlay, J. E., Band, D., Seward, F., Leahy, D., Weisskopf, M. C., and Marshall, F. E. 1984, *Ap. J.*, **277**, 286, ***.

Hall, R., and Davelaar, J. 1983, *IAU Circ.*, No. 3872.

Harding, A. K., Mészáros, P., Kirk, J. G., and Galloway, D. J. 1984, *Ap. J.*, **278**, 369.

Hasinger, G., Langmeier, A., Sztajno, M., Trümper, J., Lewin, W. H. G., and White, N. E. 1986, *Nature*, **319**, 469.

Henrichs, H. F. 1982, in *Accretion Driven Stellar X-ray Sources*, ed. W. H. G. Lewin and E. P. J. van den Heuvel (Cambridge: Cambridge University Press), pp. 393–429.

Hut, P., and van den Heuvel, E. P. J. 1981, *Astr. Ap.*, **94**, 327.

Johnston, M. D., Bradt, H. V., Doxsey, R. E., Griffiths, R. E., Schwartz, D. A., and Schwarz, J. 1979, *Ap. J. Letters*, **230**, L11.

Kelly, R. L., Jernigan, J. G., Levine, A., Petro, L. D., and Rappaport, S. A. 1983, *Ap. J.*, **264**, 568.

Lamb, D. Q., Lamb, F. K., Pines, D., and Shaham, J. 1975, *Ap. J. Letters*, **198**, L21.

Lamb, F. K. 1985, in *Japan–U. S. Seminar on Galactic and Extragalactic Compact X-ray Sources*, ed. Y. Tanaka and W. H. G. Lewin (Tokyo: I.S.A.S.), pp. 19–32.

Lamb, F. K. 1986, in *The Evolution of Galactic X-ray Binaries*, ed. J. Trümper, W. H. G. Lewin, and W. Brinkmann (Dordrecht: Reidel), pp. 151–171.

Lamb, F. K. 1988, in *Proc. COSPAR/IAU Symposium, The Physics of Compact Objects: Theory versus Observation*, ed. N. E. White and L. Filipov (Oxford: Pergamon), pp. 421–447.

Lamb, F. K., Shibazaki, N., Alpar, M. A., and Shaham, J. 1985, *Nature*, **317**, 681.

Lamb, R. C., Dower, R. G., and Fickle, R. K. 1979, *Ap. J. Letters*, **229**, L19.

Lang, F. L., *et al.* 1981, *Ap. J. Letters*, **246**, L21.

Leahy, D. A., Darbro, W., Elsner, R. F., Weisskopf, M. C., Sutherland, P. G., Kahn, S., and Grindlay, J. E. 1983a, *Ap. J.*, **266**, 160, ***.

Leahy, D. A., Elsner, R. F., and Weisskopf, M. C. 1983b, *Ap. J.*, **272**, 256, ***.

Lewin, W. H. G., van Paradijs, J., and van der Klis, M. 1988, *Space Sci. Reviews*, **46**, 273.

Li, F., Rappaport, S. A., and Epstein, A. 1978, *Nature*, **271**,, 37.

Makishima, K., *et al.* 1988, *Nature*, **333**, 746.

Manzo, G., Molteni, D., and Robba, N. R. 1978, *Astr. Ap.*, **70**, 317.

Mészáros, P. 1984, *Space Sci. Reviews*, **38**, 325.

Molnar, L. A. 1988, *Ap. J. Letters*, **331**, L25.

Naranan, S., Elsner, R. F., Darbro, W., Hardee, P. E., Ramsey, B. D., Leahy, D. A., Weisskopf, M. C., Williams, A. C., Sutherland, P. G., and Grindlay, J. E. 1985, *Ap. J.*. **290**, 487, ***.

Priedhorsky, W. 1986, *Ap. J. Letters*, **306**, L97.

Priedhorsky, W. C., and Holt, S. S. 1987, *Space Sci. Reviews*, **45**, 291.

Roberts, W. J. 1974, *Ap. J.*, **187**, 575.

Sanduleak, N., and Philip, A. G. D. 1977, *IAU Circ.*, No. 3023.

Seward, F. D., and Harnden, F. R., Jr. 1982, *Ap. J. Letters*, **256**, L45.

Seward, F. D., Harnden, F. R., Jr., and Helfand, D. J. 1984, *Ap. J. Letters*, **287**, L19.

Seward, F. D., Harnden, F. R., Jr., and Elsner, R. F. 1985, in *The Crab Nebula and Related Supernova Remnants*, ed. M. C. Kafatos and R. B. C. Henry (Cambridge: Cambridge University Press), pp. 165–172, ***.

Shibazaki, N., Elsner, R. F., and Weisskopf, M. C. 1987, *Ap. J.*, **322**, 831, ***.

Shibazaki, N., Elsner, R. F., Bussard, R. W., Ebisuzaki, T., and Weisskopf, M. C. 1988, *Ap. J.*, **331**, 247, ***.

Skinner, G. 1980, *Nature*, **288**, 141.

Skinner, G. *et al.* 1980, *Ap. J.*, **240**, 619.

Skinner, G., Bedford, B. K., Elsner, R. F., Leahy, D., Weisskopf, M. C., and Grindlay, J. 1982, *Nature*, **297**, 568, ***.

Soong, Y., Gruber, D. E., and Rothschild, R. E. 1987, *Ap. J. Letters*, **319**, L77.

Stella, L., White, N. E., and Rosner, R. 1986, *Ap. J.*, **308**, 669.

Thompson, I. 1987, *IAU Circ.*, No. 4519.

Trümper, J., Kahabka, P., Ögelman, H., Pietsch, W., and Voges, W. 1986, *Ap. J. Letters*, **300**, L63.

van der Klis, M., Jansen, F., van Paradijs, J., Lewin, W. H. G., van den Heuvel, E. P. J., Trümper, J. E., Sztajno, M. 1985, *Nature*, **316**, 225.

van der Klis, M., and Bonnet-Bidaud, J. M. 1988, *Astr. Ap.*, submitted.

Weisskopf, M. C., Elsner, R., Darbro, W., Ghosh, P., Sutherland, P. G., and Grindlay J. 1980, *NASA-TM-78273*, ***.

Weisskopf, M. C., Elsner, R. F., Sutherland, P. G., and Grindlay, J. E. 1981, *Astrophys. Letters*, **22**, 179, ***.

Weisskopf, M. C., Elsner, R. F., Darbro, W., Leahy, D., Naranan, S., Sutherland, P. G., Grindlay, J. E., Harnden, F. R., Jr., and Seward, F. D. 1983a, *Ap. J.*, **267**, 711, ***.

Weisskopf, M. C., Kahn, S. M., Darbro, W. A., Elsner, R. F., Grindlay, J. E., Naranan, S., Sutherland, P. G., and Williams, A. C. 1983b, *Ap. J. Letters*, **274**, L65, ***.

Weisskopf, M. C., Elsner, R. F., Darbro, W., Naranan, S., Weisskopf, V. J., Williams, A. C., White, N. E., Grindlay, J. E., and Sutherland, P. G. 1984, *Ap. J.*, **278**, 711, ***.

White, N. E., and Carpenter, G. F. 1978, *M.N.R.A.S.*, **183**, 11p.

White, N. E., and Stella, L. 1988, *M.N.R.A.S.*, **231**, 325.

White, N. E., Swank, J. H, and Holt, S. S. 1983, *Ap. J.*, **270**, 711.

# VARIABILITY OF THE X-RAY SOURCES IN M33 AND M31

**G. Peres, F. Reale**

Istituto di Astronomia, Universita di Palermo, Palermo, Italy

**A. Collura**

IAIF-CNR, Palermo, Italy

**G. Fabbiano**

Harvard-Smithsonian Center for Astrophysics, Cambridge, MA

## 1. INTRODUCTION

We report on the variability of x-ray sources in M33 and in M31. This work is part of an on-going systematic search for variability of the most intense compact sources in the nearby galaxies.

We have looked for variability within individual observations, analyzing the sequence of photon arrival times with the Kolmogorov-Smirnov method, the Cramer-Smirnov-Von Mises method and an upgraded version of the $\chi^2$ method (Collura *et al.* 1987, henceforth $\overline{\chi^2}$). We have checked whether the average count rates of different observations of the same source are consistent with the hypothesis of constant emission, using the method of Maccacaro, Garilli and Mereghetti (1987). We have also studied the light curves of the sources found variable. Given the relatively low number of photons for any detection, the light curves consist of data points obtained integrating at least over one continuous section of the observation (HUT). For all the analysis we have adopted the 99.73 % confidence level as threshold for variability.

## 2. M33: AN ECLIPSING BINARY AND AN 'X-RAY ONLY' AGN

The analysis of the variability of the sources in M33 has been completed (Peres *et al.* 1989). Two of the fifteen known x-ray sources of this galaxy, namely M33 X-7 and M33 X-8 (the nucleus), are variable. The light curve of M33 X-7 exhibits a pattern of high and low states, indicative of an eclipsing binary x-ray source. In Figure 1 we show the light curve of the IPC and HRI observations joined together. The IPC and HRI data points have been cross-normalized assuming that the intensity at maximum is the same. Peres *et al.* (1989) discuss this point in more detail. The data suggest a binary period of $\sim 1.79$ days and an eclipse duration of 0.4 days; a period twice the above value is also compatible with the data.

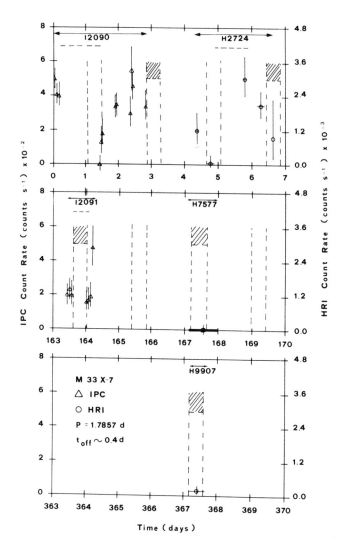

Figure 1. The IPC and HRI light curve of M33 X-7. Triangles and circles mark IPC and HRI data points, respectively. The horizontal bar of each data point indicates its time span. The abscissa is the time from the beginning of the first observation of M33. Solid and dashed horizontal lines mark the time span of each observation and the most relevant gaps within each observing sequence, respectively. The vertical dashed lines represent the recurrent minima, under the assumption that the source varies with a period P=1.7857 d. The cross-hatched minima assume a period twice as long (from Peres *et al.*, 1989).

The nuclear source, M33 X-8, varies only in the softest part of the spectrum (kT $\leq 1.2$ keV). The observations suggest variability over a time scale of $\sim 3000$ s and a rapid rising in less than three days together with variability on a longer time scale. This is further evidence in favor of its description as an 'x-ray only' mini-AGN (Markert and Rallis, 1983; Trinchieri, Fabbiano and Peres 1988).

## 3. M31: TWO BINARIES?

In a first analysis, now approaching completion (Collura, Reale, and Peres, 1989), of the M31 observations we have searched for variability within any single observation. We have studied all the 66 detections with photon counts higher than 60 (our selection threshold) up to a few thousands (for the most intense sources). We have used only the $\overline{\chi^2}$ method which, unlike the other methods, is capable of yielding the time scale and effective amplitude of variability. In this search only two sources have been found to be variable in M31. The significant variations of their light curves, shown in Figure 2 and Figure 3, can be interpreted as evidence of orbital modulations, and taken together with the optical identification by Crampton *et al.* (1984), suggest that the first source is a low mass binary and the second one a compact massive x-ray binary. The observations of M31 span, on the average, a time lapse shorter than those of M33, although their average "live time" is almost the same, and most fields have been observed only once. Detection of variability is, therefore, more difficult than for M33. Collura, Reale, and Peres (1989) determine the amplitude and time scale detectable in all sources given the available data.

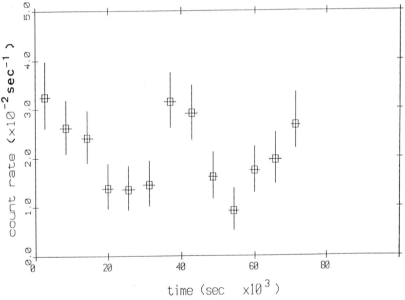

Figure 2. IPC light curve (0.8 - 3.5 keV band) of source 1. The horizontal bars yield the time span, including gaps, over which each count rate was averaged. The vertical error bars account only for statistical error. The count rates are corrected for instrumental effects and background.

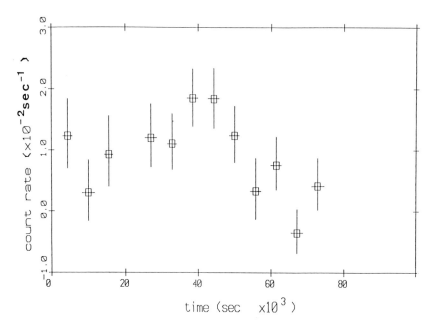

Figure 3. IPC Light curve of source 2 in the 0.2 - 3.5 keV band. Error bars and count rate corrections as in Figure 2.

## 4. AXAF PROSPECTS

The *Einstein* telescope was the first with sufficient sensitivity to detect variability of non-nuclear sources in nearby galaxies. *AXAF* will have a sensitivity higher by two orders of magnitude but also a field of view of 20' × 20', suitable for detailed studies of specific small fields and known sources. For extensive surveys of variable sources in nearby galaxies, long exposure times and a wide field of view are more effective than an increase of sensitivity of two orders of magnitude. In this respect *ROSAT*, having a sensitivity three times larger than *Einstein* and a field of view wider by approximately the same factor, should yield a dramatic improvement in the search for variable sources in nearby galaxies, at the same time preparing the path for detailed studies of specific sources and search for variable sources in outer galaxies, projects more appropriate to *AXAF*.

## REFERENCES

Crampton, D., Cowley, A. P., Hutchings, J. B., Schade, D. J., and van Speybroeck, L. P., 1984, *Ap. J.*, **284**, 663

Collura, A., Maggio, A., Sciortino, S., Serio, S., Vaiana, G. S., and Rosner, R., 1987, *Ap. J.*, **315**, 340

Collura, A., Reale, F., and Peres, G., 1989, *Ap. J.*, in preparation

Harnden, F. R. jr., Fabricant, D. G., Harris, D. E., and Schwarz, J., 1984, *Scientific Specifications of the Data Analysis System for the Einstein Observatory (HEAO-2) Imaging Proportional Counter*, Smithsonian Astrophysical Observatory Special Report n.393

Long, K. S., D'Odorico, S., Charles, P. A. and Dopita, M. A. , 1981, *Ap. J.*, **246**, L61

Maccacaro, T., Garilli, B., and Mereghetti, S., 1987, *Astron. J.*, **93**, 1484

Markert, T. H., and Rallis, A. D., 1983, *Ap. J.*, **275**, 571

Peres, G., Reale, F., Collura, A., and Fabbiano, G., 1989, *Ap. J.*, **336**, 140

Trinchieri, G., Fabbiano, G., and Peres, G., 1988, *Ap. J.*, **325**, 531

# THE SOFTEST EINSTEIN AGN

**F. A. Córdova**
Los Alamos National Laboratory

**J. Kartje**
University of Chicago

**K. O. Mason** and **J. P. D. Mittaz**
Mullard Space Science Laboratory
University College London

## 1. INTRODUCTION

We have undertaken a coarse spectral study to find the softest sources detected with the Imaging Proportional Counter (IPC) on the *Einstein Observatory*. Of the nearly 7700 IPC sources, 226 have color ratios that make them candidate "ultra-soft" sources; of these, 83 have small enough errors that we can say with confidence that they have a spectral component similar to those of the white dwarfs Sirius and HZ 43, nearby stars such as $\alpha$ Cen and Procyon, and typical "polar" cataclysmic variables. By means of catalog searches and ground-based optical and radio observations we have thus far identified 96 of the 226 candidate soft sources; 37 of them are active galactic nuclei (AGN). In the more selective subset of 83 sources, 47 have been identified, 12 of them with AGN. The list of 47 identifications is given in Córdova *et al.* (1989, hereafter Paper 1). For one QSO in our sample, E0132.8–411, we are able to fit the pulse-height data to a power-law model and obtain a best fit for the energy spectral index of $2.2^{+0.6}_{-0.4}$. For the remainder of the AGN in the higher confidence sample we are able to infer on the basis of their x-ray colors that they have a similar spectral component. Two-thirds of the AGN are detected below 0.5 keV only, while the remainder evidence a flatter spectral component in addition to the ultra-soft component.

One of the implications of our findings is that the presence of two very different spectral components in many AGN, one of them very steep, calls into question the procedure of fitting a single "average" spectrum to sources across the IPC band in order to derive global x-ray properties for AGN. In the present study we have surveyed the data base over a wider, softer energy interval (*i.e.*, 0.16 – 3.5 keV) than previous surveys (see below), allowing us to select the very softest x-ray sources. By studying the colors of the sources in *three* broad energy bands we can clearly discern two spectral components, when they are present. The simpler technique of previous surveys, which used only two broadband colors to define a spectral slope, precludes the recognition of dual components in weak sources (see, *e.g.*, Maccacaro *et al.* 1988).

A second implication is that there might be a large population of AGN that are relatively weak hard x-ray emitters and, therefore, difficult to detect at energies above 0.5 keV.

The soft component may be the high energy tail of the "big blue bump", which is often detected in the 0.6 to 0.1 $\mu$m band and thought to be emission from an accretion disk surrounding the central engine. Very soft x-ray all-sky surveys (as will be done with *ROSAT*) should uncover a large number of AGN that have substantial emission in the energy band below 0.5 keV relative to their emission between 0.5 and 1 keV. Using future x-ray telescopes such as *ROSAT*, *AXAF* and *XMM*, we may be able to fit the x-ray spectra of these sources with some precision to models for accretion disks around black holes, yielding estimates of the mass accretion rate and the mass of the central object (*cf.*, Bechtold *et al.* 1987).

## 2. THE SURVEY CRITERIA

Our first criterion for the Ultra-Soft Survey (USS) was that an *Einstein* IPC source have a signal-to-noise ratio $\geq 3$ in the energy interval 0.16 – 3.5 keV. This criterion reduced the data base to $\sim$7700 sources out of $\sim$16,000 IPC "detections"[1]. We then further required for our sample that the sources have a signal-to-noise ratio $\geq 3$ in the softer energy interval 0.16 – 0.8 keV. To select extremely soft sources, we grouped the pulse-height invariant bins into three broad energy bands: A soft energy band (C1 = 0.16 – 0.56 keV), a medium energy band (C2 = 0.56 – 1.08 keV), and a hard energy band (C3 = 1.08 – 3.5 keV). Their ratios (*i.e.*, R1 = medium/low and R2 = high/medium), when plotted against each other, form an *Einstein* color-color diagram. Examples of these diagrams are shown in Paper 1. We required for inclusion in our sample of ultra-soft sources that C1 $\geq$ 0 and R1 < 0.36. The R1 cutoff was chosen after examining the R1 values of the brightest known ultra-soft x-ray sources (see Paper 1). The number of detections satisfying these criteria are 250, of which 14 are multiple detections of the same source, and 10 are regions of probable diffuse emission associated with extended sources.

Many of the remaining 226 sources have extremely large errors in their colors. To obtain a sample of objects whose colors suggest that the objects are soft with a high probability, we further require that

$$NR1 \equiv \frac{C2 + \sigma(C2)}{C1 - \sigma(C1)} \leq 0.5 \qquad (1)$$

where $\sigma$ indicates a one sigma error on the counts in a color band. Eighty-three independent sources satisfy this criterion.

---

[1]The detection of sources is made using the standard *Einstein* IPC processing algorithms *local detect* and *map detect* in the "soft" and "hard" energy bands 0.16 – 0.81 keV and 0.81 – 3.5 keV (see Harnden *et al.* 1984), which are different than the soft, medium, and hard color bands defined presently.

This technique gives reliable, if crude, estimates of the spectra of weak sources since (1) the brightest sources can be fitted to models utilizing their full spectral resolution, and then used as 'calibrators' for the color-color diagram (an example of this is given in Section 4 below), and (2) model spectra can be folded through the response of the IPC and similarly plotted on such a diagram for comparison with the colors of weak sources (a variant of this is depicted in Figure 3 and discussed later). In Paper 1, we give numerous examples of the published spectra of the brightest ultra-soft sources in the USS. These sources have high-gravity stellar atmospheres of a few times $10^4$ K, blackbody temperatures with kT $\leq$ 100 eV, optically-thin line spectral (*i.e.*, thermal plasma) temperatures of $\sim 10^6$ K, or steep power-law spectra with energy spectral indices, $\alpha_E$, > 2.

## 3. IDENTIFICATION OF AGN IN THE SURVEY

A program of optical and radio identification of all 226 sources in the original Survey has begun, using the Wm. Herschel telescope on La Palma, the Anglo-Australian Telescope, and the Very Large Array, and by searching published catalogs for positional coincidences. We have thus far identified 96 of the 226 USS sources. Thirty seven are associated with AGN. Nineteen of these are newly identified, and the remaining eighteen appear in Hewitt and Burbidge's Catalog of Quasars (1987). These AGN are mostly at low redshifts and have apparent visual magnitudes ranging between 17 and 20. Their fluxes between 0.16 and 3.5 keV range from $1 \times 10^{-13}$ to $4 \times 10^{-12}$ ergs cm$^{-2}$ s$^{-1}$. The USS AGN are viewed at equivalent hydrogen column densities through the Galaxy of $N_H < 6 \times 10^{20}$ cm$^{-2}$. A complete list of all the AGN and their optical spectra will be published elsewhere. Paper 1 lists the probable identifications made to date of a subset of the identified USS sources, *i.e.*, those 47 identified sources that satisfy Equation 1. Twelve AGN are among this number.

## 4. SPECTRAL FITTING

The AGN with the most counts in our Survey is the quasar E0132.8–411. It is a very weak x-ray source (0.01 IPC counts s$^{-1}$), but was observed for almost 8 hours. We fit the pulse-height data on this source with a power-law model of the form $F_\nu \propto \nu^{-\alpha_E}$. We fixed $N_H$ at $2.14 \times 10^{20}$ cm$^{-2}$, the value given by Stark *et al.* (1984). We used the value for the gain of the IPC that was appropriate to the time of the observation and the position of the source on the detector (*i.e.*, the gain varies across the face of the detector and this has been calibrated before flight; the source was viewed 30 arcminutes off-axis). The total value of $\chi^2$ is 5.6 for 8 degrees of freedom. The best-fit value for $\alpha_E$ is $2.2^{+0.6}_{-0.4}$. The gain within the pixels that the source occupies is also uncertain because of anode wire-to-wire variations. To obtain a reasonable limit on the spectral slope, therefore, we varied the gain parameter ("BAL" in the notation of Harnden *et al.* 1984) by $\pm$ 1.5. This has the effect of shifting the energy intervals corresponding to the pulse-height bins. Applying a $\chi^2$ test to the fits with the extreme gain values, and adopting the

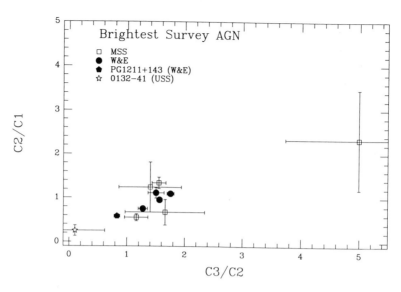

Figure 1. An IPC color-color diagram for a few of the brightest sources detected in the Ultra-Soft Survey (USS), Medium Sensitivity Survey (MSS), and Wilkes and Elvis (1987) Survey (W&E). The softest source shown is the USS quasar E0132.8–411. The energy bands, described by the notation C1, *etc.*, is given in the text.

parameter estimation scheme of Lampton, Margon, and Bowyer (1976), we find that $\alpha_E \geq 1.3$ at the 99% confidence limit. The spectral fittings allow an index as steep as 4, which was the highest value we tested.

## 5. COMPARISON WITH OTHER AGN SURVEYS

The *Einstein* Medium Sensitivity Survey (MSS) of the high-galactic latitude, serendipitous (that is, x-ray selected) sources in the IPC data base reveals that 50% of these sources are active galactic nuclei (Maccacaro *et al.* 1982 and Gioia *et al.* 1984). In a study of the IPC hardness ratios of a sample of *Einstein* serendipitous sources eight times larger than the MSS, Maccacaro *et al.* (1988) find that the distribution of power-law slopes of the AGN has a mean of $1.03^{+.05}_{-.06}$ (energy index) with an intrinsic dispersion of 0.36. By comparing a survey of serendipitous sources detected with the Channel Multiplier Arrays (CMA) on *EXOSAT* with the MSS, and by correlating the sky position of the AGN with the Galactic $N_H$, Giommi *et al.* (1988) infer that the *average* energy spectral index of the AGN detected with *EXOSAT* is steeper than 1. These surveys, as well as studies of small regions of the sky or individual AGN (see, for example, Branduardi-Raymont *et al.* 1985, Branduardi-Raymont 1986, Wilkes and Elvis 1987, Bechtold *et al.* 1987, and references therein) show that the spectra of AGN below 1 keV are considerably steeper than AGN spectra measured at energies between 2 and 10 keV (compare Mushotzky 1984). All of this work indicates that the x-ray spectra of AGN in

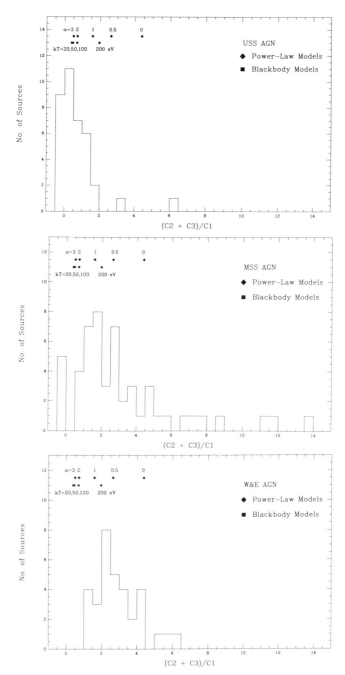

Figure 2. The distribution of AGN in the three surveys as a function of a "hardness" parameter (*i.e.*, the medium energy band plus hard energy band counts divided by the soft energy band counts). The figure illustrates that the USS AGN are much softer, on average, than the other surveys. An AGN with two prominent spectral components will appear harder on this plot than an AGN with a dominant soft spectral component.

general is quite complicated, with multiple spectral components likely.

Our sample can be compared with the MSS[2] and with the spectral survey of the brightest quasars detected with the IPC (Wilkes and Elvis 1987; hereafter, W&E).

The USS quasar E0132.8–411 has a spectrum similar to that of the soft x-ray component of PG1211+143. The spectrum of the latter has been fit with two components, $\alpha_E = 2.2 \pm 0.4$ and a flatter component with $\alpha_E \sim 0.6$ (Bechtold *et al.* 1987). PG1211+143 has a redshift of 0.085, as compared to a redshift of 0.27 for E0132.8–411, and is more than 100 times brighter in soft x-ray emission than E0132.8–411. On the color-color diagram of Figure 1, E0132.8–411 appears much softer than the brightest of the AGN in the MSS and W&E surveys. E0132.8–411 is representative of the majority of the USS AGN. The ones with additional, harder components have similar R1 values, but higher R2 values. The colors of PG1211+143 look harder than those of E0132.8–411 because of the presence of an appreciable hard component. The one-component power-law fit to its spectrum by Wilkes and Elvis (1987) gives $\alpha_E = 1.8^{+0.5}_{-0.4}$.

Figure 2 shows the distribution of (C2+C3)/C1 for the USS, MSS, and W&E AGN. These plots show unambiguously that the USS population is much softer than the AGN in the other surveys. Also shown on these plots are the (C2+C3)/C1 values of blackbody and power-law models, convolved through the response of the IPC. This gives a good indication of the distribution of spectral slopes for the different surveys.

The optical spectrum of E0132.8–411 (see Figure 3) appears typical of many low-redshift quasars. H$\alpha$, H$\beta$ and H$\gamma$ are prominent, as are the narrow forbidden lines of [OIII] 5007 Å and [OIII] 4959 Å. The redshift measured from the [OIII] lines is 0.267. The ratio of the H$\alpha$ to H$\beta$ line flux is consistent with the expected recombination value (Case B) of 2.87, suggesting that there is little or no reddening. This is consistent with the detection of this object as an ultra-soft x-ray source. The continuum slope between 4000Å and 8000Å is consistent with $F_\nu \propto \nu^{-1.1}$.

In Figure 4 the intensities of the USS and MSS AGN are plotted versus redshift. The mean intensity of both samples is 0.025 IPC counts s$^{-1}$ and the mean redshifts are 0.33 and 0.5 for the USS and the MSS, respectively (ignoring the z value of 2.36

---

[2]The USS and MSS differ in that the former includes sources anywhere in the 1° IPC field-of-view, whereas that latter includes sources only in the central 32' × 32' region. The USS includes an additional soft pulse-height bin which makes its low-energy cutoff 0.16 keV, instead of 0.33 keV as it is for the MSS sample. Finally, the USS accepts sources whose signal-to-noise ratio is ≥ 3 in *both* the soft and broad bands, whereas this limit for the MSS is 5 in the broad band alone. Two of the MSS sources are in the USS, and there may be other sources from the Extended MSS (EMSS: Maccacaro *et al.* 1988) also in the USS, but there are no published positions for the EMSS as of this date to compare with the USS positions. There are two additional sources in the MSS which are not in the USS but are also very soft. These missed being in the USS because they do not satisfy the signal-to-noise ratio criteria for both soft and broad energy bands.

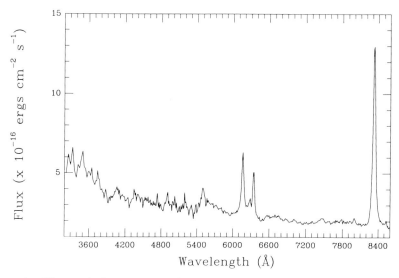

Figure 3. The optical spectrum of E0132.8–411, measured in January 1989 using the 3.9 m AAT telescope. The spectrum was split into two sections by a dichroic beamsplitter. The red section (5500 – 8300Å) was recorded using the Faint Object Red Spectrograph (FORS), while the blue section was recorded using the IPCS detector on the RGO spectrograph. The observation was made using a wide (6 arc second) slit in photometric conditions. A flux calibration standard star was observed immediately afterwards with the same instrumental setup and at a similar airmass. A bright F-type star near the quasar was also observed to calibrate atmospheric absorption features in the red part of the spectrum.

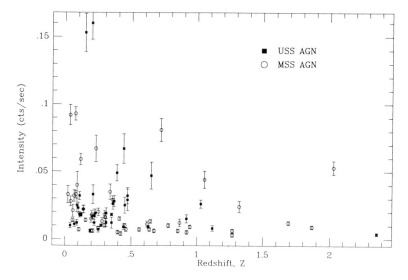

Figure 4. The intensities of the two x-ray selected samples, the USS and MSS AGN, versus their redshifts.

in the USS sample because this is highly uncertain: see Hewitt and Burbidge 1987 for references). Thus the USS samples a closer population than the MSS. An effect of this kind is expected because a soft spectral component will tend to be shifted out of the observable band at higher redshifts.

Finally, in Figure 5 we show the distribution of the three surveys as a function of the values of the equivalent hydrogen column density to the edge of the Galaxy in the appropriate direction (taken from Stark *et al.* 1984, or Elvis, Lockman, and Wilkes 1989). It is clear that the USS sample reflects lower column densities than the MSS and W&E samples.

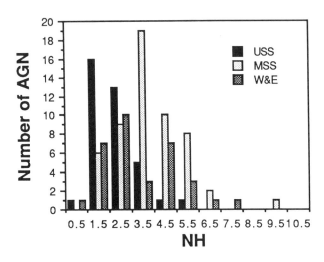

Figure 5. The distribution of the three surveys discussed in the text versus the hydrogen column density in the Galaxy along their lines of sight ($N_H$ in units of $10^{20}$ cm$^{-2}$).

## 6. INTERPRETATION

The USS results indicate that AGN with extremely steep components may be common and, in many cases, these objects evidence an additional, flatter component of approximately equal intensity in the IPC bandpass. This suggests that the adoption of a single "average" power-law to represent the spectrum (as was done by Maccacaro *et al.* (1988) for the extended MSS, will not always give a true representation of the spectrum. In particular, there will be a bias against detecting an ultra-soft component with this kind of analysis. In addition, the dispersion in the average value of the spectral slope measured by Maccacaro *et al.* (1988) will be underestimated because of ignoring the lowest energy channels where the ultra-soft sources will have most of their counts. The presence of two spectral components should not affect the *EXOSAT* results as much as the *Einstein* results because of the softer spectral response of the former telescope. The *EXOSAT* survey result, namely, finding that $\alpha_E > 1$, represents an *average* property of AGN; in the USS, on the other hand, we can select *individual* sources that are ultra-soft because of the IPC's modest spectral resolution. This capability is extremely useful for follow-up studies at other wavelengths in order to determine the overall spectra of quasars with ultra-soft components and how these compare with individual QSOs that do not show this component.

All of the ultra-soft quasars in the USS may have additional flat spectral components but these components may be variable or weak, or both, and thus not always observed. Conversely, all the brightest quasars thus far studied might have ultra-soft components but these may not be always detectable because of intrinsic variability, geometric effects, absorption by our Galaxy along their lines of sight, or dilution because of the presence of a brighter, flatter spectrum. The latter hypothesis may be applicable to the sample of Wilkes and Elvis (1987): the flatter component (or components) could be bright enough in some QSOs so that it dominates in the IPC passband and makes it difficult to measure unambiguously the spectrum of any ultrasoft component present.

We thank E. Kolb for suggesting a survey of ultra-soft sources in the *Einstein* data base. We acknowledge E. M. Puchnarewicz, G. Branduardi-Raymont, and P. G. Murdin for contributing to the optical identification program. F. R. Harnden's assistance with extracting and evaluating the x-ray data was invaluable. This research receives funding from NASA's ADP program, Los Alamos Laboratory's Institutional Supporting Research Program, the U.S. Dept. of Energy, the U.K. SERC and the Royal Society of London.

## REFERENCES

Bechtold, J., Czerny, B., Elvis, M, Fabbiano, G., and Green, R.F. 1987, *Ap.J.*, **314**, 699.

Branduardi-Raymont, G. 1986, in *The Physics of Accretion onto Compact Objects*, eds. K. O. Mason, M. G. Watson, and N. E. White [Berlin: Springer-Verlag], p. 407.

Branduardi-Raymont, G., Mason, K. O., Murdin, P. G., and Martin, C. 1985, *M.N.R.A.S.*, **216**, 1043.

Córdova, F.A., Kartje, J., Rodriguez-Bell, T., Mason, K.O., and Harnden, F.R. 1989, in *Proceedings of the Berkeley Colloquium on EUV Astronomy*, eds. R. Malina and S. Bowyer (Pergammon Press), in press. Paper 1

Elvis, M., Lockman, F., and Wilkes, B. 1989, *Ap.J.*, **97**, 777.

Elvis, M., Wilkes, B. J., and Tanabaum, H. 1985, *Ap.J.*, **292**, 357.

Gioia, I. M., *et al.* 1984, *Ap.J.*, **283**, 495.

Giommi, P., Tagliaferri, G., and Angelini, L. 1988, in *X-ray Astronomy with EXOSAT*, eds. N. E. White and R. Pallavicini, 000

Harnden, F. R., Jr., Fabricant, D. G., Harris, D. E., and Schwarz, J. 1984, "Scientific Specification of the Data Analysis System for the *Einstein Observatory (HEAO-2)* Imaging Proportional Counter", *SAO Special Report No. 393*.

Hewitt, A., and Burbidge, G. 1987, *Ap.J. Suppl.*, **63**, 1.

Kriss, G. 1982, *Ph.D. Thesis*, Massachusetts Institute of Technology.

Maccacaro, *et al.* 1982, *Ap.J.*, **253**, 504.

Stark, A.A., Heiles, C., Bally, J., and Linke R., 1984, Bell Labs, privately distributed magnetic tape.

Wilkes, B. J., and Elvis, M. 1987, *Ap.J.*, **323**, 243.

# THERMAL COMPTONIZATION IN STANDARD ACCRETION DISKS

**L. Maraschi and S. Molendi**

Dipartimento di Fisica, Università di Milano

## 1. ACCRETION DISK MODEL

According to the theory of geometrically thin accretion disks developed by Shakura and Sunyaev (1973) (where the effects of viscosity are parametrized in terms of the total pressure, viscosity parameter, $\alpha$) the innermost region of the disk, (where the pressure is due to radiation, and the main source of opacity is Thomson scattering) a disk is described by the following (vertically averaged) equations (*e.g.* Lightman 1974) :

angular momentum transfer

$$\dot{M}\,(GM/R^3)^{1/2}f(R) = 4\pi\alpha p_{rad}h \tag{1}$$

energy balance

$$Q = \frac{3}{8\pi}\frac{GM\dot{M}}{R^3}f(R) \tag{2}$$

hydrostatic equilbrium

$$\frac{p_{rad}}{\rho} = \frac{GM}{R^3}h^2 \tag{3}$$

radiative diffusion

$$Q = \frac{2cp_{rad}}{\tau_{es}} \tag{4}$$

Here $\dot{M}$ indicates the accretion rate; $M$ the mass of the central object; $R$ the distance; h the half thickness of the disk; $\rho$ the mass density; and $\tau_{es} = k_{es}\rho h$ the optical depth to Thomson scattering. The radial function, $f(R) = 1 - \left(\frac{6GM}{c^2R}\right)^{1/2}$, takes into account the boundary conditions at the innermost stable orbit. The chosen form of $f(R)$ corresponds to a Schwarzschild black hole, for which the disk is assumed to extend to $R_{min} = \frac{6GM}{c^2}$.

*It is important to stress that these four equations for the four unknowns $p_{rad}$, $\rho$, h, Q can be solved without making use of an equation for the temperature. This is not true for the other regions of the disk.*

Introducing the usual scaling, $r = \frac{Rc^2}{2GM}$, $m = \frac{M}{M_\odot}$, $\dot{m} = \frac{\dot{M}}{\dot{M}_{cr}}$ where $\dot{M}_{cr} = \frac{L_{Edd}}{\eta c^2}$ with $\eta = 0.06$, the solutions in cgs units are given by:

$$Q = 1.38\times10^{27}\dot{m}m^{-1}r^{-3}f(r) \tag{5}$$

$$p_{rad} = 2.33\times10^{15}\alpha^{-1}m^{-1}r^{-3/2} \tag{6}$$

$$\rho = 1.38\times10^{-7}\alpha^{-1}\dot{m}^{-2}m^{-1}r^{3/2}f(r)^{-2} \tag{7}$$

$$h = 3.2\times10^6\dot{m}mf(r) \tag{8}$$

When the effective optical thickness, $\tau^* = \sqrt{(\tau_{es} + \tau_{ff})\tau_{ff}} > 1$, we can write

$$p_{rad} = \frac{1}{3}aT_c^4 \qquad (9)$$

which defines the vertically averaged temperature $T_c$ .

The other relevant temperature, in fact the most interesting one, is the surface temperature, $T_s$ . This is defined by the condition that the radiated flux equals the dissipated power per unit area given by (5)

$$F(T_s, \rho, h) = Q \qquad (10)$$

When $\tau^* < 1$, $T_s > T_c$ as noted by Czerny and Elvis (1987). The inconsistency is due to the incorrect use of eq (9) which is not valid for $\tau^* < 1$. However, the set of equations (1-4) is correct for all values of $\tau^*$. In particular equation (4) remains valid as long as $\tau_{es} \gg 1$.

We will therefore use eq. (10) for the determination of the temperature, assuming that the disk is homogenous and isothermal in the vertical direction, with the density and thickness given by equations (1-4) .

## 2. COMPTONIZED BREMSSTRAHLUNG FLUX

We will assume that the mechanism of photon production is purely thermal and take into account the increase in emissivity and opacity due to lines and free-bound transitions by multiplying the free-free emissivity by a factor 30 for $10^{-3}keV < kT < 10^{-1}keV$ and by $30(kT/0.1)^{-1.5}$ for $10^{-1}keV < kT < 1keV$ (*c.f.* Cox and Tucker 1969). The same factor has been used by Czerny and Elvis (1987) in the low temperature range without allowing for a reduction at high temperatures.

The average frequency shift per scattering is, for $kT \ll m_e c^2$, (Rybicki and Lightman 1976)

$$\Delta x = \frac{kT_s(4 - x)}{m_e c^2}x \qquad (11)$$

where $x = h\nu/kT$. After $n$ scatterings the initial frequency $x_i$ will be amplified by the factor, $A(x_i, n)$, given by

$$A(x_i, n) = \frac{\exp\left(\frac{4kT_s}{m_e c^2}n\right)}{1 + \frac{x_i}{4}\left[\exp\left(\frac{4kT_s}{m_e c^2}n\right) - 1\right]} \qquad (12)$$

For $n \to \infty$ eq.(12) gives $A \to \frac{4}{x_i}$. Since the correct average frequency for thermalized photons is $x = 3$ we have set $A(x_i, n) = \frac{3}{x_i}$ whenever $A(x_i, n) \, x_i > 3$

In order to determine the amplification factor one has to estimate the number of scatterings of a photon of frequency $x$. To this end there are three relevant frequencies (Illarionov and Sunyaev 1972): $x_0$ defined by $\kappa_{es} = \kappa_{ff}(x_o)$, $x_1$, defined by $\tau^*(x_1) = 1$, and the "saturation frequency", defined by $A(x_{sat}, \tau_{es}^2)x_{sat} = 3$ (See Rybicki and Lightman 1979 pp.208-219 for complete definitions).

We estimate the radiated flux per unit area integrated over frequency with three methods of increasing complexity:

**a)** We assume that all bremsstrahlung photons with $x \geq x_{min} = max(x_{sat}, x_1)$ are shifted to $x = 3$. The flux radiated through Comptonization is approximated by:

$$F^C = 3kT_s h \int_{x_{min}}^3 n_{ff}(x)dx \qquad (13)$$

where $n_{ff}$ is the photon production rate per unit volume due to free-free, bound-free, and bound-bound processes.

**b)** We consider all photons with $x > x_o$ and for each frequency we use an average $n = \tau_{es}(z(x)) = [\kappa_{es}/\kappa_{ff}(x)]^{1/2}$ where $z(x)$ is the height in the disk for which $\tau^*(z, x) \equiv [(\kappa_{es} + \kappa_{ff}(x))\kappa_{ff}(x)]^{1/2}z(x)\rho = 1$ In this approximation the overall flux is

$$F^C = kT_s \int_{x_o}^3 n_{ff}(x)A(x, \tau_{es}^2(z, x)) \ x \ z(x)dx \qquad (14)$$

**c)** In our final approximation we compute the amplification factor using the probability distribution for the number of scatterings instead of the average number. We use the distribution derived by Sunyaev and Titarchuk (1980) for a quasi-homogeneous source $P(n, \tau) = \beta e^{-\beta n}$ where $\beta = \frac{\pi^2}{12(\tau_{es}+\frac{2}{3})^2}$.

## 3. RESULTS: TEMPERATURE PROFILES AND EMISSION SPECTRA

We determined the temperature profile numerically, solving eq (10) for $T_s$ at every radius, adopting for $F^C$ the three different approximations discussed in the previous section. The emission spectra were then computed integrating at each frequency the contributions from regions with temperatures $\frac{h\nu}{20} < kT < 40h\nu$ and energy distributions given by a modified Black Body or Wien law, depending on the dominant energy loss.

The temperature profiles obtained with the three approximations for $F^C$ are shown in Fig 1a,b for $M = 10M_\odot$ and $M = 10^8 M_\odot$ respectively, for the highest accretion rate, $\dot{m} = 1$, and viscosity parameter $\alpha = 0.5$. The figures show that the three approximations discussed above give very similar temperatures, with maximum deviations of 50%. Comptonization plays an essential role where the disk becomes optically thin. In this case the temperature is much lower than predicted by bremsstrahlung alone and much higher than derived with the modified black-body emissivity. In the thick region the correction due to Comptonization does not change the temperature significantly. Pair production and relativistic corrections have not been included. The consistency of the model, for a specific choice of parameters, can be checked *a posteriori*, by requiring that a maximum temperature, which from Svensson (1984) can be estimated as $70 keV$, is not exceeded.

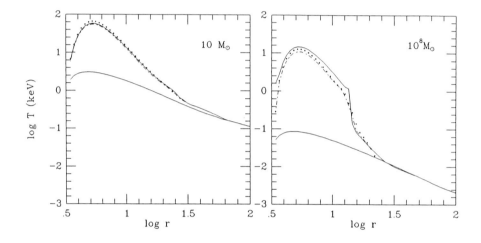

Figure 1. Temperature profiles for $\dot{m} = 1$ and $\alpha = 0.5$ for the Comptonized flux. The dotted line corresponds to approximation a), the dashed line to approximation b) and the upper solid line to approximation c). The lower continuous line represents the temperature calculated with the Modified Blackbody approximation.

The emission spectra have been calculated for the temperature profiles given by approximation b), since for numerical reasons it gives smoother profiles. The results are shown in Fig. 2a,b for the two values of the central mass, for fixed accretion rate, $\dot{m} = 1$ and different values of $\alpha$. The effect of changing $\dot{m}$ at fixed $\alpha$ is rather similar, apart from the different normalization. The most dramatic change in these spectra with respect to those obtained in the absence of Comptonization ($\alpha \ll 1$), is the presence of a hard component, as envisaged by Shakura and Sunyaev, extending from 10 to 100 keV for $10 M_\odot$ and from 1 to 30 keV for $10^8 M_\odot$. The high energy cut off, corresponding to the maximum temperature is very sensitive to the values of $\alpha$ and $\dot{m}$, however the shape within the Comptonization region is not.

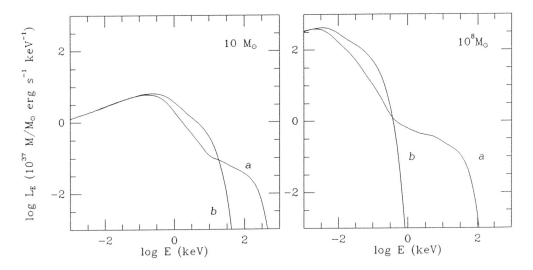

Figure 2. Emission spectra for $\dot{m} = 1$. Different curves refer to different values of the viscosity parameter $\alpha$: curve $a$ corresponds to $\alpha = 0.5$, curve b to $\alpha = 0.1$.

## 4. CONCLUSIONS

Our main result is that for large viscosity and accretion rates the inner region of a standard accretion disk model can produce quasi-power-law spectra in the x-ray range with two main components. A steep power law component originates in the transition between the optically thick and thin regions of the disk and a flatter component, which can extend to the 100 KeV range, arises in the Comptonized region. The inclusion of bound-bound and bound-free opacities decreasing with temperature in the interval $10^6 - 10^7 K$ affects the spectrum in the region intermediate between the modified blackbody and the Comptonized regime.

The model can potentially account for the x-ray spectra of black hole binaries and in particular for the existence of hard tails. However the presence of the hard component requires a nearly critical accretion rate.

In the case of large central masses, appropriate for Active Galactic Nuclei, the model can explain the UV excess and *at the same time* the power law spectra observed in x-rays. This result seems extremely promising. In particular the UV to x-ray flux ratio of our most extreme case, which translates into an effective spectral index $\alpha_{ox} = 1$ is close to the lowest observed among radio quiet QSOs where the accretion disk is likely to make an important contribution to the x-ray emission (

Kriss 1988). The model predicts that objects with $\alpha_{ox}$ close to 1 should have flat spectra in the x-ray range and should be close to the Eddington limit. Objects accreting at subcritical rates ($\dot{m} < 0.1$ ) should have much larger UV to x-ray flux ratios with steep x-ray spectra.

## REFERENCES

Cox, D.P. and Tucker, W.H. 1969, *Astrophys.J.*, **175**, 1157.

Czerny, B. and Elvis, M., 1987, *Astrophys.J.***321**, 305.

Illarionov, A.F. and Syunyaev, R.A., 1972 *Soviet Astronomy,* **16**, 45.

Lightman, A.P. 1974, *Astrophys.J.*, **194** 419.

Kriss, G.A. 1988, *Astrophys.J.*, **324**, 809.

Malkan, M.A. 1983, *Astrophys.J.*, **268** 582.

Novikov, I.D., Thorne, K.S. 1972, in *Black Holes*, Les Houches, p. 345.

Pringle, J.P. and Rees, M.J. 1972, *Astron.Astrophys.*, **21** 1.

Press, W.H., Flannery, B.P., Teukolsky, S.A. and Vetterling, W.T. 1986, *Numerical Recipes* , (Cambridge University Press).

Rybicki, G.B. and Lightman, A.P. 1979, *Radiative Processes in Astrophysics*, (New York, Wiley).

Shakura, I.N. and Sunyaev, R.A., 1973, *Astron.Astrophys.*, **24** 337.

Sunyaev, R.A. and Titarchuk, L.G. 1980, *Astron.Astrophys.*, **86** 121.

Svensson, R. 1984, *M.N.R.A.S.*, **209** 175.

Wandel, A. and Petrosian, V. 1988, *Astrophys.J. Lett.*, **329**, L11.

White, N.E., Stella, L. and Parmar, A.N., 1988, *Astrophys.J.*, **324** 363.

# COMPARISON OF VLBI RADIO CORE AND X-RAY FLUX DENSITIES OF EXTRAGALACTIC RADIO SOURCES

S.D. Bloom and A.P. Marscher

Department of Astronomy, Boston University

## 1. INTRODUCTION

The *Einstein Observatory* revealed that most quasars, selected in a variety of ways, are strong x-ray emitters (Tananbaum *et al.*1979; Ku, Helfand, and Lucy 1980; Zamorani *et al.*1981). The latter two of these papers showed that radio bright quasars are statistically more luminous in the x-ray than their radio-quiet counterparts. A much stronger connection was discovered by Owen, Helfand and Spangler (1981; see also Ledden and O'Dell 1985), who found that the 90 GHz to soft x-ray spectral index has a very small dispersion for sources selected by their strong millimeter emission. This implies a close relationship between compact radio flux density and x-ray emission. In addition Worrall *et al.*(1987), Worrall (1987), and Kembhavi, Feigelson, and Singh (1986) have found strong correlations between the arcsecond scale flux densities and soft x-ray fluxes. These authors suggest that the correlation can be explained if the soft x-rays were produced by the synchroton self-Compton (SSC) process within the compact radio emitting region.

If the x-rays are self-Compton in origin, the most likely site of production would be in the most compact components of the source. Indeed, Kembhavi *et al.*(1986) find that the correlation between radio and x-ray brightness is stronger only if the radio emission unresolved to the VLA ($\lesssim$ a few arcseconds) is included. This compact emission can be resolved on the milliarcsecond scale using VLBI, and maps thus obtained often show a complex structure (Pearson and Readhead 1988). Typically, only a fraction of the flux density in components unresolved at arcsecond resolution is contained in the true radio "core" which remains unresolved at the $\sim$ 0.3 milliarcsecond resolution of VLBI. The fraction of the flux density which remains unresolved using intercontinental VLBI at 1.3 cm ranges from nearly 100% in objects such as AO 0235+164 to $\lesssim$ 10% in 4C 39.25. In our study we test for the dominance of the self-Compton hypothesis by comparing *Einstein* x-ray flux densities to compact radio core flux densities from published VLBI maps for 49 extragalactic radio sources.

## 2. SOURCE SELECTION and DATA ANALYSIS

We used the archival *Einstein* IPC data and the program FINSPEC at the Harvard-Smithsonian Center for Astrophysics to obtain the x-ray flux densities at 2 keV. For sources with sufficiently high counts, we determined a best-fit spectrum, assuming a single power law with index $\alpha_x$ (defined such that the flux density $F_\nu \propto \nu^{-\alpha_x}$) and photoelectric absorption due to atomic hydrogen (as parameterized by the HI column densities $N_{HI}$). For sources with too few counts we assumed a value $\alpha_x = 0.7$ and extract $N_{HI}$ values from Galactic 21-cm surveys (Stark *et al.*1988). For non-detections, we take the $3\sigma$ values as upper limits to the actual counts. There are three sources with x-ray upper limits. The 2 keV flux density value for OJ 287 is derived from values quoted in Madejski and Schwartz (1988).

For the radio core flux densities, we extract 10.7 GHz VLBI maps from the literature. If there are no direct 10.7 GHz observations, we assume a radio core spectral index $\alpha_{core} = -0.5$ (*i.e.*, an inverted spectrum characteristic of a self-absorbed component containing gradients in magnetic field and relativistic electron density) to obtain the flux density at this frequency.

The flux densities were converted into luminosities by integrating over frequency (10.7–90 GHz) using the observed or assumed spectral indices (the derived luminosities are fairly insensitive to the precise values of the spectral indices) and multiplying by $4\pi d^2(1 + z)^{-1}$, where $d$ is the luminosity distance and $z$ is the redshift. For sources without 90 GHz measurements in the literature, we assumed $\alpha_{core} = -0.5$ for the integration. All luminosities assume $H_0 = 50$ km s$^{-1}$ Mpc$^{-1}$ and $q_0 = 0$. The luminosities thus derived cover different energy ranges for sources of different redshift. There is no reliable way to correct for this without assuming that the observed spectrum extends to unobserved frequencies.

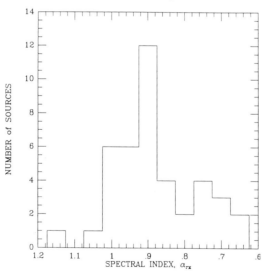

Figure 1. Histogram of spectral index $\alpha_{rx}$ between 10.7 GHz (core) and 2 keV for 41 extragalactic sources. The distribution of $\alpha_{rx}$ has a mean of 0.85 with a dispersion of 0.11. Eight sources with radio and/or x-ray upper limits are not included in this diagram. Citations of VLBI data will appear in a forthcoming journal publication (Bloom and Marscher, in preparation).

## 3. DISCUSSION

In a manner similar to that of Owen, Helfand and Spangler (1981), we have examined the distribution of radio to x-ray spectral index $\alpha_{rx}$ for all detections. The resulting $\alpha_{rx}$ histogram is shown in figure 1. We find a rather small dispersion (0.85 ± 0.11). This implies that, for a given radio core flux density, the x-ray flux density is confined to a well-defined range. This connection is consistent with the SSC model. The dispersion is not as small as for the sample of Owen, Helfand and Spangler; however, the limited size of their sample (25 sources) and selection effects may lead to a narrow distibution that would be widened if more sources had been included (Helfand, private communication).

The flux density–flux density plot (figure 2) shows a very poor one-to-one correlation. This shows that though the range of the x-ray flux density is determined by the radio emission, the exact value may depend on a variety of parameters. This too is consistent with the SSC model in which the x-ray flux is highly dependent on several radio source parameters (Marscher 1987):

$$S_\nu{}^{SC} \propto T_m{}^{3+2\alpha_r} S_m \nu_m{}^{1+\alpha_r} \delta^{-2(\alpha_r+2)}, \tag{1}$$

where $\delta$ is the relativistic Doppler factor, $\alpha_r$ is the optically thin radio spectral index, and the subscript $m$ denotes the value of the parameter at the self-absorption spectral turnover. A random scatter in the value of each parameter about the mean would account for the poor x-ray–radio flux density correlation for our sample.

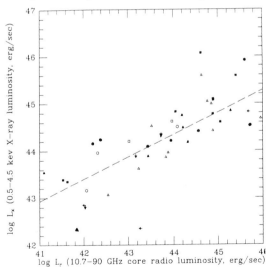

Figure 2. The 2 keV x-ray *vs.* 10.7 GHz core flux density for 49 extragalactic sources. The correlation is poor (correlation coefficient $r = 0.06$ for 41 objects.) All objects are included in the plot; however, the correlation coefficient was determined using only the detections.

The luminosity-luminosity plot (figure 3) shows a strong correlation. In general, the act of multiplying both the radio and x-ray fluxes by the same distance factor stretches the flux-flux plot (which may show a poor correlation) in a direction along a slope of unity to generate the luminosity-luminosity plot. However, the luminosity-luminosity plot has a slope of 0.48 rather than unity. This appears to be caused by a dependence of x-ray flux on redshift, with an upturn in flux at low redshift in our sample (see figure 4). Such a flux excess could be caused by lower photoelectric absorption; a very soft x-ray component which is redshifted below the IPC band in all but the lowest redshift objects (this is probably not the case since the IPC spectra would then be systematically steeper for low redshift objects); a second x-ray component which is significant only at low luminosities; or merely small number statistics and/or selection effects. The finding of Owen and Puschell (1982) that the radio–x-ray spectral index distribution has a much wider distribution for x-ray selected sources supports the notion that two or more x-ray components may exist in active galaxies.

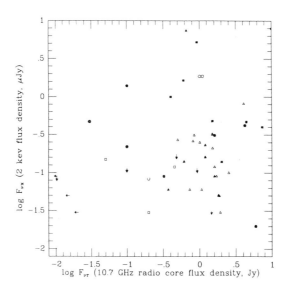

Figure 3. X-ray vs. radio core luminosity. The plot shows a very strong correlation ($r = 0.77$). Objects with upper limits to the flux densities or unknown redshift were excluded from this regression analysis (37 of 49 sources were included).

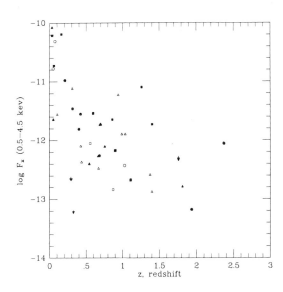

Figure 4. Redshift dependence of the 2 keV x-ray flux in the IPC band. There is also the possibility of a flux excess for low redshift objects as well as a weak correlation at higher redshifts.

Several authors (*e.g.*, Worrall 1987) find that the optical to x-ray luminosity correlation is tighter than the radio–x-ray correlation. This may be a result of the optical portion of the spectrum being closer to the x-ray in frequency rather than an indication of a closer physical connection with the x-ray emission mechanism than exists for the radio. For example, a dispersion in the radio–x-ray spectral index of 0.1 results in a radio–x-ray flux density correlation which is poorer than the corresponding optical–x-ray flux density correlation even if there is a much wider dispersion in optical–x-ray spectral index. It is apparent that the significance of such comparisons depends on how the actual emission mechanisms at various frequencies relate to each other. We are in the process of exploring this effect through theoretical simulations using various radio, optical, and x-ray emission models.

There is clearly a need to extend this sample to include more sources at low and high redshifts, in order to verify the flat radio–x-ray luminosity-luminosity slope and to ascertain the reality of the excess x-ray flux at low redshift. This will be made possible through further VLBI and x-ray observations of extragalactic radio sources.

This research was supported in part by NASA grant NAG8-671.

## REFERENCES

Bloom, S.D. and Marscher, A.P. 1989, in preparation.

Helfand, D.J., private communication.

Kembhavi, A., Feigelson, E.D., and Singh, K.P. 1986, *M.N.R.A.S.*, **220**, 51.

Ku, H.-M.K., Helfand, D.J., and Lucy, L.B. 1980, *Nature*, **288**, 323.

Ledden, J.E. and O'Dell, S.L. 1983, *Ap.J.*, **270**, 434.

Madejski, G.M. and Schwartz, D.A. 1988, *Ap.J.*, **330**, 776.

Marscher, A.P. 1987, in *Superluminal Radio Sources*, ed. J.A. Zensus and T.J. Pearson (Cambridge:Cambridge University Press), p.280.

Owen, F.N., Helfand, D.J., and Spangler, S.R. 1981, *Ap.J.(Letters)*, **250**, L55.

Owen, F.N. and Puschell, J.J. 1982, *A.J.*, **87**, 595.

Pearson, T.J. and Readhead, A.C.S. 1988, *Ap.J.*, **328**, 114.

Stark, A.A., *et al.* 1988, in preparation.

Tananbaum, H., *et al.* 1979, *Ap.J.*, **234**, L9.

Worrall, D.M. 1987, in *Superluminal Radio Sources*, ed. J.A. Zensus and T.J. Pearson (Cambridge: Cambridge University Press), p.351.

Worrall, D.M., Giommi, P., Tananbaum, H., and Zamorani, G. 1987, *Ap.J.*, **313**, 596.

Zamorani, G., *et al.* 1981, *Ap.J.*, **245**, 357.

# BL LAC OBJECTS: ONE OR TWO CLASSES?

**B. Garilli and D. Maccagni**
Istituto di Fisica Cosmica, CNR

**P. Barr, P. Giommi, and A. Pollock**
EXOSAT Observatory, Astrophysics Division
Space Science Department of ESA, ESTEC

## 1. INTRODUCTION

BL Lac objects were discovered in the optical and radio band as luminous, highly variable nuclei showing no emission lines. For a decade radio and optically selected objects were the only BL Lacs known and the properties of the class were based on these objects. The HEAO-1 and *Einstein* surveys led to the discovery o new BL Lacs in the x-ray band (Piccinotti *et al.* 1982, Gioia *et al.* 1984). It was immediately clear that the overall properties of the x-ray selected BL Lacs differ from those of radio and optically selected ones (Stocke *et al.* 1985).

We have followed up the early suggestions from *Einstein* data with an *EXOSAT* survey of BL Lacs. *EXOSAT* has pointed toward many of these sources, and a few others were serendipitously in the field of view for other pointed targets. This set of observations covers about half of the BL Lacs known and is a good tool to investigate the differences between radio or optically selected and x-ray selected objects.

## 2. THE DATA

We have considered all the BL Lac objects listed in the catalogue of Burbidge and Hewitt (1987) observed with *EXOSAT*, with the addition of 1H 1427+42, recently classified as a BL Lac by Remillard *et al.* (1988). This sample includes 35 objects, 20 of which can be classified as radio selected and 15 as x-ray or optically selected.

The sample is listed in Table 1, where in column 1 the source name is given. In column 2 the selection criterion is indicated (R for radio and X for x-ray or optical), while column 3 gives the redshift of the object (when known), together with the category of the redshift, as defined by Burbidge and Hewitt (1987): 'em' when the redshift was derived from emission lines, 'abs' when derived from absorption systems, and gal when obtained from typical galactic lines. In column 4 the total number of *EXOSAT* observations is given, and in columns 5 and 6 the number of LE and ME detections respectively. An asterisk in columns 5 or 6 indicates that counting rate variability has been detected for the object. All the observations have been reanalyzed in a uniform way so that a direct comparison is not biased by different methods of analysis.

| Source name | Selection | z-type | No. of observations | No. detections LE | ME |
|---|---|---|---|---|---|
| **Table 1** | | | | | |
| **X-ray Observations of BL Lacs** | | | | | |
| GC 0109+224 | R | — | 1 | 1 | – |
| 3C 66A | R | 0.44 em | 3 | 2 | 3 |
| AO 0235+164 | R | 0.94 em+abs | 9 | – | – |
| 4C 47.08 | R | — | 1 | 1 | – |
| 1E 0317+185 | X | 0.19 gal | 2 | 2 | 2* |
| 1H 0323+022 | X | 0.15 gal | 6 | 6* | 2* |
| 1H 0414+009 | X | — | 4 | 4 | 3* |
| PKS 0521-365 | R | 0.06 gal | 2 | 2 | 2 |
| PKS 0548-322 | X | 0.07 gal | 3 | 3 | 3* |
| PKS 0735+178 | R | 0.42 abs | 3 | 3 | – |
| OI 090.4 | R | — | 1 | 1 | 1 |
| OJ 287 | R | 0.31 em | 10 | 8* | – |
| MC 1057+100 | R | — | 1 | – | – |
| MKN 421 | X | 0.03 gal | 14 | 14* | 13* |
| MKN 180 | X | 0.05 gal | 5 | 5* | 3* |
| B2 1147+245 | R | — | 2 | 2 | – |
| 1E 1207+397 | X | 0.59 gal | 22 | 20 | – |
| ON 325 | R | — | 14 | 12* | – |
| 2A 1218+304 | X | 0.13 gal | 13 | 13* | – |
| 1E 1235+632 | X | 0.30 gal | 1 | 1 | – |
| B2 1308+326 | R | 1.00 em+abs | 8 | 8 | 3 |
| 1E 1402+042 | X | — | 4 | 4* | 3 |
| 1E 1415+259 | X | 0.24 gal | 13 | 13 | 1 |
| OQ 530 | R | — | 3 | 3 | – |
| 1H 1427+42 | X | — | 2 | 2* | 2* |
| AP Librae | R | 0.05 gal | 1 | 1 | – |
| 4C 14.60 | R | — | 1 | – | – |
| MKN 501 | X | 0.03 gal | 11 | 11* | 10* |
| I Zw 186 | X | 0.05 gal | 2 | 2 | 2 |
| 1803+78 | R | — | 4 | 4 | – |
| 3C 371 | R | 0.05 gal | 2 | 2* | 2* |
| PKS 2005-48 | R | — | 5 | 5* | 5* |
| PKS 2155-304 | X | 0.12 gal | 13 | 13* | 13* |
| BL Lac | R | 0.07 gal | 1 | 1 | – |
| 3C 446 | R | 1.40 em+abs | 13 | 11* | – |

## 3. DISCUSSION

In Figure 1 we have plotted the broad band spectral indexes $\alpha_{ro} - \alpha_{ox}$ (see Tananbaum *et al.* 1979 for the definition) for the *EXOSAT* BL Lacs. The same plot also shows the steep and flat radio spectrum QSOs observed by *Einstein*, as reported by Worrall *et al.* (1987). The spectral indices for the BL Lacs have been computed assuming an energy slope $\alpha = 0.5$, as for the QSOs. As can be seen from Figure 1, x-ray-selected BL Lacs cover a neatly separated part of the diagram from radio-selected BL Lacs. The distribution of the latter is perfectly comparable with that of radio-steep and radio-flat quasars.

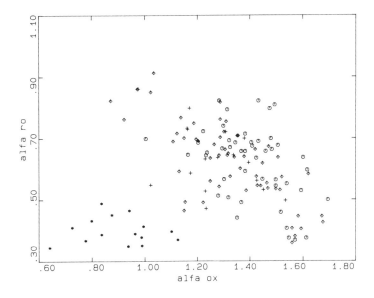

Figure 1. $\alpha_{ro}$ (radio-to-optical index) *vs.* $\alpha_{ox}$ (optical-to-x-ray index) for: $\star$ x-ray selected BL Lacs; + radio selected BL Lacs; O radio-flat QSOs; $\Diamond$ radio-steep QSOs.

This effect is hardly due to mere selection effects. A BL Lac having an $\alpha_{ro} = 0.6$ and an $\alpha_{ox} = 0.8$, with for example a radio flux as low as 5 mJy, should have an optical V magnitude around 21.5 and an x-ray flux greater than $10^{-13}$ erg cm$^{-2}$s$^{-1}$ (0.3-3.5), and thus would be detectable by present x-ray surveys. Up to now no object with these characteristics has been found, and this implies that if they exist, their number ratio to other BL Lacs is unfavorable. Given the low surface density of BL Lac objects (N($>$S) = $5 \times 10^{-2}$ per sq deg for S= $5 \times 10^{-13}$ erg cm$^{-2}$s$^{-1}$ [Giommi *et al.* 1989b, Maccacaro *et al.* 1989] in the x-rays), the number of source that would fall in the empty region of the $\alpha_{ro} - \alpha_{ox}$ diagram between the x-ray-selected BL Lacs and the QSOs and radio-selected BL Lacs is so low (even if an all sky survey with a greater limiting sensitivity than the present ones here available) that the plot in Figure 1 will not be substantially affected.

If we now accept that x-ray-selected and radio-selected BL Lacs are different, at least from the overall spectrum point of view, then we can further compare their other properties. As far as time variability is concerned, no differences were found (Giommi *et al.* 1989a, and Maccagni *et al.* 1989). It does not seem that from the point of view of polarization, the other historical feature of BL Lacs, there are differences between x-ray and radio selected sources. On the contrary, x-ray selected objects never show QSO-like emission lines in their optical spectra, as radio-selected sources sometimes do. Moreover the contrast between the nucleus and the underlying galaxy is generally higher for radio selected BL Lacs.

## 4. CONCLUSIONS

From their broad band energy distributions, BL Lacs appear to be divided into two classes, which grossly correspond to radio-selected and x-ray-selected objects. Possible intermediate objects, if they do exist, cannot be sufficiently numerous to overcome such a dichotomy.

Although the two classes have some observational similarities, radio-selected BL Lacs, especially in the optical, are more similar to radio-loud quasars, and especially to OVVs, that the x-ray-selected ones.

## REFERENCES

Burbidge, G. and Hewitt, A. 1987, *Astron.J.*, **92**, 1.

Gioia, I.M. *et al.* 1984, *Ap.J.*, **283**, 495.

Giommi, P. *et al.* 1989a, *Ap.J.*, submitted.

Giommi, P. *et al.* 1989b, Proceedings of the conference *BL Lac Objects: 10 Years After*, Como, in press.

Maccacaro, T., Gioia, I.M., Schild, R.E., wolter, A., Morris, S.L., and Stocke, J.T. 1989, Proceedings of the conference *BL Lac Objects: 10 Years After*, Como, in press.

Maccagni, D. 1989, Proceedings of the conference *BL Lac Objects: 10 Years After*, Como, in press.

Piccinotti, G. *et al.* 1982, *Ap.J.*, **253**, 485.

Remillard, R. *et al.* 1988, *Ap.J.*, in press.

Stocke, J.T. *et al.* 1985, *Ap.J.*, **298**, 619.

Tananbaum, H. *et al.* 1979, *Ap.J.(Letters)*, **234**, L9.

Worrall, D.M., Giommi, P., Tananbaum, H., and Zamorani, G. 1987, *Ap.J.*, **313**, 596.

# INTEGRITY OF HRI IMAGES

**D. E. Harris, C. P. Stern, and J. A. Biretta**
Harvard-Smithsonian Center for Astrophysics

## 1. INTRODUCTION

The High Resolution Imager (HRI) of the *Einstein Observatory* is described in Giacconi *et al.* (1978). The point response function (PRF) of the HRI-mirror combination is adequately described by two exponentials, the core and the scattering wings:

$$PRF(r) \propto [\exp(-r/1.96) + 0.01\exp(-r/12.94)]$$

where r is measured in arcsec. This expression is valid for the field center at 1.5 keV (see Henry 1979, SAO Internal Memorandum).

During the course of analyses using the PRF to subtract unresolved components of complex brightness distributions (Harris and Stern, 1987; Biretta *et al.* in preparation), we have found that apparent source structure on scales of 3 to 8 arcsec is often caused by instrumental problems. In this paper, we show two examples: a galaxy in Abell 754 for which the aspect solution is the main problem, and the jet of M87 where imperfect "gapmap" corrections cause distortions in the image.

## 2. ASPECT

Two aspect regimes are used in the standard processing: "LOCKED" and "MAP-MODE". In the first, the star trackers are locked onto their pre-assigned guide stars, and in the second, all identified stars in the field of view are used *a posteriori* to obtain a solution. Also available are the separation errors for each segment of the observation. These values give the average difference between the known and observed guide star separations or, in the case of MAPMODE segments, the average difference over all stars identified.

For example an observation of a radio galaxy in the outskirts of the cluster Abell 754, approximately $\frac{1}{3}$ of the total 6492 sec exposure was obtained during LOCKED segments. For the remainder, the average separation errors were all 2-4″ (one of the best categories for MAPMODE aspect solutions). In figure 1 we show smoothed versions of (a), the total exposure; (b), only the LOCKED data; and (c), only the MAPMODE data. It is immediately evident that the complex brightness distribution in figure 1a results from bad aspect data in MAPMODE, and the source is, in fact, unresolved (the FWHM being the same as that of the Chromium [0.57 keV] ground calibration). Furthermore, there appear to be three distinct aspect solutions: that for the locked image (L) and two (MNorth and MSouth) for the mapmode segments. The offsets of MN and MS from L are 5.9″ and 3.0″ respectively. The offsets between the x-ray and the

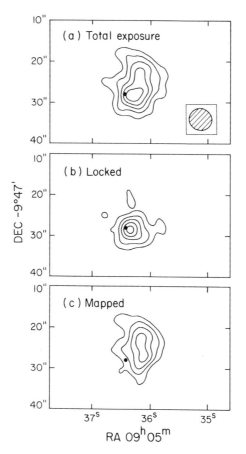

Figure 1. Smoothed maps of sequence number H7691, a radio galaxy in Abell 754. A gaussian smoothing of FWHM=3″ has been applied and a locally determined background has been subtracted. The FWHM contour of the ground calibration of Chromium (0.57 keV) is shown in the insert. The radio/optical position of the galaxy is indicated by the filled circle. Contour levels have been chosen on the basis of statistical errors so that features differing by $2\sigma$ have a 90% confidence. We specify each contour level by sigma above background and (enclosed in parentheses) percent of the peak value. Because of differing effective exposure times and differing morphologies, neither designation is strictly comparable between images.

(a) The total exposure. Contour levels are $\pm 3(9\%)$, $5(20\%)$, $7(36\%)$, $9(55\%)$, and $11(78\%)$ sigma above background (% of peak intensity).

(b) The "LOCKED" image. Contour levels are $\pm 3(8\%)$, $5(20\%)$, $7(37\%)$, $9(59\%)$, and $11(85\%)$ sigma above background (% of peak intensity).

(c) The "MAPMODE" image. Contour levels are $\pm 3(13\%)$, $5(30\%)$, $7(53\%)$, and $9(83\%)$ sigma above background (% of peak intensity).

optical/radio positions (Harris *et al.* 1984) are 2.2″(L), 6.9″(MN), and 5.1″(MS). Only for the locked data is the position within the 90% confidence radius of 3.2″ for the HRI (see Grindlay's contribution, "Astrometric Quality of HRI Positions" in Harris and Irwin, 1984).

## 3. GAPS

The gaps are regions where it is electronically impossible to assign photon positions because of the readout algorithm. Instead, these photons are assigned to permitted locations adjacent to the gaps. They occur at 1.67 arcminute intervals along both detector axes. The "gapmap" correction attempts a mapping of photons from their registered position back to their actual arrival position. However this procedure can only be completely successful for a uniform brightness distribution: the corrections were developed from observations of the bright Earth and of the Crab Nebula.

An example of gapmap imperfections is sequence H282. This is an observation of M87 lasting 339 ksec, with an on-time of 75 ksec. We divided the observation into 3000 sec segments, separated by similar length off-times (and longer inter-observation periods). For each of these 33 segments, we produced an image file with no aspect and no gapmap corrections. Since M87 is sufficiently stronger than the background so as to be visible in 3000 sec, we were able to place each segment into one of the following categories: a) Clear of the gaps - "Best", b) Mostly clear of the gaps - "Good", c) All the data rejected in (b) - "Bad", and d) Image definitely split by the gaps - "Split".

The "good" and "bad" data sets are mutually exclusive but "best" is a subset of "good", and "split" is a subset of "bad". We then made images by summing the segments in each category, with the additional constraint of accepting only LOCKED data (which comprises 32% of the data).

| Classification | Locked Percentage of 75 ksec |
|---|---|
| a) "Best" | 10% |
| b) "Good" | 15% |
| c) "Bad" | 17% |
| d) "Split" | 10% |

The results are shown in Figure 2 together with the LOCKED image. The structures of the bright eastern component (the core) and the western component (knot A) are significantly different between the "LOCKED" and "best" images. Also note that much of the north-south extended structure seen in the "LOCKED" image results from the "bad" data and is absent from the "best" image.

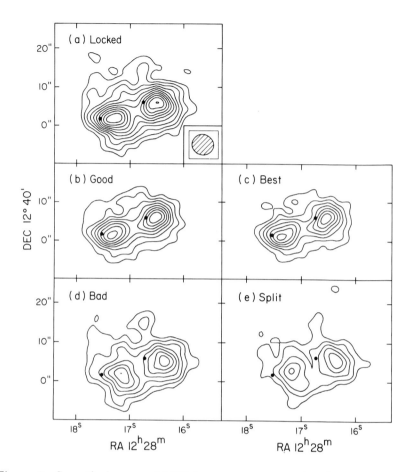

Figure 2. Smoothed maps of M87 from sequence number H282. Smoothing and contouring are the same as in figure 1.

(a) The "LOCKED" image. Contour levels are ±3(6.8%), 5(12%), 7(19%), 9(26%), 11(34%), 13(43%), 15(53%), 17(63%), 19(75%), 21(87%), and 23(99%) sigma above background (% of peak intensity).

(b) The "good" image. Contour levels are ±3(8%), 5(15%), 7(23%), 9(33%), 11(44%), 13(57%), 15(70%), and 17(85%) sigma above background (% of peak intensity).

(c) The "best" image. Contour levels are ±3(9%), 5(18%), 7(28%), 9(40%), 11(53%), 13(69%), and 15(86%) sigma above background (% of peak intensity).

(d) The "bad" image. Contour levels are ±3(10%), 5(19%), 7(29%), 9(41%), 11(55%), 13(70%), and 15(87%) sigma above background (% of peak intensity).

(e) The "split" image. Contour levels are ±3(13%), 5(25%), 7(41%), 9(58%), and 11(79%) sigma above background (% of peak intensity).

## 4. SUMMARY

Whenever source structure on the scale of 3″ to 8″ is in doubt, an evaluation of the LOCKED image should be made even if this sacrifices exposure time. This is a relatively straightforward procedure with current software. However, there is no automatic method of selecting data segments for which the target is clear of the gaps, and evaluation of imperfect gapmap corrections is therefore much more difficult.

## REFERENCES

Giacconi *et al.* 1979, *Ap.J.*, **230**, 540.

Harris, D. E., Costain, C. H., and Dewdney, P. E. 1984, *Ap.J.*, **280**, 532.

Harris, D. E., and Irwin, D. (editors) 1984 *"The Einstein Observatory Revised User's Manual"*, High Energy Division, Center for Astrophysics, Cambridge, MA.

Harris, D. E. and Stern, C. P. 1987, *Ap.J.*, **313**, 136.

# PART 3
# THE EINSTEIN
# DATABASES

# THE EINSTEIN OBSERVATORY DATA BASES

**G. Fabbiano**

Harvard-Smithsonian Center for Astrophysics

When the *Einstein Observatory* was launched the state of computer hardware available to astronomers was in general quite limited (this was before VAXes were introduced even). The SAO *Einstein* data center in particular had what now appears to be extraordinarily limited resources. In the last few years, as we all know, the situation has progressed rapidly– workstations are appearing of every desk, large amounts (~Gigabytes) of disk storage are becoming affordable, and data transfer media are moving beyond the limitations of 6250bpi tapes (which, by the by, were already a great advance. At the time of launch the standard was 1200bpi), towards CD-ROMs and other high density media.

This orders-of-magnitude change in computer resources has allowed a fresh approach to be taken to the preservation and distribution of the *Einstein* data. The SAO *Einstein* data center has made plans to use this new technology through a series of data bases which will comprise most of the *Einstein* data in forms useful to astronomers. These are described in the following articles. Each of these data bases will be available in hard copy, *via* on-line access to the SAO data bases, and *via* a computer readable version distributed on CD-ROM and/or magnetic tapes. On-line access is already available for the IPC catalog data base and is being used by astronomers in the US and abroad. We plan to implement remote on-line access to all of our data bases in the future through the NASA Astrophysics Data System, which should be available in 1990.

The largest of the data bases is the *Einstein* IPC Source Catalog which includes all IPC sources from the standard processing and maps of all the fields. The *Einstein* HRI Source Catalog will be developed next. Then there are a series of topical data bases for each of stars, galaxies and quasars, and a set of survey data bases – the Extended Medium Sensitivity Survey, the Deep Survey and the new Slew Survey. The complete MPC data base will also be put on-line.

The progress of these data bases can be monitored through the HEAO Newsletter (HEAO Newsletter, ms 6, 60 Garden St., Cambridge MA 02138, USA). Requests for assistance in accessing the data bases should be directed to either the first author of the relevant article or to Fred Seward (same address).

# THE EINSTEIN OBSERVATORY SOURCE CATALOG

**D.E. Harris (Chair), W. Forman, I.M. Gioia, J.A. Hale,**
**F.R. Harnden, Jr., C. Jones, T. Maccacaro, J.D. McSweeney,**
**F.A. Primini, J. Schwarz, and H.D. Tananbaum**
Harvard-Smithsonian Center for Astrophysics

## 1. THE IPC SOURCE CATALOG

The *Einstein Observatory* Source Catalog is computer based and contains the basic data on sources detected with the Imaging Proportional Counter (IPC); maps for each IPC observation greater than 300 sec; and attendant information concerning the observation. It will allow the recovery of upper limits for most sky positions within the observed images; the evaluation of the morphology of extended sources; and the visualization of the immediate x-ray surroundings of any location.

The catalog will be available in several different forms. We plan to publish a multi-volume set which will contain approximately 200 pages of documentation and 4000 pages of data. We will also provide a FITS version, although the medium is yet to be determined. An on-line service allowing access to the source list and the field list is already functional.

## 2. A MULTI-VOLUME SET

The "hard-copy" version of the Catalog will be published in 4 to 6 volumes which will contain approximately 4000 pages of the IPC observations, one observation per page. The accompanying text will be a separate book consisting of a description of how the catalog was constructed; details for each of the parameters presented; and various appendices which will contain procedures to convert countrate to flux, recipes to calculate upper limits for positions of interest, a complete list of unique sources, information and maps for approximately 150 merged fields, and other auxilliary information. Current projections indicate that this multi-volume set will be available early in 1990.

## 3. DESCRIPTION OF A CATALOG PAGE

An example of a catalog page is shown in Figure 1. The contour diagram is constructed from a background-subtracted array of the BROAD band (0.2-3.5keV) data smoothed with a Gaussian of FWHM=75″. Vignetting corrections (which amount to a factor of 2 at the field edge) have been applied. The first contour is at 3 sigma, where sigma is the rms fluctuation of source-free regions of the given field. Successive contours are separated by factors of 2. Units for the first contour level and the peak intensity of the map are counts per square arcmin. Negative contours are dotted. The grey scale displays the relative exposure in the image,

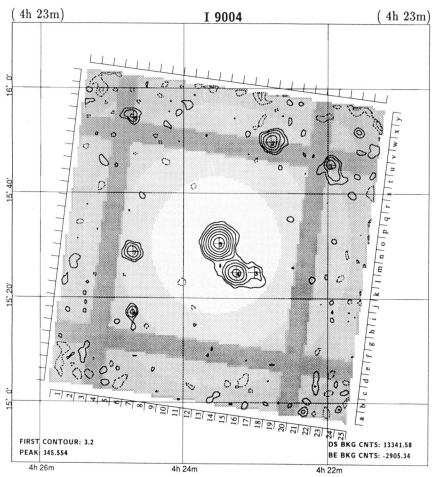

FIELD CENTER: $04^h23^m30.0^s$  $15°30'22''$ (B1950)   DATE: 1980/263 - 1980/263   NH: 1.6E+21
$04^h26^m20.8^s$  $15°37'06''$ (J2000)   LIVETIME: 10811.0s   REF/ID:
$\ell$: 180.18  $b$: -22.60   ROLL ANGLE: -82.9°   FIELD FLAGS:

| CAT # | FLD # | RA (1950) | DEC (1950) | ± '' | COUNT RATE | ± | NET CTS | BKG CTS | S/N | SIZE COR | RECO | R' | SRC FLG | ID |
|-------|-------|-----------|-----------|------|------------|------|---------|---------|------|----------|------|------|---------|-----|
| 1022 | 1 | 04 21 58.7 | 15 45 32 | 51 | *0.0094 | 0.0017 | 39.6 | 11.4 | 5.6 | 1.8 | 1409 | 26.6 | AH | * |
| 1027 | 2 | 04 22 47.2 | 15 49 46 | 48 | *0.0256 | 0.0023 | 129.5 | 11.5 | 10.9 | 1.3 | 906 | 21.9 | AH | S |
| 1030 | 3 | 04 23 00.5 | 15 24 57 | 41 | 0.0071 | 0.0012 | 51.9 | 22.1 | 6.0 | 51.5 | 0 | 8.9 | | |
| 1032 | 4 | 04 23 15.7 | 15 24 51 | 38 | 0.0397 | 0.0024 | 300.6 | 25.4 | 16.6 | 9.0 | 0 | 6.5 | AH | S |
| 1034 | 5 | 04 23 30.1 | 15 30 24 | 31 | 0.2101 | 0.0051 | 1690.8 | 23.2 | 40.8 | 1.8 | 0 | 0.2 | AH | S |
| 1045 | 6 | 04 24 42.1 | 15 54 26 | 55 | *0.0063 | 0.0016 | 24.2 | 13.8 | 3.9 | 1.0 | 1106 | 29.8 | | |
| 1046 | 7 | 04 24 42.6 | 15 28 42 | 48 | 0.0138 | 0.0017 | 83.7 | 18.3 | 8.3 | 1.2 | 0 | 17.5 | H | S |
| 1047 | 8 | 04 24 42.8 | 15 17 11 | 51 | 0.0059 | 0.0013 | 31.4 | 17.6 | 4.5 | 0.9 | 0 | 22.1 | | |

3 - 076

Figure 1

and is non-uniform because of the IPC rib shadows and vignetting. In the lower right corner are shown the "deep survey" and "bright Earth" counts which were used in constructing the background map.

Sources are selected from the standard processing if their signal-to-noise ratio of the broad-band detection exceeds 3.5. Information for the detected sources is derived from the BROAD band and includes: catalog number; field number; position (1950); corrected count rate; net counts and background counts; signal-to-noise ratio for the detection; size Cor, an intensity correction factor for extended sources; RECO, the "ribs and edges" code; distance from the field center in arcmin; a source flag, which indicates particulars for the source such as the existence of a hardness ratio or that the source has been detected in another observation; and a reference to published results which often indicates an optical identification of the source (*e.g.* "Q" for quasar, "G" for galaxy, etc).

## 4. THE CATALOG IN FITS FORMAT

The backgound-subtracted arrays used to generate the contour diagrams in the published version will be available as FITS files. Each array is 256 × 256 pixels with a 24″ separation between pixels. The data describing the observation will be included in the FITS header. As a tabular extension to each field, all sources detected in the BROAD band with a signal/noise detection > 3.5 will be listed with the same parameters as those which appear in the published version. The medium will be compact disk and/or magnetic tape. The FITS version should be available late in 1989.

## 5. THE ON-LINE VERSION OF THE CATALOG

The source list part of the catalog will be available on-line as part of the Astrophysics Master Directory, a NASA sponsored effort to provide access to databases related to space missions. The information available will be the same as that appearing in the published catalog in the source tables.

To assist scientists in preparing ROSAT proposals, the High Energy Division of the Harvard-Smithsonian Center for Astrophysics has provided access to a preliminary version of the *Einstein* IPC source list and the IPC field list. Users of the *Einstein* On-Line Service may also access various text (ASCII) files which contain important information concerning the field and source parameters as well as details about other aspects of the *Einstein* data.

# THE EINSTEIN OBSERVATORY STELLAR X-RAY DATABASE

F.R. Harnden, Jr., S. Sciortino, G. Micela, A. Maggio
G.S. Vaiana, and J.H.M.M. Schmitt

## ABSTRACT

We present the motivation for and methodology followed in constructing the *Einstein* Observatory Stellar X-ray Database from a uniform analysis of nearly 4000 Imaging Proportional Counter fields obtained during the life of this mission. This project has been implemented using the $INGRES^{TM}$ database system, so that statistical analyses of the properties of detected X-ray sources are relatively easily and flexibly accomplished. Some illustrative examples will furnish a general view both of the kind and amount of the archived information and of the statistical approach used in analyzing the global properties of the data.

## 1. INTRODUCTION

The *Einstein* Observatory was the first imaging X-ray telescope to study the properties of normal stars, and early results from *Einstein* were largely of a discovery nature, as X-ray emission was observed from virtually every type of star in the H-R diagram (Harnden *et al.* 1979, Seward *et al.* 1979, Vaiana *et al.* 1981, Helfand and Caillault 1982, Long and White 1980, Ku and Chanan 1979). Subsequent efforts focussed upon the characterization of the X-ray properties of stars, based on then-current data reduction software; the results (for reviews, see *e.g.*, Rosner, Golub and Vaiana 1985, Vaiana and Sciortino 1987) showed that, in the main, X-ray emission is indeed characteristic of most stars, principal exceptions being the A dwarf stars (Schmitt *et al.* 1985) and late-spectral type giants and supergiants (Ayres *et al.* 1981).

This paper describes a database created for the study of stellar X-ray emission, as observed by the *Einstein* Imaging Proportional Counter (IPC), and constructed from data which have been processed by a well-defined and uniformly applied analysis system (the 'Rev 1B' analysis software described by Harnden *et al.* 1984).

## 2. THE DATABASE

The database was constructed using IPC fields from the *Einstein* Data Bank and contains more than 16,000 distinct X-ray detections, some 5000 of which have been identified with other-wavelength counterparts from the *Einstein* Master Catalog (Harris and Irwin 1984), a list consisting mainly of catalogues of radio and optically identified QSO's, galaxies, and stars. Of these 5000 identifications, some 2500 are with stars.

Our stellar database was established by running specially-coded software on all *Einstein* IPC fields to (i) locate and analyze X-ray detections, (ii) search for possible identifications of detections with objects in the Master Catalog, (iii) compute upper limits for unmatched catalog objects, and (iv) port the results of the analysis to a distributed database system implemented on *SUN Workstation* and *DEC VAX* computers. This last step has allowed us to approach the formidable task of analyzing the properties of large numbers of sources ($\sim$ 2500) and upper bounds ($\sim$ 30,000) associated with stellar objects, mainly stars listed in the Smithsonian Astrophysical Observatory catalog. Within the database, the relevant information has been organized and archived into three logically-distinct sets of tables (or relations): (a) bookkeeping tables which contain general processing information, (b) scientific tables which store the X-ray or X-ray-related data, and (c) other complementary tables which contain optical stellar catalogs, such as the Yale Bright Star Catalog (Hoffleit 1982), the Woolley Catalog (Woolley 1966), the Garmany OB star catalog (Garmany, Conti and Chiosi 1981).

The scientific tables are conceptually divided in four groups, containing: (i) information on X-ray detections (*e.g.*, position, flux, spectral hardness ratio, etc.), (ii) information on the possible identification of other-wave-length counterparts of the X-ray detections, (iii) upper limits to X-ray flux for cataloged stellar objects, and (iv) general information on X-ray images (*e.g.*, exposure time, number of detections, intensity of background, etc.).

*INGRES$^{TM}$* (a trademark of Relational Technology) is a relational database management system which stores data in tables by rows and columns and provides the capability of analyzing data in sets using a high-level, non-procedural language (*i.e.*, a language in which the end user specifies only what task to do, not how to accomplish it) for the definition of data structures and their relationships to one another. Other useful capabilities include: sorting and querying on any of the fields (columns) of a given table, cross-correlating many distinct tables using any given fields as cross-reference indices, and multiple indexing for significantly-increased query rate. We have also found that the *S* interactive statistical package provides a convenient tool for performing scientific data analysis and display and is well-suited for investigations within this project.

## 3. GENERAL DATA CHARACTERIZATIONS

The full *Einstein* set of some 4,000 IPC fields covers approximately 3,500 square degrees, as shown in Figure 1 in which we plot the sky coverage achieved. Figure 2 shows the distribution of the effective exposure times of the images.

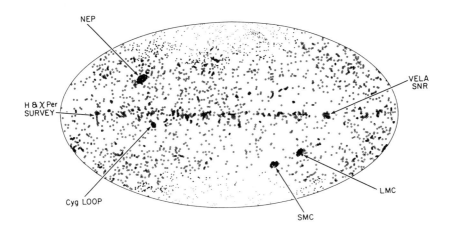

Figure 1: Galactic-coordinate sky coverage achieved with the *Einstein* IPC; each dot represents one image. Several regions subjected to repeated targeting (*e.g.*, surveys of the north ecliptic pole (NEP) and H Per and χ Per) are indicated.

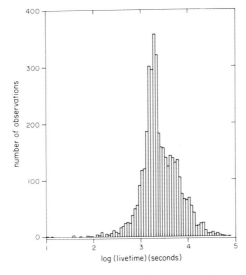

Figure 2: Distribution of image exposure times; note that a majority of images have exposure times of the order of a few thousand seconds.

## 4. EXAMPLE APPLICATIONS

One of the first applications of this database has been an extensive survey (involving 63 distinct images) of the stars in the Hyades Cluster Region (Micela *et al.* 1988), involving ∼ 150 stars, ∼ 300 X-ray detections, and ∼ 200 X-ray upper bounds. One of the major scientific results of this survey was the conclusion that the level of X-ray emission is not only a function of the stellar age but that the dependence on age itself is a function of spectral type (see Figure 3).

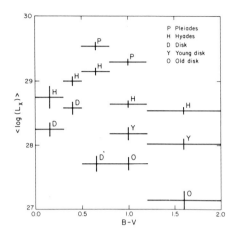

Figure 3: The behavior of mean log $L_X$ as a function of spectral type and stellar age, as deduced from volume limited sample of nearby stars, and analysis of samples of coeval stars (Micela *et al.* 1988).

In another application, we performed a survey of ~ 400 late-type giants and super-giants cataloged by Hoffleit 1982 and observed with the IPC. A major finding of this survey (Maggio *et al.* 1989) is a statistically significant decrease of X-ray luminosity level beyond spectral type K0 (cf. Figure 4).

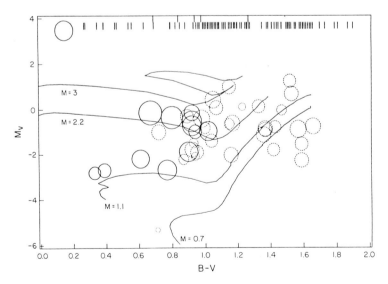

Figure 4: Bubble HR diagram of the single late-type giants and supergiants observed by the *Einstein* Observatory. The bubble dimension is proportional to log $L_X$, the solid bubbles indicate X-ray detections, while the dotted bubble indicate X-ray upper bounds. Superimposed are the evolutionary tracks (solid line, adapted from Mengel *et al.* 1979, and Sweigart and Gross 1978) for stars

of indicated masses. Note the marked reduction in the number of detections versus that of upper limits for B-V ~ 1.0 (Maggio *et al.* 1989).

Further information regarding the *Einstein* Stellar X-ray Database and its use can be obtained from the principal author.

We would like to thank Dr. R. Giacconi for encouraging this effort through his farsighted support for the initial stellar observations with *Einstein*. We would also like to thank R. Rosner for stimulating discussions. We acknowledge partial support of the Consiglio Nazionale delle Ricerche (PSN) and the Ministero Pubblica Istruzione (Italy). This work was supported in part by NASA contract NAS8-30751.

# 5. REFERENCES

Ayres, T. R., Linsky, J. L., Vaiana, G. S., Golub, L. and Rosner, R. 1981, *Ap. J.*, **250**, 293.

Garmany, C., Conti, P. and Chiosi, C. 1981, *Ap. J.*, **263**, 777.

Harnden, F. R., Jr., Branduardi, G., Elvis, M., Gorenstein, P., Grindlay, J., Pye, J. P., Rosner, R., Topka, K., and Vaiana, G. S. 1979, *Ap. J. Letters*, **234**, L51.

Harnden, F. R., Jr., Fabricant, D.G., Harris, D.E. and Schwarz, J. 1984, "Scientific Specification of the Data Analysis System for the Einstein Observatory (HEAO-2) Imaging Proportional Counter", *SAO Special Report No.* **393**.

Harris, D. E., and Irwin, D. (eds.) 1984, *Einstein Observatory Revised User's Manual*.

Helfand, D. J. and Caillault, J. 1982, *Ap. J.*, **253**, 760.

Hoffleit, D., and Jaschek, C. 1982, *The Bright Star Catalogue, Fourth Revised Edition*, (New Haven: Yale University Obs.).

Ku, W. H.-M. and Chanan, G. A. 1979, *Ap. J. Letters*, **234**, L59.

Long, K. S., and White, R. L. 1980, *Ap. J. Letters*, **239**, L65.

Maggio, A., Vaiana, G. S., Haisch, B., Stern, B., Bookbinder, J., Harnden, Jr., F. R., and Rosner, R. 1989, submitted to *Ap. J.*

Micela, G., Sciortino, S., Vaiana, G. S., Schmitt, J. H. M. M., Stern, R., Harnden, Jr., F. R., and Rosner, R. 1988, *Ap. J.*, **325**, 798.

Pallavicini, R., Golub, L., Rosner, R., Vaiana, G. S., Ayres, T., and Linsky, J. L. 1981, *Ap. J.*, **248**, 279.

Rosner, R., Golub, L., and Vaiana, G. S. 1985, *Ann. Rev. Astr. Ap.*, **23**, 413.

Schmitt, J. H. H. M., Golub, L., Harnden, Jr., F. R., Maxson, C. W., and Vaiana, G. S. 1985, *Ap. J.*, **293**, 176.

Seward, F. D., *et al.* 1979, *Ap. J. Letters*, **234**, L55.

Vaiana, G. S. and Sciortino, S. 1986, *Adv. Space Res.*, **6**, 99.

Vaiana, G., *et al.* 1981, *Ap. J.*, **244**, 163.

Woolley, R. *et al.* 1970, *Royal Obs. Ann.*, (Herstmonceaux: Royal Greenwich Observatory), **5**.

# EINSTEIN OBSERVATIONS OF SUPERNOVA REMNANTS

**F. D. Seward**

Harvard-Smithsonian Center for Astrophysics

*Einstein* was pointed at $\sim$ 80 galactic supernova remnants. 60% of these were detected and data from the imaging instruments has been gathered into a catalog.

X-ray images of 46 SNR are available as contour diagrams, grey scale plots, or as digital data on magnetic tape in FITS format. Some of these are mosaics made from as many as 50 separate observations. Some images have not yet appeared in the literature. Some SNR were observed with both HRI and IPC and some were bright enough in the IPC so that they could be divided into two energy bands. The FITS tape contains 74 images of the 46 remnants.

Counting rates for the HRI, IPC, and MPC have been derived and tabulated and are included here (Table 1). The rates given apply to the entire energy range of the IPC (pulse-height channels 1 to 15). Because many SNR are larger than the field of view, rates over the entire field of view ($\sim$ 60′ diameter) are quoted for all SNR and for all point sources. This is different from the way point-source rates are sometimes presented in other *Einstein* publications and allows direct comparison of counting rates for different SNR. The relative contributions of point sources such as internal pulsars is also easily obtained. All rates have been corrected for vignetting, dead time, and scattering from the telescope mirror.

HRI coverage of the larger remnants was usually not complete (*e.g.* Cyg Loop, IC 443). For seven remnants, with coverage from 35% to 90%, an HRI rate for the entire SNR has been calculated by using the observed IPC data and assuming the ratio of HRI to IPC rates was the same for the observed and the unobserved parts of the remnant. Such rates are footnoted.

A sample illustration of the remnant PKS 1209-52 is included here (Figure 1). The pictures, counting rates, and references will be submitted to Ap.J. Supp. These data and FITS tapes are also available upon request from the author.

Table 1 page 1

| Table 1 | | | | | | | |
|---------|--|--|--|--|--|--|--|
| **Count rates in Einstein Detectors** | | | | | | | |
| SNR No. | Remnant | Common Name | IPC Rate $(cs^{-1})$ | HRI Rate $(cs^{-1})$ (1) | MPC Rate $(cs^{-1})$ | Type | Comments |
| 1 | 4.5+6.8 | Kepler | 7.3±0.4 | 1.73±.05 | 4.50±.05 | S | SN 1604 |
| 2 | 6.4-0.1 | W 28 | 3.2±0.4 | 0.5±0.1 (2) | 3.47±0.2 | F, CO | |
| 2A | | 1E1757-233 | NRS | 0.008±.001 | NRS | | |
| 3 | 11.2-0.3 | | 1.00±0.1 | 0.11±.008 | 1.8±0.2 | S | |
| 4 | 21.5-0.9 | | 0.49±.05 | 0.029±.003 | 4.15±0.1 | F | |
| 5 | 21.8-0.6 | Kes 69 | >0.1 | ND | NRB | F? | INC |
| 6 | 27.4+0.0 | Kes 73 | 1.07±0.1 | 0.078±.008 | 4.66±0.1 | F?,CO | |
| 6A | | 1E1838-049 | NRS | 0.016±.002 | NRS | NB | |
| 7 | 29.7-0.2 | Kes 75 | 0.22±.03 | 0.0067±.0017 | 2.39±0.1 | F | |
| 8 | 31.9+0.0 | 3C 391 | 0.24±.03 | 0.048±.012 | 1.40±.15 | IR | |
| 9 | 33.6+0.0 | Kes 79 | 0.44±.05 | ND | 1.66±0.2 | S? | |
| 10 | 34.7-0.4 | W 44 | 3.3±0.3 | 0.35±.1 (2) | 4.13±0.2 | F | internal radio pulsar |
| 11 | 39.2-0.3 | 3C 396 | .06±.01 | ND | 0.6±0.2 | IR | |
| 12 | 39.7-2.0 | W 50 | 1.6 ave. | ND | NRB | IR,CO | |
| 12A | | SS433 | 1.2 ave. | 0.10 ave. | 7.0 ave | NB | variable |
| 13 | 41.1-0.3 | 3C397 | 0.75±0.1 | 0.11±.025 | 1.76±0.1 | IR | |
| 14 | 43.3-0.2 | W 49B | 0.67±.06 | 0.041±.006 | 4.7±0.2 | F | |
| 15 | 49.2-0.7 | W 51 | 0.7±0.2 | INC | 0.89±.06 | IR | |
| 16 | 53.6-2.2 | 3C 400.2 | 0.80±0.1 | 0.25±.05 | 0.20±.06 | F | |
| 17 | 54.1+0.3 | | 0.016±0.004 | ND | NRB | | |
| 18 | 65.3+5.7 | GKP | >1.7±0.3 | ND | >0.21±.12 | | INC |
| 18A | | 1E1928+313 | 0.042±.006 | ND | NRS | | |
| 19 | 69.0+2.7 | CTB 80 | 0.025±.05 | 0.025±.004 | 0.81±.04 | F,CO | |
| 19A | | 1E1951+327 | 0.14±.01 | 0.009±.001 | NRS | NI | "age" 100,000 yr |
| 20 | 74.3-8.5 | Cyg Loop | 660±60 | 177±16 (2) | 14.5±2.2 | S | INC (HRI only) |
| 21 | 74.9+1.2 | CTB 87 | .040±.01 | ND | NRB | F | |
| 22 | 78.2+2.1 | W 66 | >0.7±.15 | INC | 1.3±0.2 | | INC |
| 23 | 82.2+5.3 | W 63 | 0.4±0.1 | ND | 0.0±0.15 | IR | |
| 24 | 89.0+4.7 | HB 21 | >0.2 | ND | >0.1±0.1 | IR | INC |
| 25 | 109.1-1.0 | CTB 109 | 5.2±0.4 | ND | 3.1±0.3 | S,CO | |
| 25A | | 1E2259+586 | 1.1±0.1 | 0.09±.01 | NRS | NB? | 7s period |
| 26 | 111.7-2.1 | Cas A | 61±2 | 6.28±.13 | 109±1 | S | |
| 27 | 119.5+10.2 | CTA 1 | 0.75±.15 | ND | 0.75±0.15 | F | |
| 27A | | 1E0000+726 | 0.025±.004 | ND | NRS | | |

Table 1 page 2

| Table 1 Count rates in Einstein Detectors | | | | | | | |
|---|---|---|---|---|---|---|---|
| SNR No. | Remnant | Common Name | IPC Rate (cs$^{-1}$) | HRI Rate (cs$^{-1}$) (1) | MPC Rate (cs$^{-1}$) | Type | Comments |
| 28 | 120.1+1.4 | Tycho | 22.3±1 | 3.36±.08 | 27.5±0.3 | S | |
| 29 | 130.7+3.1 | 3C 58 | 0.35±.04 | 0.051±.008 | 0.61±.05 | F,CO | |
| 29A | | 1E0201+645 | NRS | 0.0039±.0008 | NRS | | |
| 30 | 132.7+1.3 | HB 3 | 2.1±0.4 | ND | 0.0±.15 | IR | |
| 31 | 160.4+2.8 | HB 9 | >1.6±0.5 | ND | >0.6±0.2 | IR,CO | INC |
| 31A | | 4C+46.09 | .049±.01 | ND | NRS | | |
| 32 | 184.6-5.8 | Crab | 684±35 | 120±10 | 1383±10 | F,CO | SN 1054 |
| 32A | | PSR 0531+21 | NRS | 5.2±0.3 | NRS | NI | "age" 1240 yr |
| 33 | 189.1+3.0 | IC 443 | 12.4±0.6 | 2.55±.35 (2) | 6.5±0.6 | IR | |
| 34 | 260.4-3.4 | Pup A | 250±20 | 79±2 | 36±4 | S | |
| 35 | 263.5-2.7 | Vela SNR | 520±25 | ND | 54±8 | IR,F,CO | |
| 35A | | PSR0833-45 | 2.15±0.15 (3) | 0.69±.03 (3) | 7.6±0.1 (3) | NI | "age" 13,000 yr |
| 36 | 290.1-0.8 | MSH 11-61A | 0.47±0.1 | 0.13±.03 | 0.92±.05 | F | |
| 37 | 291.0-0.1 | MSH 11-62 | 0.22±.05 | ND | 0.80±.07 | F | |
| 38 | 292.0+1.8 | MSH 11-54 | 9.1±1.0 | 1.41±.05 | 3.72±0.1 | IR | |
| 39 | 296.1-0.5 | | 3.2±0.3 | 1.2±0.2 (2) | 2.3±0.3 | S | |
| 40 | 296.5+10.0 | PKS 1209-52 | 3.6±0.3 | ND | 0.0±0.2 | S,CO | |
| 40A | | 1E1207-521 | 0.082±0.08 | 0.028±.004 | NRS | NI | No opt. Counterpart |
| 41 | 315.4-2.3 | RCW 86 | 8.5±1.0 | 1.9±.1 (2) | 6.0±1.0 | S | |
| 42 | 320.4-1.2 | MSH 15-52 | 2.40±0.2 | 0.18±.05 | 9.3±0.6 | S,F,CO | |
| 42A | | PSR 1509-58 | 0.30±.04 (3) | 0.011±.001 | NRS | NI | "age" 1550 yr |
| 43 | 326.3-1.8 | MSH 15-56 | 1.10±.10 | INC | 1.5±0.2 | IR | |
| 44 | 327.1-1.1 | | .085±.02 | 0.035±.012 | NRB | IR | |
| 45 | 327.4+0.4 | Kes 27 | 0.40±0.1 | ND | 1.18±0.1 | F | |
| 46 | 327.6+14.6 | SNR 1006 | 11.1±1.0 | 5.6±.7 (2) | 3.4±0.2 | S | |
| 47 | 332.4-0.4 | RCW 103 | 9.3±1.0 | 2.17±.08 | 4.35±0.1 | S,CO | |
| 47A | | 1E 1613-509 | NRS | 0.007±.001 | NRS | NI | No opt. Counterpart |

| | | |
|---|---|---|
| NRS | = | not resolved from other parts of SNR |
| INC | = | incomplete data, much of SNR not observed |
| NRB | = | not resolved from nearby bright sources |
| ND | = | no data, not observed |
| S | = | shell |
| F | = | filled center, plerionic |
| IR | = | irregular |
| CO | = | central object |
| NI | = | neutron star, isolated |
| NB | = | neutron star in binary system |
| (1) | = | rates corrected to 1 Jan. 1979. |
| (2) | = | estimated from partial HRI observation and HRI/IPC ratio. |
| (3) | = | includes surrounding diffuse emission. |

**296.5+10.0**      **PKS 1209-52**

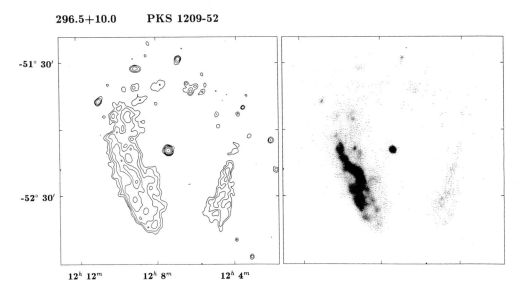

IPC, contour intervals are a factor of 1.5 in brightness.

Figure 1. Sample page of hard copy Supernova Remnant catalog.

# THE EINSTEIN NORMAL GALAXY DATABASE

**G. Fabbiano, G. Trinchieri, S. Hazelton, and D.-W. Kim**
Harvard Smithsonian Center for Astrophysics

## ABSTRACT

An X-ray Catalog and a data base for normal galaxies surveyed with the *Einstein* Observatory are under construction and will be made available for use by the astrophysical community at large. Galaxies are extended and complex X-ray sources: Careful and non-standard analysis is needed to extract their X-ray parameters. The catalog for detected galaxies consists of observational parameters, contour maps, radial profiles, spectral parameters, variability and fluxes. The X-ray database contains information for all the galaxies including detections and upper limits. The data reduction method is basically the same as described in Trinchieri *et al.* (1986) and Fabbiano *et al.* (1987) and is described below.

## 1. DATA PRODUCTS

The main data products of this project are an X-ray catalog for $\sim$150 detected normal and morderately active galaxies and a data base for $\sim$350 bright galaxies ($B_T \lesssim 13^m$) which have been surveyed with the *Einstein* Observatory, including detections and upper limits. The data base and catalog will contain the following information, and a sample page of the catalog is given in Figure 1.

- X-ray observation parameters for each galaxy (*e.g.*, date; duration).

- X-ray fluxes or upper limits within a *total* galaxy radius.

- Basic optical, IR and radio parameters.

- Isointensity contour plots of the X-ray emission superposed onto an optical image for the detected galaxies. X-ray maps in different energy bands (0.2–0.8 keV; 0.8–3.5 keV; 0.2–3.5 keV) will be produced from the IPC images.

- Radial distribution of the X-ray surface brightness, and comparison with the distribution of the emission in other wavebands

- X-ray spectral parameters

- Fluxes and variability information for each individual X-ray sources detected in a given galaxy.

(A sample page of the catalog)

# NGC 4472

### Map

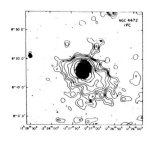

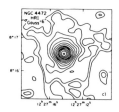

Galaxy position
$\alpha = 12^h 27^m 14^s \ \delta = 8° 16^m 42^s$
Einstein Observation
IPC : I4308 (July 1979)
HRI : H7068 (Dec 1979 and Jan 1981)

### Radial Profile

Isodensity X-ray contours for NGC 4472. The lowest contour level is a 2 $\sigma$ level over the background. Values for 3, 4 ... $\sigma$ are then plotted.

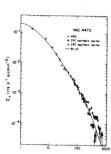

X-ray and optical surface brightness profile for NGC 4472. The scale for y-axis refers to the HRI data. For the IPC points, the value on the y-axis should be multiplied by a factor of 6.

### Spectral Analysis

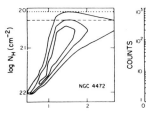

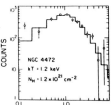

**Parameters**

IPC gain = 13.9
PHA channel = 2 – 11
Energy range = (0.2 – 5.4 keV)
KT = 1.2 keV (0.5 – 2 keV)
$N_H$ = 1.2 x $10^{21}$ cm$^{-2}$ (2.6 – 12 x $10^{21}$)

### Flux

$F_x = 1.2 \ x \ 10^{-11} \pm 2.6 \ x \ 10^{-13}$ erg sec$^{-1}$ cm$^{-2}$
$L_x = 5.6 \ x \ 10^{41} \pm 1.2 \ x \ 10^{40}$ erg sec$^{-1}$

Within radius of 780″ and energy range of 0.2 to 4 keV and at distance of 20 Mpc, and assuming thermal spectrum with kT=1 keV.

For detailed data analysis and discussion, see Trinchieri, Fabbiano and Canizares (*Ap. J.* 1986, **310**, 637)

**Figure 1. Sample page from the hardcopy form of the Normal Galaxy Catalog.**

## 2. DATA REDUCTION

The procedures of the IPC data reduction are described below. The HRI data reduction is similar, though simpler, since there is no energy information and the field background can be evaluated locally for each field.

1) IPC Background Subtraction

The IPC background is azimuthally symmetric, but depends on the distance from the field center (Harnden *et al.* 1984). The standard IPC data processing (Rev 1B) produced background templates for each field; however their normalizations may be erroneous in some cases because of diffuse emission in the field. Azimuthally-averaged radial distributions of the surface density for both the 'galaxy' image and the 'background' image centered at the source center are compared at large radii, in source-free regions. When necessary, the background template is *renormalized* to match the 'source' profile. For confused, complex fields, the image is split in sectors to obtain clean regions. The same procedure is applied to the three *standard energy bands*: Soft (0.2-0.8 keV); Hard (0.8-3.5 keV); and Broad (0.2-3.5 keV). The template is then subtracted from the raw image, and a radial profile for the galaxy emission is produced.

2) X-ray Contour Maps

The background subtracted data are then *smoothed* with a gaussian function, typically of $\sim 35''$, comparable to the IPC Point Spread Function. Iso-intensity contour plots are produced and *overlayed* onto an optical picture of the galaxy. The same procedure is applied to the three standard bands. If multiple observations of the same object are available, the images are merged together and the above procedure is applied to the resulting image.

3) Flux and Luminosity

The net counts within the optical radius or out to where there is detectable emission are obtained from the broad band and converted into fluxes assuming a spectrum appropriate for the galaxy type. Typically, for S and Irr, the model we used is a thermal bremsstrahlung with kT=5 keV, and for E and S0 an optically thin plasma with emission lines and kT=1 keV. An appropriate low energy cutoff is estimated from the line of sight HI column density (Stark *et al.* 1984). In the presence of multiple observations, the count rates in the images are derived separately and compared to search for source variability.

4) Spectral Parameters

The IPC pulse height spectra with a total number of net counts exceeding $\sim$ 200 will be fitted to optically thin plasma and thermal bremsstrahlung models with absorption cutoff at the low energy. Spectral distribution and confidence contours for the two interesting parameters, temperature $kT$ and equivalent column density $N_H$, will be presented. If a point nuclear source is dominated, a power law model will also be used.

The Catalog for detected galaxies will be available in a printed form and in the FITS format on a tape/CDROM. The data base of bright galaxies surveyed with the *Einstein* Observatory, implemented using the *INGRES*$^{TM}$ (a trademark of Relational Technology) database system, will reside in the *SAO HEAD SUN-LAN*, and will be remotely accessible. Moreover, data tables will be extracted and distributed at user's request.

This work was supported under NASA Contract NAS8-30751, and under a Scholarly Studies Grant of the Smithsonian Institution. GT acknowledges the financial support of the Italian Piano Spaziale Nazionale.

## REFERENCES

Fabbiano, G., Trinchieri, G. 1987, *Ap. J.*, **315**, 46.

Harnden, F. R., *et al.* 1984, *Scientific Specificatioin of the Data Analysis System for the Einstein Observatory Imaging Proptional Counter*

Stark, A.A., Heiles, C., Bally, J., and Linke, R. 1984, *Bell Labs.*, privately distributed magnetic tape.

Trinchieri, G., Fabbiano, G., and Canizares, C. R. 1986, *Ap. J.*, **310**, 637.

# THE EINSTEIN DATABASE OF OPTICALLY AND RADIO SELECTED QUASARS

Belinda J. Wilkes, Harvey Tananbaum, Yoram Avni
M.S. Oey and D.M. Worrall
Harvard-Smithsonian Center for Astrophysics

## ABSTRACT

The *Einstein* quasar database will provide soft X-ray count rates or upper limits, fluxes, and luminosities for *all* previously known, optically- or radio-selected quasars observed with the Imaging Proportional Counter (IPC) aboard the *Einstein* observatory. Data for all the quasars will be available in published hardcopy, on-line via computer network, and in a form suitable for distribution (*e.g.* tape, CDROM).

## 1. THE SAMPLE

The quasar database is divided into two subsets corresponding to differences in the details of the analysis procedure:
**1)** IPC targets, mostly on-axis.
**2)** Serendipitous radio- and optically-selected quasars lying in IPC fields.

The target sample contains 547 objects of which 481 have been processed as of November 1988. Those remaining have specific problems which require individual attention. The serendipitous sample will contain roughly the same number of quasars; analysis of this sample is just beginning. All quasars lying in IPC fields which are included in Hewitt and Burbidge (1987, 1989) or Véron-Cetty and Véron (1987) as well as all Seyfert galaxies classified as S1 by the latter will be included in the database. As a whole this sample does not constitute a *complete* sample in any sense, although it contains a few subsets which do: the PG quasars (Tananbaum *et al.* 1986); the 3CR radio quasars (Tananbaum *et al.* 1983); and the Braccesi sample (Marshall *et al.* 1984).

## 2. THE DATABASE

The data summarized below will be available in tabular form as a published hardcopy, in machine-readable form on tape or CDROM and by direct access via computer network using an INGRES$^{TM}$ (a trademark of Relational Technology) interface.

The database will contain the following basic parameters for each quasar: co-ordinate designation (epoch 1950) and other common names; optical coordinates; redshift; galactic $N_H$ (Stark *et al.* 1984); optical magnitude and 2500Å monochromatic luminosity; and references for all of the above. The X-ray data will include:

date of *Einstein* observation; exposure time; IPC sequence number; detector gain; off-axis angle; flag for non-detection; count rate and error; broad band (0.16-3.5 keV) flux; monochromatic (2 keV, source frame) flux; monochromatic (2 keV) luminosity; broad band (0.5-4.5 keV) luminosity; and $\alpha_{ox}$ (effective 2500Å to 2 keV slope, Tananbaum *et al.* 1979). Three-$\sigma$ upper limits will be given for sources not detected with the IPC.

Following the finding of Wilkes and Elvis (1987) that the soft X-ray slopes of quasars span a range from $\sim 0.0$ to 2.0 in energy index, $\alpha_E$ ($f_\nu \propto \nu^{-\alpha_E}$), fluxes and luminosities listed are computed using $\alpha_E = 0.0$, 0.5, 1.0, 1.5 and 2.0, respectively. A Friedmann cosmology with a Hubble constant of $H_o = 50$ km s$^{-1}$Mpc$^{-1}$ and a deceleration parameter of $q_o = 0.0$ is adopted.

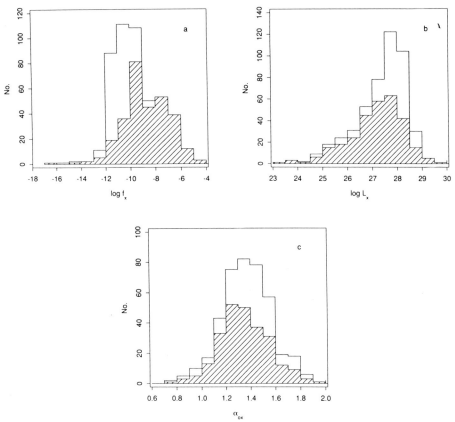

Figure 1: The distribution of X-ray properties of the quasars: (a) monochromatic (2 keV) flux distribution (keV cm$^{-2}$ s$^{-1}$keV$^{-1}$); (b) broad-band (0.16-3.5 keV) luminosity distribution (ergs s$^{-1}$); (c) $\alpha_{ox}$ distribution. The shaded regions indicate positive X-ray detections and the unshaded regions indicate upper limits for X-ray flux and luminosity and lower limits for $\alpha_{ox}$.

## 3. ANALYSIS

The quasars were analysed using a fully automated procedure to ensure uniformity and consistency. The IPC data were processed with the Rev 1B processing version and the most recent versions of the effective area table and the IPC gain calibration (Harnden *et al.* 1984) were used. Our analysis procedure employs a $2'.4 \times 2'.4$ box to determine the presence of a source and a $3'$-radius circle for flux determination, both utilizing the broad band (0.16-3.50 keV). Background subtraction was performed with reference to the background map for each image using a cell of the same size and position as that for the source image. An additional correction was made for counts (mostly low energy) that fall outside the $3'$ circle used for determining the flux. This correction factor, typically 1.8%, was determined with reference to the pulse height (PH) distribution for each source, since the point response is a function of PH and depends on the source spectrum. The effective-area table incorporates a mean correction; this extra factor reflects the difference between the source spectrum and the mean spectrum assumed in computing the table. The X-ray fluxes were corrected for absorption by gas along the line of sight using galactic $N_H$ values determined from the Bell Laboratories survey (Stark *et al.* 1984). For southern quasar positions, which are not covered by this survey, the older maps were used (Heiles and Cleary 1979).

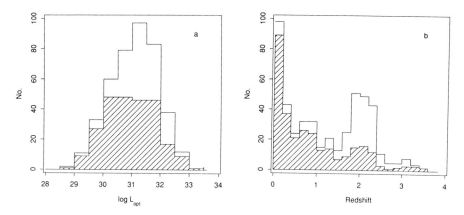

Figure 2: The distribution of optical properties of the quasars: (a) monochrommatic (2500Å) optical luminosity (erg $s^{-1}Hz^{-1}$); (b) redshift. The shaded regions indicate positive X-ray detections and the unshaded regions indicate X-ray upper limits. These distributions cannot be used as presented to infer those for quasars as a whole due to selection effects.

## 4. CURRENT STATUS

The database currently contains X-ray fluxes for 582 observations of 481 quasars from the first subset (targets). 297 of these objects (62%) are detected with the IPC. Analysis of this subset is now complete apart from 66 objects which require individual treatment due to specific problems.

Included in the database are 66 PG objects, 33 3CR objects and 35 Braccesi objects. The PG quasars were randomly selected from the original complete sample (Schmidt and Green 1983). The 3CR and Braccesi quasars are both complete samples (Schmidt 1968, Braccesi, Lynds and Sandage 1968). The remainder of the sample is a heterogeneous mixture of quasars and Seyfert 1 galaxies which were observed by many different investigators for various scientific reasons. In Figure 1a we show the distribution of the broad-band X-ray flux of the targeted quasars corrected for line of sight absorption and assuming $\alpha_E = 0.5$. Similar plots of monochrommatic X-ray luminosity and $\alpha_{ox}$ are shown in Figures 1b,c. Figures 2a,b show the distributions of optical apparent V magnitude and redshift respectively. Note that these figures serve merely to illustrate the range of parameters present in the sample. Due to the incomplete nature of this sample, they do not indicate directly the properties of quasars as a class.

Once analysis of the target sample is complete we plan to utilize this large sample to extend the work of Avni and Tananbaum (1982, 1986) on the X-ray and optical luminosity functions of quasars, to study the relations between various quasar properties and to generate maximum likelihood estimates for the luminosity distributions.

This work was supported by NASA contract NAS8-30751.

## REFERENCES

Avni, Y. and Tananbaum, H. 1982 *Ap.J.* **262**, L17.
Avni, Y. and Tananbaum, H. 1986 *Ap.J.* **305**, 83.
Braccesi, A., Lynds, R. and Sandage, A. 1968 *Ap.J.* **152**, L105.
Harnden, F.R., Fabricant, D.G., Harris, D.E. and Schwarz,J. 1984 *SAO Report* No. 393.
Heiles C. 1975 *Astr. Ap. Suppl.* **20**, 37.
Hewitt, A. and Burbidge, G. 1987 *Ap.J. Supp.* **63**, 1.
Hewitt, A. and Burbidge, G. 1989 *Ap.J. Supp.* **69**, 1.
Marshall, H. L., Avni, Y., Braccesi, A., Huchra, J. P., Tananbaum, H., Zamorani, G., and Zitelli, V., 1984, *Ap.J.*, **283**, 50.
Schmidt, M. 1968 *Ap.J.* **151**, 393.
Schmidt, M. and Green, R. F. 1983 *Ap.J.* **269**, 352.
Stark, A. A., Heiles, C., Bally, J., and Linke, R. 1984 *Bell Labs*, privately distributed magnetic tape.
Tananbaum, H. *et al.* 1979 *Ap.J.* **234**, L9.
Tananbaum, H., Wardle, J. F. C., Zamorani, G. and Avni, Y. 1983 *Ap.J.* **268**, 60.
Tananbaum, H., Avni, Y., Green, R. F., Schmidt, M. and Zamorani, G. 1986 *Ap.J.*, **305**, 57.
Véron-Cetty, M.-P. and Véron, P. 1987 *ESO Scientific Report* No. 5.
Wilkes, B.J. and Elvis M. 1987 *Ap.J.* **323**, 243.

# THE EINSTEIN ALL-SKY SLEW SURVEY DATA BASE

**Martin Elvis, D. Plummer, G. Fabbiano**

Harvard-Smithsonian Center for Astrophysics

When the *Einstein* satellite moved from pointing at one target to pointing at the next the IPC, if it was in the focal plane, was for the most part kept switched on and observing. It thus recorded a one degree wide strip of the sky, more or less on a great circle joining the two pointings. Each of these 'slews' across the sky is individually not very sensitive, $\sim$1 IPC ct s$^{-1}$, but by adding together all the slews from the whole mission we accumulate $\sim 2 \times 10^6$ s of exposure. From this we can construct an IPC All-Sky survey with an exposure of 15–100 s for 50% of sky. This implies a sensitivity of $\sim 3 \times 10^{-12}$erg cm$^{-2}$s$^{-1}$, and predicts $\sim$1000 extragalactic sources, plus stars and Galactic sources.

Although only one tenth as sensitive as the ROSAT survey, which will be carried out after the ROSAT launch in 1990, the Slew Survey should be valuable for several purposes:

- producing a *log N-log S* curve based on a single instrument from the brightest, Slew, sources through the *Einstein* Medium Survey to the Deep Surveys. This should remove some of the ambiguities in present *log N-log S* studies. The higher energy range of the *Einstein* IPC makes it less susceptible to the effects of intervening absorption.

- providing bright targets for detailed ROSAT and ASTRO-D investigations.

- comparing with the much more sensitive ROSAT all-sky survey to pick out unusual sources based on variability or spectrum

- identifying HEAO-A1 sources (in combination with HEAO-A3 diamonds), and resolving confusion problems in the high energy sky surveys.

- connecting the flux scales of sources in the hard and soft x-ray bands.

The first version of the survey contains 1024 sources (figure 1). However this still contains known aspect problems which are now being addressed. We hope to make the Slew Survey source list available by the end of 1989. More work will be needed to establish reliable time histories for sources and upper limits for non-detected objects. The source list will contain: source name (1ES....), RA, decl. (2000.0), with uncertainties, number of photons in detection, exposure time, corrected count rate and number of background counts. More information will be added later (*e.g.* dates/times observed, number of photons in each pass), including proposed identifications.

Slew2d.1 (1024  <.00001 Probability Sources) Galactic Coordinates

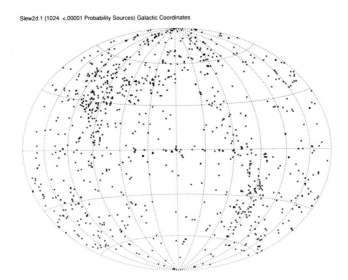

Figure 1. Distribution of sources from the first version of the Slew Survey (in Galactic coordinates).

Identification of these bright sources should be relatively easy– many will be previously cataloged (*e.g.* PG Quasars, Mrk galaxies, Abell clusters, SAO stars). Many new sources will show up in the IRAS and Green Bank surveys. We will use the Minnesota POSS digitization and the ROE UK Schmidt 'COSMOS' digitization to derive the bulk of our optical magnitudes for Slew source counterparts. The IRAS survey and the Green Bank continuum surveys from the late 300-foot telescope will also be valuable aids. An identification program has been funded by the NASA ADP program. We shall maintain a public on-line data base of sources and suggested identifications and we welcome contributions to these.

This work is supported by NASA contract NAS8-30751.

# THE EINSTEIN EXTENDED MEDIUM SENSITIVITY SURVEY DATABASE

**I.M. Gioia[1,2], T. Maccacaro[1,2], S.L. Morris[3], R.E. Schild[1], J.T. Stocke[4], and A. Wolter[1]**

1) Harvard/Smithsonian Center for Astrophysics, Cambridge, MA
2) Istituto di Radioastronomia del CNR, Bologna, Italy
3) Mount Wilson and Las Campanas Observatories, Pasadena, CA
4) Center for Astrophysics and Space Astronomy, Boulder, CO

## 1. THE SURVEY

The Extended Medium Sensitivity Survey (EMSS) is a collection of serendipitous x-ray sources detected with the Imaging Proportional Counter (IPC) on board the *Einstein* Observatory with well-defined selection criteria and well-determined sky coverage. This sample is composed of 835 x-ray sources, with fluxes in the range $6 \times 10^{-14} - 1 \times 10^{-11}$ erg cm$^{-2}$sec$^{-1}$ (0.3-3.5 keV band), selected in a total of 780 deg$^2$ of sky using the criteria outlined in the next section. Table 1 summarizess the properties of the EMSS.

| Table 1 | |
|---|---|
| Properties of the EMSS | |
| IPC images analyzed | 1435 |
| Lim. sensitivity (erg cm$^{-2}$s$^{-1}$) | $5 \times 10^{-14} - 3 \times 10^{-12}$ |
| Area for IPC field (sq.deg.) | 0.54 |
| Total area of sky (sq.deg.) | 780 |
| Sources detected | 835 |
| Flux range (erg cm$^{-2}$s$^{-1}$) | $6 \times 10^{-14} - 10^{-11}$ |
| Energy range (keV) | 0.3-3.5 |
| Significance of detection | $\geq 4\sigma$ |
| Sources identified (as of Oct. 1988) | 704 |
| Sources with radio data | 616 |

## 2. SELECTION CRITERIA

*a) IPC fields*

Fields are included in the EMSS when:

- at high galactic latitude $|b| > 20°$ (to avoid regions of high galactic absorption and high stellar density)

- with exposure time ≥800 sec (to have sufficiently good limiting sensitivity)

Fields are discarded when the **target** is:

- a very bright (>0.6 ct s$^{-1}$) source

- an extended target (e.g. SNR, rich and nearby cluster of galaxies)

- a group of targets (e.g. star cluster, group of nearby galaxies)

### b) Sources

Sources are accepted in the sample if:

- Signal to Noise Ratio ≥ 4

- physically unrelated to the target

- not obscured by the detector supporting structure

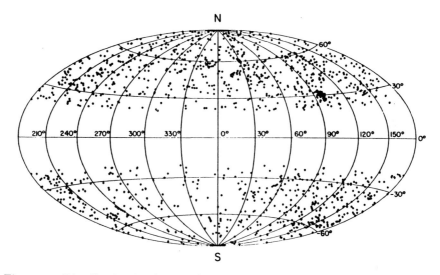

Figure 1: Distribution in the sky (in Galactic coordinates) of the 1435 IPC images used in the EMSS.

## 3. OPTICAL WORK

A program of spectroscopic identification of all the sources is underway. Spectra are taken at the following facilities:

- MMT

- MWLCO 2.5 m

- ESO 1.5 and 3.6 m at La Silla (in collaboration with D. Maccagni and P. Vettolani)

- UH 2.2 m (in collaboration with P. Henry)

CCD photometry (B, V, deep R exposures) at

- FLWO 0.61 m

The optical composition of the EMSS is shown in Table 2.

| Table 2 Optical Identification Content of EMSS | | |
|---|---|---|
| Total sources | | 835 |
| Identified | | 704 |
| | | |
| **EMSS optical ids** | | |
| Active Galactic Nuclei: | | 358 |
| BL Lacs: | | 31 |
| Clusters of galaxies: | | 87 |
| "Normal" galaxies: | | 15 |
| Stars: | | 213 |
| | | |
| **EMSS radio observations:** | Detection | Upper Limits |
| Active Galactic Nuclei: | 77 | 260 |
| BL Lacs: | 30 | 0 |
| Clusters of galaxies: | 40 | 38 |
| Normal galaxies: | 8 | 6 |

## 4. CATALOGS AND LITERATURE SEARCH

Cross-reference with IRAS point source catalog, with the SIMBAD catalog in Strasbourg, and search in the literature for previously known objects have been performed and used to assist in the identification procedure.

## 5. THE CATALOG

The EMSS catalog of 835 sources resides on the CFA HEAD local area network of SUN computers, and will be made available as a published hardcopy list (Gioia *et al.* 1990, in preparation) in machine readable form and *via* remote access through networks. Table 3 shows the present format of the catalog. The list will be kept updated, and will contain information also at other wavelengths (radio, optical).

## 6. ACKNOWLEDGEMENTS

This work has received partial financial support from the Smithsonian Scholarly Studies Grant SS48-8-84 and SS88-3-87, and NASA contracts and grants NAS8-30751, NAG8-442 and NAG8-658.

| Table 3 |
| :-- |
| EMSS Catalog Format |

| | |
| :-- | :-- |
| (1) | Source Name |
| (2) | Position: centroid of x-ray source (1950.0 coordinates). |
| (3) | Error associated to the position, in arcseconds. When a source is detected by the HRI, the uncertainty is 4″ and the source coordinates in (2) come also from HRI. |
| (4) | X-ray flux (first line) and $1\sigma$ error from photon counting statistics (erg cm$^{-2}$s$^{-1}$), in the 0.3-3.5 keV band. |
| (5) | Corrected count rate (cnts/1000 s); net counts corrected for vignetting, mirror scattering and point response function scattering. |
| (6) | Signal to Noise Ratio of detection. |
| (7) | Net counts (first line) and associated error (second line). |
| (8) | Exposure time of the IPC observation in seconds (first line); background counts (second line). |
| (9) | IPC sequence number of the image used (first line); net "extended" counts for sources resolved by the IPC (second line). |
| (10) | Hydrogen column density along the line of sight to the IPC pointing (from Stark *et al.* 1984, Heiles and Cleary 1979, Cleary, Heiles and Haslam 1979). |
| (11) | Proposed identification or classification of the x-ray source (first line). AGN = Active galactic nuclei; BL = BL Lac object; GAL = isolated "normal" galaxy; STAR = star; Class 1 = probable extragalactic object; Class 2 = probable galactic object. References to identification (second line). |

## REFERENCES

Cleary, M.N., Heiles, C., and Haslam, C.G.T. 1979, *A.A. Suppl.*, **36**, 95.
Heiles, C. and Cleary, M.N. 1979, *Aus.J.Phys.Astrophys.Suppl.*, **41**, 1.
Stark, A.A., Heiles, C., Bally, J., and Linke, R. 1984, *Bell Labs*, privately distributed magnetic tape.

# THE EINSTEIN OBSERVATORY DEEP SURVEY DATABASE

**F.A. Primini**

Harvard-Smithsonian Center for Astrophysics

## 1. INTRODUCTION

The *Einstein Observatory* Deep X-ray Survey (EDS) consists of a series of very deep X-ray exposures on selected regions of the sky at high galactic latitude. The main purposes of the survey are to investigate the nature of the extragalactic background through direct source counts at low flux levels and to study the nature of the very faint X-ray sources that contribute to the diffuse X-ray background. The survey includes 9 IPC observations and 34 HRI observations, which, in general, cover the central $32' \times 32'$ regions of the IPC fields.

In total, 178 IPC sources and 202 HRI sources were detected above thresholds set to allow ~1 false source per observation. All source detections and all relevant field data have been inserted into a relational database, using the $INGRES^{TM}$ (a trademark of Relational Technology) database management system. Data from all IPC energy bands and all HRI detect cells are included.

## 2. DATA ANALYSIS

The data analysis is largely that of the standard *Einstein* Rev. 1 HRI and Rev. 1B IPC reprocessing systems, with minor modifications as described here. For IPC observations, all good viewing geometry data have been used; the correlation between source rates and rates obtained using only the best viewing geometry data is good. Minor corrections to IPC count rates and significances have been made to account for systematic errors in the IPC analysis of long exposures.

Background estimates are provided by standard background maps, constructed from templates generated from several deep survey images, with strong sources removed. Although such maps provide a reasonable model of the detector irregularities in the background, since they are constructed from deep survey images themselves, background values used in estimating source counts for faint sources may include a contribution from the source itself. The effect depends on the relative contribution of the deep survey image to the background map template and is corrected *a posteriori*.

Count rate uncertainties are corrected to include the variance in the background. Finally, fluxes in the standard Hard and Broad bands have been modified to use values of $N_H$ appropriate to each viewing direction (Heiles and Cleary 1979; Stark *et al.* 1984), rather than the nominal value of $3.0 \times 10^{20} cm^{-2}$ used in the standard analysis. Positional uncertainties for both HRI and IPC sources have also been recalculated. For IPC sources, a systematic uncertainty that depends on source position has been included. For the hard band, such uncertainties range from $\sim 25''$ on-axis to $\sim 47''$ for sources more than $\sim 15'$ off-axis. For HRI sources, a systematic uncertainty of $3.5''$ has been added to all position errors.

## 3. DATA PRODUCTS

The basic data product is the database itself. Documentation on the organization and contents of database tables, and command files for performing simple queries on the database are also provided. The database is organized into several tables, which contain most of the information originally contained in the IPC and HRI *SDF* files. Data on the fields (*e.g.* sequence number, target position, exposure time) are contained in tables **ifield** and **hfield**. IPC source data are contained in tables **isrc** (position, counts, intensity, *etc.* for the primary *PI* band of the source), **ispec** (hardness ratio, pulse height and *PI* spectra), and **icomp** (source parameters in other *PI* bands in which the source was detected). HRI source data are contained in tables **hsrc** (position, counts, intensity, cell size), **hxtnt** (extent estimate, flux estimates), and hmatches (cross-reference of sources detected in more than one cell size). Field tables are keyed by sequence number and source tables by sequence number and source number within sequence. A cross-reference of IPC and HRI sources is provided in table **ipchri**.

Documentation for each table is contained in an ascii file, named (*table name*).doc. Command files, containing database commands in the *Ingres QUEL* query language, are also provided. Command files are identified by the file extension *.CMD*. They may be used for performing simple queries and also as examples for developing more complex queries.

## 4. USING THE DATABASE

The database is on-line on the CfA HEAD local area network. It may be accessed directly via the *INGRES$^{TM}$* query language *QUEL*, or, for standard queries, *via* pre-defined *INGRES$^{TM}$* command files.

The community may use the database by requesting information from it from the author internet address fap@cfa200.harvard.edu, SPAN address CFAROS::FAP). Requests may range from simple queries on individual sources to *ASCII* dumps of entire tables. All requests will be honored. Users with access to the CfA HEAD local area network may access the database directly.

### References

Heiles, C. and Cleary, M.N. 1979, *Aus.J.Phys.Suppl.*, **47**.
Stark, A.A., Heiles, C., Bally, J., and Linke, R. 1984, *Bell Labs*, privately distributed magnetic tape.

# The Ninth Marquess

# The
# Ninth Marquess

## JON CLEARY

WILLIAM MORROW & COMPANY, INC., NEW YORK   1972

*To Audrey and Ted Willis*

# One

John Aidan, eighth Marquess of Carmel, died, actually but un-officially, at twenty-past five on the afternoon of Saturday, the 17th of February.

"Damn!" said his son Henry, the Earl of Bayard and at that moment the new ninth Marquess. "Why couldn't he have held out till Monday?"

"He tried," said Joe Banning, helping himself to the whisky on the table beside the dead peer's bed. Glass in hand he walked to the window and looked out at the snow-flecked darkness. Snow was still falling heavily and he knew he was not going to make it home to-night. There would be a lot of work for him during the next week: patients suffering from bronchitis, pneumonia, self-pity, perhaps even another heart attack or two. He hated the English winter and wished once again, as he had so often, that he had gone out to South Africa or Australia right after the war. He turned and looked at the new Marquess, the man who, one way and another, had helped keep him here. "I've never seen anyone so reluctant to die as your father. Not that he seemed afraid to die – I don't think he cared about that."

"He didn't." Henry Aidan, called Harry, leaned against one of the posts of the four-poster bed. He must have disturbed some dust on the canopy because it floated down like thin pollen through the pale light of the bedside lamp. "If you must know, he was trying to beat the estate duties."

"Oh, I know all right. You'd have thought *he* had to pay 'em."

Harry straightened up, looked down at his father, and nodded. The flesh on the skull already seemed to have shrunk and the once ruddy skin was now pale tissue; there was no suggestion left of the strength and stubbornness that had been the character-istic of John Aidan while he had lived. The stubbornness, if not the strength, had been there last night right after the first heart attack.

"Don't worry, Harry," he had gasped when his son and Ferguson, the butler, had at last got him upstairs to his bedroom. "I'm not going yet."

"Don't talk," said his son. "You'll last longer if you shut up."

The old man had stared at him for several moments; then he had winked. Harry had been surprised by the simple expression, but then he had winked back. Thinking of it now as he looked down at his dead father, he remarked that it was the first expression of affection, if one could call it that, that had passed between them in almost sixteen years. And abruptly something of the love he had had as a boy for his father came back and he was choked with emotion.

"You'd better tell Constance," Joe Banning said. "Gently – she may break down a bit when she hears it."

"I doubt it."

"She loved the old chap." Banning looked down at his empty glass. "More than you did, Harry."

"Don't start grading the degrees of grief we should feel," Harry said, but there was no sharpness in his voice, only a slight huskiness. He moved away from the bed, away from death itself: he knew that, unlike his father, he would be afraid of it when it came for him. The room felt suddenly cooler and he crossed to the old-fashioned steam heaters and turned them up. Then he turned to the windows, pulled the heavy silk cord of the curtains, and closed them. The room became warmer, if only with colour, as the rich red drapes slid across. "Actually, I'm going to miss him."

Joe Banning debated whether to pour himself another drink and decided against it. On the winds of gossip, which never blow in a predictable direction, he had heard himself referred to as an old soak, a judgment that, drunk or sober, he would contest vehemently. He could hold his liquor as well as any man in the district and he drank much less than several of them, including the dead man on the bed. But old John Aidan had been a secret drinker and only his immediate family, Joe Banning and Ferguson, the butler, had known.

Banning put down his glass and looked across at the new Marquess, the so-so drinker. He saw a tall dark-haired man who would have been too handsome but for his broken nose: my own incompetence, thought Banning, saved him from being merely pretty. He should thank me that I was half-drunk when he came to see me after the riding accident and that I told him a visit to a London specialist would not be necessary. If he had not also been concussed, perhaps he would not have listened to me. Afterwards he had listened to Constance, who had preferred the broken nose.

8

"I suppose you will miss him. One doesn't carry on a running war with one's father for years, then just forget him as soon as he's gone."

"At least we didn't ignore each other."

His father had said exactly that in the middle of last night. Joe Banning had been sent for and somehow had managed to get his car through the snowstorm that was already blanketing the West Country. He had made the old man comfortable, chiding him for having gone out in the park that afternoon when the weather had been so threatening, then suggested that someone should sit with him throughout the night. There had been no possibility of getting a nurse to come in from either of the two nearest market towns; and there was no staff on duty other than Ferguson, the butler. Mrs Dibbens, the housekeeper, had stayed up at the house in Eaton Square to supervise some renovations; a housemaid had been supposed to be on duty but had gone home sick yesterday afternoon; the rest of the staff only worked alternate weekends during the winter. So a roster had been drawn up to sit with the Marquess: Banning, Ferguson, Harry and Constance each to do two hours. Harry had not minded the discomfort of being woken and getting out of his own warm bed to come here and sit beside his father. He was not devoid of conscience or of love and he remembered a time when he had loved his father, though it had been a long long time ago. John Aidan must have remembered it too.

"At least we didn't ignore each other. But I didn't pay as much attention to you as I should have, not when you were young – "

"You're talking again – "

"A lifelong habit. I'm not going to give it up now." But then he was silent for two or three minutes, his chest rising and falling under the blankets.

"We'll get you one of those oxygen tents first thing in the morning – "

"Oxygen tent be damned! I'm not going to die under one of those plastic contraptions. Hate plastic."

"Who said anything about dying?"

"I did." Another silence but for his rasping breath. Then: "Harry, when did you last look at the family motto?"

"I don't know. I was looking at it, or anyway thinking of it, the day we had our quarrel." He spoke without rancour, sure

9

now that his father would accept that their estrangement was over for good.

John Aidan nodded. "I was looking at it m'self that day. Must have been a different angle. But any angle, it's a good one. One of the best. Realism above all else."

He stopped again, staring at the canopy hung above the huge four-poster bed. Every morning he woke to gaze up at the coat of arms woven into the faded silk. There had been additions to the design over the centuries, but the original statement was still predominant: Roger Aidan, the first baron, had fought for the Cross against the Saracens. John Aidan knew by heart the blazon of the coat of arms: quarterly, 1 and 4, a cross engrailed argent; 2 and 3, parted per pale argent and sable, on a chaplet four fleur-de-lis counterchanged, in the centre of the quarters a crescent or, for difference. Mantling vert, double argent. Crest, upon a wreath of his liveries, a horse rampant argent; in an escroll under the shield this motto, *Veirs en tout*. It was the language of heraldry and gibberish to the uninformed, but it pained him, a traditionalist, when he had to explain in layman's terms what the description meant: in the first and fourth quarters of the shield was a silver cross drawn with a series of connected small semi-circles to form the lines of it; in the second and third quarters, which were each cut vertically into silver and black halves, were circular wreaths on which were pinned four fleur-de-lis; and in the centre of the shield, linking the four quarters, was a golden crescent. Green mantling fell away on either side from the silver helmet above the shield. Above the helmet itself, his rear hoofs planted on a wreath, a silver horse reared. Beneath the shield was a scroll bearing the motto, which, translated roughly from the Norman French, was *Realism above all else*. The proper name for the whole emblazonment was an achievement; sometimes he stared up at it and pondered how much he had ever achieved himself; it was one of his regrets that he had never been able to add anything to the family's symbol of its own pride. The canopy had been repaired countless times and the coat of arms, stitched and restitched, now looked like a splintered shield about to fall apart. The big ornamental covering did not look as if it would long survive himself. He wondered if his son would move into this bed, the head of the family's bed, when he was gone. For tradition's sake he hoped so. He had the dying wish, though without hope at all, that his son's wife would also move in here.

"It's not realistic to ignore a sign such as I've just had. One more attack like this and I'm gone."

"You could go on for years if you're careful."

"You're a fool if you think I have years to go."

"You're talking as if you're throwing in the sponge. As if you *want* to die."

The grey head rolled from side to side on the pillow. "Last thing I want to do." A thin smile fluttered on the blue lips. "A joke. Never thought I had much sense of humour, did you?"

"No, you're wrong. You have one. I just happened to think it was a bit warped, that was all."

"Perhaps." He lifted a hand and rubbed the bridge of his fine straight nose. He had once been almost as handsome as his son, but age and the secret drinking had eroded the face that had inherited the traditional Aidan good looks. The vast house was hung with portraits of handsome Aidan men going back to the oldest portrait, painted in 1488; some of the women who had married into the family had had no more beauty than that of their sizeable dowries, but somehow they had not managed to spoil the tradition of family good looks.

"I should have listened to you, y'know. About the estate, I mean. Turned it over to you when you first suggested it."

"You didn't believe me when I told you I was only thinking of the family."

"No." The once strong chest rose and fell weakly while he pondered on the suspicion and stubborn pride that had kept them apart. "No, I didn't. And I'm sorry now. Mean that, Harry." His son nodded, accepting it. "Besides, thought I was going to out-live you."

"You've had a good run."

"I'm only seventy-five," John Aidan said, as if angry at the fate that had called for him so soon. "Was going to live till I was ninety, just to spite you."

"Spite is no elixir."

"That's good. Tell 'em that where I'm going." Then he said with a regret in his voice that sounded less bearable than the pain in his chest, "We should've talked more, Harry."

"Go to sleep," he said gently, and pulled the blankets up round his father's throat. "We'll talk again to-morrow."

"Good night, Harry. And don't worry. I know down to the penny what you'll have to pay. Over five and a half million. I'll last till Monday."

But he hadn't. He had fought desperately and, Harry now mused ironically, perhaps it was the effort to stay alive that had killed him.

"You'd better go and tell Constance," said Joe Banning. "I'd do it, but I think it's your place. I'll tell Ferguson, if you like."

"No, just stay here. Do you mind?"

"I'm used to corpses." Banning looked at the whisky decanter, then helped himself to another drink. "Particularly those of old friends."

Yours is resigned indifference, Harry thought; and wondered if it was any better than his own. Whatever had happened to love? But there was no answer to that question, not in this house on this night. He went out of the room and down the Long Gallery to the east wing, feeling emptier than he had felt in – sixteen years, he supposed it was. The portraits of the previous earls and marquesses were hung here in the gallery and he was aware of their watching eyes as he passed in review beneath them. For a good part of his life the paintings had been no more than furniture at which he rarely glanced; he had then been the younger son and as such his father had never bothered to encourage him to pay homage to his ancestors. But now he was the only son, the heir, and it would be his duty to see that his father's portrait, painted ten years ago by Coldstream, would be taken out of the old man's study and hung at the end of the line. He had never had his own portrait done, but he supposed he would have to see about that soon. He wondered how a painting of himself by Bacon would look, screaming bloody blue murder.

He stopped, looked back down the Long Gallery. He was all at once aware of the silence of the great house; the ghosts were gathering again. He shivered, went on, knocked on the door of Constance's sitting-room and opened it. "I come in?"

Constance was sitting on a couch by the fire, an open book on her lap. Her suite was centrally heated, but she insisted that the heaters be turned off and that fires be kept going constantly in her sitting-room and bedroom. Harry had not been into her bathroom for years, but he knew there was a big old-fashioned Dutch porcelain stove in there and it, too, was always kept stoked. The rejection of central heating for fires was a habit that she had only developed in the last year or two, as if she was now determinedly retreating to the past, to the memories of the crumbling, unheated house in Northumberland where she had grown up.

She looked up, closing her book. "He's dead, isn't he?"

He nodded. "How did you know?"

"It's in your face."

"I must be slipping."

"Perhaps you're becoming human again." She waved him to the twin couch opposite her. "Would you like some tea?"

He nodded and sat down. As he leaned back he became aware of the exhaustion that had crept up on him; the broken night, the long day and the climax of unexpected emotion had taken their toll of him. He took the cup of tea Constance passed him and said, "I didn't come here to fight."

She hesitated, biting her lip, then nodded. "Harry, I know you didn't. I shouldn't have said that – about being human again, I mean. You've always been that – too much so, sometimes." She closed her eyes quickly, then opened them again. "Sorry. That sounded like another criticism."

"No, it sounded just like Father advised me to be. Realistic above all else." But he smiled to show there was no sting in his reply.

"Did he die quietly? I'd hate to think he went out with a lot of pain."

"He went quietly. He'd been sleeping, then he opened his eyes and said, 'Say good-bye to Constance' – " It was a lie, the old man had died without a word, but he knew she would like to hear that his father had remembered her in his last moments.

Constance said nothing for a while, her hand playing with the book beside her on the couch. He glanced at it; it was *Scandal in the Assembly*, not a recent book, and he wondered why she should be reading about the Catholic attitude towards divorce at a time like this. She saw the direction of his glance, turned the book over, and said, "So now we're the Marquess and Marchioness. I wonder how the family are going to feel about that?"

"You're harbouring old grudges – "

"Not me. Your father finally accepted me, but your sisters and your uncles and aunts and the rest of the tribe – " She shook her head. "I'm still someone who came in by the tradesmen's entrance."

It was not a subject he wanted to discuss just now, so he said, "Do you want to go up and see Father?"

She shook her head again and he saw that she was close to tears. Till now it had never occurred to him that she might have loved his father. They had always been friendly towards each other, at least for the last fifteen or twenty years, but to him they had only seemed close when they were allied against him in some common opinion. "It's silly, but I'm afraid of the dead. He would have laughed at me for that – " She was silent for a moment, regaining control of herself, then she said, "We'd better phone the boys and Robin. When will the funeral be?"

"That's what I wanted to talk to you about." He stirred his tea while he stirred the idea in his mind. "I'd like to leave it till next Thursday or Friday."

"A week? Why on earth – ?"

He said slowly, "Do you know how much we are going to have to pay the government in duties? Five and a half million."

"Do we have to talk about money so soon?" But she saw that he thought the matter urgent and important and she nodded. "I suppose we have to. Do we have that sort of money? I mean, available?"

"No. We have it in land and property and shares, but it will mean selling off a lot. If Father had made over the estate to me when I first asked him – or if he had even set up a trust – "

"Monday was to be the day, wasn't it? No, I haven't been watching the calendar – I never have been a very good house-keeper – What are you smiling at?"

"I'd never thought of it as housekeeping before."

"Well, whatever you call it. Father came to me on – Wednes-day, I think it was – and said, 'Next week, my dear, you'll be one of the richest women in England.' "

"I'll bet that wasn't what he said."

"No," she admitted. "What he said was, 'You'll be the wife of one of the richest men in England.' That's all I am, aren't I?"

"I've never denied you anything you wanted. What I meant was that Father would have been very careful of the exact order of things. He believed very strongly in the principle of primo-geniture, even though for the last sixteen years it must have nearly killed him to know everything would come to me. The eldest son, and only the eldest son, is the one with the money."

"And everyone else is dependent on him. It's a bloody awful system, one that should have gone out with Magna Carta!" He could see the temper rising in her; she had an almost Mediter-ranean tendency to erupt suddenly and violently. "God, I shudder when I think that if Roger had lived and we had needed money, we'd have had to come cap in hand and ask him for it. I'm going to do something about it when *you* die, to see that

Robin and Matthew are not dependent on Richard for handouts!"

"What, for instance?" But he did his best not to sound argumentative. "The only thing that has kept estates like ours intact has been the system. The estate would break up in no time if it were split up. And without an estate, very soon there would be no family. Or not one that counted for much." He got up, began to move about the room; he had a restless energy that tired other, more languid people more than it did himself. "England is full of families that once used to mean something but are nobodies to-day. And all because they let their estates disintegrate into nothing."

"What are you interested in – the money or the family?"

"Both." He turned back to her, paused and looked carefully at her. He knew he was going to have to tread carefully with her; careful though she was with money, she had never understood the intricacies of handling it at high levels. "We don't have five and a half million, not in liquid cash. And I'd have a great deal of trouble raising it just now."

"One of the richest men in England?" Her tone suggested she thought he was joking.

He sighed. "I could sit down and show you sums on paper that would explain it all, but even then I doubt if you'd understand."

"I know enough about money never to have overdrawn my account at the bank."

"That," he said, and he only half-intended to be humorous, "is what limits your understanding of what can happen when one is handling and is responsible for a lot of money."

"I'm not insulted," she said coolly. "Failing to understand the sort of finagling that goes on among your City friends doesn't make me a moron."

He kept his patience. "It boils down to this – at this particular time many of our holdings, the ones we could most readily sell and would not depress the market here, are worth only about forty per cent of their true value. We've had a bad drought on the properties in Australia – "

"I know that."

" – and our mining interests, both out there and in Canada, because of the fall in demand for nickel, are now down to much less than what we paid for them. They will eventually recover, both the cattle holdings and the mining shares, but it will take

15

time, perhaps a year or two, even four or five years. If we sold them now, on paper we'd be selling close to six millions' worth of property and shares to raise about two and a half million and we still wouldn't have enough. There are other factors involved in our other interests, more important ones, but that is the simplest one to explain to you."

"What do you want to do?"

He took a deep breath. "I don't want Father's death to be official till Tuesday or Wednesday. Wednesday would be better – it wouldn't look as if he had run it so close."

She shook her head. "It's disgraceful!"

"Well, if that's the way you feel about it – I can't do it without your co-operation. Yours and Joe Banning's."

"I don't mean the covering-up or whatever we'll have to do. I mean you and I here talking about money and estate duties and the family – and he isn't even dead half an hour! God, is that all his death means to you?" She looked up at him angrily, her pale face flushed with her temper. "How soon will you start dividing me up – " she gestured around her, at the only possessions she considered hers " – when I die?"

He strode towards the door. "I'll come back later. I told you I didn't come in here to fight with you – "

"No, wait a minute – " She sank back, doing her best to let the tension run out of her. Why is there still some love left for him after all this time? There's not much, but just enough to stop me from hating him. But even if I did hate him, it would still be there: hate doesn't eradicate love. Only indifference, what he feels towards me, does that. But I could never be indifferent to him. "I'm sorry, I didn't mean us to fight. But I *loved* your father, don't you understand that?"

"I'm beginning to realise it now – "

"Harry, come back and sit down." She composed herself again, arranging herself with that convent-bred neatness that had persisted from her girlhood. "What are you suggesting – that we try to defraud the government?"

"I can't think of another way of putting it." He sat down opposite her. "But be a little less blunt when we broach the subject with Joe Banning."

"Twenty-eight years of being married to you has taught me something of the uses of euphemism."

She poured herself another cup of tea, added two lumps of sugar, stirred it: the whole action as dainty and decorous as

16

that of the local parson's wife who came to the Hall for her annual tea. He knew that if ever it had been possible for him to make it to the top in politics, to have become Prime Minister as he had once dreamed of doing, Constance would have been the perfect partner. She would have handled insults from satirists, dodged thrown eggs from students, and listened to lying flattery from diplomats from unaligned countries with the same degree of unruffled grace. Yet he recognised that her routine with the tea cup was in itself satirical: she had always been able to send up herself and everything she stood for. And yet, paradoxically, she was even more dedicated to tradition than he was.

"Now tell me what you are planning to do," she said. "I don't want the estate to be broken up, any more than you do."

2

John Aidan, *sans* title, *sans* fortune, was buried at eleven o'clock on the morning of Friday, the 23rd of February. The weather was still cold, the low grey skies stretching far to the west and out into the Atlantic; the lake in the park had frozen over and waterfowl moved about on the ice with the air of angry and bewildered tenants who had been evicted. It had stopped snowing on the Wednesday, the day John Aidan's death had been announced, and most of the roads in the district had been cleared. Mourners were able to get through and the old man was buried in the family mausoleum with a suitably sized and important crowd to bid him his last farewell. All branches of the Aidan family were represented, drawn in from all over Britain, Ireland and various parts of the world like skeins wound back into the ancestral ball by the name, heritage and fate that was common to them all. The City came down from London, its representatives sleek as seals in their dress and demeanour; one property tycoon wore an astrakhan hat and even when he left was still thought to be the Russian ambassador and complimented, out of his hearing, on his Christian behaviour during the chapel service. The landed gentry came in from the surrounding counties, arriving in their Bentleys in clothes that suggested their families kept wardrobes, inherited and unrelated to fashion, for just such occasions. The two sides, like opposing football teams, took up positions on opposite sides of the Great Hall before moving into the chapel. The villagers and the workers from the estate turned up in their best clothes, their weatherbeaten faces showing more sense of loss than the faces of the newly arrived family; the eighth Marquess had been very popular with his tenants and his workers, and none of them was sure that things would not change under the ninth. Half the Shadow Cabinet and several elderly Tory peers arrived, the latter warmly wrapped against the cold and the possibility of joining the late Carmel. And Tom Brewster, the Prime Minister, put in an unexpected appearance.

"I've never forgotten that I was born in one of your villages," he told Harry as the latter greeted him. "Your father never forgot it either. Every time we met he asked me when I was going to retire and come home."

"Do you still think of the village – Skipbrook, isn't it? – as home?"

Brewster smiled, unwinding the thick woollen scarf from his thick muscular neck. After thirty-five years in politics he still looked like a farmer more interested in cows than the Commons, but he was the best politician of either party in the House. And he was also an able statesman, a not always easy accomplishment for a politician. "Home is Number Ten. They'll bury me from there. I hope you'll come."

"I'll be delighted," said Harry, smiling back. Labour had been in office two terms and all the polls were confidently predicting that when Brewster called the next election he would be returned to Downing Street for yet another term. "But thanks for coming to-day. I think my father would have been pleased."

"I hope so." Brewster looked up and around him at the Great Hall. The big chamber, almost three stories high and surrounded on three sides by galleries, was the kernel of the vast house. The walls were hung with tapestries, banners and shields, all evidence of the family's esteem for itself; suits of armour stood about like incurious security guards; only the family and the servants knew of the constant battle against woodworm in the panelling behind the panoply. Brewster, who admired tradition but never aspired to it in regard to himself, who was his own man and always had been, looked back at Harry. "Nice place you have here. Lucky you were able to keep it."

Harry kept his own face as straight as that of the Prime Minister. "Father always did cut things rather fine. It was his joke that he was born exactly nine months to the day after my grandparents were married."

"What are you going to do with the place? Keep it on?"

Harry was surprised at the question; it had not occurred to him that anyone should expect him to do otherwise. "Wouldn't you?"

Brewster shrugged. "Rose and I have never owned a house in the forty years we've been married. Possessions have never meant anything to me – they are too much of a headache. I hope the Probate boys don't make too much of a headache of this for you."

"They won't," said Harry confidently.

But he knew there would be inquiries. The bureaucrats of the Probate Registry, less subservient to tradition than to their account books, would not meekly accept the loss to them of five and a half million pounds, not when the deadline had been met so suspiciously closely.

"You're asking a lot of me," Joe Banning had said.

Harry had left Constance on Saturday evening, gone back to the west wing and locked himself in his father's bedroom with Banning. The latter, third whisky in his hand, had not looked surprised when Harry had put his idea to him.

"I know. But think of the stakes, Joe – I don't mean just the money, but the whole of the estate, the farms, the villages, everything. It'll have to be broken up to pay the duties – "

"Not necessarily," Banning interrupted. He was not argumentative, just sounded like a man ready to enter into a mild, friendly debate; he had not spent the last half-hour sitting here thinking of nothing at all. He savoured another mouthful of whisky, grateful again that the old Marquess had never served any but the best twelve-year-old stuff. "You could sell your interests in Australia - they would raise a few million, wouldn't they?"

Harry suddenly looked cautiously at the older man. "Go on."

"I've been sitting here doing a little book-keeping. Simple arithmetic, in my head. The mind reels. At least a simple mind like mine does."

"What sums did you do to knock yourself out like that?"

"Well, your Australian interests for one – the cattle stations and the sheep, and the mining leases. Then there's Canada – there's family money in lumber and mining there, isn't there? And the property in London – of course you're not in the class of the Grosvenor or Portman holdings, but you own more than a couple of semi-detacheds. And what about – the biggest of the lot, wouldn't it be? – the Aidan percentage of Anglo-Qaran Oil? All of it adds up to ten or twelve million easily, perhaps more. All of it saleable and none of it connected with the ancestral estate."

"You forgot the cottage at Montego Bay."

"No, I didn't. I happen to know that's in your own name."

"I'm surprised you took up medicine. Why didn't you become an accountant? Or a private investigator?"

"Save the sarcasm, Harry. I got all my information from him." He nodded at the dead man, face still uncovered, in the bed. "He told me all about it this morning. He put exactly the same suggestion to me, supposing that he died before Monday."

"And what did you tell him?"

"I said I'd think about it." He sipped his drink, sat back in the big faded red leather chair. He was a short thick-set man with a grey military moustache and a freckled bald scalp above wings of sandy grey hair. He looked capable and confident and only he and Harry knew of the cowardice that had made a husk of him for the last twenty-eight years. "What disappoints me, Harry, is that you have to be hypocritical with me. I thought we were two who could always be perfectly honest with each other."

"What do you mean?"

"Why not be frank? You just don't want to part with five and a half million to the government. People all over the country are trying to do the same thing every day – on a smaller scale, of course," he added with a dry smile. "You don't have to give me all that bull about trying to save the breaking up of the estate, the preservation of history – "

"A phrase I didn't use." Harry got up, began to move around. "All right, that's only part of it. The main thing is that I'll have difficulty in raising the five and a half million – that is, without suffering a pretty substantial loss. I'd suffer no loss on the sale of the London properties, but they would bring in no more than a million, perhaps a million and a half – most of it nobody would want to buy just now because the leases are coming to an end. They would rather wait a couple of years and negotiate for the freehold. If I sold everything that's immediately available I'd raise no more than about three and a half million. That would still leave me two million short."

"I'm enjoying this little flight into the stratosphere. I had to have a talk last week with my bank manager who was objecting to my overdraft of four hundred quid. Go on."

"This is a hell of a place to be having this sort of discussion." Harry looked towards his dead father. "Still, I suppose he wouldn't object."

"I'm sure he wouldn't," said Banning, and pointed up at the motto on the canopy above the bed. "Realism above all else. But what about Anglo-Qaran Oil?"

Harry shook his head. "No good. I could sell them any time I like, there are several companies who'd pay above the market price for them, but none of them are British companies. If I put them on the market now, especially with the situation as it is in the Middle East, the instant reaction would be that I had some inside information on what the government was going to do. It could mean a dozen things – deciding to back the Israelis, moving our reliance on oil from the Arabs to some other suppliers less volatile, a whole host of reasons – but the end result would be pressure on me from Whitehall to hang on to our share of Anglo-Qaran."

"One department of the government going against the other? It wouldn't be the first time it's happened, I suppose. But wouldn't the government buy the Aidan shares in Anglo-Qaran?"

"That would have the opposite effect. It would suggest the government was coming down on the side of the Arabs against the Israelis. You may find it hard to believe, Joe, with your overdraft of four hundred quid, but despite the fact that I'm a rich man, at this moment and for the next twelve months or so, I don't have any liquidity. At least not enough to satisfy the Probate men." He stopped with his back to his dead father, looked at Banning. "I know what I'm going to ask you is highly unethical and is very much against the law. You don't owe me any favours – "

"I do, actually. You've kept quiet all these years about what happened at Arnhem. You never told him, for instance." Banning nodded at the dead man.

"You'd have done the same for me."

"You wouldn't have funked it as I did." Banning finished his drink in one gulp. "Fortunately, they don't test you for cowardice at Guy's Hospital, otherwise I'd never have got my degree. Well, what do you want me to do?"

Harry felt a measure of relief at Banning's attitude. He was beginning to feel some amazement at the enormity of what he was suggesting; he would not have been angry at Banning if the doctor had refused to co-operate. "We'd have been in the clear if the bloody Socialists hadn't changed the exemption period from five to seven years back in 1968." Banning said nothing and

Harry looked at him sharply. "I've never asked you before, but you're not one of their voters, are you?"

"Allow me one or two secrets, Harry. I could be a Communist, a secret one. That's the best way – vote for the overthrow of capitalism while enjoying all its perks. Just like the students."

"I wouldn't put it past you. Anyhow – " He resumed his pacing up and down. "Stay here for the weekend, establish the fact that his heart attack was a bad one. Go home Monday, attend to your practice, come back Monday evening to visit Father. The same on Tuesday. He will die early Wednesday morning – you will stay here Tuesday night because he will have had a second attack."

"What about the servants?"

"They aren't due back till Monday morning. Constance and I can keep them out of this room, tell them you have left orders Father is not to be disturbed in any way. Thank Christ Mrs Dibbens is up in London – she's the real busybody in this house. That leaves only Ferguson. I think he will be on our side – I once heard him tell my father that his loyalty was to the family, not the government."

"There's one other snag – the undertaker. As soon as he sees it he'll know the body has been dead longer than twenty-four hours."

"I've thought of that. We'll use Davis, the estate carpenter, to make the coffin – he's also a cabinet-maker. There's no law, is there, that says we have to use an undertaker?"

"None that I know of. So long as I sign the death certificate and the body is buried privately in the family vault, you don't have to call in any outsider. Well, we'd better start now – " Banning got up, crossed to the steam heaters. He appeared to have thrown off his reluctance and looked almost enthusiastic. "We're fortunate the weather is as cold as it is. We'll keep the heating off and just cover him with a sheet – it will be as good as keeping him in a freezing chamber."

Harry felt a sudden revulsion, a wave of disgust at how they were going to use his father; Banning's cold professionalism all at once pointed up the callousness of what they were about to do. He got up and moved across to the bed, stood looking down at his father. He was surprised that the old man already looked a stranger.

"Having second thoughts?" Banning was not insensitive.

Harry hesitated, then shook his head. "No. One of the last

22

things he said to me was stick by the family motto. Realism above all else."

"A sensible suggestion. Five and a half million is a nice bit of realism."

Anger flared up in Harry, but he quickly doused it. He could not afford to be indignant: Joe had stated the truth of the situation. But: "Joe, I have the feeling that you'd like, as the Americans say, a piece of the action."

It was Banning's turn to shake his head. "If I wanted a piece it would have to be a big piece, half a million at least. Otherwise why hold you to ransom? If one is going to be a crook, always be a big one. No, Harry, I wouldn't know what to do with a lot of money, not now. Twenty-five years ago it might have been different." He paused, opening his eyes in a wide stare as if trying to look back down the years to an horizon that had disappeared forever. "Just send me a case of Chivas Regal every Christmas. Each year on the anniversary of your father's death I'll toast our success at getting away with it."

And so far, though it was still early, they had got away with it. Ferguson, as Harry had expected, had proved his loyalty was in the right place; there was no one more Tory than a butler and any conspiracy to defraud a Labour government was an honourable pursuit in his eyes. The fact that the country itself was being defrauded did not appear to occur to him and Harry did not feel it was his place to point it out.

Now John Aidan was ready to be buried, to be laid to rest among his ancestors, another link ready to rust on the long family chain. The local parish priest, Father Helvick, still looking like the Rugger forward he had once been, conducted the service in the chapel where Cardinal Wolsey had once said Mass; at John Aidan's request there were no eulogies and he was buried quietly without fuss but with grief and regret. Harry, Constance and their three children sat by themselves in the front pew; Harry's three sisters sat behind them with their husbands; the rest of the Aidans, uncles, aunts, cousins, crammed themselves like peak hour commuters into the other pews. The other mourners, including the Prime Minister, sat out in the Great Hall, murmuring among themselves like constituents waiting to see their Member of Parliament, an analogy that struck Brewster and which he remarked to his neighbour, Charles Wolfe, the Opposition leader.

"That's all past for Harry," croaked Wolfe. Pink with the cold and the two double whiskies that, no matter where he might be, were his substitute for the morning coffee break, he sat huddled in his chair, a bullfrog of a man who ran his party with a pudgy hand of iron. "He'll be resigning his seat now. A pity we're going to lose him to the Lords."

"He won't like it. There's never any excitement in that place. How would you like me to recommend you for an earldom?"

"I'd shoot you if you did," growled Wolfe. "Here they come. They're a breed apart, aren't they?"

"The Aidans?" Brewster stood up as the mourners began to file out of the chapel into the Great Hall.

"Our aristocracy. As if the same dose of sperm has been used over and over again."

"I don't think Harry looks watered down."

"A full red-blooded one pops up occasionally. But don't quote me."

"Do we have to traipse outside to this mausoleum?" Brewster, who could take or leave the aristocracy and was never worried by them, had been surprised by Wolfe's seeming contempt for the traditional pillars of his party.

"Afraid so. Damned inconsiderate of old John, dying this time of year. Might at least have waited till May or June."

"I'll bet Harry wishes he had," said Brewster, but he had turned away and Wolfe could not read if there was something hidden behind the words.

The women and the more elderly male mourners did not venture out into the snow. Constance moved about the Great Hall greeting relatives, friends and strangers, all of whom were already fortifying themselves with the coffee and biscuits that Mrs Dibbens, dignified in new black, had summoned from the kitchen on trays carried by footmen and maids. She ducked her head as Constance went past, nodding approvingly as if to say it was a very nice funeral, indeed.

"She's a snob," twenty-year-old Robin whispered in her mother's ear.

"Your father thinks the same," said Constance. "But just now you are supposed to be missing your dead grandfather, not criticising the servants."

"I honestly don't miss him, not now. But I shall. He was a dear old man, even if he never knew how to talk to girls my age. God, why do people have to die just when you're bridging the gap, getting to know them?"

Robin had inherited the looks of both her mother and her father: her mother's blonde hair and pale complexion, her father's strong features softened to womanly beauty. She was beautiful and knew it and had a constant inner struggle against vanity; she laughed at the *Vogue* captions of her as the most beautiful debutante, but secretly she was pleased and would sneak second looks at the photographs of herself. She was the lovely daughter of a very rich peer: she had everything. But then her mother had had everything up till a few years ago; and what had she now? Which warning was the one thing that stopped Robin from succumbing completely to frivolous conceit. She did not know what had caused the break-up of her parents' marriage, but she knew that when it came time to choose her own husband, even if she were desperately in love, there would be doubts that would rack her like an inherited disease.

"Go over and speak to the Duke of Swindon," said Constance. "He's an old man who knows how to talk to young girls."

"He's an old lecher," said Robin, but dutifully made her way towards the elderly duke sitting in a chair by the foot of the wide stairs. His eyes lit up as he saw her approaching and he seemed to shed years, even if his hormones did not increase proportionately.

Outside the house, up the path that led from the mausoleum discreetly hidden among a small grove of beeches, Harry and his two sons were returning from the burial. The other mourners walked behind them, leaving the Aidan men to their private moment of grief.

"We'll stay in mourning for a month," said Harry.

"All that's a bit old hat, isn't it?" Richard was slightly shorter than his father and thinner; he was handsome but the effect was spoiled by a gauntness that made him look older than his twenty-two years. His dark hair was long, down past his ears, and though he wore a suit he had over it a dufflecoat instead of an overcoat. He looked what he was, a university student interested in radical causes. "Nobody goes in for that sort of thing these days."

"We do," said Harry. "And you'll respect my wishes."

Richard shrugged, shoved his hands deep into the pockets of his dufflecoat and looked at his brother. "I'll leave it to you. You can go to daily Mass for both of us. You're the religious one."

Matthew ignored him, turning his head to look at some red deer as they skirted the frozen lake. He was seventeen and looked like none of the family, a throwback to the blunt-faced hefty Northumbrian men who had been his mother's forbears. He hated school, had no time for the family town house in London, and grabbed every opportunity he could to come back here to the estate where, unselfconsciously, he did his best to have himself accepted by the farm workers as one of themselves. He loved the countryside and already knew the working of the estate better than his father or brother ever would.

Harry, aware of the polar difference between his sons, had only now begun to appreciate the difficulties that lay ahead of him, not immediately but when it came time for him to prepare for his own small journey to the mausoleum back there in the beeches. The whole estate would go to Richard, and unless the boy radically altered his beliefs in the meantime the family's wealth, influence and tradition could crumble abruptly in the next generation. Harry, half-way between the mausoleum and the house, decided he would have to look into the possibilities of a trust.

An hour later most of the mourners, with the exception of those of the family who were staying overnight, had departed. Charles Wolfe, breath smelling of another double whisky that he had managed instead of coffee, took Harry aside.

"Can you come up and see me some time next week? Now you'll be resigning your seat I'd like to talk to you before I have to send someone down to talk to the local committee."

"I'm hoping you'll use me when I go into the Lords. I have no intention of becoming just a backwoodsman."

"Of course we'll use you, dear boy," said Wolfe, pink cheeks ballooning as he smiled placatingly. "But it's a pity you have to move to the Other Place. We still need chaps like you in the Commons."

"I could renounce the title."

Wolfe's eyes twinkled. "You wouldn't do that, Harry. You just wouldn't know how to act as a plain mister. Arrogance like yours can't be wiped out with a stroke of the pen. Come and see me next week. Nice funeral."

Then he left, ducking his head in courtly little bows to the Aidan women as he passed them, assuring them silently if hypocritically that he was the leader of their cause in the place where it counted, the Commons. Harry watched him go, envying him: Wolfe's was the power he wanted but could never have.

He turned to go back and join Constance who he could see was suffering under the barrage of family condolences. As he did so a young man, who was faintly familiar, nodded to him. He stopped, focusing his attention on the young man. "John Ferguson! I'm sorry I didn't recognise you."

"You have enough on your mind this morning."

"Where have you been?"

"Out in the Lebanon, teaching. Or I was. Actually, I've been back a year, but I missed seeing you each time I came down to see my father. I'm at the London School of Economics now." He was a hefty young man who would quite obviously never be the butler his father was; he had an air of impatience about him that suggested he would cut a five-course dinner down to a quick snack. He had dark brown hair, worn unfashionably short, and a square bony face that missed being good-looking by a wide margin yet had its own attractiveness when he smiled. "I don't suppose my father mentioned that?"

"Hardly." It occurred to Harry that he had never once inquired of Ferguson where his son was. Why am I so careless of the existence of people? he wondered. The landscape of his memory was almost barren of figures; he could remember events with startling clarity, but people came and went in his life like shadows. This young man had once been a boy with whom he had had closer communication than with his own children. "I should imagine your father thinks the L.S.E. is the British equivalent of the Kremlin. Is it?"

"Not quite, sir." John Ferguson's face sobered. "May I say how sorry I am your father has gone? He was the one who paid for me to go to university."

"I know." He winced inwardly as a memory pricked him. It was he who had once promised to help the boy through university and had then forgotten all about it. "How did *he* feel about your finishing up at the L.S.E.?"

John Ferguson grinned. "I was never game to ask him. Give my sympathy to Lady Bayard – sorry, Lady Carmel." As he turned away he bumped into Robin. "Hello, Lady Robin."

"*Lady* Robin?"

"Well, it's three years – " He looked her up and down, while Harry watched them both with careful interest. "I must say you look all right, even in black. Have you been doing anything useful lately?"

"Not *useful*. I thought it was enough just to be decorative."

"Still the same old vanity." Then he looked at Harry. "Sorry, sir. We're not used to purely decorative girls at the L.S.E. I'd better go before I start singing the 'Red Flag'."

He grinned at Robin, then headed quickly for the front doors. He walked with long strides, another contrast to his father whose measured tread was that of an archbishop on his way to his own enthronement. Robin looked after him, glowering.

"He hasn't improved his manners."

"Perhaps he was afraid of you snubbing him," said Richard, coming up to his father and sister as an escape from his overpowering relatives. "You're old enough now to get back at him."

"I think he could look after himself," said Harry. "Now, I'd better join your mother. She's ready to show the white flag."

"How does one acquire such a dreadful bag of relatives?" he heard Richard say as he moved away.

"Lack of birth control," said Robin.

Constance looked at Harry with a mixture of relief and annoyance as he approached her. "Welcome to Mafeking. I thought I was going to be left to cope with all this horde on my own."

Harry managed a graceful smile, taking her hand in his, and the watchful sisters, aunts and maiden cousins, alert for any sign of a reconciliation, nodded hopefully. Perhaps Harry and Constance would get together again, open up Aidan Hall once more as the splendorous social focal point it had once been. The countless bedrooms could be opened and used, the State ballroom thronged with the best in the land as used to occur in the thirties when Harry's mother had been mistress of the house, the park thrown open to the guns and rods of weekend house parties. A hundred thousand public visitors at thirty new pence a head did not, in the womenfolk's view, constitute a social season. In their eyes Aidan Hall was still shuttered and closed to its proper use, the entertainment of the family and their friends; but Har now that he had the family fortune, might spread some of around for their benefit. Seeing the new Marquess and Marchioness standing close together, smiling at each other, the Aidan women in turn smiled at each other like a roomful of mirrors catching a common light, and mentally began checking off dates in their diaries for the coming summer.

"I think they are already writing out their acceptances for the invitations they hope we'll send them," said Harry behind his smile.

"Nothing doing," said Constance, who had long ago learned how to keep the family at arm's length. "I'm going upstairs. Are you coming?"

They began to ascend the stairs, but Lady Cavanreagh,

Harry's eldest sister Elizabeth, halted them with a stifled shriek. She was five years older than Harry, thinly handsome, formidable and married to a fool of an Irish peer who had too much national pride to batten on his rich English relatives, a fact she had discovered too late.

"My dears, you're not deserting us!" She clutched at her hat, a floppy felt artefact that she had had for years but which now put her in fashion with her teenage daughters. "There's so much we all want to talk about – "

Constance's voice was polite but had the chill of a North-umberland wind through it. "Harry and I would like to say a few private prayers."

"Of course, of course." Elizabeth Cavanreagh backed off; she spent half her time in prayer, asking God to grant her windfalls. "But we'll see you at luncheon?"

"Yes," said Harry, feeling Constance's hand turning to a claw in his own. "But it will only be pot luck."

"For thirty-four?" Constance had already counted them.

Elizabeth waved a stringy hand, careless of inconvenience: she had learned *something* from the Irish, Constance thought. "Grand-father used to come here with a hundred, drop them on Grand-mother without a word of warning. Mother was the same – always managed to cope – "

"The good old days," said Constance, and went on up the stairs, Harry keeping pace with her, his hand still in hers. They walked along the gallery, nodding down at those gazing up at them from the floor of the Great Hall, and came to the corridor that connected their respective suites. Constance let go Harry's hand and looked at him steadily. "Do you still want to keep up the pretence?"

He saw that she was on the point of collapse, that all her marvellous control was on a rein worn thin by the morning. "Lie down till lunchtime. I'll knock on your door just before one and we'll go down together."

"Just as we used to," she said, turning away. "The good old days."

# Two

Harry had first met Constance when he had returned wounded from Arnhem at the end of September 1944. He was twenty-two then, a lieutenant in one of the parachute battalions of the 1st Airborne Division. When he had gone into the army as a second-lieutenant the family had been disappointed that he had not gone into the Grenadiers, but he had deliberately avoided the Guards.

"I'm too junior," he had told his father, "and I'm not going to spend the war tripping over bloody uncles and cousins and saluting them. And there's Roger – he's the last one I'd be saluting."

He had never got on with his elder brother. Roger, heir-apparent to the title, had always seemed to Harry like something out of the history books; he believed that he came of a class that was a chosen few, he took his privileges as his right and he sub-scribed to the ideal of *noblesse oblige*. He was very much the favourite son, educated to his responsibilities and the value of his heritage, and Harry had always been neglected, the younger son who, like the daughters of the family, had been looked upon as something of a nuisance by his father. Harry thought of his brother as a prig and that he might be posted to serve under him was a circumstance more dampening than the possibility of being captured by the Germans.

He had joined the local Yeomanry. He had expected to be posted to Italy, but to his chagrin his battalion was sent to the Persian Gulf, where he spent a year leading patrols up and down the Trucial Oman coast and inland to the desert, never firing a shot at anything more menacing than a wild camel. It gave him little comfort to know that he was helping guard part of the family fortune, the five per cent Aidan return on Qaran oil; his father was an old friend of the Sheikh of Qaran and the latter had greeted him as if he were a long-lost son; but an engorgement of sheep's eyeballs, roast camel and fly-blown dates had soon made him a willing volunteer for the endless dull patrols. At the end of a year he had had his father pull some strings and he had returned to England and transferred to the paratroopers.

He had gone to Arnhem and come back suffering less from his wound than from the shock of discovering that a man he had known all his life had proved a coward. Joe Banning had then been thirty-five and the Aidan family doctor; he had taken over the practice from his father who had brought all the Aidan children, beginning with Elizabeth, into the world. Joe had attended Harry in his illnesses and accidents during the latter's teens, and the boy had always admired the doctor's cool wry approach to everything and, still at school then, had written to wish the doctor luck when, at the outbreak of war, he had joined the Royal Medical Corps. Banning had served as R.M.O. for four years to a unit, stationed in the Orkneys, that had never seen a shot fired in anger. Home on leave, bored and fed up with his part in the war, it had not taken much persuasion on Harry's part for him to transfer to the parachute battalion. The parachute training itself, the dropping from planes and the tough, bone-grinding drill, had found no chink in him; and on 17th September, when the battalion took off for Arnhem, he was as cool and confident as the best of the other officers. Six hours later he let three men die because he did not have enough courage to go out under fire and attend to them. Harry had argued with him and threatened, but he had sat with his back against a shattered stone wall, shaking his head, flinching with every mortar bomb that landed nearby, unable to say anything but, "I'm sorry, I'm sorry, I'm sorry." Finally Harry himself had crawled out to the men, but all three had been too badly wounded for him to move them. Then he himself had been wounded and only with the greatest difficulty and luck had he been able to crawl back to the stone wall where Joe Banning still sat hunched and desolate.

"For Christ's sake, do something for me, then," Harry had snarled. He had seen cowardice before but never so abject as this, and the sight of it sickened him as much as the pain in his thigh. He was too emotionally upset to feel pity yet; instead he reacted with anger that he would later regret. "We're not going to get out of this in a hurry, so don't just sit there weeping like a bloody woman!"

Banning had said nothing. He had forced himself away from the wall and crawled towards Harry, dragging a field medical case with him. Machine-gun fire clipped the top of the wall, showering them both with dust and stone chips. Banning dropped flat, burying his face, and Harry looked down at him with contempt. He eased his own back against the rough comfort of the wall, took out his knife and cut the leg of his trousers. The sight of his bloodied thigh turned all his attention to himself.

"Get on with it!"

Banning had carefully cleaned and dressed the wound. The bullet had gone right through Harry's thigh, just grazing the bone; later evidence would show that Banning had proved himself a good doctor if not a brave one. The German mortar fire had eased off, but the occasional bomb was still landing on the other side of the road, and Banning kept flinching and ducking all the time he was attending to Harry. The latter, now only semi-conscious with the pain of the wound, had given up abusing the older man and lay wondering how soon he and Banning would be dead. Nine dead men, all that had got this far of his platoon, lay about them; the three men on the other side of the stone wall were probably already dead. Darkness was closing in and Harry, in a twilight of pain and shock and disappointment, doubted that he and Banning would see the sun rise in the morning.

But during the night another platoon had rescued them and they were withdrawn to a safer position. A casualty clearing station had been set up, but no one could be cleared from it because there was no way out. It was another week before Harry and the rest of the wounded could be evacuated. In that time he saw Joe Banning work himself to the point of collapse, though he still avoided going out into the open and exposing himself to fire. Harry said nothing of what had happened on the first night of action, and on the day he was to be evacuated Banning came to him.

"Are you going to put in a report on me?" Banning seemed to have aged ten years in the past week. His eyes were bloodshot, his jaw was covered with a ginger bristle of beard, and every minute or two he would shiver as if he were suffering from some fever.

"No."

Banning had been looking at his hands grappling nervously with each other, but now he glanced up in surprise. "Why not? You're entitled to."

"I don't think I'm entitled to make judgments on other people." The week in the makeshift hospital, watching men die with varying degrees of fortitude and fear, seeing some go out with calm acceptance and others crying out with fear of the unknown, had begun his education in the wide geography of human nature.

"You've already made your judgment. You made it pretty clear that night."

"All right, let's say that was a private one. That's as far as it goes."

Banning stood up. "Thanks. I'm sorry that I let you down. But it won't be the last time you'll be disappointed in people."

Two days later Harry was landed back at Dover on his way to a hospital somewhere in England. He was lying on a stretcher on the wharf, watching the clouds scudding low overhead and wondering if someone would come to rescue him before it rained, when an ambulance backed up to him, almost running over him. He yelled loudly and the ambulance pulled up with its back step only inches from his head. He saw a good pair of legs get down from the driver's seat and come round till their owner stood immediately above him.

"Don't look up my skirt, darling. It's against regulations when I'm on duty." Then a face, inverted to his own so that he could not recognise it, leaned over him. "My God, it's you, darling!"

The face fell on him, showering him with kisses from the wrong angle; its chin rubbed against his nose and once slid into his right eye. At last the girl straightened up, slid round till he could see her properly. "Darling, are you dreadfully wounded?"

"It's a bit late to be asking, isn't it?" He recognised Rosemary Bassett-Mason; he should have known it would be her. "How would that look in *The Times*? Smothered in action by an ambulance driver."

"Darling, you'd love it! Oh, this is Constance Grainger." Harry now saw the second girl standing above him. "My relief driver. She drives while I'm in the back looking after you dear boys. Don't look so reproving, darling," she said to the other girl. "This is Harry Aidan, an old friend who understands me."

"Obviously," said Constance Grainger coolly.

"She's a mad Catholic just like you," said Rosemary, to whom religion was on a par with vegetarianism. "Frightfully Popish, believes in the Virgin Mary and all that. Her family are still fighting the Reformation."

"It's been a long fight," said Harry, gazing up at this blonde who had an air of repose about her that Rosemary would never achieve even under sedation.

"We're not beaten yet." She had a cool soft voice laced with

33

a distinct trace of an accent that was strange to Harry's ear.

"You mad Catholics never know when to call it quits," said Rosemary.

Her accent had the drawling fluting note still left over from London society in the thirties. She was a short dark girl with a bosom a little too prominent for her slim waist and hips. There was a hint of gipsy in her looks, and there was a rumour that one of her female ancestors had been tinkered with by a passing Romany. She had been a schoolgirl in the late thirties, but her mother, the Duchess of Swindon, a devout believer in Neville Chamberlain and confident even in the summer of 1939 that peace in our time would last forever, was already preparing her for her debut. Harry had known her as a child but only in his last year at school had they seen each other regularly. Each had broken the other's virginity, she revelling in the occasion, he writhing with guilt as much as with passion. The affair had been an intermittent one, like partners meeting occasionally for a bridge game or a tennis match; neither he nor Rosemary had been interested in going steady and the arrangement had suited them both. The further Harry had got away from the Jesuits who had taught him, the more his guilt had diminished, and he had discovered the need for sex was one of his greater hungers; Rosemary, a one-woman volunteer organisation, had been more than willing to see that he did not starve. But eventually he had grown bored with her frivolous approach to everything, including the war, and he had not seen her for six months before this encounter on the docks at Dover.

"Where do you come from?" he asked Constance as Rosemary moved away to greet three other stretcher cases as they were brought up.

"Northumberland."

"No wonder we never met."

Constance glanced at Rosemary, who was showing her legs to one of the stretcher cases. "Perhaps it was just as well. I'm not the sort to compete against someone like Rosemary."

"Social life up in Northumberland sounds as if it could be pretty frigid."

"The blood runs warm occasionally. But not for every Tom, Dick and – " She stopped.

"What about Harry?"

"You're the first I've met."

She turned away before he could advance on that remark. A

34

few minutes later he and the three other stretcher cases, all officers, were in the back of the ambulance. Both girls were up front, a fact that Harry was grateful for: he could not have borne three or four hours of Rosemary's flirtatious chatter. Now that he was back in England, safe though wounded, he was beginning to suffer a reaction. Even while he had listened to Rosemary's twaddling tongue there had been an echo behind it, the anguished voice of Joe Banning. He lay with his eyes closed, listening to the conversation of the other three men.

"That dark one would be a man-eater," said the lieutenant who had introduced himself as Cowper. "That's what I've been fighting the war for. For God, England and nymphos."

"Her father is the Duke of Swindon." The plump balding man on the stretcher above him, Captain Simmons, had a high voice that seemed always on the edge of a giggle. "I understand the highest percentage of nymphomaniacs is in the titled classes."

"Where'd you hear that? What's your unit?"

"Matter of fact, I'm in the Statistics Section at SHAEF."

"What brought you home? A slipped statistic?"

Simmons did giggle, coyly. "I fell out of a colonel's bed, broke my tootsies."

The third man, who had not introduced himself, growled, "What's the percentage of pansies at SHAEF?"

Simmons made a hissing noise. Cowper raised his eyebrows, turned his head and looked at Harry, who also was on a bottom stretcher.

"I saw you talking to the two girls. Hope they're not friends of yours?"

"I used to go out with one of them." Harry opened his eyes, but did not turn his head.

"Oh, really? Which one?"

"The dark one."

"Oh." Cowper whistled softly. "Okay, I shan't ask if the statistics are right."

"No," said Harry, closing his eyes again. "Don't."

"The blonde one," said Simmons, as if to prove he was also heterosexual, "is the one who'd prove more interesting."

"She's also a friend of mine," said Harry, still with his eyes closed. "If you have to talk, let's hear some gossip about your colonel friend at SHAEF."

Again there was the hissing noise from the upper stretcher, but Simmons said no more and the interior of the ambulance lapsed into silence. The men dozed off and did not waken till the ambulance pulled up and Rosemary and Constance came round with tea and sandwiches.

"Only another hour or so, darlings," said Rosemary.

"Where are you taking us?" said Cowper. "Edinburgh?"

"Can't tell you, darling. It's a military secret."

It was another two hours before the ambulance came to its final halt. Rosemary opened the doors and carolled, "Surprise, surprise!"

"I knew all the time," said Harry, gazing out at the wide façade of Aidan Hall. "I'd arranged it before I left Ostend."

"Damn you," said Rosemary. "I never could surprise you."

Harry looked around at the other men and at Constance. "Welcome to my home."

Cowper was sitting up on his stretcher. "You mean the lot is yours? It's not council flats or something?"

"It's the family's – or it is in peacetime. Just now it's a hospital. My family live in one of the cottages on the estate."

"I'm glad to see it's being put to some use." The third man was also sitting up. He was big and dark and in his late thirties, and there was a slight trace of Cockney in his growling voice. He had a major's crown on his battle jacket and beneath the blanket that covered him there was the shape of only one leg. "When we get back after the war we'll continue to find some use for it."

"Who's we?" said Cowper. "You're not a bloody Communist, I hope?"

"The name's Wardle. I'm the Labour M.P. for East Dawsbury."

"I'm sure Lord Harry's family have as much sense of responsibility as you, Major Wardle," said Constance. "Don't start another war until this one is over."

Good for you, thought Harry. He grinned at her and winked, but there was no answering smile. She turned away as several orderlies arrived to unload the stretchers; and Rosemary, for once chatterless, followed her. Harry looked across at Wardle as the latter was laid on the ground beside him.

"My father will probably call in on you. He usually comes down from London when the House isn't sitting."

"Did you pull strings to have us brought all this way in an ambulance?"

"Of course. Otherwise it would have meant two or three train

36

changes. They were going to send us to a hospital in Yorkshire. We – I mean, people like me – we might as well use our influence while we still have it."

Wardle suddenly grinned, his big dark face breaking up like a crumbling Spanish mask. "Privilege has its advantages. In a way I suppose I'll be sorry to see you lot go. They tell me there's a Tory hidden deep inside every Socialist."

The Marquess of Carmel came to visit his son that evening. "Does this wound mean you're out of the war from now on?"

"I hope not."

"Don't be too heroic. I hear on the grapevine you've been recommended for an M. C. You've done your bit. Now get yourself properly well again before you go rushing back in. Enthusiasts are a damn' nuisance in the army. This fellow Montgomery and the American – Patterson, Patton, whatever his name is – always sound as if they're running some damn' football team."

"I'm not really in their class."

"All the same, wait till the army sends you back. Don't volunteer. How's Joe Banning?"

Harry's voice was steady. "He's all right. Working like a dog when I left."

"A good man. Hope he comes through all right. Who's this pretty gel?"

Constance had come to the foot of the bed. She was still in uniform but without her cap; her pale blonde hair was cut short with a fringe so that it framed her face like a silver cap. Harry introduced her to his father, and the Marquess, still a virile forty-seven and his wife dead ten years, nodded appreciatively with a frank look in his eye that his son had never seen before.

"Can't remember the ambulances having drivers as pretty as you in my war. Saw the pair of you to-day, you and that gel Rosemary, Bobby Swindon's daughter. You a friend of hers?"

"Miss Grainger comes from Northumberland," said Harry.

"Another kingdom," said Constance. "I don't think people in the south ever give us a thought."

"Probably right," said the Marquess. "Disraeli, wasn't it, said there were two nations here. Sounds like something he would have said. The Jews and the Irish always make the best remarks about us. Well, have to look in on a fellow named Wardle. Heard him speak in the Commons once. He wants to abolish the Lords. Might try and abolish him."

He went off and Harry and Constance were left alone. The

ward of four beds had been established in one of the numerous bedrooms, a room that Harry could not remember having seen before. The other three beds were occupied by walking patients, all of whom were now downstairs in the hospital lounge. Male orderlies were trundling trolleys up and down the corridor outside, but for the moment Harry and Constance were as alone as a couple could be in a military hospital.

"You're quite beautiful." Harry had been looking at her carefully while his father had been talking to her.

He was pleased to see she neither blushed nor acted coy. "Thank you. Are you usually as articulate as that when you're complimenting a girl? Most men never get past calling a girl jolly nice. Some of the older ones, the middle-aged schoolboys, might say 'spiffing'. But I've never been called beautiful before. May I?" She sat down on the end of his bed and looked just as carefully at him. "You're handsome. Too handsome. You should break your nose or something."

"I'll see what I can do." Which was why five years later, when he did break his nose, he did not worry when he realised Joe Banning had set it badly. "Tell me about yourself."

"There's nothing to tell. I'm an only child, my mother is dead, my father is a farmer. I know all about you."

"From Rosemary?"

"Partly. And I read all about your family in a history of Catholicism in England. It's in the library here. I've been looking it up."

"You don't seem the sort who'd be a snob about family and all that sort of thing."

"I'm not. I was just filling in the background on you." She stood up. "Rosemary and I are going back to London first thing in the morning."

"Where can I get in touch with you?"

She produced a slip of paper from her pocket, laid it on his bedside table. "I was hoping you'd ask me, so I wrote it down."

He laughed, shaking his head. "You're as bad as Rosemary in your own quiet way."

"No," she said soberly, even though she returned his smile. "I'm selective. Rosemary isn't. If you look me up when you eventually come to London, it won't be one, two, three and into bed. I think it's only fair to warn you."

He picked up the slip of paper, tucked it into the book he had been reading. "I may feel in the mood for a platonic evening."

38

"You might like the change," she said.

Two months later they were married and she was still a virgin up to their wedding night. The fact caused no complications: she proved to be as passionate and uninhibited as her husband.


2

The Aidan family name first appeared in local records in the West Country towards the end of the thirteenth century. Roger Aidan, a gentleman and farmer, was created a baron and granted permission to bear arms by Edward I after following that king, as prince, in the ninth and last Crusade. A prolific sire, he died of a surfeit of children: crossing the Bayard River one day with nine of his children the overloaded boat sank: all the children were saved but Roger was drowned. His eldest son, the first Henry, proved to be as adept in the acquiring of real estate as the abbots who had taught him; by the time of his death in 1310 the Aidan holdings were well established. *His* son, John, did not last long: availing himself of the legal and encouraged right to beat his wife, he did so, expecting no more reply than a few abusive words: instead she answered him with six inches of poniard and left that evening with an incelibate Benedictine friar. Richard, the heir, lived till he was seventy, a great age in those days, and bequeathed the estate and title to his son Roderick, who was both pious and a snob, a combination not uncommon in families that wear their religion as if God Himself were a relative. He asked Edward III for something a little more exclusive than mere Baron Aidan and was granted the Earldom of Bayard; he promptly added twenty rooms to Aidan Hall as a memorial to himself. The family survived the Reformation because the two earls current during its course were neither particularly religious nor politically stupid: they knew an ill-wind when it blew. When abbey lands were offered to them they bought eagerly, voluntarily paying premium prices, a gesture that was appreciated at Court; by the time the anti-clerical ferment had settled down the Aidans were the third biggest landowners in the West Country. In the late seventeenth century they began to move freely in Court circles, wielding influence with the quiet cunning born of long practice; it was then that the eleventh Earl married Sarah Pack, whose dowry was a sizeable number of acres just north of Holborn. In 1715 the twelfth Earl was created Marquess of Carmel, in a year when

Catholics in Northumberland, the Graingers among them, were rising in another Pilgrimage of Grace. Charles, the first Marquess, promptly went abroad, making a special trip to Hanover to show George I, the new King, that his heart and his mind were in the right places. From that time on the Aidans never looked back, except to admire their breeding; though their influence declined at Court, they got richer and richer, and accordingly their influence was turned towards the Tory party. Their political power only began to decline after World War I, but that was due more to the seventh Marquess, who found it easier to talk to horses and hounds than to politicians, rather than to a disinclination of the party to listen any longer to the Aidans. The eighth Marquess was just beginning to reassert the family's power when World War II broke out.

All this Constance, doing her own study, learned in the first month of their marriage. "Harry, what are you going to do when the war is over? Go into politics?"

"Not if I can help it. Up in Northumberland, do they usually talk politics in bed?"

"I don't know. I've never been in bed with a politician. Don't put your hand *there*. I'm trying to talk seriously with you."

"Just keeping my finger on the pulse of things – an old political expression. I've heard Roger use it. He and Father are the politicians in the family."

They were in bed in the Aidan town house in Eaton Square. It had become a *pied-à-terre* for all the Service members of the Aidan family whenever they were in London; but Harry and Constance, the latter already showing that she was not going to be brow-beaten by the family, had insisted that the front room on the third floor was theirs and theirs alone. Constance was stationed at Gravesend and came up to London at every opportunity: Harry, his leg still unfit for active duty, had been posted to the War Office as a commissioned messenger boy. Wrapt in each other, they were having a lovely war.

"Darling, let's talk for a while. We have a whole weekend for *that*. What are we going to do when the war is over? Nobody thinks it can last another year or eighteen months."

Harry rolled over on his back, moved his hand from her thigh and absentmindedly stroked the scar on his own. He was fed up with life at the War Office where the most action he saw was when someone opened a window and the wind blew the papers off his desk. Now that he was married he was no longer keen to go back to the Continent; as his father had said, he had done his bit; and he did not want to be separated from Constance for perhaps a year or even a month. The war was going well in Europe, though the Germans were said to be regrouping, and it was only a matter of time before the fighting there was finished. After that, of course, there were the Japanese; but that was another war. In the meantime he would like another posting, perhaps to a training camp. But how to get Constance posted near him, unless he got her pregnant and she was discharged from the Service?

"I could get you pregnant."

"What?"

"Oh. Sorry. I was thinking of something else."

"I thought you didn't want children just yet?"

That had been one of the few things on which they had come close to an argument. Individual conscience was something that most Catholics had then never heard of; contraception was a sin and something to be confessed on every visit to the confessional. It was a sin that did not particularly worry Harry; like most young men in wartime, he carried condoms as part of his uniform. Only because she loved him so much and was so hungry for him had Constance allowed him to use the contraceptives. She was racked with guilt each time afterwards, but she said nothing to him, just went to Brompton Oratory and there in the darkness of the booth, blushing at having to talk to a strange man about her most intimate experience, confessed her sin to the priest. Going each month, choosing a different priest each time, she was aware of the hypocrisy of her confession – "and I firmly resolve, by Thy holy grace, never more to offend Thee" – but knew she could not deny what Harry, and she, craved so much. She wanted children, but she understood, or told herself she did, Harry's viewpoint that it would be better to wait till the war was over.

"I don't want them – not yet." He changed the subject before they got too deeply into it; the thought of being a father just yet appalled him. "There's plenty I can do when I get out of the army. I can go back and help run the estate – "

"That will mean living close to your family – " She shook her head. "They'll all be back there, won't they? You said the place was overrun with them before the war. I couldn't put up with them, darling."

Her marriage to Harry had not been popular, not even with

Harry's father. The Marquess had admired her as a pretty girl, but he had been shocked and angry when Harry had announced he was going to marry her. The family had had other girls in mind for him; sensing that there might be a change in social conditions after the war, they had hoped for a consolidation of the old families. Constance's family could trace their heritage back three or four centuries, but they had never amounted to anything; they were middle-class farmers with a very modest holding and no social position, not even in Northumberland. George Grainger, Constance's father, had once had his name in *The Times*, but only at the bottom of a letter protesting against blood sports, a moral stance that won him no favour among the Aidans when they heard of it. Constance, no fortune- or title-hunter, only deeply in love with the only man she had ever loved, had been hurt but not defeated by the family's opposition to her. As Harry and the family were to learn, if she ever surrendered to any circumstance it would be only on her own terms.

"How would you like to go to Australia?" The question was as unexpected to him as it was to her.

"Australia? Good heavens, do we have to go *that* far?"

Having suggested it, his enthusiasm for the idea was non-existent. "I suppose not. It was just that we have some holdings out there – "

But she was running her tongue along her lips, a habit she had when she was thinking. "Perhaps it might be worth a try. We shouldn't have to stay there forever – would we have enough money to come home for trips occasionally?" She had no idea what their financial position would be when the war was over; she guessed that Harry was allowed an income by his father but she did not know what it was or was likely to be in the future. She had married into a rich family, but money was never discussed; the Aidan women seemed to look on it as an unavoidable evil, like garbage, and Harry had never even bothered to ask her if she needed any money. She still used her own Service pay to buy whatever personal things she needed. "I shouldn't fancy living in exile forever."

He laughed. "Dear girl, Australia isn't Siberia. But let's talk about after the war, after the war."

"Careful – you're pinching me. Aren't you ever satisfied?"

"No. Are you?"

"No." She yielded herself to him, his family and Australia, forgotten under the drug of his hand. "Do we have to go through

42

all that messy act again? Darling, I've been reading up on rhythm – "

"Too risky. Stop talking."

"This is too frustrating. It's like washing your feet with your stockings on."

"Where on earth did you get that?"

"I heard Rosemary say it. What's the matter?"

He had fallen back from her. "Do we have to have *her* in bed?"

"Sorry. I said it without thinking." She kissed him, began her seduction of him. She was jealous of Rosemary and any other girls Harry might have had; her jealousy was so consuming she was sometimes afraid of it; she actively had to fight against it. Her reaction each time was to be even more loving and abandoned, though she was never quite sure whether she was trying to prove to herself or to him that she could offer more than Rosemary or any of the others had.

"Don't stop," he said. "This is something one should always do without thinking."


3

Harry and Constance did not go to Australia. The war in Europe finished without Harry seeing any more action; he was kept on at the War Office, promoted to captain, and relieved of the duty of delivering messages below the classification of Top Secret. Constance was moved up from Gravesend to Ealing and managed to get home to Eaton Square almost every night, her only risk being the increased opportunities for pregnancy. They made love, talked in a desultory way of going to Australia and wondered, in an equally desultory way, how much longer the war in the Pacific would go on.

Then a General Election was announced. Roger came home from Germany and told the family that he was putting his name forward for nomination as the Conservative candidate for the local seat. The Conservative Association, delighted to have once more an Earl of Bayard representing the Bayard constituency, suggested to the incumbent Member that he might like to retire, and he, no illiterate when it came to reading political writing on the wall, graciously decided he had reasons of health for standing aside. Roger was nominated and at once plunged into his campaign for election.

In his speeches he talked of a Brave New World – "Not the

43

sort proposed by Mr Aldous Huxley. The sensible British people will never stand for such a society. (*Hear, hear.*) We are renowned for our commonsense. (*Hear, hear.*) We may not have much imagination – (*Silence*) – but we do have commonsense. (*Hear, hear.*) No, we shall have a Brave New World in which we can all start again with the spirit that has bound us together throughout these last five and a half dreadful years, a world in which there will be opportunities for all regardless of class – "

"Dammitall, you're going a bit far." The Marquess had sat in village halls and stood in market squares listening to the speeches with mounting irritation. "You're a Conservative, not a confounded Socialist."

"I don't think you know the mood of the voters," said Roger. "I think I'll hold this seat – it has never returned anyone but a Tory – but I'll give you two to one that the Party loses the election."

It was election eve and he, Harry and their father were sitting round the dinner table after a meal of cold consomme, rabbit pie, and an anonymous pudding made from powdered egg and stale cake. Harry, already a lover of good living though he had experienced none of it in his adult life, hoped that the Brave New World would spread pate de foie gras and pheasant among everyone regardless of class. But that might take imagination, something Roger had denied the English possessed.

"They'd never throw out Winston," said the Marquess. "He'll take us back with him."

Roger shook his head. "I was listening to the men in my regiment. They don't want us, Father. We're part of the past."

"What was wrong with our past? We had a great Empire – we still have it – "

"Do we? I doubt it. In any event, I don't think the ordinary man in the street, or in uniform, cares about the Empire any more. My batman was one of the Jarrow Marchers. He once told me all about it. Two hundred men walking three hundred miles to London just to ask for work for their town. They didn't get even an honest answer, Father. He stayed in London after the march and joined the army because he couldn't find a job outside it. He wasn't interested in the Empire and he doesn't remember Baldwin and the government with any sense of loyalty at all. I had him up on a charge of being drunk on duty and he told me then that it would be *their* turn when the war was over."

44

"What did you do to him?"

"I let him off with a caution. He was entitled to be drunk, on or off duty. It was V-E Day."

The Marquess grunted, remarking that things would have been different in *his* war. He passed the port to Harry and sat back, looking at his two sons and wondering which of them would prove the bigger disappointment. One of them was beginning to sound near-as-damn-it to a Socialist and the other looked as if his only occupation when peace came would be that of a married playboy. He would never admit it, thought the Marquess, but perhaps the old order was almost finished. And the port turned to vinegar in his mouth.

"How have you found them at the War Office?" Roger asked.

Harry shrugged. "Nobody ever talks politics there, unless it's to abuse *all* politicians."

The two brothers were very much alike, though Roger had a more solid look to him, both physically and in demeanour. He was twenty-seven and the war had had its effect on him; where he had once taken his class and privileges for granted, he had abruptly become a politically-minded man with a social conscience. He was still unmarried and had told his sisters bluntly and rudely that he would choose a wife in his own good time, which would not be till after he was settled in as an M.P. He was the one member of the family who had openly welcomed Constance, privately thinking that Harry had been lucky to get such a splendid girl.

Harry and Constance had come down from London to Aidan Hall for three days' leave. The Hall was still being used as a hospital and the Marquess's home was the gamekeeper's cottage that had been converted at the beginning of the war. Ferguson, the butler, was in the army, a mess sergeant at SHAEF headquarters who made even field-marshals feel inferior; the rest of the domestic staff had either been called up or had gone off to better paying and more patriotic jobs in munitions factories. But Mrs Dibbens, the assistant cook, had volunteered to stay on and came in each day from the village of Skipbrook to cook and do the housework. Constance, more alert than the menfolk, had at once recognised Mrs Dibbens as a woman with her eye on the main chance. *Mr* Dibbens, whoever he was, was never mentioned, and Constance never did find out if there had been such a partner.

Constance, tired out after a week of almost continuous duty, had excused herself and gone up to bed as soon as dinner was finished. But she was still awake when Harry came up. She had been about to drop off to sleep when she had heard raised voices in the sitting-room downstairs. For the last hour she had been lying here in the dark warm room wondering what the Aidan men had been arguing about and worrying, as an outsider, that it might have been about her.

She said nothing as Harry came quietly into the room, undressed in the dark and got into bed beside her. She could feel the tension in him as he lay on his back, and she put a tentative hand on his thigh. Through the thin cotton of his pyjamas she could feel the ridge of his scar, but she left her hand where it was.

"What's wrong, darling?"

Harry was not surprised to find her awake nor that she had guessed something had upset him; he was too concerned with his own anger at the proposals – no, the *directives* – that had been put to him downstairs. "Damn them, we're going to Australia, whether they like it or not!"

"Are we?" She was relieved to find the argument had not been about her, relieved not for her own sake but for Harry's. Going out to Australia after the war had been only an idle thought, its only attraction being that she would be well removed from the influence of the family, and it had been her impression that Harry had been equally lukewarm in his talk of emigration. "Why?"

"They want me to go into the City – God, I can't think of anything more bloody boring! I'm not a banker or an accountant – "

"Why do they – who's *they*? Your father and Roger?" She felt rather than saw him nod. "Why do they want you to go into the City?"

"Because if Roger is elected to-morrow – and he will be – he says he won't have time to attend to our interests up there. He's going to be a full-time politician and his relaxation – that's what he calls it, relaxation – will be to give Father an occasional hand running the estate down here. I'm to be the every-day family representative on Anglo-Qaran Oil, Aidan-Canada Lumber, and all the other things we have our money in – but that's what they think!"

"Would it all be so boring?"

"Bloody."

She said slowly, taking her hand away from him, "I don't think I want to go to Australia."

He emerged from his own self-concern enough to sense that she had made a decision of some sort. Their relationship so far had been such an intermittent one, squeezed in when both of them could escape from duty at the same time, that they were still discovering each other after eight months of marriage. Sexually their discovery of each other had been almost instant, their compatibility only marred by the disagreement about contraception; emotionally they were still strangers, neither of them yet completely alert to the nuances of the other's reactions. It was only just coming home to Harry that marriage was an even more complex business than he had realised, that there were secrets to Constance that he might never solve.

"Why not?" He was not sure himself that he wanted to go to Australia, but just now he could not think of another alternative to the City.

"I don't think it would be something that either of us would settle for permanently. I don't want ours to be a shiftless sort of marriage. I want to set up a home, have children, have some feeling of security – "

She was stating all the middle class values by which she had been brought up, but he only dimly recognised them as such. His mother had died when he was twelve and ever since he had actively rebelled against any social education by his sisters. At school the Jesuits had done nothing to lessen his sense of privilege and his father had aided and abetted the self-deception by never denying him anything he had asked for. Security, in its social sense, was a word as foreign to the Aidans as an Urdu oath: there was just never any necessity for discussion of it. Four centuries of wealth, power and influence provided their own feeling of security.

"Try the City for a year, darling. We'll set up a house or a flat in London, whatever we can get – "

With something like horror he saw, for the first time, the pattern of his future life. He was not a stupid young man, but going straight from school into the army there had been no time nor indeed necessity to consider what his life would be like after the war. He had had some vague idea that he would enjoy himself for several years to make up for the lost time: do some travelling, go the party round, perhaps ride during the steeple-chasing season. He had supposed he would eventually marry, but he had never given a thought to what married life would be like.

"My love, I'm not the type for settling down. Not like *that* – "

"How do you know? You haven't tried it." She changed her tone, realising she would have to plead with him. "Darling, I don't want you to become a dull businessman – that's the last thing I want as a husband. But I don't want us to live a butterfly existence – one can't build a marriage on that. Sooner or later you'll have to face responsibilities – "

"I didn't realise I'd married a schoolmistress."

She sighed, lying on her back and gazing at the dark ceiling. She had refrained from asking him what type he considered himself, what he thought he was if he did not think he was a banker or an accountant: those questions had seemed petty at the moment. Out in the park an owl hooted, then there was the sound of an ambulance going up the long drive to the Hall. How many men up there in the hospital, she wondered, are lying awake to-night worrying about their future? Those without limbs, without money, without jobs to go back to? Suddenly she was angry at Harry's selfishness and abruptly she turned away from him, swinging over on to her hip with a thump that creaked the bed.

"What's the matter?"

"Work it out for yourself!"

"Connie – "

But she kept her back to him, firmly brushing his hand away as he tried to touch her. The rest of the night was, with the exception of the first night at Arnhem, the most troubled he had ever spent. He was selfish, but he had sense enough not to respond with obstinate temper. He lay awake analysing himself, her and their relationship, facing some truths that were bitter, and wondering, with some fear, if the gulf between them, because of their different sets of values, would widen. Just before he at last fell asleep he thought of Roger's warning that England would be a different place after the war, perhaps even from to-morrow.

In the morning he said, "All right, you win – "

"No, darling, I don't win. Don't ever do anything for me because you think I've *won*."

His education in her was already improving; he recognised that she could be a diplomat when she chose, that she did not demand total victory. He stared at her for a moment, thinking how beautiful she looked first thing in the morning; his limited experience had already taught him that there were few girls who were twenty-four-hour beauties. Then he tentatively leaned towards her and kissed her. At once she grabbed him and pulled him down on her.

"Do you think they will hear us downstairs?"

"Who cares?"

"Oh, I love you noble lords! To hell with what the neighbours think!"

"Noble lords don't have neighbours."

"Don't be a prig – Ah! Don't ever let's quarrel again, darling."

If the Marquess, Roger and Mrs Dibbens heard what went on in the bedroom, they gave no indication of it. All the talk at the breakfast table was of Election Day; as soon as he had finished breakfast, Roger went off on a tour of the constituency's voting centres. Harry and Constance cycled into the village of Bayard and voted; it was the first time the villagers had seen the new Lady Harry Aidan and Constance felt that she, too, was being voted upon. She and Harry spent the rest of the morning wandering through the woods surrounding the park. She looked at the Hall and the 600-acre park with something less than a dispassionate eye; she knew it would never be their home and for the first time she wished Harry was heir to the title. The thought of being a Marchioness instead of just Lady Aidan did not occur to her; she had middle class values but never that class's valuation on a title. All at once she wanted to contribute to the continuation of the Aidan name, to be part of the heritage of which Harry seemed so careless. Tradition infected her and suddenly she wanted to be part of history, no matter how local and, in tomorrow's world, unimportant.

In the afternoon, with the cottage to themselves, they made love again. Once she cried out in wild ecstasy and a quarter of a mile away a bird-watcher, one of the patients from the hospital, sat waiting worriedly for a repeat cry from a bird that must be foreign to these parts. In the evening the Marquess, using some of his precious petrol in the Austin Seven he had bought when rationing began, drove them to the constituency returning office. By midnight Roger had been declared the elected Member.

Next morning the bad news began to come in from all over the country. The result was not declared immediately because so many of the troops' votes had to be counted, but within a week it was evident that *their* turn *had* come. On 26th July Mr Attlee presented himself to King George VI, both of them looking a little surprised at who was to be the new Prime Minister; and all over Britain the die-hards among the aristocrats waited for the

49

rumble of the tumbrils. Only Roger, among the Aidans, took the election result with equanimity.

"No heads are going to roll. Matter of fact, I don't think they'll get round to even thinking about people like us for at least a couple of years. Most of them are going to take that long to get over their surprise at getting in."

"It will take me even longer," said the Marquess. "God knows what dreadful schemes they have in mind."

"You should thank God Attlee isn't Cromwell," said Harry.

"Don't be facetious," his father snapped. "These fellows could turn out to be worse even than the Roundheads."

Two weeks after the election earthquake in Britain the atom bomb was dropped on Hiroshima. Harry, like most men in uniform, suffered no moral qualms; anything that finished the war quickly was welcome. Two months later he was out of the army and, after he and Constance had had a second honeymoon in Ireland, he started work in the offices of Anglo-Qaran Oil in the City. He wondered if it were some sort of omen for the future of the Aidans that the only building standing in the bomb-devastated block in Moorgate was Anglo-Qaran House.

# Three

The Persian Gulf slid up into the plane's window. Harry turned his head and looked out at the sea stretching away to the sun-bleached sky. After the eye-aching monotony of the view of the last few hours the sea offered no relief: it was no more than an extension of the desert over which they had been flying. The nude dunes of the desert gave way to a narrow strip of white beach, a sliver of bone protruding from brown dead flesh; there was the pale green of the shallow water which gradually darkened into the vast blue emptiness; then finally there was the same blinding sky that he had been looking at ever since he had left Cairo two days ago. The plane, an RAF Anson still in wartime camouflage paint, banked sharply and suddenly there in the window was Qaran.

"Darling, be careful out there," Constance had said when he had left London.

"My love, the Sheikh considers me one of his sons. The worst that can happen to me is some tummy trouble from eating roast camel. *You* be careful." He had patted her stomach.

"*Please* – the char's there on the stairs."

He had said nothing, wondering when she would grow out of this habit of always being watchful of the servants. He had been brought up to believe that servants were part of the furniture of a house, not to be knocked about but not to be noticed either. He had never carried his disregard of them as far as had his ancestor, the second Marquess, who in summer used to move from room to room of Aidan Hall stark naked; but he had never allowed their presence to inhibit his behaviour towards his family. He found it perfectly natural to pat his wife's pregnant belly in front of the charlady or anyone else.

"If I can manage it, I'll stop off in Jerusalem on the way back, pick up Roger and come home with him."

"I wish I were going with you. I'd love to see Jerusalem and Bethlehem and all the other Holy Places."

"But not Qaran? That's a Holy Place to Anglo-Qaran Oil."

"Roast camel and sheep's eyeballs are no diet for a fragile

mother-to-be. I'll stay here and have all the furnishings finished by the time you come back."

The housing shortage in bomb-shattered London had strangled any ideas they had had of having a house of their own. Constance had wanted to escape entirely from the family environment, but in the end she had acknowledged the practicality of the family motto and had decided to be realistic above all else. Harry's father offered them the third floor of the Eaton Square house as a self-contained flat and they accepted it.

Harry thought longingly now of the flat in London as the plane came in to land on the airfield on the edge of the town of Qaran. The glare of the yellow-brown earth, the stark white cube of the fort, the dun-coloured houses clustered round the shallow curve of the bay: he hated the sight of it all and the memories it brought back. He had once heard his platoon sergeant in the Yeomanry refer to Qaran as the arse-end of the world; the phrase was the ultimate in aptness. Now, in July 1946, he had come back too soon; wartime memories, like whisky, are better for age. He got out of the plane and cursed his father for the latter's bright idea at the last board meeting.

"Now they've started on the new pipe line, I think a visit from one of us would be appreciated out there – just to show them London *cares*."

There had been a murmur of laughter round the board table. No one had been laughing at the joke, Harry had remarked to himself: it was the laughter of smugness, of being in a position where the only discomfort was of being caught in disagreement with the chairman. He had looked with reluctant admiration at his father who could so dominate these men that, though there were nine on the board and totally they represented ninety per cent of the share holding, all the decisions were his. If ever the day came when he himself was chairman he hoped he would not be surrounded by so many yes-men. He doubted if he would have his father's arrogant courage to risk a wrong decision.

"I propose that Harry should go out. He knows the Sheikh – old Abu sent me a letter that he considers Harry his ninth – or was it his tenth? – son. Two or three of us will have to go out at a later date, when the pipe line is finished, but Harry will do for now. No offence, Harry."

Not much, thought Harry; but he smiled. He had been a director for exactly one month, the result of the sudden death of another member of the board, and he was beginning to enjoy the position. He had already come to recognise the Old Boy network in the City and it gave him cynical amusement to watch it at work. He was learning the wickerwork of power and influence in business, the interlocking directorships that could control an industry without the public's ever becoming aware of it. He had no desire for power himself, still being lazily inclined towards pleasure, but it amused him to be a spectator of the uses of it. The City, when one was on the inside, was not such a dull place as he had imagined it.

"You can leave next week," said his father. "Take the British Overseas Airways aeroplane to Cairo and I've arranged for the RAF to fly you on from there out to Qaran. Freddie Wilmott, at the Air Ministry, tells me you will probably have to change aeroplanes several times, but that should make it more interesting. Eh?"

"Oh, undoubtedly," said Harry, but his father had the autocrat's tin ear for irony.

Now here he was, after a trip as uninteresting as it had been long and uncomfortable, and coming across to meet him were Lucas, the company manager here in Qaran, and Hazza, the eldest of Sheikh Abu's sons. Harry thanked the RAF pilot who had flown him out, then arranged his face into a semblance of pleasure at arriving as Hazza swooped on him.

"Greetings, Lord Harry, old chap." Hazza had been taught English by the wife of the British Agent in Qaran, an ex-schoolmistress from Cheltenham; at times his voice even had a woman's inflections, though there was nothing womanly about his appearance. He was a tall ugly young man with a face as fierce as that of the falcons with which he hunted. "It is absolutely delightful to see you. Charming."

"Nice to be back," said Harry, and saw Lucas, behind Hazza, raise his eyebrows quizzically. "How are things going, Lucas?"

"Fair enough." Lucas was a short, thickset Australian with fair curly hair and a flat dry voice squeezed out through teeth clenched continually on a pipe. He was in his mid-thirties and his face was already showing the effect of too many years lived beneath a baking sun. Harry had met him fleetingly when he had come to London in March, but it would take more than a fleeting acquaintance to get to know the phlegmatic Australian. "Hazza saw we got a good mob of workers."

"Absolutely the best," said Hazza, leading the way towards

the huge pre-war Rolls-Royce in which he and Lucas had arrived. "We never have any trouble in Qaran."

No, Harry thought, slave labour never does offer any trouble. He knew the low wages that Anglo-Qaran were paying the Arab workers and he had idly wondered what the trades union members of the Labour government in London thought of the exploitation of workers in this part of the world. But perhaps, now they were in power, they had come to appreciate that empires could not be built on justice. Britain needed the Qaran oil and at a price it could afford.

"Do you like my car?" Hazza settled in behind the steering wheel, primly upright, reminding Harry of one of his aunts who drove her Rolls in the same stiff manner. "It was ordered for one of the deposed kings of Europe, I forget which one, but he never collected it. I convinced my father that he is now a king, an oil king, and he rides in this now instead of on a camel."

"Very appropriate," said Harry, and glanced back at Lucas in the back seat. The Australian lit his pipe, then slowly closed one eye through the smoke.

That night Harry and Lucas went to dinner at the Sheikh's palace, a collection of houses that had congealed together and been surrounded by a high wall. The whole lot had been given a coat of whitewash and a flagpole, from which no flag ever flew, had been erected in the scrubby garden in front of the palace.

Over dinner of roast mutton, British Army tinned vegetables, and tinned plum pudding, washed down with orangeade and thick coffee, Hazza again mentioned the lack of trouble in Qaran. "Extremely different from what the English are experiencing in Palestine, Lord Harry, old chap."

Harry wondered if the words had any other meaning behind that wide smile. Did Hazza know that Roger, on his own initiative, was at that moment in Palestine, had allied himself with the Zionists and was there to provide himself with background to fight their cause in the Commons? Roger had become a rebel in both his party and the family, and one of the reasons Harry had been chosen to come out here for Anglo-Qaran was to offset any impression that all the Aidans thought as he did.

He glanced round at the other men seated cross-legged on the thick carpet that was the dinner table. There were about twenty of them: the old Sheikh, his other sons, some minor sheikhs from the interior, and Lucas. All the Arabs were smiling at him as if they understood what Hazza had said, though none of them spoke

English; their smiles stretched away, an infinity of teeth, in the mirrors that formed the walls of the room. Lucas gave him a covert wink, then went back to eating, reaching across to pick up some crumbling white nougat from a large brass bowl. Harry sipped his coffee, feeling it coating his teeth like sandy mud.

The old Sheikh, a desert buzzard with a white beard and eyes almost blind with cataracts, said something and Hazza nodded. "My father says it will be a bad day for us Arabs if the Jews are allowed to form their country of Israel. Do you agree, Lord Harry?"

When he had been here during the war Harry had tried without success to get Hazza to call him something else but Lord Harry. But Hazza, with a mixture of formal courtesy and hinted derision, had persisted and now Harry, though he had not heard the form of address for over two years, accepted it as if he had been accustomed to it all his life.

"My own father asked me to assure you that nothing will spoil the relations between Britain and Qaran. We are friends forever."

"Does he speak for Mr Attlee and Mr Bevin?"

"Not officially – he does not belong to Mr Attlee's government. But I am sure they feel as my father does."

"And your brother?"

Harry's teeth gritted on the coffee grounds. "I'm afraid my brother speaks only for himself."

"But some day he will take your father's title, he will be head of Anglo-Qaran Oil – "

"Not necessarily." Harry cleaned his teeth with his tongue, invested in a lie: "My father is grooming me for that position."

Lucas, a piece of nougat half-way to his mouth, froze for a moment and looked at him. Harry stared back, wanting to wink that it was a politically motivated lie; but all the smiling faces were turned towards him, their eyes sharp as dagger points. Then Lucas nodded, popped the nougat into his mouth, and chewed on it ruminatively.

"That will be marvellous, Lord Harry." Hazza translated to his father and the old man beamed a five-tooth smile and said something in reply. "My father is pleased. He says you understand our country and he will be thrilled to have his ninth son as chairman of the board of Anglo-Qaran Oil."

An hour later, after elaborate farewells, Harry and Lucas left the palace. As they drove away Lucas said, "If ever you become board chairman – "

"There's no chance of that."

Lucas glanced at him, waiting for an elaboration, but Harry said no more. Though outwardly he had become more confident, even a little flamboyant, inwardly he had developed a certain reticence, a result of both his marriage and his joining Anglo-Qaran. He was learning that people who exposed themselves fully might be lovable but were vulnerable and were frequently fools. He felt safer with Constance when he held part of himself back, though he was not quite sure why: it was not a question of trust but of having something of himself in reserve for future occasions: love, he was beginning to suspect, was not sweeter for making one vulnerable. His reticence of opinion in business was only a recognition that it was another way of surviving in what was another form of guerrilla warfare. What the world saw, a handsome, slightly unconventional young lord, was camouflage for the thoughtful man beneath.

Lucas drove in silence for a while, then said bluntly, "They know your brother is in Palestine."

"How do they know? He was careful to keep it quiet."

"Search me," said Lucas laconically. He had been born in an Australian bushfire, his mother dying from burns and shock, and such a dangerous entry into the world seemed to have conditioned him to any emergency that might present itself. "But they know, all right."

Lucas lived on the far side of town from the palace, out on a rocky point where work had begun on a jetty that would jut a mile out over the shallow waters to meet the tankers. Qaran was really no more than a large village, a collection of mud dwellings stacked on either side of narrow unpaved streets; in the bright moonlight it looked like a low escarpment cut by dark wadis, a barrier between the sea and the dunes stretching inland. Lucas circled the town, driving his pre-war Packard at a reckless speed along a track that wound through patches of camel scrub, then slowed as the car bumped up the rocky road to his house.

"I'll move from here soon as the jetty is finished," he said as they got out of the car. "Too bloody noisy."

Below them, on a lower road, trucks were grinding their way out towards the jetty where the nightshift was working. Arc lights lit up the scene, turning it into a piece of artificial theatre, and Harry could see the Arab workers moving slowly about their

56

tasks. Though it was eleven o'clock the heat was still oppressive, and he remembered his own lethargy during the summer months when he had been stationed here. He could see two European supervisors, both of them moving more briskly than the Arabs, and he marvelled at their energy. Then he remembered how much they were paid and guessed that money generated its own vitality.

"Do you think we'll ever have any trouble from the Wogs? The workers, I mean. Demanding more money, unions, that sort of thing?"

"You're likely to have more trouble if they hear you calling them Wogs."

Harry looked carefully at the other man, but the moonlight was not bright enough to illuminate his expression. "You can be a snotty blighter when you wish, can't you?"

"Just like you Poms."

"Where would you go if you suddenly found yourself out of this job?"

"I have three offers in my desk from the Yanks. I'm more useful than you out here, Lord Harry."

Harry suddenly grinned and capitulated; he knew when there was more to be gained by surrender. "Righto. Sorry. Perhaps we'd better start again." He put out his hand. "And don't call me Lord Harry, please."

Lucas sucked on his pipe a moment, then put out his own hand. "Okay, Harry. Out here we'll do better if we forget our ranks. I don't mean your title – I mean you being a director and me the local manager. As far as the Arabs are concerned, from old Sheikh Abu down to those blokes down there on the jetty, you and I are outsiders. Forget the bull about you being the Sheikh's ninth son – Hazza and the other sons will have something to say about that. One day they're gunna put the bite on us, my word they are. There is a whisper of something called Arab nationalism starting up out here in the Middle East. It's only a whisper so far and you can't put your finger on it as being in any one particular place. But it's here and one day it's gunna be a bloody great explosion. And when that day comes we outsiders had better all stick together or we'll be out on our arses, probably with our throats cut into the bargain."

Harry looked back across the town. From the slight altitude of the point he could see beyond the town to the desert. On the horizon there was the glow, a sunrise that never grew, of the gas flares at the nearest oil field. A trail of lights led towards it, the spoor of the nightshift working on the new pipe line. The pipe would stretch on into the desert for two hundred miles to the farthest field, and all at once Harry had the impression that he was looking at a long fuse being laid. Nationalism, greed, xenophobia: something, some day, would put a match to it.

"One day," he said, "the old Sheikh is going to be worth millions. How do you think he'll handle it?"

"Search me. Danvers, the Political Agent, is trying to advise him to spend some of it on schools and hospitals, but the old bloke doesn't want to spoil his children, as he calls them. He's a cunning old bastard – he knows that education breeds curiosity, and curiosity is one of the germs of being discontented." He led the way into his house, an old two-storied building that had been bought at a ransom price from a local merchant, given a coat of whitewash, and turned into a combination of home and office. "It will be the young blokes, Hazza and his brothers, who'll get to use the money. They'll either turn this place into an artificial city or they'll piss off to Europe and blow the lot on cars and women. Your guess is as good as mine. Like a beer before you go up to bed?"

Harry sipped the ice-cold beer, washing the taste of orangeade and coffee from his mouth. He looked around the living-room in which they sat, at the cheap garish furniture shipped down from Baghdad, the Persian rugs on the floor, the books stacked like brightly-coloured stones against the walls. "Where's your home, Sid?"

"Here." Lucas waved a hand about him.

"You're not married? I mean, no family back in Australia?"

Lucas shook his head. "I left Australia as soon as I got out of university, back in 1935. I've been knocking around this part of the world ever since. No woman would look at me, not for more than a night or two. If I can get a lift on a plane, I fly up to Beirut when I want to get the dirty water off my chest. But no woman would ever want to come out here and set up a home."

"Don't you get lonely?"

"What do you think?" He drank from his glass, emptying it in one long swallow. "I'll stick it out for another ten years. I'll be forty-five then and I reckon I'll have enough stashed away to give me an easy time for the rest of my life, till I run out of steam. I'll come to England, find a woman who won't mind a beat-up, middle-aged joker like myself, buy a cottage in the country and settle down."

58

"'Why England? Why not Australia?'"

"I've had enough bloody sun to do me forever. I won't care if it rains every day. *I am the Poem of the Earth, said the voice of the rain* – that's Walt Whitman. When you've lived out here for ten dry years, you know what he meant." He put down his glass. "You married?"

"Yes."

"Lucky bastard. Good night." And abruptly he got up, went out of the room and upstairs.

Harry leisurely finished his drink, stood up and went out into the small garden that fronted the house. Lucas evidently had made some attempt to cultivate roses; Harry could smell their perfume even above the salt smell of the sea. Work had stopped out on the jetty for a meal break and the night had that stillness one finds only in treeless places; from somewhere at the back of the house came the thin poop-poop of a water pump. In the shadows of the town a donkey brayed, a hoarse and painful moan; and out on the jetty some of the workers laughed, a high-pitched and feminine sound. A dhow slid in past the breakwater like a giant shark's fin, and a moment later a rattling came across the water as the sail was brought down. All the small noises only heightened the immense silence, the sound of loneliness.

A lucky bastard, Harry thought. Well, perhaps I am; but why not? Even in the egalitarian society that the Socialists were trying to force on Britain there would always and inevitably be a few lucky ones. Chance was one of God's whims: he believed in God enough to believe in that. He was sorry for Lucas, but he was not going to be ashamed of his own good fortune.

He looked up at the stars, said a prayer for Constance and their unborn child, then went inside and up to bed. Now that he had seen the extreme from his own, of how other men had to live, the lonely educated Lucas and the underpaid illiterate Arab workers, he appreciated what his own life had to offer. He would do all that he could to see that the tenor of it was not disturbed.

2

"We are supposed to be an elite," said Roger. "Therefore it's up to us to set an example."

"But why the Jews?" Harry asked. "Why make them your concern?"

"I saw Dachau. That was enough to make me feel ashamed for

the whole human race." Roger paused, blinked slowly, wiping out a memory, then looked across at Harry. "You think I'm a soft-hearted, soft-headed charity worker, don't you?"

"If you are," said Harry, hedging, "you'll be the first the family has ever had."

Roger nodded morosely. "I suppose so. But I like to think I'm being practical. Hard-headed. We – not just Bevin and the Socialists but our party too – we're double-crossing the Jews. Now is the time, in this new world, if you like to call it that, now is the time for us to establish that our word can be trusted."

"I always thought an Englishman's word was his bond."

"It was an Englishman who said that. I've never found any corroborative evidence from a foreigner. The French don't call us perfidious Albion just out of bloody-mindedness."

"After what happened in 1940 the French have no right to call anyone names."

"Perhaps. But that doesn't alter our position. We earned a lot of respect for what we did in the war and now is the time to build on it. We're finished as a first-class power – "

"A remark like that could get you thrown out of the Conservatives."

Roger smiled and nodded. In the two days the brothers had been together it had surprised Harry how well the two of them had got on together. He had been picked up in Qaran by another RAF plane and flown to Amman. From there he had driven up to Jerusalem in a British Army staff car, and now he was sharing a room with Roger in the King David Hotel. In all their life, not even on the visits to the town house in Eaton Square, could Harry remember their having shared a room; at Aidan Hall each of the family had always had his or her own room, though Harry could remember as a child how afraid he had been of being alone in the big dark room. The intimacy of this hotel bedroom, impersonally furnished though it was, had almost, as it were, introduced the brothers to each other for the first time. Harry did not know what the war had done to himself, but it had changed Roger and he liked the change in him.

"Just what have you achieved by coming out here?" Harry re-wrapped in cotton-wool the silver bracelet he had bought that morning for Constance. It was the first piece of jewellery he had bought for her, and it only occurred to him now that he did not know if she cared for jewellery. He had never seen her wearing anything but the two rings, one set with diamonds and the other a plain gold band, that he had given her when they married.

"Enough to know we are going about the problem the wrong way. The Jews no longer trust us."

"Neither do the Arabs," said Harry quietly.

Roger again nodded morosely. In London, after his election to Parliament, he had always been electric with enthusiasm, seeming to relish the opportunity to attack the problems that he had made his own concern. But yesterday, within half an hour of meeting him, Harry had been acutely aware of the depression that now gripped his brother. Uninvolved himself, he had been a little shocked by Roger's deep commitment. Brought up to the tradition of always showing a front of cool detachment, no matter how partisan one might feel, Roger's zeal was entirely foreign to the character of him that Harry had known.

"That's the curse of it. The Army brass tends to favour the Arabs – they all see themselves in the image of Lawrence of Arabia, every one of them wanting to be a public school Bedouin. But I don't think the Arabs are fooled. They know all the decisions are going to be made at Westminster, not here at G.H.Q. And I'm afraid we'll try for another of our famous compromises."

"A compromise may be the only answer."

"Perhaps." Roger looked out the window and his face took on an expression of abstraction; he had been taught to recognise history, and he saw it now. "Funny I should be here trying to do this. You might say our family history began here. The first Roger could have stood right on this spot."

"He wasn't seeking any compromises. From what I've read, the Crusaders never turned the other cheek unless it was to get the other chap to lower his sword. Then they chopped his head off."

"You're too cynical. I don't think Roger was like that – " He spoke as if he had known the first Roger; there was no picture of the long-dead knight, but he had his own image of the man. "Of course, he might not understand why I'm here trying to help the Jews."

Then there was a knock at the door. Roger went to it, opened it. A small thin man in his early thirties stood there, his bad teeth half-exposed in a shy smile. He was almost bald, his hair combed in strands across his narrow skull, and he wore gold-rimmed spectacles; dressed in an open-neck white shirt, dark trousers and sandals, he looked like a clerk on his day off. The shyness of his manner contributed to the impression that he was a man who would always be under the thumb of others.

"I came to say good-bye, Lord Bayard." His English was faintly tinged with a German accent. "But I would suggest you leave the hotel immediately."

"But we don't leave until to-morrow," Roger protested. "Come in, Mr Roth. Harry, this is Mordecai Roth. My brother, Lord Aidan. Mr Roth works for the Jewish Agency – he has been my guide while I've been here."

Roth smiled nervously at Harry, then looked quickly back at Roger. "Lord Bayard, *please* – there is no time to lose. You really must leave the hotel – "

Roger wrinkled his brows. "Leave the hotel? Why?"

Harry turned away, feeling that some sort of embarrassing scene might be building up. Jerusalem lay baked under the July sun; he stared at it from the bedroom window. The domes, minarets and the square white houses seemed only an extended erosion of the bald white hills, the dark exclamation marks of the cypresses only accentuating the lunar look of the city. It was just after midday and the streets were almost deserted, even mad dogs and the English soldiers showing sense and retreating in out of the heat.

"Why leave the hotel now?" Roger repeated.

"There – " Harry turned back as Roth suddenly heaved a loud sigh and spread his hands; he looked a caricature of all the Jews Harry had seen in films and on the stage. "There is going to be an explosion – "

Even as he spoke there was a loud explosion somewhere outside. Harry, still standing by the window, whirled round and saw the dark smoke billowing up like a flowering bush beneath a tree some distance from the hotel. The tree itself also seemed to explode, leaves and branches spraying into the air like grotesque birds taking flight; then it leaned drunkenly into the smoke climbing round it. Harry looked down and saw the small group of men in Arab dress come out of the hotel entrance on the run; at the same time another group came round a corner and began shooting at the ground floor of the south wing of the hotel. Watching from his fifth-floor window, looking down on the scene with all perspective distorted, Harry at first did not grasp the actuality of what he saw. He watched with the slightly puzzled detachment of a man who had been unexpectedly shown a newsreel that meant nothing to him. The first group of Arabs

62

ran towards a taxi parked in the street; they tumbled in with comical haste, one man falling and being dragged in by his companions; then the taxi sped away with a burst of oily smoke from its exhaust. Harry, still detached, thought: *How odd – a taxi service for terrorists.* The second group finished their shooting and withdrew, Harry losing sight of them as they ran past the corner of the south wing. The whole action took only seconds and Harry was still bent forward, trying to catch another glimpse of the retreating second group, when Roger pushed into the open window beside him.

"What the hell's going on?"

"Some Arabs shooting up the Secretariat – "

"Not Arabs, gentlemen," said Mordecai Roth behind them. "Jews. Those men were from *Irgun Zvei Leumi.*"

Both Aidan men turned round. Roth stood with his arms hanging limply by his side, utterly dejected; his eyes, behind the glinting panes of his spectacles, were on the verge of tears. But it struck Harry that the little Jew was no longer nervous, only sad.

"I am sorry it has happened. But I am glad you escaped unhurt – I feared the explosion would be much bigger."

"Roth, did you know this was going to happen?" Roger's voice had a hard edge to it.

Roth nodded. "But only twenty minutes ago. I got a warning – no matter who told me." He shook his head as if he were telling Roger not to ask the obvious question. "I came here at once. I did not want to see a friend of ours hurt – "

"Did you warn anyone downstairs?"

Again Roth shook his head. "Lord Bayard, I was taking a risk coming here to warn *you*. The Agency and *Irgun* do not always agree on the methods to be used against the British. I did not agree on what was planned this morning and I told them so."

The phone rang and Harry picked it up. "For you, Roger. It's that major we had dinner with last night, the one from Intelligence."

Roger took the phone and Harry, standing close by him, could hear the crackling voice coming over the line. "Lord Bayard? Vesper heah. Thought I'd better put you in the picture. Some of your friends – " even Harry caught the sarcastic note in the voice; Roger raised his eyebrows resignedly – "have just shot one of our chaps. There's been some damage over the Secretariat too, but don't know what's what there just yet. Damned bad show, I must say. Just a moment – " There was silence for almost a minute.

The two brothers gazed at each other, neither of them saying anything. Roth had moved to the window and stood looking down at the flurry of activity now taking place outside the hotel; twice he shook his head as if in disapproval, and once he murmured something in Hebrew that could have been a prayer. Then Major Vesper came back on the line: "Sorry. We've just had some sort of warning they're going to blow up the place. Told us to get out."

"Had you not better do that?" Roger said mildly, the first words he had spoken into the phone.

"Christ – sorry. You're a Catholic too, aren't you?"

"Yes," said Roger, still mild and patient, amused at the other's effort, no matter how ridiculous, not to offend in religious matters. Harry, listening, wondered if the major was as circumspect when it came to a Jew's religion. "Yes, I'm a Catholic, if that has anything to do with it. But go ahead, Christ or no Christ."

"Sorry. Didn't mean to offend. But we aren't here to let the bloody Jews give us orders. Telling us to evacuate the place – Christ, it's just like their bloody cheek!"

"Is that all, Major?" Roger didn't wait for an answer, but hung up, only the way he slammed down the receiver showing his anger. He turned back to Roth. "Is there going to be another explosion?"

Roth came away from the window, sad thin face hanging boneless beneath his spectacles. "There could be. I just do not understand the way an *Irgun* mind works. I saw enough violence – " He shrugged and spread his hands again. God, thought Harry, don't be so *Jewish*! Then at once was ashamed for the thought. Why shouldn't the man be what he was? Had not the major on the phone sounded stupidly English? "I think, gentlemen, we should leave the hotel, if only for a little while. It is almost time for lunch. If you would be my guests, I know a place run by a French Jew – "

Roger hesitated. "I have a meeting this afternoon. I was going to spend the next couple of hours making up some notes – " Then he nodded. "No, we'll come with you. Perhaps you can help my brother get a clearer picture of what's happening out here. So far he has heard only the Arab side."

"Don't put me on the defensive," Harry said. "I haven't spoken to an Arab since I landed in Jerusalem."

64

"If you have spoken to the English officers," said Roth, and amazingly there seemed no bitterness in his voice; he spoke with the candour of a man who knew the exact truth of the circumstances in which he lived, "you have heard the Arab side."

Harry glanced at Roger, but the latter was looking at his watch. "It's half-past twelve. Do you mind if we go now?"

"The sooner the better," said Roth.

Harry, tieless and jacketless, followed the shirt-sleeved Roth out of the room. Roger did not immediately follow them; when he came out and closed the door behind him he had put on his regimental tie and a jacket. Ah, Roger, Harry thought, the stuffed shirt rebel. But he smiled affectionately at his brother, and Roger, not understanding the smile but pleased at this new relationship between them, smiled back.

Roth led the way along the corridor to the lift. He looked relieved now they were on their way, but when they reached the lift he pressed the button impatiently.

"With these lift drivers, sometimes it is quicker to walk. But we are on the fifth floor."

He pressed the button again and at that moment a tremendous roar came up the lift well. The doors blew out and Harry, in the instant before he instinctively shut his eyes, saw the lift shoot past up the well, a blur of white frightened faces congealed into one mass behind its grille. The floor of the corridor rocked violently, then abruptly tilted. Harry opened his eyes and saw that he was standing at the top of a steep tiled slope; Roger and Roth stood beside him, all three lined up like starters in a ski run. Below them the wall of the building was peeling slowly backwards into the sunlight; then the whole corridor, floor, walls and ceiling, a moving tube, began to slide after the disappearing outer wall. Harry flung himself sideways, clutching at the knob of a door that, still set firmly in its jamb, was tilted at an acute angle. He saw the number on the door, 57, and the card flapping wildly on the knob beneath his hand, *Do Not Disturb*. Then palms, pots, bricks, mortar and dust were hurtling by him. Out of the corner of his eye he saw Roger go plunging in slow motion down the now perpendicular corridor into a grey surf of mortar and dust. The door suddenly came out of its jamb and he went down the disintegrating corridor, the door held above his head like some ridiculous oblong umbrella. The dust rose up to meet him and he shut his eyes and mouth. Then something hit him from below, the door came down on his head and the world went abruptly black.

### 3

Harry was buried for three hours. He came to some time after he had fallen out of the collapsing hotel wing, though he would never know how long he had been unconscious. He lay in a pile of bricks, stone and thick dust, unable to move; the door, supported on two large slabs, was three inches from his face, holding back whatever was above him. A tiny ray of light came in from somewhere behind his head; the minute opening gave him enough air to breathe. His face was masked with thick dust, granules of mortar resting in the corners of his eyes; he moved his lids, trying to look back towards the pencil of light, but the granules slid into his eyes and he shut them quickly again. He could feel the coating on his lips and he resisted the temptation to open his mouth and yell; something told him he would choke to death on his own cry for help. His legs and arms were trapped; he tried to move them and stopped at once as he felt the broken bone in his leg stick into his flesh. Panic whipped through him and his whole body trembled; he knew suddenly that he was going to die, slowly and with no one knowing he was buried here. Sweat broke out on him and for the first time he became aware of the almost intolerable heat of this oven of rubble in which he was enclosed. With paradoxical desperation he tried for calmness; involuntarily he began to pray. Slowly he relaxed, felt the rubble slide a little as the stiffness went out of his body and he lay back, mentally if not physically, in his death bed.

He spent the next two hours slipping between grey coma and clear sharp thought. He prayed, promising everything that God might ask of him; but even as he prayed one part of his mind was conscious of the hypocrisy that the fear of death could expose; God had his uses, but only in times of desperation. He thought of Constance and their unborn child, of Roger and Mordecai Roth; he prayed for all of them, trying to dispel the abject pleading of his prayers for himself. But he was unconscious when the army sappers finally dug him out, and he did not wake till he was safe in a hospital bed.

The young army doctor, plump and red-faced, feeling the July heat, nodded encouragingly at him. "You were very lucky, Lord Aidan. You should have been dead by rights."

"Thank you," said Harry dryly, and the doctor looked pleased: anyone who still had his sense of humour would be sure to survive. He was a simple young man who believed in simple remedies and he was perfectly at home as an army doctor.

"How is my brother?"

The plump red face folded downwards. "I'm afraid he's dead. He was one of the first they dug out. Rather an awful mess, I'm afraid."

Harry felt sick, retched and turned away. An orderly shoved a bowl under his chin, but it wasn't necessary: there was nothing inside Harry but an aching emptiness. He and Roger had never been close and it came as a shock to him that he was going to miss his brother, that the last two days had proved that of all the family Roger was the one of whom he might have grown fondest. He lay with his eyes closed for a while, then he opened them and looked up at the doctor.

"There was another man with us, chap named Roth – ?"

"He's all right. A few scratches and bruises, nothing more. He was the one who found you. He's outside waiting to see you. Shall I have him sent in?"

Harry hesitated, wondering if he really wanted to see Roth. The man meant nothing to him and just now he wanted to be left alone, to be involved in nothing but his grief for Roger. But then he nodded, feeling he owed it to Roger to see that Roth was all right. "I'll see him."

Roth, clothes dusty and torn, came into the ward. He had lost his spectacles and he picked his way carefully among the stretcher cases on the crowded floor, his short-sighted eyes peering at each bed as he looked for Harry. Then he stood at the foot of Harry's bed, grasping the white iron rail and leaning nervously forward like a man standing in the dock of a criminal court. Harry, the image at once in his mind, waited for Roth to plead Not Guilty.

"They have told you? About your brother?" Harry nodded. Roth's hands tightened on the rail and he dropped his head. "What can I say? To kill one of our few friends – "

"You've killed damned near a hundred others," said the doctor, face getting redder; then he nodded at the stretcher cases who were still being brought in. "And injured all these."

Harry, not yet feeling any sympathy for Roth, all at once was angry at the young doctor. "That'll be all, Doctor. I'll call you when I need you."

The doctor looked at him in surprise, bright blue eyes popping.

"Who do you think you are? This is a military hospital – your bloody title hasn't any rank in here – "

Roth looked acutely embarrassed, but Harry just gazed steadily at the doctor, his own anger now under control as the other's temper rose. Though he was not aware of it himself, for the first time he showed a touch of arrogant authority, was his father's son in manner and cold confidence. "I haven't mentioned my title, you stupid clot – you're the only one who has done that. I'm just asking for my rights as a patient. Patients have rights, even in a military hospital – this isn't the first one I've been in. For the record I'm still on the Reserve of Officers, with the substantive rank of lieutenant. Now go and attend to some of the injured instead of standing there dispensing your doses of anti-Semitism."

The doctor glared at him, his face almost ready to burst. For a moment it looked as if he might actually attack Harry; then there was a snicker from the man in the next bed. That was enough: the doctor abruptly wheeled about and went striding down the ward, threatening with every step to put his foot down on one of the stretcher cases lying on the floor.

"Don't take any notice of him," said the man in the next bed, peering with one eye from behind a mask of bandages. "He's not only anti-Semitic, he's anti-every-bloody-thing. The bugger needed to be put in his place."

Harry thanked him for his comment, then looked back at Roth. The little Jew had not moved from his place at the foot of the bed, still stood with his hands gripping the rail. Harry, suddenly not wanting to be in the position of a judge, impatiently waved him to the stool that stood beside the bed. "Perhaps you can do something for me. Get in touch with my wife and my father – can you get through from here to London by phone?"

"I don't know. I – I'd rather send a cable, if you don't mind. I'm a stranger to your father – I should not like to be the one to tell him directly that his son is dead. Especially since I am a Jew."

Harry nodded sympathetically. "I understand. All right, send cables, one to my father and one to my wife. Don't be too terse – use enough words to tell them I'm perfectly all right. My wallet should be with my clothes, wherever they are – "

Roth held up a hand. "It is little enough for me to do – I shall pay for the cables. And I shall be as explicit as possible. The one to your wife will be a pleasure to compose, to tell her you are safe. But the other – " He blinked, as if his eyes had suddenly become sore without his glasses. "Even on such short acquaintance I thought of Lord Bayard as my friend – " He stopped, blinked again and looked at Harry. "But you are Lord Bayard now, are you not? Isn't that how your system works?"

It had not yet occurred to Harry that his identity had abruptly and decisively changed. In other circumstances he would have been aware of the implications of Roger's death, but to-day the thought of his own narrow escape and the urgent wish to let Constance know that he was still alive had pushed all other considerations out of his mind. He sank back into his pillow, as exhausted by the prospect of the future as he was by the ordeal of the past few hours. He was now the Earl of Bayard, heir to his father's title, inheritor of responsibilities from which the accident of birth had exempted him and with which an accident of death had now burdened him. It was as if he were buried again, this time beneath the rubble of tradition. God, that sardonic opponent, had answered his prayer but had offered His account. His life had been saved but he was expected to pay for it. The tenor of his living, that he had been so thankful for only a few days ago in Qaran, was wrecked and might never be the same again.

"Perhaps I should go now," said Roth. "You do not look well. I shall send the cables and come back this evening."

Roth came back that evening and every day for the next two weeks. Gradually Harry came to appreciate the thin little man's qualities: the quiet strength beneath the nervousness, the compassion for others, the pride in being a Jew. On the fifth day they discussed, calmly and amicably, the letter put out the previous day by General Barker, the GOC in Palestine. Though marked Restricted and intended only for army officers, copies at once had found their way into other hands. *Irgun* copied it on to a poster and was distributing it throughout the country. The stark anti-Semitism of the letter threatened another explosion as great as the one it was protesting.

"It is difficult to imagine a man in his position making such a cheap jibe," said Roth. He read from the copy of the poster he held: " – *will be punishing the Jews in a way the race dislikes as much as any – by striking at their pockets and showing our contempt for them.*" He folded the poster, put it carefully away in his pocket. "Does he think all Jews condone what *Irgun* did? My father has spent four whole days in prayer as atonement for what has been done. Why do people hate us so much, Harry?"

"I just don't know." Harry had had little experience of Jews, but he remembered with shame that there had been times in the past when he had shared the general attitude towards them. His own class, he knew, were among the worst offenders; he remembered some of his father's remarks about the Jewish peers in the Lords. Each time Roth came to visit him he felt the tension in the atmosphere of the ward as the little Jew came nervously down the aisle between the beds. No one said anything directly to Harry, but he had seen the whispering in the more distant beds, and each day he waited for some remark and wondered what his own answer would be. The young red-faced doctor had been replaced by another, more tolerant man, but some of the orderlies did nothing to hide their feelings about a patient whose most regular visitor was a Jew.

"All countries forgive their enemies sooner or later," said Roth. "You watch, in twenty or twenty-five years' time, the Germans and the Italians and the Japanese will have been forgiven. And why? Because of business, trade – because they will be needed. But no one ever forgives the Jews. And why? Because we are so good at business."

"Perhaps you should not be so successful."

"Personally, a talent for business, I don't have. But let us not discuss my woes. Have you heard from your wife?"

"She is well, thank you. Unfortunately the cables you sent were delayed – they didn't reach her and my father until the evening of the day after the explosion. They weren't in London but down at our home in the country."

"They must have been worried, hearing the news on the wireless and reading it in the newspapers. But I am glad she is well. And your father also, I hope?"

"He wanted to fly out here, but I sent him another cable to stay where he was. He wanted to take my brother's body back to England to be buried."

"How did you explain it in your cable? That there was very little left of Roger?"

"I didn't attempt to." Harry doodled on the back of the pad he was holding; he had been writing to Constance when Roth had come in. "Mordecai, if the local Conservative Association will have me, I am going into politics when I get back home. I am going to take over my brother's seat. But I can't promise you that I shall take up the banner for Israel as he did."

"Will you be against us?" Roth had new spectacles, horn-rimmed ones that gave him an owlish expression, especially when he cocked his head on one side as he did now.

"No. I promise you *that*. I'll do everything I can to help you, but don't ask me to be a crusader."

Roth smiled. "A crusader for the Jews? But I suppose there have been allies just as odd."

Harry was flown home with several other casualties of the King David explosion. An RAF Stirling, converted to an ambuance plane, flew them to an airfield in Wiltshire, and they were whisked away to their various destinations before the Press knew of their arrival in England. The Marquess and Joe Banning came up from the estate to meet him, and as Harry hobbled out of the aircraft on crutches and saw his father waiting for him on the tarmac, he wondered how emotional their meeting would be.

But the older man's self-control was marvellous. He put a gentle arm about his son's shoulders and said quietly, "Damned good to see you, Harry. A pity Roger – " He paused and bit his lip, but it was only a momentary lapse. "Well, it was God's will. He would have thought of it that way himself, I'm sure."

Harry nodded, shook hands with Banning, then hobbled across to the car. "The Bentley? That's a bit extravagant, isn't it?"

"They gave me an extra ration of petrol," said the Marquess. "Said you were a deserving case."

They drove home through the gentle English sunset. Harry looked out at the peaceful fields, mauve and green in the soft slanting light, and for the first time fell in love with the nature of his country. They passed a yellow hill where harvesters, working late, were walking down a zig-zag track behind a high-laden wagon; it suddenly came to Harry's mind that they could be pilgrims, believers in the countryside that they worked. He was surprised at his own reaction to the landscape as he drove through it, and now he understood why Lucas should want to retire here.

"Why didn't Constance come up with you?" He had been disappointed when he had got off the plane and not seen her there waiting for him. And piqued too, he confessed to himself.

"I've ordered her to rest," said Banning. "No, she's all right, no need to worry. But we have some bad news for you."

He and Harry were in the back of the car, the Marquess in front at the wheel. Banning looked at the Marquess, the latter caught the look in the driving-mirror and nodded.

"She lost the child," said Banning. "The day after the news

71

of the explosion, before we heard that you were safe, she had a miscarriage."

"Damned pity," said the Marquess, driving the Bentley with the cold haughty recklessness that was characteristic of him whenever he got behind the wheel, his Boadicea complex as Roger had called it. "I was looking forward to having a grandson. Your sisters have given me nothing but gels."

"There was no guarantee – " Then Harry glanced at Banning. "Or did you know what it was?"

Banning nodded. "It was a boy, formed enough for us to know what it was. That's why Constance needs to rest. She had a pretty bad time, the miscarriage coming so late."

"How will she be? I mean about more children?"

"No worry at all, I should think, though you may have to wait a while. There's nothing physically wrong with her. She's just had a terrible emotional shock. We all did." He looked again at the Marquess and again the latter nodded. "We got the news about Roger before we heard that you were safe. For a while it looked as if you might have gone too."

"Dreadful business," said the Marquess. "What are they doing about the Jews out there?"

"Looking for revenge."

Harry turned and looked out the car window, closing the subject. He saw the slight stiffening of his father's neck and beside him he felt Banning look at him; but he said nothing, just kept looking out the window and thinking about Constance. He had always thought of her as stronger than himself in character and temperament and it frightened him to think that she had gone so to pieces at the thought that he had been killed.

He was further shocked when he saw her. Aidan Hall was no longer a military hospital and had been handed back to the family. The Marquess, accompanied by Ferguson, Mrs Dibbens, a cook and a maid, had moved back into one wing of the house. Furniture, under dustsheets, was piled up in the Great Hall and the corridors like piles of grey rock, but the inhabited wing had been made comfortable. Constance came out on to the terrace to meet Harry, then with hardly a word to her father-in-law and Joe Banning, led him into the house and their bedroom.

"I'm sorry, darling, there'll be no loving. Not for a while."

Standing awkwardly because of the cast on his leg, leaning on her for support, he held her to him and felt the bones of her in every part of her body. She had never been a big fleshy girl but she had never been skinny and fragile as she was now. Her face was gaunt and the fingers she put up to stroke his cheek were as thin and weak as an old woman's. He could have wept for her and, indeed, felt the tears come to his eyes.

"I was ready to die," she whispered. "I *wanted* to. Then the cable came – "

He held her while she wept silently against his shoulder. "I'm sorry you lost the baby – "

"I haven't thought about it. All I've thought of is you – I haven't been able to eat – Mrs Dibbens and Joe Banning have been cross with me – " She looked up at him, her face made even more gaunt by her weeping. "I shouldn't want to live if anything happened to you."

He was both humbled and frightened by such devotion. No man had a right to such love; love should not demand a life. "You mustn't think like that. One person shouldn't be the whole world for another."

"You are for me." She dried her eyes and tried to smile. "No argument, please."

That night they lay in bed, both on their backs, the only way he could be comfortable, hands clasped like modest virginal lovers.

"What a way to celebrate our reunion," she said.

"It's been only three and a half weeks."

"The longest we've ever been without it. How long is your leg going to be in that cast?"

"I'll keep it on till you're really well again, put back all the weight you've lost. I'm not as sex-mad as you seem to think."

"Haw-haw."

Already she seemed to have regained her vitality if not her strength. At dinner she had been bubbling over, charming the men with her good humour; even Ferguson and Mrs Dibbens had remarked upon the improvement in her and had beamed approval. In the past when in the family environment her demeanour had always been a trifle cool and withdrawn, she was always conscious that she was an outsider; to-night she had been completely relaxed in her happiness, and the Marquess, seeing a new side to his daughter-in-law, had at last fully and warmly accepted her. And she had responded to his acceptance of her.

Joe Banning, when leaving, had said to Harry, "Some day she will be the Marchioness. It will be the nicest accident that's ever happened to this family. Maybe to the whole bloody aristocracy."

"I think I'll have to get rid of you. It's dangerous having a doctor in love with one's wife."

Banning smiled. "I've been thinking of going out to Australia or South Africa."

"Seriously?"

"Seriously. There's not going to be much future for a country G.P. when this National Health scheme comes in. I'll barely make a living."

"Do you want to leave England?"

Banning shrugged. "Not particularly. But it's not just what the Socialists are doing – actually, I'm in favour of some of their ideas. But don't quote me to your father. No, there's something else – "

Harry waited for him to go on, but Banning seemed to think he had said enough. "What, for instance?"

Banning said reluctantly, "Well, I'd like to make a new start, put a few things behind me."

"Such as what happened at Arnhem? Is that still on your mind?"

"Isn't it on yours occasionally?"

"Oh, for Christ's sake! Joe, that's over and done with. Forget it. Forget about going to Australia or South Africa too. Look, if you're going to lose money when this new scheme comes in, I'll talk to Father. We'll come to some arrangement. If you moved out, who's left? Old Peterson, who's useless, wouldn't know a hernia from tonsilitis. You've been here too long, Joe – you're as much part of this region as I am –"

"I'll think about it," said Banning, cursing himself for his weakness. He did not want to leave England, but something told him the price of not going would be his soul. "But don't fool yourself – neither of us will ever forget what happened at Arnhem."

Then Constance had come out of the drawing-room and gone out with Banning to his car. When she came back she said, "You're in Joe's debt, you know. If it had not been for him, I might not be here."

"It was as close as that?" It was a further shock to him to know she had actually come so close to dying.

She nodded. "But that's behind us, darling. Just let's think about the future from now on."

Now, an hour later, in bed, she said, "Your father is delighted that you're going into politics."

"How do you feel about it?"

"I'm happy, if that's what you want to do."

"My love, let's stop building both our lives around *me* – Are you listening?"

"Yes. I was just thinking – how could I have once thought you were selfish?"

He did not answer that because he knew, or suspected, the selfishness that still lay within him. "If I'm elected it will mean extra work for you. Being an M.P.'s wife may not be your idea of fun."

"When do you think you'll be elected?"

"If the local Association will have me – "

"I'll bet that's already arranged. I heard your father on the phone talking to someone. No, *telling* them."

Harry smiled to himself in the darkness. "That would be Father – he thinks of the local Association as his own little band of faithful followers. Anyhow, supposing I'm nominated, the by-election can be rushed through – a constituency like this does not require much organising – and I could be in the Commons before Christmas."

"I'll be pregnant again by then."

"You will not. We're going to wait at least a year before we try again. Even if I have to keep this cast on my leg to hold you off."

4

In the morning Harry was sitting out on the terrace, his leg propped up in front of him, when his father came out and sat down beside him. The Marquess pulled a chair round till he was facing the same way as his son: they sat together looking down across the wide expanse of the park, at the trees pregnant with summer and the gold-topped grass reflecting the bright hot sun.

"Wanted to talk with you." Harry understood now why his father had sat down *beside* him. The Old Man did not want a face-to-face talk: whatever he was going to say it was going to be difficult for him. "Afraid I've neglected you. You're the heir now and you know damn-all of what's expected of you. My fault entirely." Harry said nothing and his father shot a quick glance at him. "You agree?"

"I agree."

"There are responsibilities – you'll inherit them some day,

75

along with all this." He waved a hand that took in the house, the park and the surrounding fields and hills. "There's more than just the four farms and the cottages and shops and land in the six villages. There are the people. Your responsibility too. Or will be."

"The people can look after themselves these days. This isn't the Middle Ages."

"Wrong." The Marquess was still speaking in the drawl fashionable with his generation and class, but he was minimising his wordage; even at the best of times he had never quite known how to talk to his younger son, and this was not the best of times. "It's our duty to look after them. We have the money and the privileges. They don't."

"Father – " Harry exploded in good humoured exasperation. "They don't *want* us interfering in their lives. People want to be left alone. So long as we employ them at a fair wage, or house them at a fair rent, they couldn't care less about our responsibilities towards them. They'd laugh if we mentioned *noblesse oblige* to them. They'd think it was some French horse come over for the Derby."

"Perhaps. Doesn't alter the fact from our viewpoint. You would think differently if I'd brought you up to it. My fault."

"You said that."

His father turned his head, looked directly at him for the first time. "Do you care for people?"

"Some. I care for Constance. And you." He hoped the hesitation before the last sentence had not been too apparent. He did not want to offend his father, not while he was still grieving for Roger.

"Thank you. No, I mean people in general. Them." He waved his hand again, as if the park and the hills were full of watching people.

"I don't know. I haven't really thought about that."

"Learn to, then. We'd be nothing without them. A man's no leader who has no one to lead."

Harry's eyebrows shot up and he laughed in embarrassment at his father's seriousness. "I'm not going to be a leader!"

"Perhaps not. Roger – " Then he stopped, realising just a little too late that from now on there must be no comparisons with the dead son; among all the responsibilities he had been speaking of, his biggest was to the son who sat beside him. But out of the corner of his eye he had seen the stiffening of Harry's good leg.

"Roger might have done the trick." Harry's voice was flat and toneless.

"Perhaps." Backed by the huge house, he looked out at the park, at the long drive leading down to the tall iron gates that carried the rusted coat of arms on their railings. "But people will always need someone to look up to. You're in a better position than anyone else round here to some day be that someone."

# Four

"God, as we all know, is an Englishman." Harry paused and looked around the House. The actual chamber was that of the House of Lords, but the peers had moved out of it and the Commons members had moved in while their own bombed-out chamber was being rebuilt. The red leather benches on both sides were packed; the Whips had not had to go cracking this afternoon. "I see some nods of agreement even from the Government side – "

"Withdraw!" Wardle, the Member for East Dawsbury, stood up awkwardly; the Member beside him shifted to make room as Wardle swung his tin leg round. "Mr Speaker, are we to allow blasphemy from the Opposition? We know the Honourable Member for Bayard comes of a class that considers itself on a par with the Divine – "

"Withdraw!" The Opposition beat the backs of the benches in front of them with order papers and the palms of their hands.

The Speaker, face strictly neutral beneath his heavy wig, wondered why grown men had to indulge in frivolities on an occasion as serious as this. An old army man, he knew light machine-gun fire often preceded a big battle, but this was just grapeshot. "I am sure the Honourable Member for Bayard had no blasphemous intent."

"None at all," said Harry. "But if I withdraw the remark am I not then, by implication, denying that God *is* an Englishman? Would the Honourable Member for East Dawsbury then accuse me of devaluing our status throughout the world, as indeed his own Party has done by its monetary action last week?"

"Hear, hear," cried the Tory Members. God, they all knew, would have been seated on their side of the House if He had decided to run for election.

Constance, crushed between two burly men in the crowded Gallery, felt the mixed reception around her. Parliament, recalled on this September afternoon in 1949 after Sir Stafford Cripps had announced the devaluation of the pound, was in a sour disillusioned mood. Spectators, expecting a heated debate, had packed the Gallery, and only the brusque efforts of one of the ushers who admired such a good-looking woman as the Countess of Bayard had got Constance a seat. The spectators were about evenly divided between Labour and Tory supporters, and Constance sensed that the feeling in the Gallery would be more inflammable than that down in the Chamber.

"What bloody right has he got to speak?" a man behind Constance whispered hoarsely to his neighbour. "He personally could be devalued by fifty per cent and he'd never feel it."

"He can't have much left after taxes," said the other man mildly.

"His kind know how to beat taxes."

Harry, unexpectedly, had been the only backbencher chosen to speak for the Opposition. He had made several notable speeches in his almost three years in the Commons, but no one had thought he would be given a place in to-day's debate. Constance, aware of Harry's youth in terms of parliamentary age and conscious of the handicap in the Commons of his title, had wondered if he should accept the invitation to speak; but Harry, confident and delighted, had brushed aside her doubts, pointing out that he would be speaking as the *Member* for Bayard and not as the Earl of Bayard, and anyone who forgot that fact would be forgetting how democratic the House of Commons was.

"I'm not a peer yet," he had said. "The title is a courtesy one, in the Commons it doesn't mean any more than a plain knight-hood – "

"Darling, you don't have to explain the system to me. I've done my homework in Burke and Debrett. But I still have a nice middle-class mind and I think I know how middle class – and working class – minds think. When the Earl of Bayard gets up in the House and says something, I don't think a great percentage of them are of the opinion that you know what you're talking about. I mean in relation to *them, their* problems."

"My love, I have enough sense not to get up and speak about everyday things that affect the man in the street. Deep in their hearts the British public likes to be governed by its betters – I'm sure we'll get back into power at the next elections – but just to show its independence it draws the line where we can have an opinion. For instance, I've never spoken on food rationing – who would believe you and I are affected by it?"

"I say an Act of Contrition every time you have that black market meat delivered."

"But you eat it."

"Except on Fridays," she said with mock virtue.

"The stomach is like a stiff cock – it has no conscience."

"Don't be vulgar. At least not in public."

"We're not in public. We're in bed, the proper place for vulgarity. How is your stomach?" He ran a gentle affectionate hand over her pregnant belly. "Do you think you should come to the House to-morrow when you're like that?"

"He may be going into the House himself one day. I think I should get him accustomed to the atmosphere as soon as possible."

After her miscarriage in 1946 she had waited a year before she had suggested they should try again for a child. But by then Harry had been caught up in politics, business and a social life that involved both those pursuits; Constance had found herself a hostess who was kept as busy as her husband. She had not disliked the life, had found it exciting and even glamorous, a word she had never thought she would use, after the dull girlhood she had spent in Northumberland; but she had hungered for a family, and now was the time to start, while she and Harry were both young. But Harry had reverted to his old attitude towards children, that they would only tie down Constance and himself, that they should establish the pattern of their own lives before fitting children into it. There had been several arguments, one of them very fierce, but Constance had always given in. She had gone to a Harley Street gynaecologist and had had him fit a diaphragm, and for two and a half years it had worked safely. Then six months ago, one night after coming home from a party where they had both been gay and happy and physically hungry for each other, she had removed the diaphragm without telling Harry. When she had been sure she was pregnant and had told him, he had not been angry and disappointed as she had expected; he had been philosophical and, she suspected, secretly happy. But she had never told him that the coming baby had been planned deliberately by her. It was the only secret she had from him and it tormented her, but something told her it would be best kept from him. She was beginning to discover more about Harry than he had about her.

She looked down on him now as he got into his stride in his speech. Though she knew nothing of international economics even she began to appreciate the brilliance of his argument: all she had to do was look at the discomfort on the faces of the Government Members opposite him.

"He knows his facts," muttered one of the two men behind her.

"Why shouldn't he?" growled his companion. "He's got his bloody money spread all round the globe. I shouldn't be surprised if he could quote off the cuff the currency rate of exchange for every country in the world."

"He doesn't look as if he has a penny to his name."

"That's only camouflage. The ones with the old money are all the same these days – they do their best to hide it."

Constance smiled to herself as she looked down at Harry in his rumpled suit bought off the rack. It was not camouflage on his part that made him dress like that but a genuine disinterest in clothes; among the well-dressed men on his side of the House, many of them in striped trousers and black jackets, he looked like someone sent up from the labour exchange in search of a job. He wore a yellow shirt and a tie of a regiment that was not his own: God knew where he had got it and sooner or later he would be challenged by someone who was fussy about such things. When clothes rationing had ended six months ago she had gone out and bought him a new wardrobe at Hawkes in Savile Row; she had not been able to persuade him to come with her and she had had to resort to some elaborate lies to explain to the tailors why Lord Bayard, instead of coming to them for a fitting, wanted six ready-to-wear suits and two tweed jackets off the rack in a hurry. But each day when he dressed he put on whatever was closest to hand, and this morning it had been the five-guinea suit he had bought when he had first got out of the army. It had not occurred to him that the occasion, his most important day since entering the Commons, called for any window-dressing. Constance, when she had seen him come down for lunch after working all morning brushing up his speech, had only shrugged resignedly. In a family that could trace its lineage back six hundred years, clothes quite obviously were not needed to make the man.

Harry finished his speech and sat down. Several Tory Members leaned forward and patted him on the back; on the Front Bench Churchill and Butler turned and nodded approvingly. Over on the Treasury Bench Stafford Cripps, Puritan face stiffly composed behind his rimless glasses, an austere man perfectly cast for a period of austerity, rose to reply to some of Harry's points. Constance, aware again now of the crush of spectators on either side of her, got up and made her way out of the Gallery. She glared at the two men behind her, but they were unconscious of her hostility; all their attention was on Cripps, the saint who had let them down. She looked across towards the Peers' Gallery, saw her father-in-law nod to her and rise from his seat.

They met down in the Central Lobby, the only common ground between the two Houses. "Splendid, eh?" The Marquess nodded his head enthusiastically. "Really gave them what-for, eh? Care for some tea? Oh, here's Elizabeth."

Elizabeth Cavanreagh came gliding across the floor of the Lobby, her beautiful face smiling out at them from between the picture hat she wore and the mound of old wool that Constance supposed was her dress. Elizabeth had the unconscious talent of looking even more derelict than her brother. "My God," said her father, who was vain about his own elegance, "why does she always look as if she has been dressed at a church bazaar? Her mother was the same."

"What time does he speak?" said Elizabeth as she coasted to a stop in front of them. All her movements were graceful, but her voice was loud and hoarse and gave the overall impression that she was harsh and awkward. "Damned buses – never run to time as they used to – "

"He's spoken," said the Marquess. "Damned good, too."

"Blast," said Elizabeth, her voice suiting the word; several ushers looked at her, recognised the Marquess, and decided against cautioning quiet. Elizabeth, unaware of anyone's attention, looked at Constance. "So you're pregnant again."

Constance looked down at her prominent stomach. "I think so."

Her father-in-law's mouth twitched at the corners, but all he said to Elizabeth was, "You should keep in touch more."

"Bloody wilds of Ireland. No news ever gets through there."

Constance, a twice-weekly letter-writer to her father, had been surprised at how little the Aidans kept in touch with each other. As far as she knew the Marquess never wrote to any of his three daughters, and their only notes to him would be to announce their impending arrival on a visit. Elizabeth lived in Connemara; and Anne and Grace, the other two sisters, lived in Norfolk where their husbands, both peers who never bothered to come to the House of Lords, owned neighbouring estates. Yet for all their lack of communication with their father and Harry, the Aidan sisters always seemed remarkably well informed of what went on at the core of the family. Constance could only suppose that there was some grapevine that she had not yet discovered.

The Marquess took Constance and Elizabeth into tea in the

dining-room. They were served ham paste sandwiches and caraway seed cake which Constance thought went well with the waitress who served them. She looked around the dining-room which had the funereal chic of a men's club, which in effect was what it was; then, for want of something better to do, gave her attention to the conversation from which the Aidans, father and daughter, had excluded her.

"I want to move back to the Hall," Elizabeth was saying.

"How does Patrick feel about that?"

"Stubborn. But that's the Irish for you. But I can't survive any longer, Father, in the bloody bogs of Ireland." Several people looked up from their tea and cake, but Elizabeth was oblivious of the fact that anyone might overhear her. "I'll talk Patrick into sanity."

"Well, we'll have to speak to Harry about it."

P.S., thought Constance, and Constance too.

"And Constance, too, of course," said the Marquess, and smiled at her as if he had meant to include her all along. "It's their home as well as mine."

Elizabeth looked at her sister-in-law as if puzzled as to why it was necessary to include her. "We shouldn't disturb you. The house is big enough for half a dozen families."

"Who else is coming?" asked Constance innocently.

Elizabeth recognised an enemy when she was shot at. Her expression did not change, but she lowered her voice. "After all, it is *our* home."

Constance looked at her father-in-law for some support, but the old swine had conveniently chosen that moment to turn away and exchange greetings with a fellow peer. Men, she thought disgustedly: they always duck out on women's disagreements. Only this was more than a disagreement: Elizabeth was about to declare war.

Constance, despising herself, beat a strategic retreat: "I'm afraid you'll have to discuss it with Harry."

"I intended to all along," said Elizabeth, and bit into a slice of cake as if it were Constance's throat.

The Marquess came back from his temporary neutrality. "There will be the Ministry of Works too. Can't upset them, eh?"

"What the hell have they got to do with it?" Elizabeth spat out a caraway seed as if it were a B-B shot; Constance instinctively leaned back out of range.

So the grapevine does fall down occasionally, she thought; and said, "Didn't you know? Your father had to approach them, otherwise we should have to close most of it down. It costs the earth to run, what with central heating and the rest of it."

"What are you going to do with it?" Elizabeth addressed her father, ignoring Constance.

"First, we're going to have everything repaired and refurbished. Nothing's been done to it for ten years. Lots of dry rot, death watch beetle, stuff like that. Roof's leaking too. The Ministry is going to match me pound for pound on what it costs. Will cost a small fortune."

"What do the Ministry want in return for their money?"

"Place has to be thrown open to the public for a certain number of days in the year. We'll charge them, of course, a shilling or two a head – it will help cut down the maintenance expenses. Damned nuisance, of course, but it can't be helped."

"I'm damned if I'm going to live in a house overrun by bloody strangers!"

"That solves that, then, doesn't it?" said the Marquess; and turned away to order another pot of tea. You cunning old blighter, thought Constance, you know exactly how to handle them all. And determined to study him thoroughly in case the day came when, for some reason, he tried to handle her too.

Then an usher came in to tell her that His Lordship was waiting for her in the Central Lobby. She rose to go. "Will you be coming over for the hunting in November?"

"Of course," said Eliazbeth. "That's why I'm in London now. Ordering outfits for Patrick and myself. Tell Harry we'll need horses."

Constance just stopped herself from saying, *Yes, ma'am*. She left the Aidan father and daughter and went out to the one Aidan she truly loved. Half a dozen Tory Members were congregated about him, patting him on the back, but he was taking their congratulations with an equanimity only a degree short of smugness. Another Tory Member went by, just nodding curtly to him, the resentment showing as plainly on his face as the dark stubble on his jowls: Constance recognised him as Michael Argus. He was six or seven years older than Harry, a black-haired, black-jowled man who had inherited a small newspaper chain in the North and who was tipped as a possible Chancellor of the Exchequer in the future. He had been the back-bencher expected to speak for the Tories in the devaluation debate, and some of the political commentators in this morning's newspapers had openly

84

wondered if the substitution of Harry in the debate meant that Argus was now thought less promising.

Constance, waiting for Harry to escape from his admirers, saw him glance at Argus. The latter paused in his stride for just a moment; the men's gazes locked like swords caught by their hilts. Constance felt herself draw in her breath: Harry, darling, don't make enemies. But whatever Harry thought, Argus had declared himself. He strode on across the Lobby, his thick neck over his collar dark with temper. Harry detached himself from the group and came across to Constance.

"Hello, my love." He kissed her on the cheek. "What did you think of it?"

"Very good," she said, but fed his ego no more than that. "Darling, why does Michael Argus dislike you so much?"

He smiled. "Your beautiful blue eyes don't miss much, do they? He has it in for me because I did the dirty on him. He was waiting to be asked to speak in the debate. And I didn't – wait to be asked, that is. I sent a copy of what I wanted to say to someone on the Front Bench, he showed it to the others, and that was it."

"Is it usual to do that – I mean, let someone see your speech in advance when you haven't even been invited to speak?"

"No. But gentlemanly conduct never got anyone anywhere in politics. Argus isn't a gentleman, but he made the mistake of trying to act like one."

"Are you a gentleman?"

"Of course. But one doesn't have to go on proving it all the time."

2

Bayard was a large village, and unlike most large villages it had not been spoiled by the encroaching blight of later years. The local urban council, another of the Marquess's fiefs though they did not like to think of themselves as such, had been careful that any development fitted in with the local scene. A few Nissen huts had been erected during the war years, but immediately peace same they had been dismantled and sold to local farmers as storage sheds. So the village, built mostly of local limestone with a few half-timber houses thrown in for relief, with thatched or slate roofs, continued to look much as it had for centuries. Even the garage at the end of the High Street, with its yellow petrol bowsers, looked as if it were only a conversion from a staging inn for the coaches that used to run through the village from Bath to Portsmouth. There was a church at either end of the winding High Street, an Anglican and a Roman Catholic, with the R.C. church, due to the Aidan patronage, in better repair than the C. of E. Between the churches were three pubs, all in better repair than the churches. All the shops were locally owned, with the exception of the Boots chemist shop, under whose thatched roof were sold Glaxo, Aspros, Laxettes, condoms and other ancient potions and preventives. An air of serenity, centuries old, hung above the fold in the hills in which the village lay.

The Bayard Hunt, a mixture of pink and black coats and tweed jackets, came clop-clopping up the High Street, headed towards Aidan Hall. The Marquess was Master of Foxhounds and the hunt at its first meet always started from in front of the Hall; it delighted his eye to see the fifty or sixty riders coming up the drive in a long line, and it had become an unwritten rule that no one arrived at the Hall independently. This was the first meet of the season and everyone was looking forward to it. Except the Marquess, who had woken that morning with an attack of gout so bad that he could not pull on a boot.

The hunt, horses and riders all champing at the bit, came to the end of the High Street opposite the Roman Catholic church. A large pre-war Daimler was parked across the roadway and on either side of it, effectively blocking any passage, stood half a dozen people carrying placards. Leaning against the car, arms folded across her breast, was a good-looking redheaded woman. The hunt clattered to a halt, horses rearing and milling about as their riders drew rein at this unscheduled hurdle.

Five minutes later the phone rang at the Hall. Harry took the phone. "Lord Bayard? Could you come down to the village? Afraid we've run into a spot of bother. One of those bloody anti-blood sports groups."

Harry drove down to the village with his father and Constance. The Marquess, his swollen feet encased in felt slippers, sat in the back of the Bentley and cursed all busybodies who interfered in the pleasure of others. "Everyone enjoys it – riders, horses, hounds, people who follow to watch – everyone!"

"Everyone but the fox," said Harry. The laurels beside the road were beginning to lose the last of their yellow leaves and up on a hill a line of chestnuts added a bronze postscript to the cloud-streaked page of sky. It was a beautiful day for hunting

86

and he had been looking forward to it with great enthusiasm. But he was realistic. "These people think of themselves as spokesmen for the fox."

"Don't give them that concession when you talk to them," snapped his father. "Tell the hunt to ride right over them!"

Harry winked at Constance, turned the car out of the side road round the graveyard of the Catholic church and pulled up on the opposite side of the Daimler to where the huntsmen stood with the impatient horses and hounds. The protesters, feeling they were being unfairly attacked from the rear, spun round, placards at the ready. The redheaded woman came round the car to face Harry as he got out of the Bentley.

Behind her Constance heard the Marquess snort in angry shock. "Good God, it's the Kilminster gel!"

"Who's she?"

"Freddie Kilminster's daughter. Been away for years – married an American or something. What's she doing around here?"

"Doesn't she belong here?"

"Comes from down in Dorset – leastways, that's where her father had his place. He's out in Kenya now, done a bunk from the Socialists. Used to go down to Dorset and hunt with him occasionally. Janet, that's her name, used to hunt with us. Just a schoolgirl in those days."

"How do you recognise her after all these years?"

"Saw her picture in *The Tatler*," said the Marquess, then added defensively: "Always look at it. Like to see the pretty gels."

"She's pretty enough. And sexy, too," said Constance, looking at the redheaded woman with new interest. "Harry seems to think so. He hasn't yet ordered the hunt to ride over her."

Harry had never met the redheaded woman before. Leaning back against the mudguard of her car, her arms once more folded across her breast, she looked coolly at him. "Are you the Master?"

"For to-day, yes. Who are you?"

"Janet Buck. You're Harry Aidan, aren't you?"

That told Harry he was dealing with someone from his own class and he looked more guardedly at the woman. She was a couple of years older than himself, he guessed; about thirty, though like most men his guesses at a woman's age were usually wide misses. She was undeniably good-looking though not strictly beautiful, with a broad faintly freckled face under the combed-back red hair. In the tweed suit she wore it was impossible to tell whether her figure was good or just ordinary, but he noticed that her legs were extraordinarily shapely, the sort of legs that he thought of as sexy.

"You don't belong here, do you?"

"I do as of this week." There was just the faintest American intonation in her voice. "I've just bought Four Orchards Farm."

"I thought it had been sold to some company."

"I am the company. Buck Investments. And I understand you plan to hunt across my fields."

"If the fox runs that way – yes."

She shook her head and there was a loud murmur of dissent from the other demonstrators. The members of the hunt, bound to their horses, were all still on the other side of the Daimler parked across the roadway. Behind them half a dozen cars and a delivery van had pulled up, their drivers getting out to watch the fun. Old Father Regan, Irish and a horse-lover, stood at the door of his church, silently blessing the hunt and cursing the busy-bodies who tried to follow in the footsteps of St Francis of Assisi, and thanked the Lord *he* wasn't alive to-day. Behind Harry the Marquess had poked his head out of the Bentley and was advising the huntsmen to ride right over the demonstrators.

"Don't waste time on 'em, Harry! Shove 'em out of the way!"

"Morning, Lord Carmel," said Janet Buck. "How's your apoplexy these days?"

Harry kept his face straight as he heard his father utter a strangled cry. "Insults aren't going to help, Miss Buck," he said.

"Mrs." She showed him the ring on her finger.

"Is your husband here? Perhaps he will show some reason."

"He's in America. And he was never a very reasonable man at the best of times. You are the one who should show some reason. Chasing after a defenceless fox with a pack of dogs – is that a sane humane sport for a grown man? Why not sit at home and pull wings off butterflies?"

"You hunted as a young gel!" The Marquess still had his head hung out of the window of the Bentley. "I remember you – you're Kilminster's daughter!"

Harry understood now why she knew him and recognised him as an equal. Her father was the Earl of Kilminster, of a family almost as old as the Aidans, and she had her own title of Lady Janet, though she had probably given it up when she had married the American.

"So you're a convert to them?" He nodded at the other demonstrators. "Converts are always the worst."

"Possibly. But that doesn't change the immorality of what you're doing."

At that Harry laughed: protestors always took themselves too seriously. "*Immorality?* Chasing a fox is immoral? Chasing a chap or a girl – that could be immoral, especially if he were married. But a *fox* – !"

A shadow crossed her face and her jaw tightened. "Keep your remarks to yourself! And stay off my land!"

She wheeled about, dragged open the door of the Daimler and jumped in. She started up the car, surprising the demonstrators as much as the huntsmen, swung it round, narrowly missing several horses, and sped off down the High Street. The protesters, suddenly lacking a leader, abruptly folded up; placards came down like surrendering flags and the group withdrew into the Catholic graveyard, whence they were soon hustled out by Father Regan in his own demonstration of Christian charity. Harry told the hunt to continue on up to the Hall, then came back and got into the Bentley.

"Strange woman. One would have thought I'd threatened to shoot her or something."

"Well, you can have your hunt now," said Constance. "Just let's hope the fox doesn't make for Four Orchards Farm."

An hour later that was exactly what the fox did. Harry, at the head of the strung-out riders, came up a slope on his favourite big black and saw the hounds turn at an abrupt angle along the top of the ridge. They vanished into a small copse, their baying fading for a moment, then they burst into sight again, a moving frieze along the long downward curving sweep of the ridge. Harry swung the black to the right, lifted it over a low blackthorn hedge, and led the hunt slantwise up the slope.

He could feel the black labouring beneath him and he did not push it too hard; this was the first run of the season and the big horse was still lacking condition. But it had galloped and jumped well and Harry had enjoyed every minute of the ride. On these outings, astride a willing horse, he was possessed by a feeling of – *escape*, he supposed it was. All his concentration was on his riding, with occasional attention to anything admirable that caught his eye, the line of hill against sky, the blur of colour of trees mourning the dying year, the graceful flight of a horse in front of him as it took a hedge without losing stride; his mind freed itself from all else in the world, even Constance, and he experienced a sense of aloneness that was almost akin to drunkenness. There was also something sexual in the exhilaration he felt and he had often arrived back from a ride to take Constance immediately to bed with him. There would be none of that to-day: she never let him touch her after the sixth month of her pregnancy.

The black was slowing as it came up to the crest of the ridge, but the other horses had fallen even further behind. Harry eased the horse back as he stood up in the stirrups to sight the hounds. He saw them come out from a screen of alders, scramble through a gate, cross a narrow lane, filter through another gate, then lope on up a long stubble field. Then he saw the fox, a small brown blur half-way up the field and beginning to slow.

He dug in his heels and set the horse down the slope of the ridge. He was fifty yards from the first gate when he caught a glimpse of the Daimler coming up the lane. Up till then he had not been really concerned which way the hunt was heading; now he recognised he was already on Four Orchards land and across the lane the fox and the pack were heading up towards the main house, now showing above the trees bordering the far field. He drew rein for just an instant as he saw the car put on speed; then he dug in his heels again and urged the horse on. He would be over the gates and into the second field before the car could intercept him. The hounds were now going to get the fox anyway and there was nothing the damned Mrs Buck could do about it.

He was ten yards from the first gate when out of the corner of his eye he saw he had misjudged the speed of the car: it was much closer than he had reckoned. He hurried the jump, lifting the black too soon; even as they took off he knew they were not going to make it. He saw the gate beneath them, felt the horse hit it, then he was going over the horse's head, both of them falling in a wildly threshing heap, and the ground was coming up at him so swiftly there was not even time for him to turn his face away. He felt a dreadful pain go right through his skull from front to back, then everything was black and silent.


3

"No man is entitled to such luck," said Joe Banning. "You should have broken your neck, falling that way."

Harry, face swathed in bandages, tried to ease himself up in bed. "How long do I have to stay here?" He winced as he felt some of the bruises. "I wanted to go back to London for tomorrow's sitting."

"That's out. By rights I should have kept you in hospital – "

"I'm just as comfortable here. There's only this broken nose – "

"And concussion and that torn shoulder. You should stay here for at least a week. I'll come in every day and bring a nurse with me to change the dressings."

"Never mind the nurse – she's an expense we can do without." He saw Banning shake his head and he snapped, "What's the matter now?"

"You're a tight man with a penny. You'll spend quids on your horses – What do you think you're going to do with all the pennies you save?"

"All right, bring the nurse in!" He knew he was inclined to economise on small ridiculous things; he remembered his grandfather who was always going about switching off lights so that the Hall at night had been in a perpetual state of semi-darkness. He guessed it must be part of the conscience of the rich: he never thought of money except when it had to be expended in small amounts, and then he always thought of reasons why it should not be spent. "How is my nose going to be?"

Banning had his back to Harry as he closed his medical bag. "It'll be all right, I think. You can have an ENT man look at it later. He might want to re-set it."

"When I came to, yesterday morning, I thought I'd woken up in a distillery. Was that Friday night's breath I smelled or had you been drinking so early in the day?"

"You couldn't smell a damned thing!" Banning's voice was a little too sharp. "You didn't know where you were till we got you to the hospital. I had to manipulate the bone in your nose so the blood could run out instead of back into your sinuses. I've told you – if you think I haven't done a good enough job, you can always get a specialist to work on it."

"Simmer down, Joe. I have no complaints – yet. But you had been drinking yesterday morning – I didn't need to smell it. They would have noticed it at the hospital too. Hit the bottle as much as you wish at night, but keep away from it during the day."

"Listen – "

"No, *you* listen. My father is paying you an extra thousand a year to stay on here as the local G.P. Whether you recognise it or not, that makes you the estate doctor. There's no contract, but

I'll see there is one if it's the only way of getting you to do your job properly."

Banning picked up his bag, jammed his cap on his head. "You don't own me, y'know. I can earn double the money I make here, out in South Africa – "

"You'll never go, Joe. The effort would be too much. And you'd drink just as much out there – perhaps more. Just see your drinking sessions take place when you're off duty. I'm telling you that as much for your own good as for that of your patients. Otherwise one day you may lose one of your patients and you'll find yourself in court."

Banning, boiling with anger, had a lot he wanted to say; but he knew none of it would be any real answer. So he was glad when the door opened and Constance came into the bedroom. When Harry's accident had occurred yesterday Banning's first concern had been for her. Following the hunt with the Marquess by car, she had arrived on the scene before Banning himself, while Harry, bloodied and unconscious, still lay in the soft earth in which he had fallen. But there had been no evidence in her of uncontrolled shock and this morning, before coming up to see Harry, he had examined her and been relieved to find she was suffering no delayed reaction. The baby was only a week or two away and he hated to think of the effect on her if she should have a second miscarriage. He looked at her with love, said a curt good-bye to both of them, and left.

"What's the matter with Joe?"

"A little difference of opinion, nothing important."

"Janet Buck is downstairs. Do you want to see her?"

Harry's head and, it seemed, all his body were aching; but now might be the best time to deal with Mrs Buck. Constance went out to the gallery and called down into the Great Hall. A few moments later she came back into the bedroom with Janet Buck.

"I spoke to Dr Banning on his way out. He told me you had nothing seriously wrong. I'm glad of that. When I saw you hit the ground yesterday – " Then Janet Buck looked at Constance. "Sorry. I didn't mean to upset you."

Constance, fat and ungainly, lowered herself into a chair. "It's all right. When one is hunting one expects the possibility of an accident. And the way my husband rides – "

"He did try to make the gate much too fast. How is the horse?"

"We had to destroy it."

"Pity."

"Yes. Good hunters are not easy to come by these days."

Then both women turned to look at Harry again. Masked by his bandages his expression was hidden, but there was no mistaking the dry sarcasm in his voice. "Good men are no more easily available, either. If you hadn't tried to race me, Mrs Buck, I'd have taken that gate in my own good time."

"You were on my land, where your own good time doesn't count." She was dressed this morning in a mustard-coloured suit that went well with her red hair, a suit that was less bulky than the one she had worn yesterday and which showed that her figure was indeed a good one. "I'm sorry you were hurt – and your horse had to be destroyed – but don't blame me for your accident. You were trespassing and that's all there is to it."

"What have you got against fox-hunting?"

"The barbarity of it. Your hounds caught the fox yesterday right outside my back door and tore it to pieces."

"That wouldn't have happened if I hadn't fallen. I'd have been there in time."

"Perhaps. But I didn't come here to argue that. I just wanted to be sure that you weren't seriously hurt. I hope your face is not disfigured behind all those bandages." She looked at Constance. "Good-looking husbands are in short supply too."

Constance smiled. "I don't think I'll have him destroyed."

Janet Buck returned the smile, nodded. "Well, I'll be going. No, don't bother – I can find my own way out." She motioned Constance not to rise. "Good luck with the baby. I hope there are no complications as a result of this."

"It doesn't seem to have taken any notice of it."

"Good." She paused at the bedroom door, looked back at Harry. "My farm is out of bounds from now on. You had better tell the Hunt."

She went out. Constance said, "She doesn't beat about the bush. But I think I could like her."

"There's something wrong with her. She's bitchy against the world for something that it has done to her."

"Spoken like a true man. A woman voices some principle and at once she's branded as just getting her own back on someone. Why is it only men that have honourable principles?"

"Whose side are you on?"

"Here we go again – choosing sides. Darling, I'm on *your* side – whose else? But be fair to Janet Buck. It's her farm and she can do what she likes with it. And evidently you hit her a pretty low blow yesterday – when you said in front of all those people that it was immoral to chase a man. That was why she went off in such a rush."

"What are you talking about?"

"Joe Banning gave me the gossip this morning. He wasn't here during the war, but he seemed to know all about it. Perhaps that's how he's paid off now under the National Health – with gossip."

"Get to the point," Harry said patiently.

"Well, Janet Buck evidently set her cap at an American colonel – Colonel Buck – who was stationed near her place, and she chased after him like you chase after the fox. He was married, but that didn't appear to deter her. He took her back to America with him, had his wife divorce him and he and Janet were married. Now she has divorced him – the rumour is that he has gone back to his first wife. But Joe wasn't sure of that."

"A pity. Some patient must have short-changed him."

"Don't be smug. How would you politicians function without gossip? Isn't it the oil that keeps the wheels working?"

"You've been reading those smart political columnists again."

"Haven't you?" She picked up the Sunday newspapers that were lying on the foot of the bed. On the Thursday of that week he had made another notable speech in the House, his first on foreign affairs, and the reaction in his own Party had been just as favourable as his September speech in the devaluation debate. "The *Sunday Times* and *The Observer* each has a paragraph on you. And the *Express* is positively ecstatic – but then they are always ecstatic about anyone with a title, so long as he isn't a Socialist. I presume your father has read them all to you. Do you want to hear them again?"

"Is there anything in the papers about my accident?"

Constance nodded. "Shall I read those too? *The Observer* paints you as a barbarian who got his just deserts."

Then the Marquess and Elizabeth knocked on the door and came into the room. Elizabeth had ridden in the hunt yesterday but had fallen at an early fence and had brought her limping horse back by float. Harry's accident and the Marquess's attack of gout had cancelled the dinner-party that was to have been held last night, and all in all she had had an abortive weekend. She had come over for only three days and now she was having to return to Connemara with nothing to show for the trip.

"Suggested to Constance I'd stay on for a week or two, give

her a hand till you're back on your feet. But she says she can cope."

"No trouble at all," said Constance, looking imploringly at Harry not to suggest otherwise.

"I'll help, if needs be," said the Marquess, hobbling about on slippered feet. He sat down heavily on a chair and rested both hands on the silver-topped walking-stick he carried. "But better get well soon, Harry. Now's the time for us to strike, while you are in the news again."

"Strike what?" Harry asked.

"The Party, of course. Set up a position and consolidate it. In eight to ten years' time we might be able to nominate our own Prime Minister. We did it once before, with the help of the Stanleys."

"I thought it was the other way round – we helped them."

"That's the way the Stanleys *would* tell it," said Elizabeth. "They always did hog credit."

"Winston won't last forever," said the Marquess, impatient at the interruptions. "Eden's a good man, but he's waited so long a lot of the new Members aren't going to take to a fellow who's worn out before he starts."

"We have to put the Socialists out first."

The Marquess waved a hand of dismissal. "They won't last another twelve months. No, the fifties are going to be *our* time. And we – us, the Aidans – we can have a big part in it. It's time we got back to government by elite. It's the only way."

"Whom have you got picked out to be our P.M.? I mean *our* one?"

"Got my eye on several. But they'll have to prove themselves. Never fear, the right one will turn up when the time is ripe."

"What about me?"

"You as Prime Minister?" Elizabeth snorted. "You're much too young. The country will never accept another Pitt. Neither would I, not even you. We want a mature man."

"I'm not talking about now. Eight to ten years' time, that's what Father said." He looked at Constance. "What do you think?"

Constance was not accustomed to ambition, even on a small scale. Her attitude towards life had always been positive, much more so than Harry's when they had first met, but purpose rather than ambition had been the force in her; life on the Northumberland farm had not been conducive to a drifter's outlook. Her religion, too, had conditioned her; she had adhered to its strict discipline and only lately had she begun to query some of its more trivial rules. Five years of marriage had not exposed her to ambition. It was not a characteristic of the upper classes, indeed it was looked upon as a vice; content on the peak of their own sense of superiority there was no cause for them to aim higher. Now, confronted by Harry's question, she was almost inclined to tell him he must be joking. But a note in his voice warned her that would be the wrong thing to do.

"What about the title?" she hedged.

"She's right." The Marquess had been unusually thoughtful, chin resting on his hands on his stick and his lips pursed. He had quickly examined the possibilities of his son's unexpected suggestion, but reasons had soon doused the quick fire of enthusiasm he had felt. "You would have to renounce the title, any claim to it. The Commons will never take another peer as P.M. In any case, we don't want it. Much better to run things from behind the scenes. That way one gets all the advantages and none of the blame."

"I may be in the Commons for another thirty years before you die. I don't want to be a back-bencher all that time." He was no longer lightheaded: he recognised the logic of what his father had said and he woke from the dream before it had really begun.

"There's nothing to stop you having a Cabinet post. Foreign Secretary – how about that? Start specialising in foreign affairs now and in a few years' time the job could be yours. Eden was only thirty-eight when they gave him the F.O. You could beat his record."

Constance watched the bandaged face lying among the pillows. Only the mouth and jaw showed, but there was no hint of expression there; yet somehow she knew that she stood with Harry at a fork in his life's road. She had arrived here without warning and something told her that whichever fork he took their life would never be the same again. He had tasted success over the past couple of months and had succumbed to the drug of it. It had surprised her to find that someone who seemingly had everything was not immune to it.

Then a gong rang downstairs. "Luncheon," said Elizabeth, rising with a Pavlovan reflex; food was her drug and she ate as if on the verge of starvation. "I wonder what Mrs Dibbens has for us to-day?"

"Some of what we should have had at the dinner last night," said Constance. "We can't afford to waste it."

Elizabeth glided out of the room and the Marquess got to his sore feet. "Damned Banning has put me on a diet for a week. Hardly worth living if you can't eat and drink what you want."

"Well, don't pop off for a while. I'm not ready for the Lords just yet."

"Not popping off. You'll be an old man before I make way for you. Coming, Constance?"

"In a moment, Father." She waited till the Marquess had gone out of the room, then she turned back to Harry. "Would you really like to be Prime Minister?"

"Why not?"

"That's no answer."

He was silent for a moment or two, sorting out a true answer for himself as much as for her. Lately he had begun to investigate himself, as an outsider might; not looking for faults, though he knew there were many, but for talents that could be used. He had discovered he had a need for prospects, where an outsider would have thought that one of his class and wealth was beyond such considerations. "I want to be someone more than the Earl of Bayard. Or the Marquess of Carmel. It isn't enough. In our system a lot of people think a title *is* enough – but it isn't. A man shouldn't be judged by his circumstances – and a title is just a part of those."

"Are you trying to tell me you are against the hereditary system?"

"Of course not. But because a man is born with all advantages, why shouldn't he take advantage of them?"

Pregnant, burdened with her own prospects though not unhappy with them, she felt suddenly too weary to debate with him. Whatever he wanted would satisfy her, so long as he took her with him. She stood up slowly, feeling the child stir slightly in her womb: what circumstances lay ahead of it? she wondered.

"Just be careful, darling Whatever you want, don't chase after it as you did after the fox. Or Janet Buck after her colonel."

He smiled beneath the bandages. "I came a cropper there. And I suppose she did too."

She picked up his hand and laid it against her cheek. "It will never happen to us."

Then she went out of the room and he lay in bed looking out the large mullioned windows at the park. He could see the yew-lined drive curving down round the slope above the lake to the distant gates. A small herd of deer moved slowly down the slope like drifting red shrubs that had come loose from their roots; they had had to be rounded up yesterday before the hounds had been let loose. The lake glimmered silver in the pearl-coloured day; there might be rain this evening. On the small island in the lake the gazebo built by Capability Brown for the second Marquess had once more begun to assume shape. For ten years, since the beginning of the war, the island had been allowed to be overgrown with weeds and shrubs; this week the Marquess had started workmen on clearing it. The gazebo, decorative but useless, was a symbol of what the Aidans might have become.

Ferguson came into the room carrying a tray. Like a good butler, he never knocked on the doors of his two masters, the Marquess and the Earl; this was his house as much as theirs. He knocked on Constance's door, but only out of deference to her modesty, not out of respect for her authority. He had his own authority, his own independence, and the Marquess and Harry knew by instinct where the line was drawn. Constance, brought up in a home where the only servant had been a woman who came in daily to cook and tidy up, was still in private awe of Ferguson, though she was not sure if the butler suspected.

"Soup, sir, and a little junket. Dr Banning has recommended that you have nothing that needs chewing."

"Ferguson, what would you be doing if you were not butling?"

"Occasionally, during my time in the army, I gave the matter some thought. I decided there were no attractive alternatives. Though I should not wish to be butler to *anyone*, sir."

"Do you think you might be a snob?"

"The possibility is there, sir. I don't consider it reprehensible. I like to think I should be an honest snob." His eyes smiled, though his face was still sober. "There would not be many of us."

Harry, who was no snob, agreed. "Ferguson, perhaps you and your father and your grandfather were more pillars of the Aidan establishment than members of the family like myself. At least butlers are more constant. What happens when you retire? Are you already training your boy? How old is he now? Seven, eight?"

"Seven. Unless he changes, sir, I'm afraid he will never be a butler. I tell my wife he has too much of her father in him." Mrs Ferguson, a cheery dark little woman, did all the household sewing, maintaining her independence of Mrs Dibbens by only taking orders from Ferguson. "My wife has plans for him."

"Is she an ambitious woman?"

"Not for herself, sir. But for John – yes, I'm afraid so."

"Why afraid so?"

"Ambition is not for us, sir." *Us* being the Fergusons and the Aidans. He's impregnable in his smugness, Harry thought, he is more a brother to Elizabeth and Anne and Grace than I am. "Will there be anything else you want, sir?"

Only tolerance for my own ambition. "No, Ferguson, that will be all."


4

"I suppose you wonder what I'm doing here," said Janet Buck.

"Frankly, yes," said Harry.

"I came with Dr Banning. The poor man had no partner and I volunteered. You might say I'm gate-crashing."

"How did you know he had no partner?"

"I asked him. He came out to see me – I'm having a little trouble with my insides – "

"How interesting." I'll have a word with that bloody Banning.

"Not really. But Dr Banning mentioned he was coming to the ball and I asked him was his wife looking forward to it. Then he told me he had no wife – "

"And one word led to another and here you are. Why? Hoping to see what sort of blood sports we get up to at night?"

"No. I wanted to see what you had done with the house. This is one of the first of the big houses re-opened since the war. I must say you've done it all magnificently."

"The credit is due to my wife and my father."

"And the National Trust – or was it the Ministry of Public Works?"

"Watch your step. I mean your dancing – you almost trod on my toes then."

She smiled up at him as she leaned back from him. She was not a good dancer, not as good as he, and she had a habit of dancing with her stomach pressed against him. If she was meaning to be provocative he thought she was being much too obvious. "I like your nose the way it is now. Makes you more interesting-looking. You were rather a pretty boy before."

Harry said nothing, but looked away, bored by her. She

certainly looked attractive this evening, sexually attractive, he had to admit; but he was not interested in any invitation she might offer and he would tell Joe Banning that she was not to be brought to the house again. He had had this dance with her so that he could find out how she had been invited, but now he could lose her for the rest of the evening.

The ball was being given by the Marquess to re-open Aidan Hall and to celebrate the birth of his first grandson. Richard Roger, Lord Aidan, was now two months old and to-night was sound asleep in his nursery, an unwitting excuse for the thumb-nosing fling against austerity now taking place in the Blue Ballroom and the State Gallery. This was the first ball held at the Hall in over ten years, the last having taken place in June 1939, and the Marquess had thrown money around as if devaluation had convinced him there was no point in holding on to it any longer. An eight-piece band had been brought down from London and a six-piece combination brought up from Torquay. Caterers and decorators had been brought down, the caterers also thumbing their noses at austerity by somehow producing salmon seemingly by the school, beef Stroganoff (or so they claimed, though Constance thought it might have been horse Stroganoff), cherries Jubilee, ice-cream, and a variety of cheeses that the Marquess had not seen since before the war. The caterers also managed to produce champagne, not vintage stuff but good enough to satisfy the majority of palates which had not tasted champagne for much too long. The decorators, with few flowers available to them at the end of February, had used ferns and other greenery from the Hall's conservatory; for the rest they had used ribbons of silver and black, the predominant colours in the family's coat of arms. The Blue Ballroom had been a later addition to the original Hall, built on by the second Marquess; it had been designed by William Kent and he had succeeded admirably in linking it to the rest of the house while retaining his own style. During the war it had been the main ward of the military hospital, and since the war it had been shut up completely. To-night it looked as it might have looked at its opening ball in 1740.

A few newspapermen had come down from London to report on how the top two per cent went about its pleasures while the other ninety-eight per cent was limited to Bob Hope at the local Odeon and a Mars bar at intermission. But the estate workers, with the perverse loyalty that the Czars used to find among their serfs, proved themselves an efficient security force, and with the enthusiasm of tipsy rugby forwards threw out of the park a man from the *Daily Express*, one from the *Daily Herald*, a girl from *Vogue*, an elderly dowager correspondent from *The Tatler*, and a correspondent from Tass, thereby proving the catholicity of their rejection of the Fourth Estate.

The guests, three hundred of them, were the best decoration of the evening. Most of the women from London had come in their Diors, Balmains and Hartnells; the local women came in their Pyms, Smiths and Snodgrasses, the products of dressmakers in nearby towns and villages. Tiaras glittered on several heads, and pearls, diamonds and rubies were as common as the coffee crystals on the supper tables. Constance, her figure regained except for the bosom which was still full of milk for young Richard ("the only unrationed one in the house," Harry had commented) was wearing a Balenciaga, a present from Harry. He, in turn, was wearing a new suit of tails, a present from her and the only way she had been able to prevent him from appearing in the pre-war dinner suit that had been his only formal evening attire.

The Marquess, elegant as a Harrods or Fortnum and Mason's shopwalker, the handsomest man at the ball, gazed at Constance admiringly as he two-stepped her round the huge room. His dancing style had not changed since the twenties, and even in those days it had not been good; he had a peculiar trick of negotiating a turn with one foot stuck out at an angle; Constance had the feeling she was being *steered*, as a speedway rider would handle his motor cycle round a bend. But the Marquess was enjoying himself.

"Best-looking gel in the room. Harry shouldn't let you out of his sight."

"How about having an exchange with him? Swap me for that Buck woman who's trying to rape him."

"Rape? What? Where? Oh, *her*. Must say she does dance damned peculiarly." He stuck out a leg and they negotiated another turn at full speed; Constance looked down to see if chips flew off the floor. "Not jealous of her, are you, m'dear?"

Constance laughed; her jealousy did not worry her to-night. Three weeks ago she and Harry had resumed their love-making; both of them had been so insatiable that it amused her to think that Harry had enough energy left to be interested in another woman. They had made love this evening before they had bathed and dressed and she could still feel him in her. "I don't think so. If ever I'm going to be jealous, it will be of politics. They take up more of his time than any woman ever will."

"Do you mind?"

"A little. But not enough to make an issue of it. In a way I'm glad he's found politics so interesting. For a while – don't tell ever him I told you so – I thought I'd married a playboy. If he'd turned out to be that, then sooner or later I should have been jealous of other women. But politicians can't get up to much misbehaviour, can they?"

"Don't be naïve, m'dear. Lloyd George had more mistresses than Ministers."

"Then we'll have to see that Harry doesn't become P.M., shan't we?"

Then Joe Banning, only one whisky aboard even though it was almost eleven o'clock, tapped the Marquess on the shoulder. "Sir, you have your privileges, but can another man claim a dance with the best-looking woman here?"

"Exactly what I called her. That gives *you* the privilege, Doctor." The Marquess would never have invited Joe Banning's father to a ball, even though Robert Banning had been a more respected man than his son; but times were changing, and when Harry and Constance had added Joe's name to the guest list the Marquess had not objected. But it went against his grain to entertain socially a man he had to employ professionally. The family lawyer was not here to-night, nor would the Marquess have thought of inviting him any more than he would have his gamekeeper. Harry and Joe Banning were not close friends, but there was some understanding between them and the Marquess often wondered what it was. They had not been together long enough in the army for it to be anything to do with the war. But he would never ask Harry: privacy, the Marquess believed, was the most precious thing a man could own. He handed over Constance as if he were presenting a prize to Banning. "She's a tip-top dancer too."

"How would he know?" Banning said as he moved Constance away. "He's the worst dancer in the West Country."

"Joe, do you mind if we sit this one out? It's a long time since I danced and I'm out of training."

They edged their way round the crowd, crossed the State Gallery and went out into the Great Hall. But as they came out of the ballroom Constance paused and looked back at the beautifully proportioned room with its re-upholstered chairs and settees and its rich new curtains.

"It's wonderful, isn't it?"

"The room or the occasion?"

"Both."

Banning nodded. "England – and the world, I suppose – needs some of this." He looked both ways down the State Gallery, the long room that divided the Blue Ballroom from the Great Hall; the deep mirrors on the carved walls reflected the dancers and the music echoed back from the high ceiling; the evening of pleasure seemed doubled and magnified. "The trouble is, most of them take it all for granted."

"Who?"

"You know who I mean. The Marquess and his lot. Look at them all – treating it as if it were no more than a Saturday night dance in the village hall."

"You're wrong, Joe. That's their public face – they don't know how to put on another face in public."

"You sound as if you've been making a study of them."

"I need to know them – I'm going to spend the rest of my life with them." They had come out into the Great Hall and now she stopped and looked quizzically at him. "How is it I'm confiding so easily in you?"

"Women often confide in their doctors. Things they don't want to discuss with their priests or their parsons. Another thing, I'm an outsider, like you."

"You were *born* here."

"Certainly I was. But not in the right class. You were lucky – you married into it." Then he smiled. "Sorry, I'm talking too much. And I'm stone cold sober. Perhaps there's something to be said for drink."

"Why do you drink so much?"

"Thirst. And I have a condition known as a parched ego. Quite a few bachelors suffer from it."

"Your ego shouldn't be too parched to-night. Not with Janet Buck on your arm."

"Ah yes, I wondered how you and Harry would react to her. But it was either her or come alone. In which case I might have drunk more and probably disgraced myself. The nobles can commit murder and rape among themselves, but they never forgive unseemly behaviour in lesser mortals."

"I think this topic has gone far enough."

"I'm sorry." He sounded genuinely contrite. "I keep forgetting you belong with them now. And you're better than the whole damned lot of them."

Abruptly he wheeled and went off, leaving her alone at the bottom of the wide stairs. Then her father said at her shoulder, "Who's that rude chap?"

"My doctor. And he's not rude, just truthful."

"Amounts to the same thing sometimes."

George Grainger was a lean, grey-haired man with an air of composure about him that his daughter had inherited. He had once played rugby for his county and he still had an athlete's easy movements; he worked as hard on his farm as his four farmhands and he neither drank nor smoked. He still loved and missed his wife who had died when Constance was ten years old; he had occasional affairs with women he met on holiday or perhaps some woman from a neighbouring county, but he always kept his slate clean in Northumberland. He practised his religion, but more out of habit than conviction; his whole life had become a habit. He was saved from being a dull man only by his occasional unexpected adoption of causes.

"What are you crusading for now?" Constance put her arm in her father's; she loved him and welcomed his visits.

He did a tattoo with his knuckles on one of the two suits of armour that flanked the bottom of the stairs; the sound was hollow and tinny. "I sometimes wonder how much a crusade achieves. I get very dispirited sometimes. I wonder if the chaps who wore these outfits ever got the same feeling, that it wasn't worthwhile buckling on all this stuff?"

"You're starting to sound *old*!"

He grinned ruefully. "Perhaps that's it, one shouldn't be a crusader after one turns thirty."

They began to move back towards the ballroom. "What did you think of your grandson?"

"All children look alike to me up till they're teenagers. Can't say I'm looking forward to the day when I'll say to someone, 'Meet my grandson, Lord Aidan.' Or the Earl of Bayard, if I last that long."

"You sound like one of those perverse – or do I mean reverse? – snobs who think people shouldn't marry above them."

"Perhaps. Though I don't consciously think that way." They had entered the ballroom. The band was playing "Buttons and Bows" and some of the older guests went jigging past in self-conscious abandon. "But I'll always be only on the fringe. Even our friends up home will always think of me as an outsider to my own grandson's class. It's the way we English think."

"I think that will all change." Constance looked at the dancers whirling past; for a moment she had the image of a lake of bright colours vortexing down into a bottomless abyss. "Sometimes I wonder if this sort of life will survive till Richard is grown up."

"Will you regret it if it doesn't?"

"I don't know. I'm happy just being Harry's wife. I'm not so much one of *them*, though Dr Banning seems to think I am. I don't know that I ever will be."

Then Harry came up to them. "Look at Millie Swindon. She dances like one of those Lipizzaner horses from the Spanish Riding School in Vienna."

The Duchess of Swindon, laughing equinely, whinnying with delight, went by with her husband. Her daughter Rosemary, bosom bouncing, followed in the arms of a lean, balding young man who looked exhausted. She threw a kiss towards Harry and Constance and bounced on, while over her shoulder the young man raised his eyebrows at them in an expression of resignation.

"Was that kiss for you or me?" said Constance; then looked at her father. "She was Harry's first girl."

"My ninth, actually," said Harry.

George Grainger smiled warmly at his son-in-law. He had liked Harry from their first meeting, though he had had early doubts of him, thinking that he might amount to no more than a wealthy loafer. But Harry had changed, and though Grainger himself was a Liberal voter he had lately begun to take pride in his son-in-law's promise in the Tories. "I think I remember her from your wedding. Rosemary Something-or-Other."

"Her name's Kidman now. That's her husband she's dancing with. Ian Kidman, he's in the Foreign Office."

Then the Aidan sisters went past, all looking remarkably alike, each in the arms of her husband, none of whom looked alike. The Aidan sisters had been famous beauties in the thirties, but now they looked out of date and, as so often happens when out of fashion, even a little plain. Each of them had married a farming man with no interest at all in what went on in London, and life in East Anglia and the wilds of Ireland seemed to have stopped the clock on them. They were all wearing new dresses, they were very much aware of the fact that they were the Aidan girls back in their own setting, yet they suggested nothing but a dusty, out-of-focus print of the past. Harry, looking at them, suddenly felt sorry for them.

"Why so serious?" Constance took his arm as they moved towards one of the adjoining rooms where supper was being served.

"I was just thinking how fortunate I am to be as young as I am. Ten or fifteen years older and I might have been as dull as my sisters and brothers-in-law are."

Then an anything but dull figure appeared beside them. "Lord Harry, what a splendid occasion! An English Arabian night!"

"No English belly dancers, unfortunately. I was sorry, Hazza, I couldn't get out for your father's burial. I was caught up that week at Westminster."

"My father would have forgiven you." Hazza waved a be-ringed hand that flashed like a battery of semaphore lamps. He had always been something of a dandy, but to-night, with his pale blue silk burnous and his jewellery, he put some of the women to shame, not least the Aidan sisters.

"What are you doing in England?" Constance asked.

"Such food!" exclaimed Hazza, helping himself to a little of practically everything that was laid out on the table. Take a lesson, thought Harry, and turn on something better than roast camel next time I'm out in Qaran. "I am here, Lady Bayard, to talk with the Foreign Office. I am not only Sheikh and Prime Minister of Qaran, I am also my Foreign Minister."

"What about your brothers?" Harry said.

The semaphore messages flashed from the fingers. "Could I trust one of them to come abroad? I leave them at home to fight among themselves. I have one loyal brother – I shan't tell you which one – and he reports to me on the other seven. What do you call this delightful concoction?"

"Marinated sheep's balls," said Harry, "an old Elizabethan dish."

"They are cherries Jubilee," said Constance.

Hazza's eyes twinkled. "Lord Harry has his little joke. But things are changing in Qaran, my dear chap. I am going to bring it up to date. Including good English food."

"Not necessarily English," said Harry. "Why don't you import a French chef? You can afford it."

"I may do that. But I shall have to find one my brothers can't bribe."

He moved on out of the room, his plate piled high, and Constance said, "Does he mean his brothers might try to poison him?"

"Qaran's share of what Anglo-Qaran takes out in oil is thirty million pounds a year and some day it is going to be double that. And whoever is Sheikh owns the lot of it. That's worth a little poisoning, isn't it?"

"But they're brothers!"

"Lloyd George once said there were no friends at the top. He might also have said there were no brothers. There goes one of the Seymours – " He nodded at a girl who went past in the arms of the Marquess. "One of her ancestors nominated his own brother for the block and Edward the Sixth obliged him. Some salmon?"

"You've just destroyed my appetite."

"My love, you are much too trusting. Most men have to struggle to be honest – they are more corrupt than you imagine. Do have some salmon. Some mayonnaise?"

"Darling, don't ever become corrupt."

He smiled and winked at her. "We'll fight it together."

"I think married couples who are so much in love in public are absolutely disgusting," said Rosemary Kidman at their elbow.

"She should talk," said her husband. "She's just been trying to undo the studs in my dress shirt. Every room is a bedroom to her."

"Really?" said Constance. "She looks so innocent and – is virginal the word I want?"

"What on earth is that?" said Rosemary unabashed.

Ian Kidman smiled resignedly at Harry and began to fill his supper plate. He was a personable young man with a thin intelligent face and an air of diffidence that disguised a mind that knew exactly what it wanted and how best to go about getting it. He was not ruthless, but knew the limits of his talents and how best to use them. As far as Harry was aware, Kidman knew nothing of his and Rosemary's relationship in the past. Kidman was a late-comer to Rosemary's world: his father was in insurance and though the family was well-off it was neither rich nor well-connected. Kidman had gone to a good public school, to Oxford, where he had distinguished himself, then into the Foreign Office where, Harry had heard, he was looked upon as one of the coming young men. Knowing Rosemary, Harry sometimes wondered if the marriage would last.

"I'm being posted," Kidman said. "To Madrid."

"How's your Spanish?"

"Terrible. My French is good and I took Russian at Oxford. But the F.O. is like the army – they always send you in the direction opposite to what you've trained for."

"I'm looking forward to the bullfights. It'll be such a change after watching cricket. When he asked me to marry him, he didn't tell me he was a cricket lover."

"It was out of season," said Kidman. "If you'd been patient enough to wait till the summer, you'd have known."

"How could we wait? I thought I was pregnant. Fortunately I wasn't."

She and her husband moved on down the supper table. Harry looked after them. "Ian's an intelligent man. How could he ever have proposed to her?"

Three hours later the last of the guests not staying at the Hall had departed into the winter night. Joe Banning and Janet Buck were among the last to leave. Janet Buck, warm with mink and gin, gave Harry her hand, pressed his, smiled and moved on to Constance. "This was a real welcome home to England for me. I met so many people I hadn't seen in years."

"That's how we planned it," said Constance sweetly. "I hope everyone made you welcome, especially the men."

Janet Buck's smile lost none of its luminosity. "Especially the men. I had forgotten how nice English men can be."

She and Banning went on out the door and Harry said, "You were a bit bitchy then, weren't you?"

"At three o'clock in the morning I'm not going to play the Angel of Mercy – I'm too tired. Poor Joe. I hope he doesn't become involved with her. I think she'd be worse than Rosemary used to be."

He nodded. "Rosemary was an innocent nymphomaniac compared to her."

"An innocent nymphomaniac – that's a new one. Quite a feat, isn't it?"

"I wouldn't know. I'm only going on gossip. I think you're still jealous of Rosemary."

"Of course I am. And those other eight girls you never told me about."

The Great Hall was emptying as the guests went up to bed. The Aidan sisters, flushed with memories, went up the wide stairs together. Their husbands trailed them like pets: an English sheep dog, a beagle and a rather emaciated and drunken Irish wolfhound. The Marquess and Charles Wolfe, a junior Shadow Cabinet Minister, mounted the stairs, heads close together: they looked as if they were talking politics but actually they were talking women. Both of them widowers, each of them, against his better judgment, had succumbed to the flirtatious Janet Buck and she had raised the sap in both of them. George Grainger, another widower, had retired to bed before Janet Buck had got round to him. When everyone had disappeared upstairs Harry and Constance prepared to go up to their own bedroom.

At the foot of the stairs Ferguson said good night to them. "A most successful evening, m'lord. Like old times."

"You enjoyed it?" Constance asked.

"The whole staff did, m'lady. The junior maids were in ecstasies. They are too young ever to have seen anything like this before."

I'd never seen it either, thought Constance. "We must do it again."

"It does wonders for the morale," said Ferguson, and went off to bed and Mrs Ferguson, fortified in the knowledge that the elite were still alive and capable of enjoying themselves.

As Harry and Constance, despite the lateness of the hour, enjoyed themselves in a manner common to elite and proletariat when they got to bed. Then they fell asleep, secure in each other's arms, and the smug comfort that their happiness was impregnable.

# Five

In October 1951 the Conservative Party went back to West-minster as the Government. Harry was returned as the Member for Bayard with a majority of 9000 over his Labour and Liberal opponents. When Churchill, once more Prime Minister after six years of what he looked upon as a rude interruption, announced his Government, Harry found himself Parliamentary Private Secretary to Charles Wolfe at the Colonial Office.

"It will do for the time being," said his father. "But tread carefully. Don't get your name linked with any calls for inde-pendence. Doesn't go down well with the rank and file in the Party, this breaking up the Empire. Often get letters from chaps who served in India, complaining about what the Indians and the Pakistanis have done to the country since we gave it back to them. There's one of them here in the district, chap called Bedley, colonel in the Gurkhas, you should talk to him some time."

"Retired Indian Army colonels don't make up the rank and file of the Party. I think you're a bit out of touch, Father."

They were walking in the formal gardens at the west side of the Hall. The gardens had been planned by the eighth Earl, one of the few Aidans to show any artistic or aesthetic talent, and they had been left intact by Capability Brown when he had come to design the rest of the park. The late autumn sun turned the sculptured privet hedges to a red gold; the colour was seemingly reflected by the gold carp swimming close to the surface of the big ornamental pool in the centre of the garden. The house, built of Ham Hill stone, also had a golden hue to it in this light; there was a delicateness to it that suggested it was made of spun sugar, an effect heightened by the finials that flowered at the tops of the numerous tall delicate chimneys. Here at the side of the house the park sloped away steeply; in the distance they could see the River Bayard glinting like a bent and rusted sword driven into the bosom of a hill as it cut its way towards the Severn. From up on the rear terrace there came the high laughter of Richard, now almost two, and an occasional thin bawl from his sister Robin, aged three months. Their nanny, trying to hush them, would look anxiously down at the two men in the garden, but Harry and the Marquess, neither wishing to impress the other as being intolerant of children, had by unspoken agreement moved to the far end of the garden.

"Nonsense! You're talking like one of those damned Socialists."

"Just what you said to Roger one time. Don't be afraid. I'm not as liberal and rebellious as he was. The days are past when a rebel will get anywhere in politics – you chaps in charge of the party machine will see to that. I don't want my head chopped off, not when I'm making such steady progress."

"Still dreaming of the Foreign Office?"

"That should come eventually," Harry said, and his father took his arrogant confidence for granted; like father, like son: but at that moment there was no outsider there to comment on it. "No, I'm dreaming of more than that."

"Being P.M.? I still think that will never be on. *That* way, I agree, the country is changing. A pity. An elite still makes the best government, no argument about it. Equality never begat quality." A thin diffident smile flickered on his lips. "Made that up myself. Or think I did. Sounds original."

"If ever I become P.M. I'll borrow it. There's been no one in Downing Street since Disraeli who has coined epigrams."

Over the next year or so, as he entered into the work of the Colonial Office and enjoyed it, he said no more about his dream of becoming P.M. He saw the growing impatience of Anthony Eden as the latter waited for the reluctant Churchill to stand down; he saw the growing list of hopefuls who waited in turn for Eden to retire; and he saw the queue, led by Michael Argus, gathering for the more distant future. Pragmatic as always, he put the dream away, as he might have some memento from his youth. He turned thirty, still a youth in Parliament, but he no longer felt a young man. He was still slim and in condition, energetic in his work and virile in bed; he still hunted with the same reckless dash and he played squash and tennis with a hustling abandon that ran most of his opponents off the court. But his outlook had changed, he had the feeling he was approaching closer to his horizons. He never thought of death and decay, but all innocence, if he had ever had any, was gone forever.

The Korean War, at one time feared to be the beginning of World War III, had developed into a stalemate. Harry, like so many in Britain, had lost interest in it; for them it became an American war, and though Harry was not guilty of such pettiness there were many who were delighted to see the Americans getting their first sour taste of what it was like to be the world's Number One power; the worst offenders were those who bemoaned the loss of the Empire. The Festival of Britain, which had caused more debate in British newspapers than the war in Korea, turned out to be a qualified success; but the Tories, as soon as they were back in power, moved in and had the site of the Festival flattened, as if future generations might mistake it as a monument to the departed Socialists. Food rationing finally petered out and the English no longer had an excuse for the bad cooking that had become endemic. Political columnists wrote off Hong Kong as likely to be swamped by Communist China within the next couple of years; the State of Israel went quickly through its infancy and surprised the British by just as quickly forgetting its old bitterness against them; World War II, six years after the last shot was fired, was formally ended by peace treaty, governments once again proving their talent for declaring war quicker than they could declare peace; and Constance became pregnant again. The last fact was the one important event in the year for her.

"I'm a dreadful politician's wife, darling. I really should care more about what's going on in the world, but as soon as I know I'm pregnant nothing else matters."

He made a show of enthusiasm. He loved the two children they already had, but he was not excited by the prospect of any more. He was the sort of man who considered children women's things; they would become his when they were old enough to be treated as young adults. But he did his best to hide this failing of his from Constance. "I must get you pregnant every time an election is coming up. It wins the hearts of women voters."

"That would make me a real political wife, wouldn't it? I don't fancy that. Do you want any more children after this one?"

"Do you?" He wondered if it were better to let her guess at his lack of enthusiasm or be frank about it. Pregnant women were more touchy about their condition than politicians in danger of losing their seats.

"Another two, I think," said Constance, who was *not* touchy about her condition; for her it was something natural and she enjoyed every moment of it. But Harry had never asked her how she felt and she had never broached it; she took it for granted that he could see her complete happiness in her pregnancy. She did not even mind the last three months of it when she would not allow him to touch her: it made his return to her after the baby was born so much more satisfying. "We'll stop when we have five."

"I might decide at fifty that I'd like a large family."

"You'll have another think coming. I'll be forty-eight, well into my menopause, I hope, and I'll be damned if I'll start adding to the family then." She moved towards him, kissed him. "What questions do you have to answer in the House to-night?"

"One on the Mau Mau in Kenya."

"What do you know about it?"

"Not very much. Secret societies wouldn't be secret if we knew everything about them. But this one seems pretty horrible. Much more savage than anything we had to put up with in India or Palestine. It's one of the wages of Empire, I suppose – though Father wouldn't think of it like that."

2

Winter came and went and early in March 1953 Harry was asked to go out to East Africa. Charles Wolfe, the Secretary of State for the Colonies, an eighteenth-century title that made him uncomfortable, was in his late forties, a broad-beamed teddy-bear of a man with the bite of a tiger: several people, from his own Party as well as from the Opposition, had made the mistake of stroking him the wrong way. He had never seen any of the Colonies till a year ago when he had been handed the Cabinet post, but that had been considered no handicap: better to have no knowledge than too much, which might have suggested he was a professional. He *was* a professional, but the cult of the amateur had not yet completely died in the Tory Party. He had proved a success in the job and Harry had a great deal of respect for him.

"We want no publicity on this, Harry – there'll be an official Commission going out later. Just go out quietly, as if on private business. The Governor has been warned you are coming, but he's not to put on any ceremony for you." He tugged at the thick black horn of an eyebrow. "Just get the real word on how strong these Mau Mau are. Talk to everyone who might help – the white settlers, the district officers, the black leaders. You have a month."

"What sort of report are you looking for?"

"An honest one. I should kick you in the behind for asking such a question."

"Sorry. But you know there is a pretty strong group in the Party who don't want a good word said for the Africans."

"You may find it pretty hard to say a good word for the Mau Mau. But if they have the rest of the blacks as scared as I'm told, we may have to change our tactics. The gun isn't always the answer."

After Harry had told Constance where he was going she said, "Damn! Why do you have to choose now, while I'm pregnant? I'd love to see Africa."

"I didn't choose to go. I'm being sent."

"You won't be in any danger, will you?"

"The Mau Mau wouldn't know what a P.P.S. was, let alone want to kill one."

"Janet Buck is out there. In Kenya."

"I wonder if she is protesting to the Mau Mau about *their* blood sports?"

"Don't be horrible. I just felt the baby heave when you said that."

"What's she doing out in Kenya? And how did you know?"

"Joe Banning told me. Her father died and she's been out there three or four months now, settling the estate."

"I'll do my best to avoid her. She's a bore."

"You will be careful, won't you? Of the Mau Mau, I mean."

"My love, they only attack outlying farms – I'll be in the towns most of the time. Don't get yourself upset as you did that time I was in Jerusalem. I shouldn't want that to happen to you again."

He flew out to Kenya by Comet. It was his first flight by jet and he marvelled at the speed and comfort of it. He was not technologically minded, but he could not escape the awareness that the world had suddenly become a smaller place, that man was creating pressures he might some day find difficult to combat. He flew over the vast sunbaked continent of Africa (from 30,000 feet it was not the Dark Continent: he wondered what Park, Speke, Livingstone and Stanley would have thought of it from this height, if they would have been as fascinated by it) and came down to the pleasant little city of Nairobi. The Governor's A.D.C. met him at the airport and he was driven to Government House.

"H.E. understands you want no fuss, sir." Burgess, the A.D.C., was a slim young lieutenant from the African Rifles. He looked as if he might be competent as a soldier, but he was nervous of protocol. "Or do I call you My lord? I've only been in this job a week."

"Please yourself," said Harry. "Have you seen any action out here?"

"You mean against the Mau Mau? Yes, sir."

"What are they like?"

"Murdering animal bastards, sir."

"How do you suggest they be dealt with?"

"Catch them and hang them, sir. Every one of them, women as well as men."

This is Palestine all over again, only more primitive. "Why haven't you done that?"

Burgess flushed. "We can't catch them. It's like chasing ghosts."

Sir Humphrey Livingstone, the Governor, was an old Africa hand who had never quite filled the stuffed shirt image of so many Colonial governors. He was a small trim man in his late fifties, with a hooked nose, a grey military moustache and bright blue eyes. He greeted Harry dressed in desert boots, long white socks, white shorts and a faded blue shirt. "London said you wanted no ceremony, right? Glad to have you out here. Get so damned few opportunities to talk to anyone from London. Want to sit out here or too hot for you?"

"I've just come from winter. I'll take all the sun I can get."

"Well, we'll see you get that. I've had forty years of it and I must say I still love it. I'll retire here. Could never go back to the Old Country. With a name like mine I belong here, eh? Actually, I'm not related to the Doc. Got used to the jokes now, but they were a bloody bore when I first came out. Jokers were always popping in on me, introducing themselves as Stanley. One Colonial Secretary once sent me a Navy chap named Stanley as my A.D.C. Must have been killing themselves with laughter in Whitehall. I sent him packing, didn't even give him time to finish his greeting salute. Young blighter was grinning all over his face as he walked into my office."

Over the next few days Harry came to like and admire Livingstone. The little man had a much more difficult job than had been appreciated in Whitehall, but he did not complain when he talked with Harry. "How could you know in London what it's like out here? The climate of opinion here is as mixed as an African storm. Among the whites we have diehards, moderates, liberals, don't-cares and fly-by-nights who are only here to make a quick quid. And the blacks – well, they're the same, only not quite so articulate and organised. Except for the Mau Mau. *They're* organised, indeed they are."

"I want to get out of Nairobi now. Can you arrange for me to stay with someone somewhere farther out?"

"Better go to Aswami. The District Commissioner there is a chap named Ryder – knows the country like the back of his hand. Knows the blacks, too, which is more important."

"How do I get up there?"

"There's a woman coming to luncheon to-day. She can drive you back when she goes this afternoon. Name's Buck. Good-looking girl. Should be good company for you."

Harry was about to find some reason for not wanting Janet Buck's good company, but at that moment Betty Livingstone, a bright-eyed, fluffy-haired poodle of a woman, came out on to the terrace where the two men sat. She and her husband had an electric quality about them that the years of colonial life had not dulled; it communicated itself to all their guests, and Harry was no exception. Janet Buck fell out of his mind as the Livingstones entertained him with stories of their years in Africa.

"The wives are the worst, of course," said Livingstone. "I suppose it's unfair they should be sent out here at all. Most of 'em are nobodies back home, they come out here and suddenly they are mistresses with half a dozen servants. Start acting like Catherine the Great. I remember one woman – well, tell you another time. Here comes our guest."

Harry had caught only fleeting glimpses of Janet Buck since the night of the ball at Aidan Hall three years ago, but he had not spoken to her and had once or twice had Constance decline invitations to dinners at which she was to have been a fellow guest. Gradually he had forgotten her as she had retreated more and more from the county scene; she still owned Four Orchards Farm but over the past year she had spent very little time there. He looked at her with a mixture of interest and dislike as she got out of her Land-Rover and came up the steps to greet Livingstone who had gone forward to meet her.

"Janet is one woman Humphrey does approve of," Betty Livingstone whispered to Harry. "If I didn't know him better, I'd suspect him of having an affair with her."

"She doesn't look the type," said Harry dryly.

Betty Livingstone looked at him sharply: was Lord Bayard,

underneath, a pansy? "Oh, she's the type all right. But she's sweet, for all that." She stood up as Janet Buck came along the terrace. "Janet, darling, you look so well! This is the Earl of Bayard."

"We've met," said Janet, smiling coolly at Harry. "We're neighbours back home."

"You didn't tell me!" Betty Livingstone looked severely at Harry.

"Lady Janet and I never jump the fence that separates us. You might say we're in opposite camps." But Harry held out his hand to Janet and gave her a smile as cool as her own. "But you do look well. Life out here must agree with you."

She nodded. "I may even stay."

"Risky," said Livingstone. "You need a man up there on that farm with you. Especially now."

"Well, don't let's spoil our lunch," said Betty Livingstone. "At the risk of being frivolous, let's talk of brighter things."

Three hours later Harry and Janet Buck were on the main road leading out of Nairobi. Soon they turned off on to a side road and began to climb past farms where cattle and their black herdsmen gazed after them with the same dark placid stare. Thorn trees stood in paddocks like scarecrow umbrellas, and in one paddock a moving grey mound suddenly rose into the air, vultures rising from the carcase of a zebra.

"The zebras are a pest in this district. They break down the fences and ruin the grazing. And they bring the ticks with them. Rinderpest, that sort of thing."

"You sound quite knowledgeable." She had softened from the woman he had met in England, was less obvious in her personality. At luncheon he had found his dislike of her gradually being diluted by her charm and affability. She had even been a little offhanded towards him, but he had welcomed that: it meant she would be much easier to deal with when they were alone on the drive up to Aswami.

She nodded, as if she took the compliment for granted. "I think I've at last found the place where I want to settle down."

"You may have left it too late. What will you do if Kenya becomes independent?"

"You think it will?" But she nodded in answer to her own question. "Of course it will, in time. But I think I'll be able to live with the Africans. I didn't bring any prejudices out here with me, not like some people already here."

"What about a husband? Doesn't a woman need a man in this country?"

"I'll find one eventually, I suppose. I'll just be a little more selective than I was with the last one."

It was almost dark when they drew up before the District Commissioner's house on the outskirts of Aswami. The town itself was just one wide street flanked by mud-walled, iron-roofed stores, a native school, an open-air cinema and a couple of dozen bungalows. The D.C.'s house was the biggest in town, a bungalow with wide verandas all round, a flagpole in the middle of the half-acre of brown lawn that fronted the house, and a hedge of castor oil bush that separated it from the paddocks at the back. A marabou stork took off from the roof ridge as the Land-Rover pulled into the red gravel drive, went flapping away into the last light of day like a giant bird of evil.

A native policeman, folding the flag he had just pulled down from the flagpole, came towards them. "Bwana Ryder not here. Truck break down at Kukuru. Him not come back till morning."

"Where's Bwana Hartley? That's his assistant," she explained to Harry.

"Him away in mountains chasing Mau Mau."

Janet looked at Harry. "You can stay here, they'll look after you all right, though I can tell you the food is terrible. Or you can come up to my place and I'll bring you back in the morning." There was no invitation in her voice or her manner: he could take it or leave it. "There are no whites in the town. Only Indians and blacks."

Harry looked out at the dusk creeping in like armies of dark ghosts, stealing up out of the narrow valleys between the hills. On the still air there was the smell that he had come to recognise as that of East Africa: the thick semi-sweetness of dried cow dung, of grass tinged with the odour of animals' bodies, of wood burning, and of the smell of Africans themselves, that of the policeman standing by the Land-Rover, thick, musky and slightly ammoniac. The air was unmoving and sounds, strange to his ear, came gliding in on it: music played on a pipe, the high-pitched yell of a herdsman urging his cattle home, the beat of heavy wings as a vulture took off from behind the castor oil bushes. Because the sounds and smells were unfamiliar they increased his awareness of his alienness in this lonely village. He turned to the one familiar identification he knew.

"Do you mind putting me up for the night?"

"Not at all."

The Kilminster farm was eight miles out of town. It was dark by the time they reached there. As they drove up the gravel drive past a line of flame trees doused to ashes by the night, two arc lights suddenly blazed from the house.

"Just a precaution," Janet said.

"Are you afraid you might be attacked some night?"

"It could happen." She did not sound afraid. "But I hope that the Mau Mau know by now that I'm not one of the diehards."

"What was your father?"

"I'm afraid he was very much one of those," she said ruefully. "He believed in the whip for the natives. Perhaps he died just in time."

Dinner was better than any of the meals he had had at Government House. They had a South African wine with it and when the meal was finished the houseboy wheeled in a tray loaded with a mixture of liqueurs. "You live well."

"A legacy from my father. The cook-boy is very intelligent." She looked up at the houseboy. "Samuel, tell Maranga to come in."

The cook-boy, in spotless white singlet and apron, came silently into the room, dark intelligent eyes glancing quickly at the tall white man sitting opposite the missus. "Something wrong, missus?"

"No, Maranga. Lord Bayard just wanted to tell you what a beautiful dinner you cooked for us."

Harry took his cue. "Excellent, Maranga. The best dinner I've had in Kenya."

"Thank you, bwana." Maranga smiled. "I try to do better to-morrow."

"Lord Bayard won't be here to-morrow. He is going to stay with Bwana Ryder."

"The bwana is guvmint man?" Maranga looked at Harry.

"In a way, yes."

Maranga nodded, smiling again. "Very nice to cook for guvmint. Thank you, bwana. Good night, missus."

When he and the houseboy had gone, Harry said, "What was the point of that?"

"Bringing him in so you could compliment him? I'm trying to make up for all the indignities my father made him suffer. I'm told my father used to whip him every time his cooking didn't come up to scratch."

119

"Your father sounds an old swine."

"He was, even back home. He was the main reason I ran away to America."

"The American colonel was just the means to an end?"

"No. I was in love with him. His added attraction was that if we married we shouldn't have to live in England."

Relaxed by the dinner, the wine and her easy company, he said, "Why did you try so hard when you came back from America? I mean to be noticed?"

"I am against blood sports, if that's what you mean."

"You know that's not what I mean."

She looked at him in silence for a while. The table lamp behind the couch on which she sat burnished her red hair to a brilliant shine; light flowed down over her cashmere-clad shoulders and was lost in the shadows beneath her deep bosom. One arm rested along the back of the couch and the huge emerald on her hand flashed a green flame as her fingers stroked the upholstery. She had been as relaxed and composed as he, but now there was a sudden tenseness in her. He knew at once he had said the wrong thing.

"I made a mistake about you, that was all." Her voice was steady but low.

"Yes," he said, and decided to retreat. "Well, I'd better turn in. I'd like to be down at Ryder's first thing in the morning."

The house was a big rambling one made of stone. She led him out of the living-room and down a long wide hall to a bedroom as large as his and Constance's room at Aidan Hall. The windows were shuttered and a fire burned in the big stone fireplace. "It's a bit stuffy, but you can open the windows. Don't open the shutters, though, just in case."

"Where's your room?" She looked at him and he said, "That's not a leading question."

"I'm just across the hall. This was my father's room. He brought all the furniture out from England when he came." She moved back to the door. "I'm always up early. I'll call you at seven."

He undressed, got into bed and looked round the room before he put out the light. Kilminster had brought as much of England with him as he could: the heavy Victorian furniture, including the glass-fronted gun cabinet beside the fireplace; the sporting prints on the walls; the painting of a hunter by Stubbs over the fireplace; the bound copies of *Punch* on the shelf beside the bed.

Harry wondered where Kilminster had kept his whip, the colonial piece of furniture to which he had become so attached.

He switched out the light, turned over and tried to go to sleep. And discovered he was wide awake, aware of the breathing house all around him, of the strange noises outside, of Janet Buck in the room across the hall. He did not sit up but just turned his head on the pillow, unsurprised, when the door opened and she came in, closing it behind her. In the glow from the fire he could see she wore a green quilted dressing-gown buttoned to her neck.

She came and stood by the bed. "Was I mistaken about you?"

He made her wait before he said, "No."

She went to the fireplace, put more logs on the fire, came back and took off her robe. She was naked. "I like a warm room. I don't like doing it under the bedclothes." She pulled the bedclothes down from him, touched him. "No, I wasn't mistaken. A man with your eyes *has* to be randy at times."

She had a professional's skill and an amateur's enthusiasm, moaning all the time like an animal. She bit and clawed at him as if fighting him, but never letting him go; he would only feel the pain from her teeth and nails in the morning. She did not let him rest but brought him back to her each time he rolled away from her exhausted. She was mounted on him, clawing at his shoulders, panting like a madwoman, when the bullet slammed through the shutter and hit the wall just above her head. Then the other shots rang out and behind the shots was the sound of whistles being blown.

"It's them! They always blow whistles!" She fell off him, hurting him. "The guns, quick! They're over here!"

He caught a confused glimpse of her as, still naked, she darted across the room. He heard the splintering of glass as she smashed the front of the gun cabinet, then saw her pulling a rifle from it. She ran back, threw the rifle on the bed, grabbed up her robe and wrenched open the door. "I've got ammunition in my room!"

Still aching in the groin, he found his pyjama pants and dragged them on. He shoved his feet into his slippers, picked up the rifle and ran across the hall into her room. She threw him a carton of ammunition. "They must have shot out the lights outside! I can't see a bloody thing out there!"

Firing was still going on outside and the whistles were still being blown. Dogs were barking, a cow was bellowing; then there came a horrible blood-chilling scream. Harry, jamming bullets into the magazine of the rifle, stopped and looked across at Janet. She had switched on a small bedside lamp as he had come into the room, to give him his bearings, but now she switched it off.

"That's from the servants' quarters!"

"Are you on the phone?"

"Yes. It's out in the living-room. But whom do we ring?" Bare feet ran along the veranda outside and she spun round. While the light had been on he had seen she carried a heavy pistol; he sensed rather than saw her bring it up now. "Ryder's away – "

"The police boys will be a help – "

"They'd never get here in time – " Suddenly there was a blaze from outside; the shutters became thick black bars and thin red streaks of light. "They're setting fire to the house!"

He was still confused, still exhausted from the last hour's other battle, bewildered by the geography of a strange house. He snapped the magazine home, glad that the rifle was a Lee Enfield .303: at least *that* was familiar. He picked up the carton of ammunition, not quite sure what he was going to do but knowing he was not going to *wait*, not with the house likely to burn down over his head. He could hear the flames crackling outside as the fire took hold of the veranda; there was a thump, thump, thump as bare feet ran across the corrugated iron roof. Smoke was seeping into the room through the shutters and Janet coughed and backed away from the windows. Then he heard the axes tearing at the front door and the splintering of wood.

Glad of the opportunity for action, to be able to shoot at someone instead of just listening to them, he ran out of the room and down the long hall. The whistles were still being blown and screams were still coming from the servants' quarters. He ran into the living-room, heading for the entrance lobby at the far end. And fell headlong over something bulky lying on the floor.

He hit a heavy coffee table, gasping with pain as a corner of it stabbed into his collarbone, bounced off it and fell against a couch. He twisted his head, looked back at what had tripped him. In the red glow from the dying fire in the fireplace he saw the body of Samuel, the houseboy. The head was a yard or so away from the body, lying on its side, the eyes alive only with the reflection of the red ashes of the fire. Harry stared at it uncomprehending for a moment. Who had cut off the houseboy's head if the Mau Mau were still outside trying to get in?

Then the front door splintered and burst in. Harry swung round, still behind the couch. Against the glare of the burning veranda he saw the five men come in the doorway. He got up on one knee, levelled the rifle and fired. The shot crashed throughout the room, bouncing off the stone walls; he fired four more times in rapid succession, turning the room into a chamber of deafening sound. The five men went down, one of them getting half-way across the room towards him. The thunderous noise died away and then, as if the shout had been there in his ears from the moment the attackers had burst in the door, he heard:

"Git de guvmint man!"

For a moment he thought he had imagined it, that it was a trick of hearing brought on by the crashing blast of the five shots. Then in a horrifying flash of comprehension he knew who had cut off the head of the houseboy. The Mau Mau man inside the house was Maranga, the cook-boy. He swung round, half-expecting the cook-boy to come at him out of the shadows of the room; and at that moment a piercing shriek, like that of an animal or a bird dying, came down the hall. He jumped up, frantically reloaded the rifle, leapt across the dead natives and ran out into the hall. Maranga, a heavy machete raised above his head, came running down the long corridor, running with the high knee action of the African through wide swathes of red light that came in through the open doors of the rooms backing on to the burning veranda. Harry went down on one knee, took aim and waited. Maranga came swiftly down through the divisions of red and black, growing bigger each time he flashed past a doorway; when he was ten yards away Harry shot him, aiming low, feeling an animal satisfaction as he squeezed the trigger. Maranga broke stride, then pitched forward, hurling the machete with a snarl of rage. Harry flung himself sideways and the heavy weapon hissed by his head, clanged off the wall and skidded along the floor. Maranga fell into a doorway, clawed at the jamb, then rolled over on his back, his legs writhing in horrible spasms below his shattered pelvis.

Harry stood up, crossed to the African and held the muzzle of the rifle against the twitching purple face. The eyes, full of agony and hate but no fear or plea, stared up at him. Abruptly Harry snapped on the safety catch, reversed the rifle and hit Maranga across the head with the butt. Then he turned and ran down the hall to Janet's room.

She lay in the middle of the room, naked as she had been with him, the green dressing-gown wrenched from her and lying beside her in the blood that was still spilling out of her. Her face was untouched but her body was unrecognisable as anything human. He went past her, ran out across the blazing veranda, careless of the flames and the falling debris, and into the garden. He saw the last of the attackers, four of them, running down the drive, no longer blowing their whistles. He snapped off the safety catch, held the rifle at his hip and fired, jerking each time as it kicked in his hands. He saw one man fling up his hands and go down; then the others had disappeared into the darkness.

He turned back to the house, now blazing furiously as the flames ate their way inside. He bent over, leaning on the rifle, and was violently ill. He straightened up, shivering and weak. Then he remembered why he had knocked Maranga unconscious instead of shooting him. He dropped the rifle, ran down towards the front door where the fire had not yet taken hold, raced into the house and down the smoke-filled corridor. Maranga still lay where he had fallen.

Harry grabbed one long black arm and, hauling the African as if he were an animal carcase, staggered back along the hall, across the veranda and out into the garden again. Servants, eyes big with horror and fear, were coming out of the fire-threatened darkness.

"Where missus, bwana?" An old grey-haired man twisted his toothless mouth at Harry.

Harry nodded at the pyre that had been the house. "It is better to leave her there."


3

"You mustn't blame yourself," said Livingstone. "My dear chap, it was just one of those unfortunate coincidences. They'd probably been planning to attack her place for weeks – "

"No." Harry shook his head slowly and deliberately. "I heard them yell they wanted the government man. Maranga, the cook-boy, made a particular point of asking if I was from the government. They were after me, not Janet. Or perhaps both of us," he conceded. "But I'm sure if I hadn't been at her house last night they wouldn't have attacked it."

Livingstone pursed his lips, then nodded. They were in his office and this afternoon he was more formally dressed in a light grey suit, with a regimental tie hanging neatly from the stiff collar of his white shirt. Harry had lost his bags and all his clothing in the fire and now wore an ill-fitting tussore silk suit that had been bought that morning in an Indian store in Aswami. His body was bruised and sore from the falls he had taken last night and the burns he had suffered when he had run across the blazing veranda; but nothing hurt as much as Janet's teeth and nail marks on him. The Indian doctor who had attended to him in Aswami had looked curiously at the marks, dabbed them with spirit, but had said nothing. He was not to know that the spirit was no protection against the infection of guilt.

"The Press want to interview you. Do you feel up to it?"

"No, I don't want to see them."

"They've already made you a hero, y'know. Killing those six Mau Mau, then capturing this chappie Maranga. You'll be an even bigger hero when we have to announce what Maranga has told us. He's given us the names of all the leaders up there at Aswami and the neighbouring districts. The Security chaps are up there now rounding them up."

"I'm glad some good has come out of it."

There was a knock on the door and Burgess came in. "Lord Bayard, your wife is on the phone from London. I've had them put the call through to my office."

Harry went into the small adjoining office, closed the door and picked up the phone. "Harry? Darling, I was at the doctor's when you phoned. How are you? Are you hurt at all? The news is in all the newspapers and on the B.B.C. Are you hurt?"

Nothing that I can tell you about. "No, I'm all right, nothing to worry about. How are you?"

"You mean has it upset me? Of course, darling. But I must be getting tough in my old age – the baby is as placid as can be. There's no fear of my losing him, darling." She was silent for a moment, the line crackling in his ear; then from a long way away, and the distance to London had nothing to do with it, she said, "I'm glad I didn't lose you."

"I was lucky."

"But poor Janet Buck wasn't?"

Nuances were lost at six thousand miles on a bad line: it was impossible to tell if the question had another one unspoken behind it. "No."

"How soon will you be home?"

"I'll be here at least another week. My love, I love you."

"Of course, darling." What the hell did she mean by that? "And I love you. Never forget that."

When he hung up he sat on in the small office for a few minutes, staring out the window at a native gardener sweeping up petals from a flame tree. They looked like glowing embers; and he abruptly swung round from the window. Was he from now on going to find reminders of last night everywhere he looked? The old Catholic sense of sin came back: he had thought he had lost it, but the bloody Jesuits had been right: once it was implanted in you you could never be rid of it. Christ, but what punishment for a bit of adultery! The whore mutilated and burned: it was like something out of the Old Testament. But even as he thought it he only half-believed it. Reason fought with superstition: that was all bloody religion was. But conscience wasn't superstition; and he was stuck with that. And how would he assuage his guilt, where would he look for forgiveness? Janet? She was dead and past caring. Constance? Would she forgive him for what he had done to Janet? Might she not think he had betrayed her just as much? God? Did He care whether one was forgiven? He hadn't forgiven Janet for whatever she had done with her life.

"I beg your pardon, sir?" Burgess, pink and innocent as a small child, not a care in the world but the hunting and killing of Mau Mau, stood in the doorway.

"What?"

"I thought you said something. Did you have a good connection to London?"

"Not very good."

"Sometimes it's dreadful. One of these days Whitehall will phone H.E. in an emergency and he'll get the message wrong. I don't mean it will be H.E.'s fault," he added hastily. "You'd think in this day and age communications would be better, wouldn't you?"

"There's always the human factor."

"What? Oh yes, of course." But Burgess knew he had missed something. His Lordship sounded as if he were talking to himself, as if he were still in a daze from last night.

Harry spent another ten days in East Africa, moving from Kenya down to Tanganyika, up to Uganda, then back to Nyasaland, finally returning to Nairobi to catch the plane for home.

Constance, pregnant and placid, was at Heathrow to meet him. They sat holding hands in the back of the Bentley while Lambert, the chauffeur, drove them back along the A4 towards London and Eaton Square.

"Where are the children?"

"Down at the Hall with Nanny."

"You look well." But on a second look he wasn't sure that she did; now they were in the car there was an air of strain about her. "I'm sorry I took so long back there."

"I thought those reporters were never going to let you go. But you're a hero, aren't you?"

"Not really. That night it was the Mau Mau or me. A man isn't a hero just because he wants to stay alive."

"Never mind, you're a hero to me." Was her tone sardonic? He couldn't tell. He saw Lambert, Welsh and curious, glance at them in the driving mirror, but Constance seemed oblivious of him. She turned away to look out at the suburbs gliding by, the houses neat, prim-faced and anonymous in the grey rain. He turned to look out at them too. He never glanced at the suburbs when he travelled through them, they were the Great British Desert to him, but now, gazing at the buttoned-up houses, he wondered what went on behind those chastely curtained windows. How many adulterers lived in Osterley Park, how many men in Chiswick had been responsible for a woman's death?

"You must miss the sunshine you had the past few weeks."

What was the matter with her? She sounded so – so fatuous, a word he had never thought to apply to her. "I suppose so."

An hour later, after the welcome home by his father – "Hope you learned a few things out there. Proves they're just down out of the trees, doesn't it? All this talk of independence – tommy-rot! And damned dangerous too. You proved that." – and the pompous but sincere murmur of relief by Ferguson that he had returned safely, they were in their own sitting-room adjoining their bedroom. He stood at the window while she went into the other room and took off her hat. The rain had stopped and out in the square a bowler-hatted man waved his furled umbrella like a black sword at passing taxis; none of them answered his challenge and even from this distance Harry could see the man's fury growing. You're a victim of a lack of communication, old man, Harry thought. Join the club; so am I. But with my own wife, not with taxi-drivers.

Constance came back into the room and at once contradicted him. "I suppose you think I'm bitchy?"

"I'm not quite sure what to think. Something's obviously worrying you."

"Good God!" She shook her head in wonder. "*Something's obviously worrying me.* Are you really so dense? Or does conscience fog up your thinking processes?"

He wasn't so dense, not really. "I gather we're going to talk about Janet Buck."

"Oh, don't be so bloody pompous! You sound like a cross between Ferguson and the Speaker in the House." He smiled at the description of himself and she misunderstood his smile. "I'm not joking!"

"Neither am I. That's the last thing I want to do. Would you believe me if I told you nothing happened between me and her?" He did not suggest the lie to save himself but to save her; but he knew at once that she would not understand. Something told him that women did not want to be saved from pain, not when they were in the right.

"No," she said bitterly; then added against her will, "I'd like to, but I know you'd be lying. *Why?* Why did you have to choose her to spend the night with?"

"I didn't choose. It just happened – I was supposed to spend the night with a District Commissioner, but he wasn't there. It was just circumstances – "

"You must have known what would happen. Or was it as you once said – a stiff cock has no conscience?" She made the words sound more vulgar than they had ever sounded in his own mouth; and he was angry at her for cheapening herself. She must have tasted the words, because her mouth twisted.

"Look, don't you think I've been punished enough? If I had not been at her house that night, they wouldn't have attacked the place and killed her."

"I'm not interested in Divine retribution, if that's what you're trying to tell me. I'm sorry she died – especially in that horrible way. And I don't say she deserved it because she got you into her bed. She's gone now, she's – *irrelevant.* No, I don't mean that. No one is irrelevant, not even a dead whore. But you and I – " She sat down, not suddenly but carefully; all at once he remembered the baby was due within the next week or so. Christ, how selfishly preoccupied could one be? He instinctively moved towards her, but she waved him away. "No, don't come near me."

Then she began to weep. Like so many men he was rendered

useless and superfluous by a woman's tears. He was intelligent enough not to be angry by them; but all he could offer was the unintelligent advice, "Don't upset yourself."

She began to giggle through her weeping, and that made him angry. "*Don't upset myself!* That's just like you – blaming me for upsetting myself!" She looked up at him and he saw that her laughter was not that of amusement; she was on the verge of hysteria, something that was as frightening as it was unexpected. "Get out – get out and don't come back till I send for you! Out! Right out!"

Her face was contorted with her weeping and her fury. He felt weak, useless, castrated: they could cut out your balls with the knife of their tears, if you had any conscience. And despite her sneer, he did have a conscience, would have one forever, he told himself. He abruptly turned away from her and went out of the room. He was half-way down the stairs when he heard her cry out. He hesitated, then ran back up the stairs. She was lying stretched out in the chair, both hands holding her stomach.

"Quick! The doctor – !"

There was no time to move her to the hospital. Matthew Henry Aidan was born twenty minutes later in the bedroom. The gynaecologist, caught fortunately in his rooms in Harley Street, arrived just in time, all teeth and comfort.

"Not to worry." He smiled continually, like a comedian sharing his jokes with an appreciative audience. *A funny thing happened to me on the way to the labour ward . . .* Harry wanted to hit him, wished that Joe Banning instead was here, half-drunk, blunt but honest. "There'll be no complications. It's just reaction to your homecoming. Her happiness will probably communicate itself to the baby. Shouldn't be surprised if it's born laughing." He stretched his mouth, exposing every tooth in his head. "Not to worry."

Matthew was born without complications, but not laughing. Harry stood in the sitting-room and behind the closed door of the bedroom ("Care to watch?" the doctor had asked. "No? I understand. Makes one feel sort of useless, even a little inferior. No good for a man's ego. They're the ones with all the pain. We have it damned easy, you know.") he heard the thin cry as the baby was slapped (what a way to enter the world) into life. The Marquess came up from downstairs as the doctor came out of the bedroom.

"Another boy, eh? Good, good. Prefer boys, m'self." The

Marquess patted Harry on the shoulder, then looked back at the doctor. "How's the mother?"

"Quiet." The doctor looked puzzled. Unsmiling, his teeth hidden, he was another man, a stranger.

"But she'll be all right?" Harry said anxiously.

"Oh yes, no physical damage, nothing like that. But she seemed – disappointed." He screwed up his face as if that were not the word he wanted. Then he salvaged his teeth, displayed them again. "Perhaps she just wanted a girl, that's all. She'll get over it."

Later, when he went in to see her, Harry asked Constance, "Are you disappointed it's not a girl?"

"No. Girls are too vulnerable, don't you think?" She was wan and exhausted, her face almost as white as the pillow beneath her head. He had seen her immediately after the birth of the other two children and on each occasion she had been almost boisterously gay, almost ready to welcome him into the bed with her then and there. But not to-day: "Leave me alone for a while, I have to think."

"I have to go to the office. They've just telephoned – Stalin is dead. It will be in the news later."

"I suppose that's good news?"

"Stalin's being dead? Yes, I think so. But it's hard to believe – one always thought of him as somehow immortal."

"Perhaps the Cold War will die out now. I'd like to think the children could grow up in a peaceful world."

He would have liked to think the same way, but he knew the world would not be peaceful for a long time yet. "They may," he lied. He reached for her hand, lifted it, kissed it, as much as he dared do just now. "I'm glad you're all right."

She stared up at him, her eyes big and dark in her pale face. "Things will never be the same again. Don't expect them to be."

This was not the time to argue with her. How could he explain that the going to bed with Janet Buck had left him untouched? Even the physical memories were already gone; Janet Buck forever now would not be someone with whom he had had a casual affair but a woman whose death he had unwittingly caused. But perhaps that was it. Was Constance, while hating her, also sympathetic towards Janet Buck for the cause of her death? But he could not ask that question now.

"I'll be back in time for dinner. We'll have it up here together. Do you feel up to some champagne?"

"I'll see. Tell your father I'm going to sleep. I don't want to see anyone for at least a couple of hours. I'll be thinking about us."

"Simply forgive me and we'll start over – "

"It won't be that simple. The trouble is, Harry, I don't think you've ever realised just how much I loved you. To share you with anyone, even for a night, just isn't in my nature."

# Six

I

Over the next few years Harry had time and cause to examine his own nature and that of the world in which he lived. News became history, gossip turned into memoirs; time did its job, part of which is to spread a patina of respectability over suspicion, lies and calumny, all of which are ingredients in the stew of politics. Harry was capable of both suspicion and lies: he learned that complete honesty in politics, like partial virginity, was a contradiction in terms. But calumny was beyond him; he could not bring himself to character assassination, a treasured weapon of men in power from a time even before there had been voters. In one way and another he discovered the weaknesses and moral errors of other men, but he never used them to his own advantage. And because he never commented on or passed on the slanders other men brought to him he came to have a reputation for probity and circumspection. Slanderers respect a man of integrity: it usually means he won't be a witness against them in a court action.

Though his report to Charles Wolfe was never published, his one-night battle against the Mau Mau made him a hero in the Party. Privately he had recommended to Wolfe that independence should be encouraged, but publicly he said nothing: if independence should prove a failure he did not feel that a P.P.S. should bear any of the blame or censure that was sure to come from the Right Wing. That was a Minister's burden, not his. He was not going to spend the rest of his political life in the Colonial Ministry.

Britain, in the Fifties, slowly began to change – for better or worse, depending upon one's age, class and prejudices. All rationing had finally disappeared: people went back to ruining their teeth with an abundance of sweets, fumes from unlimited petrol supplies began to pollute the air: Britain once more was as good a place as any in which to die. Travel allowances were increased and Englishmen went abroad again to discover that the French were still as rude and profiteering as ever, that the Italians were happy and lovable and couldn't possibly have been genuine enemies during the war, and that the Spanish, though Fascist, made you and your quid very welcome on something called the Costa Brava. American tourists began to arrive in Britain in greater numbers, looking for all that tradition and those cute Cockney bus conductors: the British took their money with a polite smile and privately asked themselves what right the bloody Yanks had to ask for service, whatever that was. Banana boats came in from the West Indies and unloaded bananas and something called immigrants, a term that in the past an Englishman had thought was only a synonym for Irishmen: a few blacks might be rather fun, brighten up the place, thought some liberals in Hampstead and Chelsea, two places where the black immigrants could not afford to settle. Dickie Valentine sang "Three Coins in A Fountain," Tommy Trinder told his audiences they were lucky people, and at the local Odeon Diana Dors proved how easy it was for a golden-hearted girl to get into trouble. People said how glad they were that the war and its after-effects were at last over and went out and bought hundreds of thousands of copies of books by generals, prison camp escapees, and ladies who had given their all and had not resisted in the interests of the wartime Resistance.

King George the Sixth died and his daughter came to the throne. Britons woke up to find they were the New Elizabethans; slightly bewildered and, not sure what they were supposed to do in their new image, they rushed out and climbed mountains, set out in tiny boats to sail to America instead of the Isle of Wight. Harry, his reputation in the Party growing, was transferred to the Foreign Office as a Parliamentary Under-Secretary, and in the middle of 1954 he was a member of the British delegation that went to Geneva for the signing of the treaty that ended the war in Indo-China.

When he came back from Geneva he had to take out a map and show Constance where the war had been. "But it doesn't really matter. Unless the Americans can't resist shoving their noses in there, we'll never hear of Vietnam again. It will be like Patagonia, a place known only to stamp collectors."

"You sounded anti-American then."

"Did I? I didn't mean to. But they are going to make a lot of mistakes while Foster Dulles is their Secretary of State. He seems to think the world is a Presbyterian parish."

"And what do you think it is?"

"At the level we deal at, a civilised jungle."

"It's dreadful to think that all you men who, well, control our destinies, that you're all so cynical."

"My love – "

Which she still was. The tragedy of Janet Buck's death had receded into the background of his mind, though he would never completely forget it: each time he heard a mention of East Africa or the Mau Mau there would be the sudden migraine of a dreadful memory. Constance never mentioned Janet at all. She had moved down to Aidan Hall immediately after the birth of Matthew, and Harry, accepting her temporary banishment of him, optimistic in his male conceit that it would not be permanent, immersed himself in business at Westminster and in the City. He stayed at Eaton Square from Monday to Friday and went down to Aidan Hall at the weekend. She would not let him sleep with her and he moved into the bedroom next to theirs in their suite in the east wing. The Marquess had retreated to the west wing, though he usually had dinner with Constance and Harry. The rest of the house had finally been restored and was open to the public Wednesday afternoons and all day Saturday from Easter till the middle of November. Neither Harry nor the Marquess ever became accustomed to strangers gawking their way through their home, but each of them would remember the motto carved beneath the stone coat of arms over the front doors and would shrug resignedly. One had to be realistic about expenses, and by co-operating with the Ministry of Works they were saving themselves thousands of pounds. If the Ministry considered it a patriotic duty for them to expose the house, and sometimes, inadvertently, themselves, to the public gaze it was not to their advantage to argue. Not if it meant losing the Ministry's pound for pound.

If the Marquess noticed the rift between Harry and Constance he never mentioned it. Of the servants, Ferguson and Mrs Dibbens, having more experience of the upper classes, minded their own business: what were extra bedrooms for if not to be used? The two maids, one of whom doubled as Constance's lady's maid, thought that something must be wrong when His Lordship moved into the adjoining bedroom; they were both young and full of juice and as they confided to each other in the bedroom *they* shared they would not have kicked His Nibs out of bed. Their suspicions that something had been wrong were confirmed when Harry moved back in with Constance after three months' banishment.

It had not been an easy decision for Constance to forgive Harry; but she loved him too much not to. She had been shocked as much as hurt when she had learned he had been to bed with Janet Buck; completely happy in her sex life, it had never occurred to her that he might want another woman. She believed simplistically in the vows of marriage; she admitted to herself that she might think otherwise if love should die. But love hadn't died, neither on her part nor, she was certain, on his. So to discover that he had been unfaithful had been almost as great a shock as if she had learned he had committed murder. It frightened her, too, to discover that she did not know him as completely as she had thought.

She did not forgive him abruptly. She found the uses of calculated patience; she yearned for him physically again less than a month after their row and the birth of Matthew, but she denied herself as well as him. She was cool to him for several weeks, giving him no more gestures of affection than were necessary to keep up appearances in front of the Marquess. There were times when she wept, but no one, not even the children, ever saw her in those moments; she was afraid she wept as much out of self-pity as anything else and that was an indulgence of which she was ashamed. She had not lost Harry entirely, only part of his thoughts for no more than an hour or two of his whole married life with her up till now: she knew that many men and even a lot of women would find it impossible to think of that as any loss at all. There would be some, she supposed, who would consider her a selfish prig, but they would never know: at least Harry was not the juvenile sort of man who would boast about his affairs. Affairs? In the *plural*? She felt a chill of apprehension, then shook it away. There would be no more women in his life, not unless she drove him to them.

She began to distribute her forgiveness in small hand-outs, her own rationing system. A smile that was allowed to linger, interested attention when he told her about the Commons, the revival of an old joke or two that was private to them: gradually she allowed him back into the intimacy of herself. But she did not allow him to make love to her; not till time, if it didn't heal, at least put a layer of colodion over the wound. Then one warm morning early in July 1953 she had woken up and knew that before the day was out she and Harry would be together again in their bed in the house in Eaton Square.

It was Wimbledon fortnight, something they had attended

regularly since the end of the war. Harry was a member of the All-England Club and tennis and squash were his only sports other than hunting; he played at Wimbledon with other club members, at the indoor courts at Queens Club when the weather was bad, and there was a lawn court at Aidan Hall where he played at weekends with house guests or, sometimes, Joe Banning. He was an average player whose athleticism made up for his lack of finesse; he thumped the ball with malicious force, as if on the court he could rid himself of certain primeval urges such as wanting to belt the heads off obstructive and obtuse political opponents. Constance played a good woman's game and both of them knew enough about the finer points of tennis to be able to appreciate all they saw on the courts in Wimbledon fortnight. Which was more than could be said for the majority of spectators, most of whom looked on it as a social occasion and were more concerned with the quality of this year's strawberries and what sort of hats would be worn in the Royal Box.

Constance left the children in the country and drove up to London. Harry had bought her a Rover the year before and she handled it with the same cool competence as she had ambulances during the war. She had had a secret yen for an Aston-Martin, to go speeding along the country roads with the fields whipping by on either side and the wind tearing at her hair and face; but she had recognised that, though Harry would have bought the Aston-Martin for her, she would not have shown much common sense in asking for it. Young mothers should not run the risk of having their children blown out of speeding sports cars. Still, when she was alone she drove the Rover at a much faster pace than did the usually sedate drivers of such a car.

She and Harry ate strawberries and cream among the flower-hatted dowagers of Wimbledon, watched Vic Seixas and Maureen Connolly and a couple of Australian youngsters named Rosewall and Hoad, greeted friends, relatives and acquaintances: did everything that made Wimbledon the occasion that it was every year. But this year Harry sensed there was something more to Constance's enjoyment of the day. He had been patient with her banishment of him, appreciating that the hurried birth of Matthew had probably contributed to her attitude towards him; but as the weeks had gone by he had become annoyed by her encouragement and then her stopping short of inviting him to sleep with her again. Once or twice while alone in London, feeling

136

both randy and angry, he had been tempted to follow up the unspoken invitations from several women in their circle. But he had resisted the temptation, partly out of a sense of morality, partly because he knew that if ever Constance discovered he had been unfaithful to her a second time she would never forgive him and never have him back in her bed again. Then to-day, in the least likely setting for seduction, among the afternoon teas, the gossiping matrons and the elderly male spectators with creaky memories of yesterday's champions, he realised he was being seduced by his wife.

They went to bed as soon as they returned to Eaton Square, with the curtains half-drawn and the room lit by a pink twilight. He had intended to be gentle with her at first, but she was not interested in gentleness. She clawed at him with an abandon that, for one of those moments when the mind is cold in the midst of boiling blood, he had a frightening memory of the last time he had made love. He shut out the memory by becoming as violent as she: they fought out their love in their bed, snarling obscenities that were endearments at each other, and when they finally died the small death at the climax each had surrendered and everything was as it once was.

"I don't think I want any more children," she said later as she lay in his arms, her legs entwined in his and her hand rediscovering his body.

"What made you change your mind? You wanted five or six."

"I don't know. I just decided three will be enough to love and raise properly." In the cool aftermath of love-making there was often the desire to hurt, when the other side of love, hate, showed itself in a lie or, worse, even the truth. But she had become wiser in her understanding of him, perhaps wiser than he would ever become in his interpretation of her. She knew now that he had less conscience about sex than she had, that having been unfaithful once he could be again. So she would give him as few opportunities as possible; if she became pregnant again she would be presenting him with another opportunity. But she kept that truth to herself.

"Perhaps I shouldn't ask this – will you have an hysterectomy?"

"No," she said gently but firmly. "But you don't have to worry. I've had a new diaphragm fitted that the doctor tells me is ninety-nine per cent safe. What a pity they can't invent a pill or something one can take. I was reading the other day about a herb that some Indians in South America chew and it makes the women sterile for a certain period."

"I'll send to South America and have a shipload sent over."

"What's happening down *here*?"

"I think I'm rising to the occasion."

"Oh darling, I love you! Will you love me forever?"

"Forever, my love. Now let go or I'm going to finish up with it in splints."

So they went through the rest of 1953 and through 1954 with nothing to jar their happiness in each other. In January 1955 Harry went out to Australia on a business trip; on his way back he and Constance were to meet at Klosters in Switzerland for a week's ski-ing. They had never been to Klosters before and they chose it because they hoped to avoid the people they usually met when they went to St Moritz.

Harry arrived in the mountain-ringed village a day ahead of schedule, having picked up a BOAC Constellation that brought him direct to Zurich. When he checked into the hotel where he and Constance had booked he was fortunate to find their room already available. He rang Constance in London and felt a leaden disappointment when she said, "I can't come, darling. Richard looks as if he is building up for the measles. There has been quite a lot of it going around – "

"I'll come home to-morrow."

"Don't be silly. What can you do? You've paid for the room, so stay and enjoy it."

"That's good middle-class sense."

"Nuts to you, milord."

"But it won't be the same without you." He looked out the window, at the mountains pushing back the salmon-coloured sky. Skiers were making their last runs down the slopes, riding their shadows down out of the last bright snow of day into the blue snow of the coming evening. "The ski-ing *does* look good – "

"Then stay there. Get the exercise and come home all tanned and fit. I'll imagine you're one of those sexy ski instructors they're always talking about. I love you, darling."

Next morning he went out ski-ing on the Parsenn. He was a good natural skier and normally he would have attempted something more difficult than the Parsenn, but they told him at the hotel it was foolish to ski on the Drostobel or any of the other more difficult slopes in January. If he had gone ski-ing anywhere but on the Parsenn he would probably not have met Nicoletta Parmi.

138

She was the sort of skier who looked best when standing still; as soon as she started to move, what appeared to be a graceful figure became a mad semaphore of arms and legs. She cut right across Harry's path, falling awkwardly, as he came down at full speed and he had to lift himself over her in a soaring jump to avoid impaling her on his skis. He pulled up, went back and abused her roundly in unparliamentary and ungentlemanly language.

She sat in the snow till he had finished, then she got to her feet, smiled and said, "I like angry men and you have a right to be angry."

"Thank you." His sarcasm had more bite to it than the wind now beginning to blow down from the mountains.

"I am not a very good skier, as you can see."

"I saw it almost too late. You should never have left the nursery slopes."

"Perhaps you will allow me to buy you a drink as an apology?"

Harry looked back up the slope, saw the snow beginning to lift off the top of the ridges under the rising wind, then looked back at the girl. She was dark, beautiful and, though probably empty-headed, might prove good protection against the groups of hearty drinkers at the hotel who had been trying to get him to join them.

She told him all about herself in their first ten minutes in the bar in the basement of the hotel. She had been born in Lerici, had gone to Milan to become a model, had married a man older than herself who manufactured refrigerators and had left him last year – "He was colder than his own refrigerators." It was a remark she had obviously made before, probably to her husband; it slipped off her tongue like a well-oiled cliche. "Now I am planning a leisurely trip round the world to see all the *nice* places. I do not like to see misery, so I shall not go to India or any of the poor places."

He looked at the gorgeous body under the expensive blue sweater and ski pants and nodded. "No, somehow I don't think you'd feel at home in the poor places."

He took her to dinner and knew he was doing the wrong thing. She held his hand and kept asking him for advice on what *nice* places she should visit.

"I think you should come to England," he said. "To Bognor Regis. One of the nicest places you could find. No misery there."

"Is it exciting? I have never heard of it."

"Tremendously exciting. And very very *chic* – that's why you haven't heard of it."

"I'll come. Do you go there?"

"All the time. To the casino."

"I didn't know they allowed casinos in England."

"They don't. That's part of the excitement." I'm out of my mind, he thought. I'm playing up to this empty-headed beauty, telling her outrageous lies like some sixteen-year-old sophisticate, and for what? Because I came back from a month's trip to Australia randy for my wife and my wife wasn't here to satisfy me. I should have gone home last night as soon as she told me she was not coming to Klosters.

"What are you thinking about?"

"Thinking I had better go to bed." Her eyes opened a little with what might have been expectation, but he kept his face and his voice straight: "I think the long flight from Australia is just starting to catch up with me."

"I am so sorry!" She jumped up, swinging her bosom at him. "But perhaps an early night will do me good too."

They went up to their rooms, which happened to be on the same floor. At her door he kissed her good night and for a moment was tempted: she could not have made her body more inviting: he thought of all the *nice* places he would find in her. Then he mentally shook his head, gave her a second quick kiss, and left her.

She was waiting for him next day when he came in from a good morning's ski-ing. She was in red to-day, another expensive outfit, and looked even more beautiful than yesterday. Other men in the bar looked enviously at him as he sat down beside her and he felt the natural pride of possession that any man feels when he has the company of a woman other men desire. He knew he would be safer out on the ski runs, but while he had to be in the hotel Nicoletta was a better companion than anyone else he had seen.

He went to bed with her that night and next day when he came back from ski-ing, Constance was waiting for him in the lobby of the hotel.

"I thought I'd surprise you. I didn't tell you when you phoned yesterday evening, but Richard's suspected measles turned out to be a false alarm. He just had some sort of upset, but he's perfectly all right now. So I caught a flight to Zurich, got the train up and here I am."

140

He could not have been happier to see her. He did not give a thought to Nicoletta: she and their love-making last night had not crossed his mind all day while he had been out on the slopes; he had not come back to the hotel for lunch and so had not seen her since he had left her bed at breakfast time. Having had everything she was capable of giving, which was no more than her body, he had dismissed her from his mind and had intended leaving that evening to return to London. Another night might mean too much involvement with her.

Then while he was standing with Constance at the lobby desk, Nicoletta said behind him, "Harry darling, where have you been all day?"

He turned round, feeling perfectly at ease except for a sudden twinge of conscience. "I've been out ski-ing. Miss Parmi, may I present my wife?"

Nicoletta blinked, went pale and gave everything away in an instant. Constance looked at the body under the revealing yellow shirt and pants, read the empty mind behind the beautiful face and knew at once that she had arrived just in time. Or had she?

Later, in their room, she said, "*Harry darling* – do all the girls you meet anywhere in the world call you that?"

He smiled, deciding the best way to counter her suspicions was not to treat them seriously. He was genuinely contrite for what had happened last night, but it would not help to make an act of contrition. It had occurred to him before that it was easier to talk to God than to one's wife on certain occasions.

"A girl like that would have called Stalin *Joey darling*."

"I'd have trusted Stalin with her. How far did it go?"

"Nowhere near as far as you think. Come here and we'll go a lot further."

But Constance knew in her heart that she had *not* arrived in time but, for once cowardly, she did not want confirmation of her fears. In the following months she made no further reference to Nicoletta Parmi, and the girl, like so many other people, dropped completely out of Harry's memory.

Harry moved up in the hierarchy of Anglo-Qaran Oil and when he came back from Klosters became deputy chairman to his father; there were some grumblings about nepotism, but the board, most of them elderly men who did not want the discomfort and inconvenience of having to leave London, knew the value of Harry's twice-yearly trips to Qaran and acknowledged that when the Marquess died or retired they would need a chairman who knew what went on out in the sunbaked, fly-blown waste whence came Anglo-Qaran's money.

### 2

"I went to school at Auschwitz," said Deborah Roth.

There was at that moment a lull in the conversation and the remark was heard clearly throughout the room: Mrs Roth could not have created a worse effect even by belching. The other guests glanced at her, then looked quickly away, diving into each other's conversation to escape her tasteless joke.

"These people," said Charles Wolfe, making sure the people within earshot of him were not Jews, "they are always so vulgarly ostentatious about everything, even their persecutions."

"I suppose they did have rather a bad time of it," said the Duchess of Swindon, her goitre struggling to free itself from her pearl choker. "But they do rather go on about it. How's your gout these days, Charles?"

The Marquess, on the other side of the room, had also heard Mrs Roth's remark and his chin had gone up. He would have denied that he was anti-Semitic but would have admitted that he did not like Jews, just as he didn't like Indians, Chinese or blacks; prejudice was not malicious if one spread it around. His father had been violently anti-Semitic and *his* father before him; but they had just been part of the atmosphere of the House of Lords of their day; the Lords had also been anti-Catholic and an Anglo-Irish peer had once called his grandfather a Papist lackey. The Marquess always tried to avoid calling anyone names, at least in their hearing. But he wished now that he had not stayed home for this dinner, had gone instead to dine with the French Ambassador. The Frogs at least kept the conversation on a civilised level.

When Mordecai Roth, balder and plumper after ten years, had arrived to-night with the small dark woman on his arm, Harry had found it hard to believe that she was his wife. Deborah Roth, without being beautiful, was much too good-looking for her husband; he wondered what she saw in Mordecai. Roth had not presented her to Harry and Constance with any pride, as plain men often do when they have a good-looking wife, as if such a possession were an achievement in itself. He had been offhanded and when Constance had suggested he might like to meet Lord Bromberg had immediately gone off with her, leaving his wife standing beside Harry like a new girl at school left with the headmaster by a parent glad to be rid of her.

"Your first time in London?" Harry had said.

"Yes." She had looked around the room as if marking in her mind the escape routes.

"Find it appeals to you?" He had little patience with silent, unco-operative women.

"Yes."

"Your first time out of Israel?" He gave her one last chance; another monosyllabic answer and he would abandon her.

"No." Then her eyes twinkled though she remained unsmiling. "I was born in Poland."

He caught the smile in her eyes. "I'm glad you elaborated. I was beginning to wonder if you had any other English but *yes* and *no*."

"I understood that a woman in England wasn't required to know any more."

He looked at her with new interest. "Your English is good. Where did you go to school?"

"I went to school in Auschwitz." Deborah Roth was aware of the silence into which her remark fell, but it did not worry her. Embarrassment was a minor emotion that had been scoured out of her in the horrifying indignities of life in a concentration camp. 'There was an English Jewess there, a professor from the university in Cracow. She educated me."

"How did she manage to survive?"

"She was attractive. She didn't believe there was any fate worse than death, so she did not fight when the guards picked her out to sleep with them." She shook her head, reading his unspoken question. "No, they never picked on me. That was part of her bargain with them. You see, she appointed herself my foster mother. My own mother was killed by the S.S."

"Did your foster mother survive?"

"Yes. She lives here in London. She is the main reason I came on this visit with my husband. We haven't seen each other in ten years."

The women of Harry's acquaintance had never exposed themselves so fully and quickly, not even Janet Buck. In less than five minutes he had learned as much about Deborah Roth as he was ever likely to know; or so he thought. He suddenly did not want to be burdened with any more knowledge of the woman, and when his father went past he grabbed him with relief.

"Father, this is Mrs Roth."

The Marquess, though he did not like Jews, liked pretty women. "Your first time in London?"

"Yes."

"Find it interesting?"

"Yes."

Harry left them, but not before he had seen the twinkle again in Deborah Roth's eyes.

Mordecai Roth appeared in front of him. The little man was no longer meek and apologetic, but he still did not exude confidence; he seemed uncomfortable in the dinner jacket that was a size too large for him and he kept jerking his arms as if trying to find his hands beneath the cuffs that were continually slipping down to cover them. But he was now a very senior official in the Israeli Foreign Ministry and Harry knew enough about the Israelis to know they did not promote incompetents.

"Lord Bayard, do you realise it is almost ten years to the very month since we last met? I said good-bye to you in August 1946 and here it is July '56. To coin a cliché, how time flies."

"I'm glad to see time has treated you so well, Mr Roth."

"And you, too."

Then they looked at each other again and smiled. Harry said, "Mordecai, do we have to talk like stuffed shirt diplomats? What are you doing here in London anyway? It was only by accident I found out – "

"You'd think we'd be seeing you people, wouldn't you? But I'm afraid your Foreign Office is so pro-Arab they think of us as just a nuisance. I hope you're at least neutral, Harry."

"I try to be. But I'm suspect with your Embassy chaps here – they can't forget I'm with Anglo-Qaran."

"I've followed your career in Parliament. You did what you promised – you never became pro-Zionist."

"How could I? It was a pity you lost my brother to speak for you."

"How will you speak if your friend Nasser goes on the way he is threatening to?"

"He's not my friend. I have enough on my plate looking after our friends in Qaran. But you still haven't told me what you're here for. Or is that an undiplomatic question?"

"No comment," said Roth with his gentle smile. "Another cliché."

During dinner Deborah Roth sat on Harry's right and took in

her first experience of the British power class and the British rich. She had never sat at a table as discreetly extravagant as this one: the Sèvres china, the sterling silver, the lace cloth: the Aidans did not keep their treasures in glass cabinets but used them as their ancestors had. She watched the butler as he supervised the serving of the meal by several footmen; he seemed like something out of English films, a character actor employed for the evening. The food was excellent, better than she had been led to expect in an English home, and the wine was certainly better than anything she had tasted in Israel. The Aidans, Mordecai had told her, were millionaires, but they wore their wealth lightly and discreetly. The china, the silver and the expensive food were all taken for granted.

There were several peers at the dinner, a Cabinet Minister, two M.P.s and some rich British Zionists; she could not make up her mind whether their womenfolk contributed anything to the power structure. There were several at the table, men and women, who her sensitive antenna had told her were anti-Semitic: not actively so, perhaps, but whose antipathy was enough to prevent them from ever raising a hand to help Israel. At first she had wondered why Lord Bayard had invited them, but as dinner went on she realised it was a planned cross-section of the sort of opinion her husband was likely to encounter if he was here in England on official business. She looked down the table at Mordecai with something approaching pity; then she turned to Harry with admiration. This man intrigued her more than any she had ever met.

"We can't trust the Americans if anything blows up with Nasser." That was Wolfe, the big plump man on her right. "This is an election year for them. Eisenhower is a play safe man in politics."

"Can't trust the French either." The woman opposite Deborah was Lady Crespin, a thin angular woman who had touched up her faded cheeks and lips with rouge and lipstick that made her look as if she were getting over a bad wounding. She had a lorgnette, something Deborah had always thought was only a stage prop, which she kept raising to her eyes every time she wanted to look at someone farther down the table.

"Why not the French?" asked Deborah.

"Just unreliable, that's all. Look what they did in 1940. What's this?" She looked at her plate as a footman placed another course in front of her.

"I think I heard my wife say it was Alose au Four Gourmandine," said Harry. "But think of it as fish and chips if you suspect the French."

"Can't always trust 'em even on their food, the things they put into it," said Lady Crespin, but her appetite overcame her suspicion and she at once attacked her plate.

"Whom do you mistrust, Mrs Roth?" said Harry.

"Everyone." But she smiled; she had no cause to make enemies here. She had made enemies in Tel Aviv with her frankness and she knew she had sometimes been a handicap to Mordecai, poor dear. She had discovered that in Israel frankness was only a virtue if practised by oneself; it was a vice when other people used it. Open-neck shirt government, the elimination of the frills of formality, had not reduced the human aversion to blunt truth. The English, she had learned from Mordecai, were marvels at camouflaging the truth of anything and she wondered how much he trusted *them*.

"I thought you'd be at the dinner for King Feisal and Nuri-whatever-his-name-is, the Iraq fellow," said Lady Crespin. Her husband specialised in foreign affairs in the Lords, but she had never managed to remember any foreign names unless they had a simple English ring to them. "I gather Anthony was putting on a full show for them."

"I thought my absence would be more appreciated than my presence," said Harry. "I'm associated with Qaran and Qaran is a place Iraq would like under its wing. They're afraid that some day the Persians may move in ahead of them."

"Are you afraid the Persians may do that?" asked Deborah.

"In the Middle East," said Harry, "anyone may do anything."

The women had retired to the drawing-room and the men were sitting over the port when Ferguson brought in an envelope on a tray. "From the Foreign Office, m'lord. The messenger is waiting outside."

Harry tore open the envelope, read the two-line message. "Have the messenger tell the duty officer I'll be there at once." He waited till Ferguson had gone out, then he announced to the other men, "Gentlemen, our friend Colonel Nasser has just nationalised the Suez Canal."

There was silence for a moment, then the Marquess said, "Poppycock! He wouldn't dare!"

"I think he would," said Charles Wolfe, gulping down his

Drambuie and reaching to pour himself another. "He has nothing to lose."

"He could lose his head," said the Marquess. "We could kick him out and replace him with someone else."

"Those days are over," said Harry. "We don't have the troops there any more." He was watching Mordecai Roth, who had not looked up from his brandy goblet from the moment Harry had announced the news. "How do you feel, Mordecai?"

Roth looked up now. "Pleased." Then he rose, escaping the unsprung traps of a dozen unspoken questions. "I must go. I think I may be needed at our embassy."

"I'll give you a lift," said Harry.

"It's out of your way – "

"Not really. I have a feeling that over the next few weeks there'll be no direct routes to anywhere. least of all to a solution to the problem of Colonel Nasser."

The dinner-party broke up with no complaints from any of the guests: they were accustomed to the demands of politics; and with relief from the servants: it meant an early night. The women, having just reached the point of exchanging gossip, went home frustrated; the men hurried off to their clubs or to Westminster in search of rumour. Harry, aware of Roth's impatience, left Constance to see off all the guests.

"See Mrs Roth gets back to their hotel. Lambert can take her in the Bentley. I'll take your car." He kissed her cheek. "I may have to bunk down in my office to-night."

"I'll have Ferguson bring over breakfast for you."

Harry said good night to Deborah Roth. "This may be good-bye, in case Mordecai has to whisk you back to Israel in a hurry." Then he added with a social hypocrisy he usually avoided, "Next time you are in London I hope you'll look us up." He felt rather than saw Constance's look of disapproval; she and Deborah Roth had not taken to each other. "Let's hope things are not made too uncomfortable for you in Israel."

Driving the Rover through the twilit streets, he said to Mordecai Roth, "Is your wife as dedicated to Zionism as you are?"

Roth, now he was in the car and on his way, was less impatient. He took off his glasses and polished them, exposing a face that seemed to have added twenty years in the ten years since Harry had last seen it. "I don't know that I care about it myself any more, not as a concept. I'm an Israeli now and all I want is to see that our country survives." He looked out at the Londoners strolling through the summer evening and envied them the way their country had always managed to survive. They looked at peace, ignorant of and unconcerned with Nasser and what went on in the Middle East: it didn't matter to them whether Israel survived or not. But he could not hate them: he was not a Jew who felt the world owed him a living. He said doggedly, as much to the passing crowd as to Harry, "We shall survive."

"You still don't want to tell me why you're here? Were you people expecting this to happen?"

"Didn't you consider it among the possibilities?"

Harry took the car up Kensington Gore. A concert had just finished at the Albert Hall; the crowd oozed out from under it like an un-set pudding from beneath an upturned bowl. The last of the lovers, marked with grass and blushes, were being ushered out of Kensington Gardens by the park keepers. A coach was pulled up outside an hotel, a load of tourists spilling out of it, their exhaustion making them all one big unhappy family; the manager of the hotel stood in the doorway, his teeth looking as if they had been dragged out of his gums to make a welcoming smile. Roth continued to look out at the passing scene.

"Do you think your ordinary people will back Eden in whatever he does?"

"What do you expect him to do?"

"Who knows?" Roth shrugged; but Harry had the feeling it was a cover-up, not a gesture of ignorance. "Will you back him, Harry?"

"No comment," said Harry. "Here's your embassy. Telephone me to-morrow."


3

Harry did not see Mordecai Roth again that summer. The country simmered with conflicting opinions of what should be done about the Suez Canal. At Westminster several M.P.s developed severe cases of split vision as they tried to look east and west at the same time, at the common enemies Nasser and Dulles. In what came to be known as the "Munich reflex," swords were rattled like a brigade of tambourines: there was going to be no appeasement of this Gyppo dictator. Anti-Semites in both Houses suddenly began shouting the praises of brave little Israel; someone suggested it be invited to join the

148

British Commonwealth. The French all at once became honest and trustworthy and the flights between London and Paris became packed with diplomats and politicians from both countries, the traffic becoming so urgent that several times parties that were supposed to meet in one or other of the capitals passed each other in flight.

Harry, following the lead of several others, put forward a suggestion at a Party meeting that the matter should be taken to the United Nations. It evoked more hostility than he had expected and he decided to seek more information before he spoke again. He commissioned Sid Lucas to gather what intelligence he could from Middle East capitals, and in the meantime he read everything he could of what came in from the British embassies in the Middle East.

And for the first time a division between Harry and his father became a serious one.

"Now is when we should all be completely loyal," said the Marquess. "And you are being nothing but damned *dis*loyal!"

"I'm being realistic," said Harry.

They were driving down to the country from London. That week, to Lambert's unspoken delight, the Marquess had bought a new Bentley and this was its first journey out of town. But the Marquess ignored the power and comfort of the car to concentrate on his argument: "Dammit, we can't let the Gyppos get away with it. If he knows the country isn't one hundred per cent behind Eden – "

"It isn't. For the first time in years the chap in the street is enjoying all the comforts he's been waiting for – he doesn't want to go to war again – "

"Who said anything about going to war? This fellow Nasser is bluffing. A few troops on his doorstep, couple of warships at either end of the Canal, and he'd soon know when it was time to use some commonsense."

"What about the Russians?"

"Bluffing too. They're not going to drop an atom bomb just to help out some jumped-up Wog dictator. No, if we all pull together, show we're behind Eden, diplomacy will do the rest."

"Diplomacy and the big stick, you mean."

"If you like. It's worked before, it'll work again."

Harry shook his head, but said no more. His father lived in the past; he was a firm believer in Palmerston's principle that half-civilised governments all required a dressing-down every eight or ten years to keep them in order. Harry well knew that his father was not an unintelligent man, but as the Marquess had reached late middle age, had begun to feel the decline of his own physical powers as Britain (though he always called it England) had begun to feel the decline of her political power, he had taken to mistaking nostalgia for the past for confirmation of the present. Harry had already decided what his own course of action would be if Eden should decide on a show of force. But as the weeks dragged on it seemed that Eden was undecided what *his* course of action should be.

Now in the middle of October the two Aidan men were driving down to the Hall for the weekend. The local Conservative agent had called a meeting to discuss what was going on in the country over Suez, and Harry was expected to explain why he advocated taking the problem to the United Nations. As the car pulled in through the big gates of the park Harry saw the children, with their nanny in attendance, heading down towards the lake. On an impulse he told Lambert to pull up.

"Don't forget we're due at the meeting at eight," said the Marquess. "Dinner will be early to-night."

I'm still *his* child, Harry thought. "Punctuality is one of my few virtues."

"It should be," said the Marquess. "I taught it to you."

Harry watched the car go on up the long driveway, then he moved through the high soft grass towards the lake. In the time of his grandfather there had been enough gardeners to keep all the grass of the park neatly shorn; he could remember the men, scythes flashing like silver whips in the sun, moving slowly through the park like figures in a dreamlike ballet. Now the grass-trimming was left to the herd of deer and a few sheep, and they preferred to keep away from the driveway and its traffic of visitors' cars and coaches.

The children caught sight of him and came running back. He knew he spent too little time with them, yet he was happy that Constance had won her argument to have them when she had wanted them; as he had told Lucas, he was proud of being an Aidan and the children were the continuum of that pride. They ran at him now, throwing their arms round his legs, fighting among themselves for his attention. He dropped down into the grass and the three of them tumbled over him, shrieking with delight.

"Children!"

Harry, flat on his back, looked up through the tangle of limbs and heads, at the nanny. "It's all right, Nellie."

Nellie Purcell was prim, plain and plump: born to take care of other people's children, she exuded a sadness that Harry, on the rare occasions when he noticed her at all, found depressing. "They are much too boisterous, m'lord."

"Boisterous?" Richard was now almost seven, bright, alert and interested in words as other boys might be in trains or toy cars. "What's that?"

But Robin and Matthew, too active for words, gave Harry no time to answer; they rolled over him, enveloping him with their arms and bodies. Enclosed by them he had a sudden paradoxical feeling of release; too often, he realised, he denied himself the simple joys of being a father. He rolled around in the grass with the children, giving way to the happiness in them that had all at once swamped him, while Nellie Purcell stood above them with a smile as disapproving as a scowl frozen on her white plump face.

At last, gasping and bruised, Harry struggled out from under the children. He stood up and only then saw the thickset boy in checked shirt, corduroy trousers and gumboots who had come up from the lake's edge. He had not seen John Ferguson for almost a year, not since he had come down to go to the boy's mother's funeral. Dorothy Ferguson had had a heart condition that not even her husband had suspected and she had died suddenly. At the funeral, Harry remembered, he had seen Ferguson for the first and only time give way to emotion. The butler had been replaced by a deeply grieving husband.

"Hello, John."

"Afternoon, m'lord." John Ferguson's voice crackled half-way through the last word, but he was not embarrassed: he grinned at the sound of it.

"We want to go on the lake with John!" cried Richard, and the younger children yelped their support. "But Nanny won't let us."

"It's too unsafe, m'lord. I can't swim – if anything happened, I'd never forgive myself – "

"I'll go out with them, Nellie. They'll be safe."

"I'm sure they will." But it was obvious that Nellie Purcell thought the children would be safer and better behaved if their parents stayed out of the way. People should not interfere if they were going to employ a nanny. If there were a nannies' trade union there would be none of such nonsense. But the last thing Nellie Purcell would ever join would be a trade union. Nannies were a class, almost a race, apart, their only identification that with their charges.

Once out on the lake, with Harry stroking the oars, Richard wanted to take the tiller of the small flat-bottomed boat. As soon as he mentioned it, Robin had the same idea. They struggled to reach the tiller and the boat rocked. Harry snapped at them to sit still, but the two children either did not hear him or ignored him. Suddenly they both let out a howl and sat back holding their ears. John Ferguson had slapped them.

"Sorry, sir." He suddenly seemed to realise that the children's father was in the boat with them.

"It's all right, John. It worked."

"He does it all the time," whimpered Robin, still holding her ear.

"Only when you deserve it," said John.

"You should remember who you are," said Richard.

"So should you," said Harry, wondering who had made his son so aware of his position at such an early age. Could it be that damned nanny? She was a snob; but then so was Ferguson. And he did not appear to have made *his* son aware of his position. Harry looked at the boy carefully and the boy looked just as carefully back at him. "Do you like the children, John?"

"Yes, sir. When they don't play up."

"Are they spoiled?"

"Only a bit, sir."

"In what way?" Harry had never had a conversation with a fourteen-year-old boy before; but something told him not all fourteen-year-olds were as composed as this one. Richard, Robin and Matthew now sat quietly in the boat and all at once it struck Harry that they had moved closer to John than to himself. It was as if they felt the older boy was being attacked and suddenly they were on his side, even though they were the subject of the discussion.

For the first time John looked less than composed. He glanced sideways at the three children, who stared soberly at him, then he looked back at Harry. "Sometimes they think they're better than me, sir."

"We don't." Richard had abruptly changed sides and opinions. "You're much better at some things than us. Isn't he?" Robin and Matthew nodded enthusiastically.

"That's not what you mean, is it, John?"

"No, sir." The boy's voice cracked again, but this time he did not smile.

"Where do you go to school?"

"The grammar school in Roebuck." Roebuck was a market town ten miles from Bayard.

"That's a longish way to go every day."

"It's the best school, sir, so it's worth it."

On an impulse Harry said, "How would you like to go to a boarding school?"

"You mean a public school, sir? Like you went to?" Harry nodded. The boy thought for a moment, then shook his head. "No, sir."

"Why not?"

John hesitated. "I don't think my father would like it."

And he, Harry thought, is always going to be your problem. He knew that if he offered to send the boy to a public school, Ferguson would politely but firmly rebuff him. And probably rightly so, Harry conceded on second thoughts. The butler had achieved an almost perfect balance in the master-servant relationship, one that gave him eminent satisfaction, and he would not want the balance upset, neither by becoming indebted to his master nor by having his son possibly educated out of his class.

Harry stayed with the children for another half-hour, sculling the boat about the lake, going ashore with them on the island and climbing the small rise to look for birds' nests in the gazebo. Occasionally he looked with amused admiration at the way John Ferguson kept the children in hand, playing no favourites but, when they were climbing the rise, taking Matthew up on his back. The children obviously adored him and just as obviously shared confidences with him that Harry, the outsider though their father, would never know. When Harry finally pulled the boat back to the shore Constance was waiting for them.

Nellie, glad to have the children once more in her charge, herded them up towards the house. Harry turned to John Ferguson, who was tying up the boat and stowing the oars. "John, when you finish grammar school, what do you want to do?"

The boy looked at Constance, as if afraid that she might introduce another element into this tenuous relationship that had built up in the last half-hour between himself and His Lordship. But Constance's smile, though she did not understand what the conversation was about, reassured him. He said, "I'd like to go to university, sir. Be an economist."

Harry's eyebrows went up. "Well, when you're ready to go there, come and see me."

"Thank you, sir."

"What was that all about?" Constance asked as she and Harry began to walk up towards the house.

"I think we're seeing the beginning of a revolution in the family – that is, if you consider Ferguson a part of the family."

"I'm sure he does."

"I'm sure his son doesn't. Or won't when he's a few years older."

"That's a pity. He and the children get on so well. He's their best friend, you know. But it can't last, can it?" He looked at her. "You know what I mean. Our damned class structure."

"We don't know what our structure is going to be in ten or fifteen years' time. John is going to be one of the classless merit-ocracy that the newspapers tell us is going to be the future England."

"And where will that leave you?"

He kissed her hair. "We'll survive, my love. England may one day be a meritocracy, but personally I don't think it will ever be classless. Wherever there's a money economy there isn't a classless society anywhere in the world. Even in the Communist countries there are classes. I was talking to one of the Foreign Office chaps who has just come back from four years in Moscow. He says it shows there in lots of little ways. The senior civil servants, for instance, don't like going to the Black Sea for their holidays because that's where all the workers go. They prefer to go to Riga, savour a little of what was the good life back in pre-1917. The only classless society is that of the dead."

"Where did you get that?"

"Oh, I dreamed it up. I thought I might try to be the latter-day Chesterfield, write letters full of pithy wisdom to my sons."

"Did anyone ever publish Chesterfield's son's replies?"

"If they did, no one ever read them." He looked at his watch. "Dammit, I'll have to get a move on. I'm due at that meeting at eight."

"Watch your pithy wisdom to-night. I gather some of the local Empire Loyalists are going to take you apart."

Not only the Empire Loyalists but a majority of the meeting did indeed try to take Harry apart. The Anglican church hall in Bayard was packed; the vicar privately lamented that he could not even get a quarter of this crowd to the church next door. The local Boy Scout troop also used this hall for its meetings, and as Harry entered he saw the Scouts' badge and motto, *Be Prepared*, hung on a banner at the back of the stage. He was not prepared for the fury he had to face. He knew the country was divided in its attitude towards Nasser and the Suez problem, but this was the first time he had faced these attitudes at an open meeting. Within five minutes it was apparent that any support for his own approach was minimal. But it pleased him to see that the support he did have was led by Joe Banning.

"As Lord Bayard has consistently said, this matter should go before the United Nations – "

"To hell with the United Nations!" The leader of the Empire Loyalists was Colonel Bedley, a thin, bald-headed man who looked more like a sherry-charged accountant than an ex-Gurkha officer. "Talk, talk, talk – that's all we'll get from that crowd of spineless nincompoops!"

"Mr Chairman – " Banning looked patiently at the Marquess. "I thought I had the floor?"

"You have, Doctor," said the Marquess reluctantly, wishing Banning would take his ideas elsewhere. "Please sit down, Colonel."

"We didn't come here to listen to Dr Banning. We want to hear why our M.P. has been speaking as he has – "

"If you don't shut up and sit down," said Banning, "I'll give you a bit of what you used to give the sepoys – "

It took the Marquess, three stewards and the village policeman five minutes to quieten the mêlée. Harry sat on the stage behind the barricade of the committee table knowing that the rest of the meeting was going to be a waste of time. Emotionalism had taken over and nothing he put up in argument would be of any avail. When order was finally restored and the meeting got under way again it went exactly as Harry had predicted. No vote was taken but the atmosphere was bitter enough to assure any neutral observers, had there been any, that the local Conservatives did not back their M.P. in his attitude towards the Suez problem.

"Perhaps you should have come down to see us more frequently." Colonel Bedley had the floor at last; he addressed Harry as he might have a wayward N.C.O. "We never see you unless we specifically ask for an appointment. We did not elect you to the House of Commons for your ends, sir – you are there to represent *us*. We are the people, sir, and we expect you to speak for us!"

A few minutes later the Marquess closed the meeting. As Harry, with his father and Joe Banning, came out of the hall their way was barred by Colonel Bedley. "Sir, I want you to know that I, on a personal level, absolutely despise your stand on this. You are a traitor to your party and to England!"

Anger flared in Harry, but he controlled it. "If that were so, Colonel, I should be in the Tower and fools like you would be running the country. Fortunately the government – your party and mine – still respects the rights of individuals to have opinions different from its own. *That*, Colonel, is England – not your narrow concept of it. Now get out of my way!"

Bedley almost split apart with fury, but he stepped aside. Harry, still angry, said a much too curt good night to Joe Banning, got into the car with his father, and snapped at Lambert to get a move on. As they drove away through the sullen crowd the Marquess said, "You didn't have to insult Bedley like that."

"He insulted me."

"You're his M.P. – he thinks he's entitled to say what he likes to you. That's *his* concept of democracy – which, I presume, was what you were talking about back there."

Harry said nothing, aware of Lambert sitting stiffly in the front seat, his head tilted slightly as he listened to the conversation in the back of the car. Harry wondered what the chauffeur thought about Suez. But he guessed Lambert might be on his side, if only to oppose Ferguson; the Welsh chauffeur and the English butler had no respect for each other. The Marquess had sunk into a silence as sullen as that of the crowd outside the hall, and the rest of the drive home was suffered in an atmosphere as cold and abrasive as a sleet-laden wind. Or, as Harry remembered, as the sand-laden winter wind that could blow through the Middle East. There might be a lot of young men in the crowd back there in the village who would be experiencing that wind in the coming winter.

Once home the Marquess led the way into his study, not bothering to look back to see if Harry was following him. The latter, struggling hard to play the role of a dutiful son, did follow him, closing the study door after them. The Marquess poured two whiskies, gave one to Harry, took a long swallow from his own, then broke his silence.

"You will back the government on Suez all the way."

Harry sipped his drink slowly while he thought out an answer to this unexpected attack. He did not want an outright quarrel with his father, although the symptoms of such an eventuality had been in the air for weeks; if it was going to come it would come to-night, but he hoped it could be avoided. Their lives were too closely interwoven to allow for the latitude for disagreement that might occur between other fathers and sons. Other sons might desert their fathers and the generations, the family as a whole, could survive; eleven years ago Harry himself could have gone off and his father, and the Aidans, would have remained intact if slightly reduced. But that was before Roger's death, before Harry had been woven irretrievably into that tapestry of time, history, tradition, politics and business that was the Aidans.

He said patiently, "Father, you must leave me to my own judgment of things – "

"Your judgment has proved it isn't worth tuppence! It's time you accepted the advice of others. We can clear up this whole business just by standing firm. It may be necessary to show a bit of force, but we can do that and get away with it. It's only a matter of showing confidence, of letting Nasser and the world know we're right!"

The Marquess was pacing up and down, the whisky in his glass splashing as his hand shook. Harry recognised the symptoms: the flickering passion, brought alive in one last flaming, of an Englishman of a class and tradition that was dying. The aristocracy had not all been in the van of the Empire builders; but they had profited from it, directly and indirectly, and they, like all Englishmen, had taken on added lustre from the spread of English tradition. Harry himself was of the class and tradition and believed in them; but he was young enough not to be blinded by the cataracts of sentiment. The world did not owe England an Empire.

"You saw the letter Sid Lucas sent us – "

"What would he know of the larger issues? Does he know of the strength Eden can call upon?"

"He doesn't need to know anything about our end and he doesn't pretend to." Lucas's letter, phlegmatic as the man himself, his dry flat voice heard in every typed word (and he had typed it, badly, himself, with *bloodys* and xxxx'd-out words scattered throughout it) had been put before the board of Anglo-Qaran the previous Monday. "But he knows the Arab end and that concerns us just as much. He says they will cut the pipelines, block the Canal – "

157

"Cutting off their noses to spite their faces – they won't do it."

"They'll do it," said Harry emphatically. "They're emotional people – illogical, irrational, romantic. I sometimes think that's why they appeal so much to the F.O. – because they are the very opposite of what we try to turn out there. The worst of it is there are so many people like you – Eden, Macmillan, some of the back-benchers – who are beginning to be more emotional than the Arabs."

"Poppycock!" The Marquess did not appear to recognise that he was only illustrating Harry's charge against him. "You can't compare us with a lot of damned Wogs! Emotionalism didn't get us where we are to-day – " His step had quickened and his words with it; Harry had never seen his father like this. "You'll back up everything Eden says and does! I shan't have any more of your damned showing-off – "

Harry had been sitting on the arm of one of the leather chairs but now he stood up. He was surprised at his own icy calm; he could feel himself boiling inside but his voice had none of the trembling edge to it that shredded his father's. "Don't tell me what I can and cannot do. I'm a long way past that age. If you want to be the Member for Bayard you're welcome to it – I'll find another seat – "

"You know I can't sit in the Commons." The Marquess for a moment returned to logic.

"All you have to do is give up the title. If principles mean so much to you, then being plain John Aidan won't be much of a sacrifice." It was a spur of the moment argument that Harry had put forward, one he had never considered before, but he saw at once that it had his father ducking. The Marquess wanted to be no one else but what he was, *the* Aidan among the Aidans: Harry had a moment of clear detachment when he stood outside the family and he saw how a stranger, an egalitarian without a title, might think the matter shallow and hardly worthy of consequence against the other principles involved. The Marquess wanted to fight for England's good name, but not at the expense of his own name and its privileges.

The Marquess seemed to realise his principles had been exposed as bendable and he began to bluster, throwing up a smoke-screen of hurt pride. "I can't do that! There are too many obligations, too many responsibilities – you'll have them yourself some day. This Suez business will be over in no time – all we have to do is show 'em we'll take no nonsense from them – but being an Aidan goes on – "

All his life Harry had taken for granted the fact of being an Aidan. But it had never been an issue with him. The British aristocracy had never been faced with the challenges that had destroyed the French and Russians; one did not have to defend being an English lord. In the heat of the present argument that matter of one's identity seemed to have little weight against the larger issue involved. "Dammit, Suez will be a long bloody business if we don't handle it properly! We stand to lose a lot of money if the Arabs react as Lucas says they will – "

"Who'll lose money?"

"The country. And ourselves in particular. We'd be out on our bums from Anglo-Qaran as soon as I voted in the House – if I voted the way you want me to."

"Hazza won't get rid of us. He knows we can manage the partnership far better than he could."

"He may not have any say in it. Can't you see – if this thing blows up, it's going to be a pan-Arab thing, not just a lot of separate States making up their own minds whether they'll become involved or not. Hazza will be expected to back Nasser whether he likes it or not. He's not strong enough to do otherwise. I *know* what I'm doing is the right thing – I'm being *realistic* – and nothing you say, nor Colonel Bedley nor even Eden, will make me change my mind!"

"You'll regret it. You'll – "

But Harry had had enough. He stalked out of the room, leaving his father with glass upraised in protest; the argument, he knew, could go on all night and neither of them would win. The Aidan stubbornness would defeat them both.

Constance was in bed, hair brushed and shining in the light from her bedside table-lamp, her make-up carefully if heavily applied – her whore's make-up, as she called it – and a silk bed-jacket thrown over her naked shoulders. He knew she was naked beneath the sheets, all ready for him, but he shook his head.

"I've just had a row with Father. I don't think I could get it up to-night."

"That doesn't say much for me." She was disappointed but not unkind. She had been looking forward to to-night's love-making, was ready for him even now before he had got into bed with her. But she sensed at once that the row with his father had been a serious one.

He got into his pyjamas and sat down on the side of the bed beside her. He told her what had been said downstairs, making no attempt to justify his own attitude: it was not in his nature to do so, as she well knew. "I'm afraid things are going to get worse. Not just with me and Father, but the Suez business too. I shan't back Eden, not unless he completely reverses his attitude."

"Hand me my nightgown." She got out of bed, slipped the gown over her head and got back between the sheets. "No use getting a cold for nothing. Do you think Eden and the others will change their minds?"

He shook his head. "There's something cooking. I haven't the faintest idea what it is, but there's too much evasion every time I speak to one or two of the chaps who should be in the know. Unless they're trying to come to some sort of deal with Nasser behind everyone's back."

"Is that likely?"

"Anything's likely in the Middle East. But I'd be surprised if Nasser did an undercover deal. He's just become an Arab hero – the first they've had in a long long time. They'd never forgive him, probably shoot him, if he double-crossed them. And that seems to be the only way out of this mess, a double-cross by someone."

"Eden wouldn't do that. He's always been the soul of honour."

"There's always a first time."

He got into bed, lay back with his hands behind his head. The row with his father had upset him more than he cared to show; on both a personal and political level that had come to a division wider than any that had separated them before. He had never been his father's favourite, not as Roger had, but over the past few years, as he had made his own mark at Westminster, his father had come to accept him as more than just a substitute for Roger. The latter had annoyed their father right after the war with his liberal views, but Harry knew in his heart that even if Roger had turned Socialist his father would have found some way of forgiving him. But he knew he could not expect the same forgiveness for his own intransigence.

Suddenly needing love he turned towards Constance and took her in his arms. "Feeling better?" she said.

"No. Worse."

Her hand had slipped down to his belly, but now it came back to his hip, stayed there. "You love him, don't you?"

Love was a word that did not come naturally to his mind in

regard to his father. He supposed he would have loved his mother had she lived: Philippa Aidan had been vague and too lenient, a flower-drugged fairy godmother more than a mother, but she had been devoted to her children and they had all been her favourites. But, with the exception of Roger, John Aidan had treated all his children as if he were no more than their guardian; any untoward expression of affection for one's children was something indulged in by Italians and other emotionally unbalanced types. Love was something one expressed to one's wife, and then only when they were alone. It was difficult to feel that you might love a father who had insisted, until you rebelled when you went into the army, that you always called him "sir."

"He never encouraged me to love him. I remember once – " Then he stopped: he was afraid of sounding sentimental.

"Go on." Constance's voice was kind and encouraging.

Why not? Perhaps it had been his father's fear of being sentimental that had caused the rejection he now so vividly remembered. "We were walking in the beeches one day, I was home on mid-term from school. We passed the mausoleum and all of a sudden I stopped – I don't know why – and said, 'Do you think Mother misses us?' Then I threw my arms round him and began to cry – I don't know why I did that either. He just pushed me off, told me not to act like a baby, and walked on."

"If you ever did that to any of our children, I'd kick you." Then after a moment's pause she said, "But I think I know why he acted that way. It's that damned English habit of thinking any sort of emotion is bad taste."

"That's an easy generalisation."

"Perhaps. But I think it would be part of the reason. He might have been afraid that he would break down too. I'm sure he loved your mother very dearly."

"Am I like that – afraid of emotion?"

"No, but you might have been if it hadn't been for my influence. Daddy isn't either, thank God. He's a stickler for convention in everything but showing the world he loves his daughter and his grandchildren." She kissed him. "Be patient with your father. One of these days you may find he loves you as much as I do."

"Watch your hand."

"It wanders around of its own accord."

"It's likely to get rapped across the knuckles."

"What with? That? It would make a good rapper if it were a little stiffer."

He laughed and pulled her over on top of him. So much of their love-making had humour to it; the jokes were not very good, but they were their own appreciative audience. Laughing as she kissed him, she shed the nightgown like a loose skin, uncoiling her body and her arms with a serpentine fluidity that he always found beautiful. The sex act itself had no grace to it, but in the minutes before it a woman could be as graceful in her sensuality as she could ever be. Love would always be waiting for him in this bed and that was enough. He could live with his father without the need for love from him.


4

"If you British would give us our independence," said the visitor from Lagos, black lid closing slowly over a yellow eye-ball, "we'd stand foursquare behind you against the Wogs."

"You can't trust them," said the visitor from Accra, grinning like a black Cheshire cat. "My grandfather once ate a Wog slave trader. The blighter kept repeating on him for days."

"Do you still have cannibalism on the Gold Coast?" asked the wife of the Member for Belgravia East.

"Oh yes," said the man from Accra. "But now we have Alka-Seltzer too. One of the fringe benefits of colonialism."

"If Nasser shuts the Canal completely," said the man from the Automobile Association, "that will mean petrol rationing again. Everything falls on the poor bloody motorist."

"I've had to give up ciggies," said the ex-debutante, "now that I'm preggers."

"Her education cost me a thousand pounds a year," said her father, the Member for Belgravia East, ruefully. "And that didn't include the school fees."

"We should never have given 'em Mesopotamia," said the peer who was also a general. "That was when the rot set in."

The Commons, chamber and galleries, would be packed to-day. Harry could see that as he pushed his way through the chattering throng in the Central Lobby; he had never seen such congestion. There was an atmosphere of tension, of barely controlled bitterness that caught at one like invisible brambles: he heard his name mentioned as he passed one group and almost felt the spittle in the sound. A fellow Tory jostled him, glared at him, made no apology and passed on: he was the enemy in his own camp and the other man obviously thought he should be shot on sight. It was no consolation to know that he was not the only traitor: group treason, unlike group polygamy, was not made respectable by multiplication of numbers. He looked around for some of the others who felt as he did about Suez: Nutting, Nicolson, Boothby; but there seemed to be no recognisable faces in the crowd, only a montage of fiercely arguing strangers. Then a hand fasted on his arm.

He turned, took a moment or two to recognise the man who had stopped him. "Oh, hello, Kidman. What are you doing back here?"

"I'm just up for a few days, to report for the Ambassador." Kidman had lost some more hair, but looked lean and fit and tanned. Harry had not seen him since the night of the ball at the Hall six years ago, but he knew that Kidman was now the First Secretary at the British Embassy in Tunis. He lowered his voice. "Off the record, I think your approach is the right one. But now I'm here, I don't think you're going to win. H.E. is in for a shock when I get back and tell him nobody here in London is interested in our opinion. All of us at all the embassies in the Middle East might just as well shut up shop and go on holiday."

"It's as bad as that, is it?" Kidman nodded; then Harry shrugged. "Well, when it happens I hope the mobs don't smash too many windows in your embassy. How's Rosemary?"

"Bored." Kidman offered no more comment on Rosemary, and Harry decided it was time to move on.

"Next time you're in London, come and see me. That is, if I'm still around."

He had just left Kidman when another hand fastened firmly on his arm.

"Harry – " It was Charles Wolfe, the bags under his eyes as dark as bruises: it was not only the traitors who had had sleepless nights over the past week or two. He kept his hold on Harry's arm, pulling him out of the thick scrub of the crowd into a corner. "I hear you've been asking some awkward questions."

"Only awkward if I can't get an honest answer to them."

"Don't be priggish, dear chap. This is a time for common-sense."

They were standing by the statue of Gladstone, a man of infinite commonsense; Harry was not sure that the Old Man had not winked down at him. "Do you think that what Cabinet is planning to do is commonsense?"

Wolfe looked at him shrewdly, not to be caught like that. "How do you know what Cabinet is planning?"

Harry also was not going to be caught. "I'm only guessing. That's why I've been asking questions. Better to ask them in private than on the floor of the House."

"Is that what you're planning to do?"

"If I don't get my answers in private – yes."

"And if you get a private answer, what then?"

"That remains to be seen. You see, Charles, I belong to a dying breed – I'm a man of principle." And felt the taste of the lie like a cold penny, on his tongue.

Wolfe coughed a laugh. "Now I *know* you're bluffing. But just in case you're not – ask your questions on the floor and I promise you you'll be finished in the Party. Think it over, dear chap."

Then Wolfe slipped back into the scrub of the crowd like an old fat lion. Backs had been discreetly presented as the Minister had dealt with the rebel: Harry recognised now that Wolfe had not been the friend he had supposed but the official hatchet-man sent by – by whom? The cabal within the Cabinet who were intent on plunging suicidally into war? Or by the Party chiefs intent only on Party solidarity and disguising it under the name of patriotism?

When he had begun asking his questions this morning, tracking Ministers down by phone or confronting them in their offices, he had met with, at best, evasiveness and, at worst, blunt rudeness. Black Jack Mathis, the Earl of Hendon, had been the friendliest. His family was older even than the Aidans and had a record, dating back to the time of the Normans, of always choosing the wrong side to back. But Black Jack, craggy-faced, black hair now turned grey, a one-time wild hero now worn out by the adventures of his youth, had at last put the Mathis on the side of compromise. The family fortunes had disappeared and he needed his salary as Minister for Internal Development.

"You can't vote against the Government, Harry. I think, m'self, that Eden is sick and making all the wrong decisions, but he's made 'em – *announced* 'em, what's worse – and we have to stick with him. If Nasser sees the Government is split he'll just make rude gestures at us and we'll never get him out. You know the mood of the country – even the ordinary working fella wants to have a bash – "

"I don't believe that's true."

164

"Well, I don't think this is the time for chaps like you and me to go against the grain."

"Jack, your ancestors must be spinning in their graves listening to you."

For a moment there was a flash from the old fighter; then he shook his head. "They never won a fight, for all their bravery and devotion to a cause. No, Harry lad, stay with the Government. I hope I'm wrong, but we may be needed to help pick up the pieces."

"So you won't tell me if there's some sort of conspiracy going on with the French and the Israelis?"

"I honestly don't know," said Black Jack, and all at once Harry believed him; perhaps all the Cabinet did not know what was going on. "And I don't know where you picked up such a preposterous idea."

I could tell you, but you wouldn't think any better of me. It had been sheer luck. He had come out of the Commons yesterday afternoon and Deborah Roth had been there in the Lobby. He had not recognised her at first; she had done something to her hair, and though he usually never noticed what women wore it seemed to him that she was much more smartly dressed than when he had seen her last. She blocked his way but did not do it obviously: only later would he realise how practised she was in getting what she wanted.

"Is Mordecai with you?"

"Yes – and no. I came over from Tel Aviv a month ago on my own. Mordecai has been over once or twice since."

"I thought he might have looked me up," said Harry, surprised.

"I gather he isn't looking anyone up. Mordecai never tells me anything, but this time he's been particularly close-mouthed. But then I suppose everyone is just now. Things don't look too promising, do they?"

"War never does look promising except to stupid generals and equally stupid dictators." Israeli troops had that morning launched an attack that appeared to have the intention of cutting off Egypt from Jordan and bottling up the Egyptian forces in the Sinai peninsula. "Or is that an offensive remark to an Israeli?"

"I am a woman and to me all wars are stupid. Also, though perhaps I should not confess it, I still have to forget I'm Polish. One doesn't automatically become an Israeli just by taking out citizenship papers. Legally, yes. But not in here - " She tapped her breast.

"Mordecai has made the transition."

"He was there when Israel was born. I wasn't. You might say he was the one who experienced the birth pangs, not me."

Harry, more to be polite than anything else, said, "It's too early to offer you a drink. But some tea – ?"

She looked around her, at the Members and their visitors passing to and fro on slipstreams of whisper. Harry had noticed during his years in the Commons that the more grave the situation the softer the voices: even the loud voices of the backwoods Members, the brigadiers and squires from the hunting shires, were muted to hoarse grunts of conspiracy. Deborah Roth said, "Not here, if you don't mind. Mordecai asked me not to make myself too conspicuous. I shouldn't be here at all, except that I can't control my curiosity."

His own curiosity about Mordecai aroused, he didn't ask her about hers. "Where would you like to go?"

"I have a service flat in Curzon Street. Is that too far for you?"

She met his gaze steadily, bland as a census taker. *Too far for what? It was a long way to go for a cup of tea.* He looked at his watch. "I have a couple of hours."

"Plenty of time, then. Do you have a car?"

The Bentley was out in the Yard, with Lambert probably dozing at the wheel or reading the *New Statesman*, which he kept discreetly hidden in the glove-box when he was driving the Marquess. But Mordecai had told Deborah not to make herself too conspicuous (why? Harry told himself *that* was why he was going with her, to find out why Mordecai was paying a secret visit to London); so inconspicuous they would both be. Let Lambert, a suspected gossip, remain undisturbed in his *New Statesman*. "No. We'll take a taxi."

They sat apart in the taxi, formally polite. "Is it an Israeli government flat?"

"No, it's mine." She looked out at Pall Mall, granite-faced, a man's domain; she wondered what all those men in those clubs found to say to each other day after day. "I'm staying in London, I'm not going back to Israel. I'm leaving Mordecai. Do you belong to one of the clubs along here?"

"Not here. Up in St James."

"One where they lock out women?"

"Not lock them out. Discourage them."

"Mordecai would feel very much at home in them. He is very much in favour of the English way of life, you know. He's not a ladies' man."

"That isn't necessarily a part of the English way of life – being anti-female, I mean. I seem to remember you brought that up the first time we met."

She smiled. "I must have a complex about Englishmen."

"What about Mordecai? Do you have a complex about him?"

"No." She stopped smiling. "I just feel terribly sad for him. He doesn't have enough stamina to be both a politician and a lover."

"Yes," he said carefully, wondering if he should stop the taxi now and get out. "You need stamina if you're going to be both. Full-time, I mean."

The flat in Curzon Street was small, a well-furnished box designed for transient tenants poised on one foot for to-morrow's plane or for clandestine lovers midway between the office and home. *Why am I here?* Harry asked himself again; and Deborah Roth said, "Tea?"

"That's as good a reason as any," he said; then laughed at her puzzlement. "Sorry, I was thinking of something else."

"A drink would be better," she said, taking his agreement for granted. She brought him a whisky, with a vodka for herself, and sat down opposite him. The flat's living-room was so small that as he stretched out his legs his shoes brushed against hers; she took off her shoes and tucked her legs up under her. The room's smallness forced them into an intimacy that for the moment he found premature. She took advantage of the intimacy by saying, "I wondered if I should ever get a chance to talk to you like this."

"You'll stay on here when Mordecai goes back?"

"Not here. I'll get something more permanent. But I shan't invite you to visit me."

"No," he said, "that might only lead to complications. Are we likely to have Mordecai come home now?"

"He's over in Paris."

*Don't rush it: this is why I've come here.* "Just for the day?"

"I don't know. He's gone over three or four times. I gather he is some sort of go-between, is that what you call it?"

"It could be." He sipped his drink, careful not to appear too curious. "I've done the same job myself once or twice. How does Mordecai get on with the French?"

167

"So-so, I think. He said the other night that Jews and the French both have a natural talent for conspiracy."

"The French say the same thing about the English."

"I wouldn't know about that. So far I've never had to conspire with the English."

Her coquetry was becoming more obvious by the minute; but he did not mind. His brain now was absolutely cold, unrelated to the blood in his loins. He would make love to her in his own time: she expected it and it would be some sort of payment for the information she had given him.

"Do you have principles?" she asked unexpectedly.

"Yes. But sometimes they have to be bent. Does Mordecai have any?"

She nodded slowly. "He's an honest man – entirely. Whatever he's doing now, I don't think he likes it. He was speaking to someone on the phone last night and he said something about expediency being the lesser of two evils. But why are you so interested in him?"

He rose to go: he would not get any more worthwhile information out of her. "I think it would be best if you didn't mention you'd talked to me."

"Where are you going?" Her voice was abruptly sharp; then she softened it, half-smiled. "If you go, I think I might mention it."

"Tell him you'd brought me here to get me into bed?"

"No, I'd never hurt him that way. No, I'd tell him you invited yourself here to try and – pump me, is that the word?"

"You don't care a damn about Israel, do you?"

"Not really. I've faced the fact of myself – I'm utterly selfish. Some people came out of Auschwitz as saints. I'm afraid I didn't."

He looked at his watch. "We've been here ten minutes. How quickly do love affairs get started in Tel Aviv?"

"Ours isn't going to be an affair. I wanted a man, a *real* man, to make love to me, and I chose you."

"Honoured," he said. "Is that why this is called a service flat?"

She led him into the tiny bedroom. They both undressed deliberately and unhurriedly; he had never had a whore but he guessed this was how it would be. Deborah Roth was a sun-lover, her body completely tanned all over, her nipples like dark brown scars in the middle of her breasts; in the darkly curtained room, against the white sheets, Harry felt he could have been making love to a coloured woman, one of the cafe-au-lait girls who were just starting to appear in numbers in England. The crude eagerness of her surprised him; it was as if she had learned nothing in her years with Mordecai. When they reached their climax she did not let him go but held him to her.

"I want it again. And again. You said you had two hours."

"Is this a test of my stamina?"

"You are the most beautiful man I've ever been to bed with."

"That remark would reduce any man but an Italian to impotency."

"What should I have said?"

"That I was jolly good or rather fun."

But evidently she did not find English clichés funny. She did not smile and he wondered if she was one of those women who lost their sense of humour in bed. But then in bed with Mordecai she had perhaps had very little to smile about.

"Don't scratch me," he said.

She flattened out her fingers, stroked his back with the soft pads of them. "Are you afraid your wife will see the scratches?"

He pushed himself off her, rolled over on his back. "Don't mention her."

"All right, this is just between us." She sat up, lit a cigarette and lay back again. "We have just been using each other, haven't we?"

"Blackmailed into bed – I never thought it could happen to me." But he just said the words casually; he was more intent on forcing the thought of Constance out of his mind. He felt no pang of conscience; she would never learn of this afternoon's affair (if one could call it that), so she would never be hurt. What he had done would be no more than payment for information received. Deborah herself had thought of it as nothing more and, being a sensible woman, would not be likely to broadcast what had happened. Nonetheless, he did not feel comfortable to hear his wife mentioned in this room.

"When I first saw you," Deborah said, "I knew you were exactly what I wanted. Not because you were an English lord – though that did appeal to me, I'm a romantic – nor that you were handsome nor that you looked as if you would be good in bed – which you are – "

"Thank you. But don't recommend me to your friends. What else was it that appealed to you?"

"Your – I thought out my English carefully, even looked it

up in the dictionary – your ruthless indifference. I don't think you care about people as *people*. I knew you could get out of any woman's bed, except your wife's, and be absolutely untouched by whatever had happened in it. That was what I wanted – a man who, if I saw him again, could meet me as if I were a complete stranger."

He was a complete stranger to himself if he was as she had described him. He was like most men, reluctant to face the truth of himself; his very upbringing had been against that. The Marquess was no Lord Chesterfield nor had he wanted to be; he had almost never devoted himself to the moulding of his children's characters. Harry knew he had the fault of pride, but it was a fault in the eyes of only a few: he had never aspired to sainthood. He was pragmatic rather than completely honest; but wasn't that the family motto? He was perhaps deficient in love towards his children, but that was a common fault in his class. He supposed he had innumerable other minor faults that irritated people but that had never disturbed him: he must ask Constance to lay a charge-sheet against him. But ruthless indifference?

"There's a part of you, I think, that would always stop you from having really close friends."

"You have a certain indifference yourself – towards Mordecai."

"You're wrong, I feel terribly sorry for him. I didn't love him when I married him, but after what I had been through in Auschwitz – well, any kind man was a hero to me." She put one arm behind her head and he saw the faint number stained on the dark skin. Auschwitz would always be with her, no matter whose bed she was in. "I thought his kindness would make me love him, but it didn't. I needed more than that."

"This, for instance." He patted the bed.

"Yes. I thought for a while of finding a lover – if I had, I might have stayed with Mordecai. But love affairs are not easy in Tel Aviv – at a certain level it is really no more than a village, a kibbutz if you like, and in a kibbutz there are no secrets. I have a friend, one I love very dearly – the one I told you about at your dinner-party, the woman professor who now lives here in London."

"And no one else?"

She stubbed out her cigarette, got out of bed and put on a robe.

"No more?" he asked, though he would not have made love

to her again if she had offered him all the secrets of the Israeli Cabinet.

"No. There *is* someone else, a man I may yet fall in love with. It's all very complicated and I don't know if it will work out. But now I've thought of him, I don't feel like making love any more."

He got up, went into the bathroom, had a shower, dried himself, then looked at himself carefully in the mirror. Ruthless indifference? Perhaps: he would have to watch himself in the future. He came back into the bedroom and Mordecai Roth, in topcoat and hat, briefcase in hand, stood in the doorway.

His thin face showed more embarrassment than anger. He blinked behind his glasses at Harry's nakedness, then he abruptly turned, his briefcase banging against the door jamb, and a moment later the front door of the flat slammed shut behind him. Harry all at once felt sick; if he had ever been ruthlessly indifferent, he was not now. The stricken look on Roth's face as he had turned away had been like a blow to the stomach to Harry. He pulled on his underpants, like some Puritan hiding his sin, and sat down on the edge of the bed. In a clear moment of insight he realised that, despite the fact they had not seen each other for ten years, Mordecai Roth had looked upon him as a friend: that he had been cuckolded was bad enough, but that it had been by a friend was a terrible wounding. It was no excuse to say that Harry had never thought of Roth as a friend: one did not put premiums on another man's feelings for onself.

"I've hurt him," said Deborah. "I'll never forgive myself for that."

"Not only you." He stood up, slowly began to dress. Deborah had pulled back the heavy drapes, letting some sunlight into the room; through the nylon curtains he could see into the lighted offices across the street. A man and a woman stood at a window drinking tea; behind them another man sat reading a newspaper and a girl was painting her nails. The English way of life went on: cups of tea at regular intervals and as little work as possible: it bred the national image, equanimity, unflappability. Then as if to contradict his cynicism, especially that of a man who was spending *his* working day making love to a woman he hardly knew, the man and woman put down their cups, snapped at each other and parted; the reader dropped his newspaper in a basket, grabbed some files and hurried out of the office; the girl put away her nail polish and began to bash her typewriter furiously. He remembered now that the offices were a branch of a government department: what was on *their* minds that had so suddenly galvanised them? Then he remembered what had been on his mind till twenty minutes ago, why he had come here. He bent down in front of the dressing-table mirror, knotted his tie; he noticed for the first time that to-day he had chosen to wear his old school tie. He wondered what the Jesuits would make of this situation, how they would apply their logic. "Do you want me to talk to him?"

She sat down on the bed, gazed unseeingly at the westering sun slanting down the street: the world looked bright and golden outside. "No. Somehow I don't think he would want to talk to you."

"Probably not. But I don't think you should have to face him alone – "

She looked up at him and smiled with just a trace of mockery. "Are you trying to prove that I was wrong? That you aren't indifferent?"

"Possibly. Perhaps I'm just trying to act like a gentleman."

"You stopped acting like that when you got into bed with me. No, leave Mordecai to me. This gives him an excuse now to let me leave him. He would have fought – no, not *fought*. *Pleaded* with me if I had just told him I was leaving him because our marriage was a failure. He is afraid of failure on his own part, of being inadequate – that's why he works so hard in his job. But now he can blame you – or me – one of us, but not himself, for the break-up of our marriage."

"Is he likely to contest the divorce? I would not want my wife to know – "

"I don't think he is a vindictive man. But then perhaps I don't know him as well as I should – " She shrugged. "I'll do my best. You and I can say good-bye now and I'll tell him honestly that I never intended to see you again. That may satisfy him."

"I hope so."

Dressed and ready to go, he wondered how you said good-bye to a woman, not a whore, with whom you had just been to bed for the one and only time. To kiss her now seemed too intimate, to shake hands just ridiculous. He went out into the living-room, picked up his umbrella and the battered suède hat he had been wearing, and turned round to find her standing very close to him.

"Good-bye, Harry." It was the first time she had called him by his first name. She stood on her toes and kissed him on the cheek, not lovingly but mockingly. "We both got what we wanted, didn't we?"

"More than we wanted."

And that had been yesterday. He had gone back to the House, approached two Ministers with his question as to what was going on amongst London, Paris and Tel Aviv, and been brushed aside as if he had been no more than a muckraking journalist. He had gone home to Eaton Square, somehow glad that Constance was still down in the country and that he would not have to sleep in the same bed as her to-night. He had slept fitfully, then this morning gone looking for more answers, carrying his questions with him like an infectious disease. And getting nowhere till Charles Wolfe had held him up and demanded his loyalty or his political life.

He came out of the corner where Wolfe had left him standing, heading for the entrance to the Chamber, and found his way blocked by Mordecai Roth. The little man looked gaunt and old; the slight plumpness of several months ago seemed to have dropped from him. A picture sprang into Harry's mind of the stunned and ghastly faces he had seen in old newsreels as the gates of Dachau and Buchenwald and (yes) Auschwitz had been opened; but Mordecai had been in none of those places. He stared at Harry, his lips trembling, and for one awful moment Harry thought he was going to break into tears.

But Roth had a sense of dignity and place. He bit his lips, straightened his face and his back. "Lord Bayard, I hope you are going to be discreet in the House this afternoon."

For a moment Harry was puzzled: what had a cuckolded husband to fear from business in the House of Commons? Then he realised the ridiculousness of the question: he was facing Roth the diplomat not Mordecai the betrayed friend. "I am just looking for the truth, Roth."

No *Mordecai*, no *Harry*.

"You have a despicable way of looking for it. My wife told me the questions you asked her." He saw the look of surprise on Harry's face. "She didn't give me the information willingly. Not as willingly as she seems to have been with you – " He allowed himself a moment of personal spite, but he recovered at once. "But if ever you do find the truth you'll realise it is much more complex than you imagine."

"Is that something from the Talmud or from Ben-Gurion's Book of Aphorisms?"

"Both," said Roth, controlling his temper; he knew what the

173

English thought of Jews. It was ironic that Eden and his Ministers had to be on their side; he wondered how the Foreign Office, who knew nothing of what was going on, would explain it to their Arab friends. "But I didn't think Christians had anything to learn about the complexity of truth. They've been using it to their advantage ever since Christ, a Jew, taught it to them."

"Is that all you wanted to say to me – to swap racial insults?"

"You started it," said Roth, and Harry caught the layers of meaning behind the words. Then Roth recovered himself, became the impersonal diplomat. "I should be discreet when you go in there this afternoon, don't ask any more questions. Otherwise you may regret it."

"Tell me what you're doing and why, and I may listen to your threats."

But Roth only shook his head. "You'll learn all about it soon enough. In the meantime, as the Americans say, don't make waves."

"Are the Americans in on it?"

But Roth had said enough, more than he had intended to. He abruptly wheeled about and moved towards the staircase that led up to the galleries. Harry stood stockstill for a moment, paralysed by a sudden stroke of doubt. Was he doing the right thing by persisting with his questions? He found it difficult to believe that the Prime Minister and several of his senior Ministers were engaged in some conspiracy with the French and Israelis that they did not wish to share with even their junior Ministers let alone with their Party back-benchers. In wartime the Cabinet did not take the whole Party into its confidence; it was ridiculous to expect that it should. But this was not wartime, not yet, and the secrecy and evasion that he had met in his inquiries made him suspect that whatever was going on in the clandestine discussions presaged a major commitment that the Prime Minister wanted to present to the country as a *fait accompli*. If the Israelis were involved, then the commitment, by his own assessment, would be utterly wrong. He was not anti-Israeli nor was he pro-Arab. He was merely certain that Sid Lucas's forecast of what would happen if war broke out between Israel and Egypt, and Britain intervened, was correct; it was a forecast based upon knowledge and pragmatism, not favouritism of one side against another nor upon a dangerous nostalgia for Britain's past glory, and he was prepared to go along with Lucas's opinion all the way. But now, in the moment before entering the Commons to commit himself, there came home to him the enormity of what he was about to do. At a time of national crisis he was about to question the integrity of his Prime Minister and own Party leader before the country and the world.

"Troubled, Harry?" Bob Wardle creaked by on his tin leg. "Why don't you sit on our side for a change? Our consciences are clear."

What did Wardle and the Opposition Members know? Then Harry realised that Wardle was speaking only in general terms: no Opposition really ever had to worry about its conscience. He smiled, making a joke of it. "If I do come over, keep me a place next to you."

He followed Wardle into the Chamber, took his place where the Foreign Secretary could catch his eye if he were wanted. A Right-wing back-bencher, one of the shire knights, was on his feet uttering unctuous hypocrisy about what was going on in Hungary – "This complete contempt for another nation's independence shows the Russians in their true colours."

How ironic, Harry thought. Whatever the Prime Minister and Company plan to do, they now have the Russians on their side. If we send troops into Suez no one will take any notice of the Russians, after their Hungarian excursion, if they get up and criticise us, as they are bound to do, in the U.N. He allowed himself a moment of fantasy: was there also some collusion with the Russians? You go your way and we'll go ours and to hell with Foster Dulles and the sanctimonious United States. But that was too fantastic even in to-day's world of double-dealing.

He saw the Speaker looking at him and he realised he was next to be called. He had put his name down for a question and he would not have been surprised if his name had been lost in the shuffle. But the Speaker was an honest man, proud of his impartiality, and though he came from Harry's own Party and must know that Harry was going to embarrass the P.M. with his question, he was being scrupulously fair. Harry's name had been drawn and now he was being invited to stand up and accuse the Prime Minister of God knew what.

Harry stood up, aware of all heads turned towards him. No one had been dozing in the House this afternoon, but not all of the Members had been paying strict attention to what was being said. Now everyone sat up and looked towards the Member who, rumour said, had already been promised banishment if he became awkward. There had been some whispering on the back benches but that now died away: both sides of the House were united in a Joint Committee of Curiosity. Christ, he thought, looking at the expectant, encouraging faces on the other side of the chamber, Gaitskell and his crowd have just endorsed me as their spokesman. He looked down at his own Front Bench and suddenly felt a spasm of anger. None of the Ministers had turned to look at him; all he could see was the backs of their heads, their necks stiff with their ignoring of him. He had already been banished, had been tried and condemned before his crime was out in the open.

"The Honourable Member for Bayard — "

Harry looked down at the backs of the heads in front of him. "I should like to ask the Prime Minister if meetings have been taking place, in England and in France, between members of Her Majesty's Government and representatives of the French and Israeli Governments and if so what has been the purpose of those meetings?"

The Prime Minister stood up, still with his back to Harry, and moved towards the Dispatch Box. "Discussions have taken place with the governments mentioned by the Honourable Member, just as discussions have taken place with other countries involved in the Suez situation. We are doing everything we can to settle the dispute and discussion is, as the Honourable Member will agree, still the best, the most economical and the most peaceful way of reaching a solution."

It was an answer that was no answer, the perfect parliamentary riposte. The Prime Minister sat down to a chorus of *hear, hear* from the Front Bench which was at once taken up by the majority of back-benchers. Party loyalty, like blood, told in the end. The Front Bench still did not turn round, but a battery of fierce glares burned in Harry's direction from the back benches. He looked for a sympathetic face, found a few, but they were not enough. He settled back in his seat, perturbed but not showing it: across the chamber an Opposition Member whispered to his neighbour, "He's a cold-blooded sod, just like all his kind." The Leader of the Opposition rose to put a follow-up question to Harry's, but even the way he put the question one knew he did not expect an answer any more enlightening than Harry's had been.

Harry had just hanged himself and the Party would charge him for the use of the rope. Bored now by the debate, knowing no real information on Suez would be forthcoming but knowing also that to leave the chamber before the debate had finished would be too much of an ignominious retreat, he glanced idly up round the galleries, wondering just how many supporters he had up there. And saw Mordecai Roth rising from his seat in the Diplomats' Gallery. The Israeli paused for a moment in the aisle, looked back and down at Harry. It was impossible to read the expression on his face at that distance, but Harry suddenly had a moment of chill foreboding.

5

"Why have you told me this?" Constance asked.

"I'm not sure." Roth's voice at the other end of the line was uncertain. She heard something that could have been a sigh or a stifled sob, then he went on, "Spite, revenge, I just don't know, Lady Bayard. I – I'm sorry now. I did not stop to think of the effect on you – "

"You had plenty of time to think." She hated the man for what he had done to her, wanted to scream abuse at him; but she kept control of herself and a mile away, in a public phone box, Mordecai Roth thought, Don't the English have any feeling? "You said you found my husband and your wife together yesterday afternoon – "

"If that were all, I should not have called you. But he went to bed with my wife to pry out of her what I was doing – "

Constance laughed harshly. "Oh, come, Mr Roth! I think you are making all this up. It's like some spy story – " But she didn't believe what she said: truth was not only stranger than fiction, it was more horrifying. She wanted to crash down the telephone, but it seemed to cling to her hand like a white lobster.

"My colleagues and I have been working for weeks – " Roth sounded distant and pompous, nothing like the gentle little man she remembered from the dinner-party in the summer. "Your husband has endangered everything that has been planned – "

"*Befoul*," said a woman's voice. "What a dreadful word to use about a dog!"

Constance took the phone away from her ear and looked at it in astonishment. Another woman's voice said faintly, "One can't make the poor dears stand in the gutter, not with the traffic the way it is these days."

"Would you please get off the line?" said Constance.

"I beg your pardon?" said the first woman, and in the background a falsetto dog began to bark hysterically. "This is our line – kindly remove yourself!"

"To hell with your dogs," said Constance. "Get off the line! Are you still there, Mr Roth?"

But Roth had gone and the woman said, "Are you there, Jennifer? The manners of some people these days – "

Constance slammed down the phone, sat down heavily on the bed and stared at the suitcases, still unpacked, just inside the bedroom door. She had driven up from the country this afternoon and as she had come into the bedroom the phone had been ringing. Oh God, she thought, tell me it's all just a bad dream, that it's no more real than those women and their befouling dogs. She felt physically sick, hated the man Roth for his terrible lies. Poison pen letters had been superseded by poison phone calls: technology had made its contribution to evil and corruption. But even as she silently abused him she knew Roth was no liar.

She lay back on the bed. She closed her eyes but oblivion was not so easily come by. What was she going to say to Harry when he came home? Should she confront him with what she had been told? She had no evidence that Roth had been telling the truth – yet why should he lie about such a thing? She knew it was true; and she knew that dirt and muck, further poisoning her, would be rubbed into the wound if Harry should lie to her by denying it. She opened her eyes again, to look across at the suitcases. Should she take them down to the car and flee back to the country? But that would only postpone the inevitable and that was not her nature.

A numbness began to seep through her and she felt suddenly afraid, as if she had been left utterly alone, without maps or even the sun to guide her, in a desolate landscape without landmarks. Oh God, she cried in the silence of her skull, why have You made me so dependent on him? She could not even reach the children without him as a reference point. She loved them (more, she knew in her heart, than he did); but only because they were the manifestation of her love for him. No, not *only*. She struggled out of the numbness like someone fighting a drug. She loved them for themselves. But that didn't alter the fact that each time she looked at them she saw him in them.

She was lying still on the bed, in the dark, when Harry came home. He came into the bedroom, switched on the light, then pulled up as he saw her. "Not well, my love?"

"No." She did not open her eyes at once, not wanting to be caught at a disadvantage by blinking in the light.

The flatness of her voice warned him of what was to come.

178

There had been the uneasiness in the back of his mind that she might eventually learn of what had happened with Deborah Roth, though he was not quite sure how: perhaps a slip of the tongue on his own part at some time in the future, perhaps a piece of gossip fluff wafting by her at a party or in a beauty salon. But he had not expected her to know so soon, though he was certain now that she did know. He sat down, not on the side of the bed as he normally would have but in the cherry silk-covered chair that matched the curtains and provided the only colour in the otherwise all-white room.

"Something wrong?" He was now suddenly cautious, hoping without any real hope that they were not going to discuss Deborah Roth.

She sat up, swung her legs over the side of the bed, opened her eyes and looked directly at him without blinking. "Mr Roth rang me. He said you had been to bed with his wife yesterday afternoon." She had hoped that she might keep her voice unemotional, but she had failed. She found herself trembling with anger and hurt, and uncontrollably she burst out, "Good God, why her – a stranger? Or have you been seeing her for weeks or months? What sort of – " But then she stopped, knowing that in a moment she would be reduced to wild babbling.

He had the advantage, at least for the moment, in that his own voice was well under control. "Yesterday afternoon was the first time I saw her since the dinner-party here at the house."

"You must have got on famously, then. And so quickly, too." But sarcasm wasn't going to help her and, unable to help herself, she opened a chink in her armour: "Was he lying? Is he being spiteful for some political reason?"

"I don't know. Possibly." He thought of Roth's threat as they had met in the Central Lobby; he had not expected it to be carried out like this. For a moment he thought of developing the opening she had given him, of lying that the meeting with Deborah Roth had been entirely innocent; but he did not have the strength or the courage or whatever it was that one needed to lie to one's wife when she knew the truth, even if she did not want to believe it. He said lamely, "Their marriage is already finished. He can't blame me for that."

"And neither shall I," she said, recovering. "I couldn't care less about their marriage. It's ours I care about."

"You may not believe it, but so do I."

She shook her head, got up and began to rearrange things on

179

her dressing-table: anything to keep her hands occupied, to disguise the trembling in them. Even so she dropped a bottle of hand lotion on the glass top of the dressing-table and jumped nervously at the noise she made. "Don't you get enough at home, that you have to go out and leap into bed with any woman who wants you? It's only four days since we last had it. How many others have there been – ?"

"Don't start talking rot!" He was angry at her for allowing herself to fall into such a cheap argument. He stood up, moved towards her, but she avoided him. "My love, she doesn't mean a thing to me. When I went to her flat yesterday afternoon there was no thought in my mind it would finish up as it did. I was – paying her off, if you like – "

"Oh my God!" She tried to laugh, but it was more like the harsh cry of a stricken bird. She was glad he had not attempted to lie to her, but was not glad enough to forgive him; and now he was cheapening himself, talking about paying off the Roth woman as if he were some stud bull. It was a cheap contemptible argument and in turn she was angry at him: "What were you buying? Compliments on your looks or your equipment – ?"

Oh Christ, none of their arguments had ever got as low as this! "She gave me some information – "

"And you paid for it with that?" She gestured at his crotch.

He became angrier still: he had never seen her like this. She had never been a lady in bed – what an expression! – but she had never been a fishwife out of it. "Stop talking like some common whore!"

Her eyes opened wide at that. The row had got out of hand, words were shooting off at tangents like firecrackers; but for him to call *her* a whore – "Should I apologise?" Sarcasm slid back on her tongue; it was such an easy weapon. "I didn't mean to compare myself with the women you know – "

He turned angrily away from her, tripped over the suitcases and almost fell headlong. The argument was becoming farcical and that added to his anger. The sense of humour that occurred to them in their love-making was no help at all when they were fighting. He straightened up, ignoring the pain in his shin. He looked down at the suitcases. "You'd better go back to the country till you've simmered down and we can talk intelligently."

He went out of the room. Foolishly she picked up the suitcases, meaning to leave for the country immediately; then she saw herself in the dressing-table mirror, shoeless, coatless, her hair awry. She dropped the suitcases, sat down at the dressing-table and stared at the miserable, ghastly stranger in front of her. Then she buried her face in her arms and began to weep, great sobs that wracked her like blows from a fist.

## 6

It was another hour before she left to return to Aidan Hall. The House of Lords was still sitting and the Marquess had phoned to say he would have dinner at the House; that meant she could escape seeing him now and she would have till the weekend to sort out what was going to happen between her and Harry. She rang downstairs and O'Brien, the footman, came up and took her suitcases down and put them in the Rover. O'Brien was a young, round-faced Irishman and he had the Irish inability to catch an expression before it gave him away; he went out of the bedroom with his face swollen with gossip, almost running down the stairs like a man fleeing with the household silver. Constance shrugged: she had learned that in a big household gossip was one of the fringe benefits you had to pay to your staff.

As she came down into the entrance hall Harry came out of the drawing-room. "I'm sorry I lost my temper up there. Stay here, don't go back. We can talk this out. I have to go back to the House to-night, but I shan't stay late – "

She drew on her driving gloves, very much in command of herself now. "No, your advice was right. I need time to simmer down."

O'Brien came back in from putting the suitcases in the car and stood waiting in the hall. "That will be all," said Harry. "I'll see Lady Bayard to her car."

O'Brien had composed his face, a blank map on which nothing showed. "Yes, sir. Have a safe trip, ma'am."

Harry took Constance's arm as they went down the three steps to the pavement. She flinched, afraid of herself as much as of him, but did not draw her elbow away from his hand. He opened the door of the car, she got in and he gently closed the door with the careful deliberate action that takes hold of one when one is trying to find proper words to say. There was a sad stricken look on his face and she thought he had never looked handsomer. Yet, and she wondered at herself as the feeling possessed her, she suddenly hated him.

"We'll talk at the weekend," he said.

"Don't hurry back. You have enough to keep you occupied up here."

Meaning she had nothing to keep her occupied, that she had all the time in the world to think about what he had done to their marriage. She put the car into gear and drove off. She could not resist the temptation to look back at him in the driving mirror. He still stood on the pavement, illuminated by the light from the open front door, growing smaller and smaller as she drove away from him. The diminishing image of love, she thought; and wondered again at the coldness that possessed her. She turned the corner, driving with deliberate carefulness, as if to show him that she would get home to Aidan Hall safely, that he would not be called to her bedside in some hospital where in her pain and misery she would forgive him everything.

Harry arrived at lunchtime on the Saturday. She saw the Bentley coming up the drive and she decided she would wait till he came upstairs to their room; she did not want to face him in front of her father-in-law. But when Harry got out of the car the Marquess was not with him. He ran up the steps past Ferguson and a footman going down to collect his luggage, pausing for a moment to say something to Ferguson. Constance, watching from the bedroom window, saw the butler nod up towards her. She drew back, not wanting Harry to think she had been waiting anxiously for him, sat down in a chair and picked up a book. She glanced at the title, then hurriedly put the book face down on the table beside her. It was a copy of a recent bestseller, *Beyond Desire*, by Pierre La Mure.

On an impulse she stood up, went out and met Harry as he came up the stairs and into the corridor. He pulled up sharply as he saw her and his face wrinkled in inquiry. But she gave him no encouragement, walked past him and along to the suite where he had slept for the few months after the Janet Buck affair.

"I've had this made up for you again." In her own ears she sounded like a superior housekeeper; she might even be able to dispense with Mrs Dibbens. "Unless you'd prefer to stay in London all the time?"

He heard the footman coming up the stairs with the suitcases. He turned his back on Constance, walked along to the other bedroom, *their* bedroom. She hesitated, feeling emotion beginning to quiver again inside her, but the footman, young and not too bright, had come into the corridor and was looking at her blankly.

"Ferguson said, ma'am, the master's bags were to go in the Blue Suite."

"That's right. Put them in there."

She was not going to be bulldozed into any compromise. She went back to her bedroom, closed the door and leaned against it as Harry, legs apart, faced her like an adversary. Behind him the sunlight flooded in through the leaded windows and she could not see his face too clearly. The first of the coaches and cars were arriving with the afternoon's visitors and she could hear shouts as children began to run about.

"Don't let's be bloody stupid," he said. "I've told you the Roth woman – "

"Where is your father?" she interrupted. "Have you told him what happened?"

"Of course not," he snapped. He moved away from the window and the light now struck sideways across his face; he looked wan and tired, but there was no sympathy in her for him. He tempered his voice: "He's still caught up in the Suez business."

"And you're not any more?"

"Of course I am." He was having difficulty in keeping his voice on an even note. "Eden has got us in a hell of a mess. But I didn't come down here to talk about that. I came to talk about us. You and me. And the Roth woman, if we have to."

"Oh, I don't think we should leave her out of it, do you?" One very small part of her despised the weapon of sarcasm; but what other armoury did she have except hysterical anger? And that she was determined she would not use.

'All right, we'll talk about her. She means absolutely nothing to me. No more than if she were a woman I'd picked up in Shepherd Market or Piccadilly."

"And you think I'd overlook that – if she were just a prostitute? You men have your own ideas of how a woman thinks, haven't you?"

"To be honest, I don't know how a woman thinks. You or any woman. But I didn't mean I'd expect you to overlook it if I'd been with a prostitute. One point is that you wouldn't have known. But the main point is that it would not have meant a thing, *nothing*, to me. I'd have got in and out of her bed as I would have got in and out of a bus. It was the same with the Roth woman. What I did with her in no way touched what I feel for you."

Deborah Roth had become the Roth woman, almost as depersonalised as she had been in Auschwitz, a faint number on a bare arm against a pillow.

Constance shook her head. "No, the main point is that I *do* know. I told you three years ago I will not share you with anyone. Not even for an hour, not even with someone you don't care two hoots about. It has happened twice now. Three times – there was that Italian girl at Klosters, wasn't there? How can I be sure it won't happen again and again?"

"I swear it won't – "

"That's what you swore last time." She was not sure if that was true, but his memory would be no better than hers; better to attack with confidence, not concede a thing. She had become an expert in the politics of love. "We're supposed to be in love – "

"We are!"

She went on as if he hadn't spoken: " – but what will happen if we start to become tired of each other? It happens to lots of marriages. Your sisters are tired, fed up to the teeth with their husbands. It could happen to us. And you'd be jumping in and out of beds, a bed passenger instead of a bus passenger, to use your analogy, and I'd be stuck down here – "

"You're letting your imagination run away with you." His temper was shortening. He had come home this weekend determined to surrender completely, to be abject in asking her forgiveness; but she wasn't interested in surrender, she was breaking off all relations with him. She was acting just as Nasser had done. In the Commons he had argued, by implication, for Nasser; but now, faced with the same action on a domestic level, he could not argue for her. He was not going to be kicked out of the bed that had belonged to them both for twelve years. But he was too shrewd now to mention the analogy with Nasser. She would only laugh, as she had almost done when he had mentioned the bed and bus simile. He tried reason once again: "You can't produce one shred of evidence that we are tired of each other. We've been just as happy up till this last week as we've ever been – "

But she was not interested in reason either: it had nothing to do with love or happiness. "I don't have your sort of mind, I can't blot things out of it. There are still times when I remember you and the Buck woman – " She stopped, all at once feeling sick. She sat down on the bed, knees together, hands folded in her lap: the convent pose: be careful of men, the nuns had warned her. "Please go. I don't want to talk to you any more."

He did not move at once. He stared at her, half-angry, half-afraid; all of a sudden he did not know how to handle her. He had nothing to fall back on; self-righteousness was a hollow pose when you were in the wrong; it was against his character to attempt it. He raised a hand in a silent plea, but she just continued to stare straight ahead. He dropped the hand to his side in a gesture of defeat, went out of the room and along to the suite at the other end of the short corridor. From downstairs in the Great Hall he could hear the voice of the guide as the first group entered the house:

"The Aidan family has lived here undisturbed since the fifteenth century when the first section of the Hall was built – "

He went into the suite. The Blue Suite: it sounded like the room for honeymooners in a Brighton hotel. He slammed the door behind him and wrenched off his jacket.

Then the phone rang. It was Pickles, the local Conservative agent. "M'lord, the Association has called a special meeting for to-morrow night."

"Sunday?"

"I'm afraid so. They wanted to catch you while you were down here. Felt it was better to talk to you direct rather than have me write you. I'll phone the Marquess in London. He'll want to be there, I'm sure."

"What's it about?"

There was a hesitant clearing of the throat at the other end of the line. Then: "There's a motion of no confidence in you, m'lord. And some of them, a majority, I'm afraid, want you to resign your seat."

"I'll be there to-morrow night. In the meantime tell them to go to hell!"


7

The Marquess, elated at the British action over Suez, certain that strong measures had produced the moment of truth for Nasser, came down from London by train on the Saturday evening. "We've bombed their airfields, obliterated their air force. It's only a matter of a few days and Nasser will know he's bitten off more than he can chew. Now to-morrow night at the meeting – I'll step down from the chair, don't think it would be proper in the circumstances – all you have to do is say you'll back the Government. Tell the locals you now endorse Eden's action. Then keep your mouth shut. The war will be over in a few days and by Christmas everything will have been forgotten. Where's Constance?"

"Upstairs. She's – she's not feeling well."

"Nothing serious, I hope?"

"No. Just woman sickness."

"Ah yes. Damnable thing, woman's periods. Your mother suffered from them all the time. Twice a month. Damned unfair, what they have to put up with. Well, we shan't wait dinner for her. I'm famished. Food at the Lords these days is dreadful."

"Father, before we go in to dinner – "

"I told you, I'm famished. Let's talk at the table."

The dining parlour, as the Marquess always called it because that was what it had always been called, was no place for a talk that Harry knew was going to be as difficult as any he had ever had with his father. It was a long room with a sixteenth-century painted ceiling that depicted a stag hunt in the park outside, and with walls of dark wainscot that dated from Tudor times. The refectory table could seat thirty-two diners, but most of the chairs were stood back against the walls. Three places, Constance's at the head, had been set at one end of the table. When the three of them dined alone the Marquess, gallant as always towards his daughter-in-law, insisted that Constance should take the place at the head of the table between himself and Harry. Now the empty chair seemed to Harry to sit like a referee between himself and his father.

Ferguson and O'Brien retired after the soup had been served. The Marquess sipped loudly, nodded appreciatively, then looked across at Harry. "What did you want to say?"

Harry dipped his spoon in his soup, but he had no appetite. "I'm going to spoil your dinner."

The Marquess paused, gave him a hard stare, then took another mouthful of soup. "Unlike yours appears to be, my stomach isn't related to my emotions. Go ahead."

"First, Constance and I have had a row. A pretty dreadful one."

"What about?"

"That's our business, unless she wants to tell you about it."

The Marquess, spoon held in mid-air, nodded. "Sorry. Should not have asked. Go on."

"Pickles tells me the locals want me to resign my seat. I'm not going to."

"I'd be disappointed if you did." The Marquess did not miss

a stroke in the consumption of his soup. "What's made you change your mind? You practically threw it at me last week."

Their eyes met across the table: the Marquess was much more in control of himself than he had been the last time they had had this discussion. The Government had committed itself in the right direction, both morally and militarily, and in a day or two all intelligent men would know how right it had been. Whether his son proved intelligent or not did not concern him too much at this moment. He himself had spoken for the Aidans in the Lords and what Harry might say now was of no consequence. The family honour was safe. He knew what had been going on in the corridors of both Houses and he had discreetly let it be known that his son spoke only for himself.

Harry lowered his eyes, dipped his spoon in his soup and began to catch up with his father. He had not regained his appetite, only his composure. Spoons rose and fell, flashing like knife-blades in the glow of the one huge candelabra that lit the table; the candle-lit dinners had been Constance's idea and the Marquess had delightedly fallen in with her suggestion. "I'm looking to the future. One way or another Britain is always going to be involved in the Middle East. In five, ten, perhaps fifteen years' time the Arabs will be looking for friends again in the West. They'll never trust the Americans and they'll never love the French or the Germans. Once they both trusted and loved us – or so we like to tell ourselves. Enough books have been written about the subject, though offhand I can't remember any that were written by Arabs. But they'll turn to us again and when that time comes I want to be one of the few untainted by what's happened these past few days. I shan't vote against the Government when the vote comes up, but I'll abstain."

"It amounts to the same thing." The Marquess had finished his soup, put down his spoon. Ferguson and O'Brien at once came through from the kitchen, the latter carrying a silver tray on which rested a small roast duck. O'Brien took away the soup plates and Ferguson, at a sideboard, began to carve the duck. The Marquess had always expected deaf ears of his servants and he went on, "The constituency will be on your neck to-morrow night if they learn what you intend to do. I was hoping you had made your protest with your question in the House the other day. But you don't seem to know when enough is enough."

"Enough," said Harry, watching Ferguson pile his plate as if preparing the last meal of a man about to be executed.

"A little more for me," said the Marquess. The late March-ioness, a pecker of food, had said her husband was the only thoroughbred she knew who ate like a farm horse. "More vegetables too."

"I had a cable from Sid Lucas yesterday. He said Hazza was pleased with my attitude in the House." Harry hesitated, then went on, "Lucas was discreet, but I gather Hazza is not too pleased with you."

The Marquess waited until Ferguson and O'Brien had left the dining parlour. There was a limit to be expected to the deafness of one's servants: one didn't want oneself criticised in front of them. He looked at the empty chair between them and said, "What does Constance think of what you've done?"

"She thinks the game wasn't worth the candle." *But she was thinking in another context*: but he didn't say that.

"Sensible gel." The Marquess was silent while he attacked the duck. He savoured several mouthfuls, then said, "If you persist in your attitude, I'll have to ask you to resign your seat."

Good Christ, I'm just another one of his staff, like Ferguson or O'Brien or one of the gardeners. Well, perhaps not Ferguson: *he* was part of the Establishment, to use the new word, and he would never be asked to resign. Harry carefully put his knife and fork down beside his untouched duck, stood up.

"I'll give up my seat when the voters throw me out. There are sixteen thousand Labour voters and nine thousand Liberals in the constituency – they add up to the number of Conservatives who voted for me. For all I know I may be speaking for all those Labour and Liberal voters and some Tories too. That would mean a majority of the people in this constituency back me. I'll only know if that is true when the next election comes up. Until then I'll continue to be the M.P. for Bayard and to hell with you and your Victorian ideas of how to run the Empire we no longer have!"

He went out of the dining parlour, across the Great Hall, his footsteps echoing like the tattoo of bones on the tiled floor, and ran up the stairs. When he reached the end of the gallery he automatically turned towards the right, towards his and Con-stance's suite. Then suddenly the anger that had blinded him cleared; he pulled up only feet from Constance's door. What was she doing in the silent room? Had she heard him come upstairs, was she waiting for him to come in? But if he went in, what would he say? What would she say? Everything had been said to-day; only time was now left to close the gap between them. And there hadn't been enough time so far, not nearly enough if he had read her temper to-day. He turned quietly and went back along the corridor to the Blue Suite.

As he closed the door behind him he had the frightening, empty feeling that he was closing a door on a part of his life.

8

He did not sleep well, lying awake most of the night in an atmosphere of ghosts, memories and nearby presences that wore at him till his head ached. He got up at first light, wondering what was happening two thousand miles to the south-east; dawn was always a good time for invasion and he wondered if British troops were already ashore at Suez, telling the Wog dealer in the local bazaar that they had come back for his own good. The wars that were started in the name of peace, graveyards dug for those still living . . .

He bathed, hoping the hot water might take some of the ache and pain out of him, dressed and went downstairs. He went out on to the front terrace and from there watched the mist drifting off the lake up the slope of the island, further distorting the already grotesque shape of the gazebo. That had been his ancestor's folly, built to please a mistress who had left him six months later for a duke. He wondered what monument would be built to his own folly. Perhaps a small reproduction of the Sphinx, but with one eye closed in an enigmatic wink. But he knew in his heart who had committed the folly. God, he prayed, it's not too late. Turn them back, show them the error of their ways.

On an impulse he went along the terrace, down the side steps and round to the stables at the rear. The stables had once housed a dozen hacks and eight coach horses; now there were stalls for four hunters and the children's three ponies; the rest had been converted into garages for the cars and storerooms and offices for the estate manager. As Harry crossed the cobbled yard he saw John Ferguson come out of one of the stalls.

"You're up early, John."

"Sunday morning I do the work for Flannery, sir." Flannery was the grumpy Irishman who looked after the horses and was the general handyman around the yard. "It's his day off."

"Do you care for horses?" He stopped, glad of the opportunity to speak to the boy. Build for the future, Harry: fourteen-year-old boys may be the only ones who will talk to you hereabouts from now on.

"Not other people's, sir."

"Are you always so honest?"

The boy looked down at his gumboot tracing the shape of the cobblestone beneath it. "I don't mean to be. It just slips out."

"Never apologise for honesty, whether it slips out or not. Would you like a horse of your own?" The boy looked at him cautiously. "It's all right. I'm not going to embarrass you by offering you one."

"In that case – yes, sir. I think it would be nice to own one."

"Not *nice*. If you want to continue our conversations, never use the word *nice*. It's a word worn to a smooth nothingness by lazy tongues." But who am I to tell other people what to say? What words have I used this past week that have grated on other ears? "Do you want to hunt?"

"No, sir." The honesty did not slip out this time: it was firm and direct.

"Why not?"

"I don't agree with it, sir. I'm sorry," he added apologetically but unafraid.

I wonder what arguments go on below stairs? Does the Establishment down there, in the person of Ferguson, have as much rebellion on its hands as it has upstairs? He looked at John with a twinkle in his eye that the boy would never understand. The earl and the stableboy: who said England was divided? "Some day we must discuss it. I may be able to convince you that it is not as cruel as you think."

He turned away and then John said, "Sir – " Harry looked back. "Sir, I read in *The Times* what you said about Suez. I mean about going to the United Nations. I think you are right."

Out of the mouths of babes and innocents who read *The Times*. "Thank you," he said gravely, careful not to sound condescending. "Let us hope we are both right."

He went into the garages. There were three cars there, the Bentley, the Rover and a Land-Rover. He took the Land-Rover, wanting to use nothing to-day to which he was not entitled. Legally the Land-Rover belonged to the estate and therefore to his father, but his father never rode in it. He took it out of the yard, round the house and down the long drive. He wondered if Constance was at her bedroom window, *their* window, watching him drive away; but he did not look back, just drove steadily down the drive, out of the park and down the road towards Bayard. He was available to everyone, his wife, his father, the local voters, but he was not looking over his shoulder to plead with any of them.

He went to Mass at the church in Bayard. He had been an irregular churchgoer over the past few years, going only when Constance prevailed upon him. There was no longer anything in the Roman Catholic Church to excite or even interest him; to him it had become arthritic in its administration, stiff-necked in its doctrine, hypocritical in its compassion. He felt it was dying on its feet, but he had found no other religion to replace it, though he had not gone looking for any. He still believed in God and the Christian ethic, but he no longer believed in Rome. But going to Mass was in itself a discipline and to-day was a day that he felt should begin with some sort of self-discipline.

Though there was a private chapel at the Hall, the front pew of the village church had been reserved for the Aidan family for over five hundred years. Roger, the first Lord Aidan, a devout man, had established the tradition that there should be no cushions to protect the knees of the Aidan worshippers; the kneeling-board had been worn smooth by countless restless and aching knees. Only two Masses were said each Sunday, this early Mass and another in mid-morning. The church was half-full and the service had just begun as Harry walked down the main aisle and took his place in the front pew, adjusting his knee-caps to some grooves in the board. A new young priest, Father Helvick, had just recently taken over the parish and Harry had not yet met him. Father Helvick was an energetic young man, moving about the altar as if trying to find his place in a rugger scrum; Harry remembered hearing something that the new priest had played in an English trial. He whipped through the first part of the Mass, gabbling the Latin as if it were doggerel with which he had become too familiar and bored. Young as he is, Harry thought, he's already in the mould. Content to take the Church for granted, happy just to go through the motions. Just as the Party expects me to do.

Then Father Helvick moved up into the pulpit to make the week's announcements and deliver his sermon. There would be a christening this afternoon, a funeral to-morrow, a dance next Saturday night: birth, death and the pursuit of pleasure in the parish of Bayard. "The second collection last week amounted to four pounds three shillings and ninepence. I am deeply grateful for your generosity. I shall do my best to see that I am not profligate in the use of it."

Harry glanced up. This fellow was not going to soft-soap the natives: there was the grit of sand-soap in his voice. He looked down at Harry, no deference in his expression, then out at the congregation in general. "I am going to talk this morning about tolerance, a virtue that we English claim to have invented It includes among its various meanings a respect for the other man's opinions, even when you know, or think you know, he is wrong – "

Behind him Harry felt a stirring of discomfort among the parishioners. As he had walked down the aisle he had noticed several men and women who would be at the Conservative Association meeting to-night; their glances towards him had been as sparing of Christian charity as their contributions to last week's collection plate. But what of his own tolerance? He was candid enough with himself to acknowledge that his attitude on Suez was based more on commercial pragmatism than on a respect for international morality. The priest was talking about Suez without ever making a direct reference to it, but soon Harry was no longer listening to him. Melancholy settled on him as he thought of Constance; Suez became unimportant. Would Father Helvick give the same sermon at the later Mass, when Constance would attend with the children and the Marquess? Was tolerance to be expected from a wife who had been wronged?

". . . our house is divided," the priest was saying, "but we do not have to destroy it. To mix my metaphors, there is no profitable harvest in the seed of bitterness – "

Father Helvick, after all, was not above the use of hackneyed phrases; seminaries, Harry had long suspected, did not believe in testing the imagination of either its graduates nor their congregations. There was some muted coughing in the pews at the back and the irritation of one or two of the older parishioners prickled the air like a cold breeze off the hills to the west of the village. The congregation was prepared to be lectured on its sins, on the damage caused to its soul by fornication, lying, thieving and turning its back on the grace of the sacraments; but intolerance and prejudice were a man's own affair and what was the yardstick for such weaknesses anyway? The parishioners glared back at the priest, offering him no tolerance at all, and, becoming angry with them, he lost his own tolerance and finished his sermon in a burst of temper that called for fire and brimstone for all who ignored his warning.

Father Helvick returned to the altar and the wardens came round with the plate. Threepences and sixpences were dropped in with tiny clinks of disapproval for their recipient; they would make sure there was no profligacy in the presbytery this week. Harry had meant to put five pounds in the plate, but he had left his wallet at home and all he gave was a nod.

When Mass was over he went out through a side door, escaping those Conservative parishioners who, he knew, would be waiting for him at the front of the church with their own collection plates for services they were going to demand of him.

He drove round the countryside all day, not wanting to go home and face either or both Constance and his father. Though he had been born and reared in this district, came down to Aidan Hall almost every weekend, hunted across the fields each autumn and winter, he drove that day through a country almost lost and forgotten. He visited the villages, driving slowly down their High Streets like a tourist; he toured the estate farms but never stopped to speak to any of the farmworkers he saw. With a stranger's eye he saw for the first time, or so it seemed, the geography of home. The curve of a certain hill; a house that lurked behind a screen of trees, its windows glinting like watching binoculars; the thatched cottages of the hamlet of Skipbrook, where Brewster, the ex-Prime Minister, had been born, lying like a tumble of bracken and white rock in the cleft between two hills. Late in the afternoon he found himself in a lane that seemed vaguely familiar, though he could not remember ever having driven up it before. He slowed the Land-Rover, looked up the long slope of the field to his left, then turned his head and looked up the slope to his right. He pulled the vehicle into the side of the road, switched off the engine, and knew now why this lane was familiar. And felt the return of a sickness that had not touched him in a long time.

He was opposite the gate where he had fallen from his horse seven years ago; instinctively his hand went up to his face and stroked his broken nose. Up the hill to his right he could see the autumn-thinned trees that hid the house of Four Orchards Farm. Seven years ago to this very weekend, or almost: this was the weekend when the hunting season began, though his father had postponed the first meeting of the Bayard Hunt because of the Suez situation. The farm had been sold after Janet Buck's death and it was now owned by Colonel Bedley, the Empire Loyalist.

The irony of revenge: the skeleton of Janet Buck must be rattling with glee in its Kenyan grave at how he was being made to pay for her death. He got out of the Land-Rover, feeling he was sick enough to throw up; but he had had no breakfast and no lunch and there was nothing in him to come up. Then he heard the car come over the rise at the top of the lane and he turned away, hoping it would not be someone who would recognise him.

Fat chance. The car pulled up and Joe Banning got out. "Visiting old disaster scenes?" he said, looking at the gate on the other side of the road.

"In a way." In his own ears his voice sounded cool and he tried to make it friendlier. Joe Banning might be the only friend, other than young John Ferguson, he had in the district. "Are you coming to the meeting to-night?"

"Wouldn't miss it for worlds. I've just come from looking at a woman who is expecting a baby, but I've told her if she chooses to have it this evening between eight and eleven, she's on her own. How do you think you'll go?"

"I don't know. I've made the mistake of neglecting them down here – "

"That's true. You're not a good M.P. Not in the local sense. You don't care about the grass roots. But that's not why they'll be trying to shoot you down this evening. You are asking them to substitute practical commonsense for patriotism. This is a rural community – if we get into a war over Suez, how many of them will be called up? Very few. Most of them will be excused because they're either too old or they're in jobs that will exempt them. They won't be concerned about the bank clerks and shop assistants and street cleaners in the towns around here or in London and Manchester and Birmingham, the ones who'll be called up to fight the war. If they can't be patriotic, who else can afford to be?"

"Good God, I thought politicians were supposed to be the cynical ones!"

"You are. But not as much as doctors. There *are* some sentimentalists, but you don't find them holding office in the B.M.A."

"You're talking like a Communist."

"If I were a Communist I'd be more in favour of the National Health than I am. No, I'm a selfish, sensible G.P. who hates war." He paused, then said softly, a different man, "As you should know."

Arnhem was twelve years past: he had forgotten it. "I only

remember it, Joe, when you keep mentioning it." Then he walked over to Banning's car. "How long have you had this?"

"Nearly a year." It was a pre-war V-12 Lagonda that had been repainted and upholstered, an open car loved by bachelors of all ages and hated by women old enough to go to a hairdresser. Every part of it showed the hand of a man who had accepted it as part-substitute for a wife: rubbed with tender care, money spent on it, displayed proudly to other men not so fortunate. "We haven't seen much of each other lately, have we?"

"I'm sorry, Joe. One shouldn't neglect the grass roots of friendship either."

"I still think of you as a friend," said Banning, getting into the car. "I suppose we are?"

"I've never thought of us as anything less," said Harry, and tried not to sound like a man desperately in need of a friend at the moment.

"Well, we'll stick together this evening. But I think we're going to lose."

He pulled the Lagonda out in the roadway and drove on, the thrumming of its engine still to be heard when the car was out of sight round a bend in the lane. One should work on friendship too, Harry thought, polishing it, giving it loving care. Vintage friends were the best.

Still reluctant to go back to the Hall, he found a pub where he had a supper of cold roast beef, salad and beer. Several people in the pub recognised him, but none of them spoke to him. Then he drove to the meeting. The church hall was packed, with an overflow of people spilling out into the church graveyard. His father, sitting at the end of the committee table, sticking to his decision to relinquish the chairmanship for the evening, gave him only a curt nod as he entered and took his place on the platform. The acting chairman, the Dowager Countess Abigale, a human tank with a turret-like hat and a disposition that rode rough-shod over other people's sensibilities, rapped her gavel for the battle to commence. Colonel Bedley led the charge, on his feet before Madame Chairman had quite brought the meeting to order.

Everything went even worse than Harry had expected. He got no support at all; if there were any on his side they kept discreetly silent; he could have been Ben-Gurion trying to drum up votes around the Pyramids. He looked in vain for Joe Banning; then twenty minutes after the meeting had begun a note was slipped up to him. *Sorry. The damned baby, a pro-Eden supporter no doubt, has chosen to-night to arrive. Good luck.* But more than good luck was needed. Harry was a surrogate Wog and the meeting, led by Colonel Bedley, shot him down in short order. The Dowager Countess Abigale showed no impartiality as Madame Chairman; and each time Harry looked towards his father the Marquess was keeping his face averted. A vote was taken on the motion that the Earl of Bayard should resign his seat, and the vote was 812 to 25 in favour of the motion. Harry stood up, told the meeting he would resign if the voting went against him at the next elections in three years' time, gave a stiff bow to the Dowager Countess, did not look at his father, and walked off the platform. The meeting broke up in wild disorder, but Harry was already out of the hall, in the Land-Rover and driving quickly up the High Street and out of the village.

"What do you think?" The vicar, Peter Royden, was standing at the back of the hall with Father Helvick.

"He doesn't care a damn what any of them think or threaten to do," said Helvick. "He could have been dismissing a mob of serfs, the way he walked off that stage."

"The arrogance of knowing you're right," said Royden. "Sometimes I wish I could strike that note in the pulpit."

"We R.C.s have it all the time," said Helvick with a grin. "But right or wrong, the Earl will never care a damn what anyone thinks."

When Harry pulled into the garages behind Aidan Hall they were empty. His father had driven down to the village in the Bentley; but where was the Rover? He went into the house and Ferguson, urbane as ever, showing no sign of the upheaval in the family to which he belonged, appeared out of the shadows in the dimly-lit Great Hall. "Lady Bayard asked me to give you this, m'lord."

Lady Bayard: that formality did not augur well. Harry took the envelope. "Isn't my wife upstairs?"

"No, sir. She left the house with the children this morning."

"Before or after she had gone to church?"

"I don't believe Her Ladyship went to church this morning."

So she hadn't heard Father Helvick's sermon on tolerance. He said a curt good night to Ferguson, went upstairs and into the bedroom that had always been theirs. The bed had not been turned down; Constance's dressing-table was clear of the bottles of lotions that usually stood on it; the books she kept on her bedside table were all gone. The room was now only one of the sixty bedrooms in the house. Somehow, for the moment, she had even stripped it of memories.

He opened the letter. *Harry – I am going up to spend some time at home in Northumberland. I am taking the children with me, even though it will be interrupting their schooling. Your father knows we have separated* (separated? Was she thinking in such terms of permanency?), *but I did not give him the details why. He had the decency not to interfere, so I think you should not burden him with your reasons or excuses. I need time and isolation to think what is to be done for the future of the children. I shall write to you when I have decided what is to be done. Constance.*

He sat down on the bed, empty of feeling. The letter was a communication from a stranger, cold-blooded as a memorandum from a sub-section in the Foreign Office: *reference to your behaviour of the 2nd inst* . . . He lay back on the bed, one hand over his eyes. He was lying on *her* side of the bed. Whatever had happened to the warm-blooded woman who had lain here beneath him, who would have been utterly incapable of writing the letter he held in his hand? Had he killed her as he had killed Janet Buck?

But melodramatic thoughts like that would get him nowhere. He sat up, picked up the phone and asked for the Grainger number in Northumberland. George Grainger answered, his voice cool and distant with either a bad connection or hostility, it was difficult to tell. "Connie doesn't want to talk to you, Harry."

"Tell her it's important – "

"I'm sure it is. But when the phone rang she guessed it might be you and she is adamant she won't talk to you. And I don't think it would be a good idea if you spoke to the children either."

"How are they? Upset?"

There was a slight hesitation. "I think Richard guesses something is wrong. But the other two just think they are on an unexpected holiday." There was a pause, then George Grainger's voice sounded closer and warmer. "I'm not going to interfere, Harry. It's between you and Connie. But I hope it all blows over. Good night."

# Seven

But the trouble between Harry and Constance did not blow over. The Suez affair collapsed suddenly, British troops withdrawing ignominiously from the Canal, abused and spat upon for an action they had been forced into by politicians thousands of miles away. Diehards, ten and fifteen years later, would still be insisting that the world would have been better if the British-French-Israeli action had been carried through, but most people were relieved the confrontation fizzled out and were content to leave Egypt and the Canal to Nasser and the Egyptians. The Israelis went back to their pre-1948 attitude that one could never trust the British, and the French, who never trusted anyone, least of all the British, shrugged and went off to try and solve their own problems in Algeria. The Russians slammed the lid on the boil-over in Hungary, and at the United Nations they and the British accused each other of interfering in other nations' sovereignty. International politics went on much as before and so, indeed, did domestic politics. Harry was told he would no longer be needed as an Under-Secretary at the Foreign Office and he found himself so out in the cold he could have been the Member for North Siberia. Early in 1957 Eden, a sick man, dropped out and was not missed; Macmillan, already gathering speed, took the baton from him as if he had owned it all along, and the relay race through Downing Street went on as smoothly as ever. On the other side of the Atlantic the American elections had been held and Eisenhower, golf putter in hand, had been returned to the White House and the Burning Tree clubhouse. The hiccups of history were quietened and the world settled back to wait for the next spasm.

But things were never the same again at Aidan Hall. Constance, with the children, stayed away till Christmas. Harry wrote her a letter that pleaded forgiveness but was not abject, and in return received a note as cool as a declination of a dinner invitation. It did say, however, *I shall come back to the Hall for Christmas. I think we should spend Christmas together as a family for the children's sake. I shall write to Mrs Dibbens to prepare everything. Give my love to your father*. There was no love for him: it was just signed *Constance*.

She and the children arrived two days before Christmas. Harry had gone out to Qaran to check with Sid Lucas on the after-effects of Suez and was not due back till Christmas Eve. Constance stopped off in London for a day and she and the children went on an orgy of Christmas shopping. She had always been careful as to how much the children were given to spend, but now she placed no limit on their spending; she bought only a tie and half a dozen handkerchiefs for Harry, the sort of present a woman might give to any man friend, but she allowed the children to buy for him whatever caught their fancy. For her father-in-law, whom she had not seen or spoken to since she had left at the beginning of November, she bought a silk dressing-gown and a velvet smoking jacket.

The Marquess was waiting on the front terrace for her and the children when they got out of the car. The children ran up to him, shouting with delight, and threw their arms round him; he gathered them to him, his face suddenly excited as theirs. Constance, standing back so that she would not intrude on this welcome, all at once remembered what Harry had told her of his own attempt to be close to his father and how he had been rejected. She had a moment's sympathy (love?) for Harry, but she quickly put it out of her mind. She had not come home to be sorry for him.

The children went on inside and the Marquess turned to her and held out his hand. She was shocked at the change in him in the seven weeks since she had seen him. He had lost weight, his eyes were bloodshot, and when she leaned forward to kiss him on the cheek she smelled the whisky on his breath. But all she said was, "It's nice to be back."

From within the house came the shouts and laughter of the children magnified by the sound chamber of the Great Hall. He nodded towards the open front doors and smiled. He was still holding her hand tightly and something glimmered in his eyes, rheum or tears, as he said, "I'd never realised before how large the house is. Much too large."

"It will be full enough for Christmas." The Aidan sisters and their husbands and families were due to-morrow; it would be the first time she would be glad of their presence in the house. "You will be worn out by the time they all leave."

"Suppose so." The Marquess waited till the two footmen had gone past with the luggage and presents. It was cold out here on the terrace and Constance wanted to get inside to the warmth of the house. But the Marquess had something to say that, if he did not say it now, would never be said. He cleared his throat, forced himself to look at her: all his life he had stayed out of the domestic affairs of his kin or his friends. "How serious is it between you and Harry? This – this rift, I mean?"

She had had plenty of time to be composed for such questions, seven weeks of searching her heart and her mind for a line on the future. "Too serious to be healed, I'm afraid."

"Is it another woman?" It hurt him to ask such a question.

"Yes." She was not going to elaborate.

"Damn! Why doesn't the bloody fool appreciate what he's got? Is he thinking about a divorce or something like that?"

"We haven't discussed it."

"Would you give him one?" The Marquess did not believe in divorce, but he also did not believe in forcing his religious views on others. Political views, yes: but religion was a man's own personal need for a God and he had never aspired to speak for the Almighty, though there were several people who were willing to say he could have fooled them with such a claim to modesty.

"No."

"Good." The Marquess nodded emphatic approval; then his face softened. "But it isn't going to be easy for you, is it?"

"No." The cold wind was biting through her; she could feel the tears starting in her eyes. "Nor the children. But I think it would be worse for them if we separated."

"So you'll go on living here?"

"If it's all right with you – ?"

He took her hand again, led her into the house. "My dear, you are a part of this place now. I can't imagine it without you and the children. Harry and I, too, are – er – separated." Till his dying day the Marquess would believe that Eden had been right in his attitude towards Suez; Harry had committed treason just as much as if he had gone over to fight for Nasser and the Wogs. "The three of us will have to arrive at some sort of – er – *modus operandi*."

*Modus operandi*: the new name for a marriage. "We'll find a way, Father. After all, we're intelligent people."

"Of course. And civilised, too." He closed the doors and at once the warmth of the house wrapped itself round them. The Marquess in many ways was a relic of the past, but he believed in the comforts of the present; the house was more than adequately supplied with central heating and it was always kept comfortably warm. He rubbed his hands, looked up to where Robin and Matthew were chasing each other along the gallery of the Great Hall, their laughter echoing in the big chamber. "We'll get along. Just a matter of adjustment."

"Of course," said Constance, and went up to the bedroom that was going to be hers alone from now on.


2

"How were things in Qaran?"

"Some trouble-makers from the bazaars tried to blow up the pipe-line, but Sid Lucas heard about it and got Hazza to send out his Levies. Things are quiet now, but we are going to have to ship all our oil round the Cape. Doesn't look as if the Canal will be opened for months, perhaps years. How were things in Northumberland?"

"No trouble up there. The natives are still friendly."

They were keeping up appearances, she thought, acting out a charade for the relatives. Harry had arrived back half an hour ago and now they were having drinks in the library among the other Aidans, the last of whom had arrived a couple of hours before. It amused Constance that, though none of the relatives, sisters, brothers-in-law, nieces and nephews, were now named Aidan, the Marquess and his daughters still looked upon the body corporate as the Aidans. Lords Cavanreagh, Edgecliff and Truscott had titles that were courtesy titles only at Aidan Hall.

She had been out in the kitchen with Mrs Dibbens when Harry had arrived and he was already here in the library with the family when she had come in. She had offered him her cheek and he had kissed it lightly, then brought her her favourite dry sherry. Elizabeth, nose up on the scent, had glanced at them and Constance had wondered if she and her sisters already suspected something. She could not imagine Harry's confiding anything to his sisters and it was too soon for the Marquess to have made a slip of the tongue. She reassured herself that Elizabeth was only once again showing her resentment that an outsider was mistress of the Hall.

And that's what I'm going to stay, she thought. She moved on about the library, playing bountiful aunt to nieces and nephews, all of whom, she knew, liked her better than their mothers did. The house party was to last four days and in that time, she hoped, she and Harry would be able, in the confusion of other relationships, to slide into their *modus operandi*. How the children and the Marquess would fit into it was something that she had neither the courage nor the prescience to contemplate.

Half an hour later the group broke up to dress for dinner, a custom the Marquess still insisted upon whenever guests were in the house, though not all his guests welcomed the compliment to them. Constance was in her bedroom when there was a knock on the door. It was Harry waiting to be told he might enter.

"Do we have to go through with this charade?" he asked as he came in and closed the door behind him.

"It's no charade," she said, denying what she had been thinking downstairs. "It's what your father calls a *modus operandi*."

Harry snorted, half with laughter, half with exasperation. He felt unsure with her, as if he were paying court to a woman whom he knew only slightly and who had given him no encouragement. He had felt more at ease with her in the first weeks of their actual courtship twelve years ago than he did now. *Amor*, as well as *tempus fugit*.

"He can call it what he likes, it's bloody stupid!" He sat down on the chaise longue at the foot of the bed, an uncomfortable convenience designed for posturing mistresses rather than angry husbands. He flung an arm along the curved and curlicued back, got a sleeve button caught in one of the ornamentations and his temper increased as he struggled to free it. He had come in here for what he had hoped would be a cool civilised talk that might bring about a gradual reconciliation; he had given up hope that she would welcome him back with open arms and legs at once; but, like an offender on a good behaviour bond, he had hoped for some encouragement for the future. But things had not started auspiciously. "Do you think we can go on living like this? What about the children?"

"They are too young to query why we are not sleeping together. They don't think sex is a necessary part of marriage."

"Neither do you evidently." He had got his sleeve free and he composed himself, determined again to be cool and civilised.

She had made a slip, but she was not going to let him take advantage of it. "Not when I feel I'm just the principal shareholder in a lending library."

He winced inwardly but said calmly, "Not very original nor very good."

"But true." She picked up a brush and began to stroke her hair. "Now what did you want to talk about?"

"Us, for Christ's sake!" His temper suddenly got the better of him. He got up, began to pace about the room. "What the hell do you think I want to talk about? Good God Almighty, twelve years of marriage is being tossed out of the window, a family broken up, and you sit there like a bloody Mother Superior, a frozen prig trying to sound as if I've never been to bed with you, and ask me what I want to talk about!"

She knew exactly how she sounded, that what he said of her was true; but she had no other defence, knew the weakness in character could only be protected by going against the grain of character. If she were to be nobody but herself while with him, surrender to all the weaknesses in her, and they included love of him, she would have him back in her bed at once. But she knew, because she had already experienced it once, that as soon as he entered her there would be the picture, sharp and clear in her otherwise disordered mind, of his entering Deborah Roth and, in the earlier experience, Janet Buck. Imagination was another of her weaknesses; as was jealousy. But jealousy, perversely, was also a strength, at least a temporary one, though she knew that in the end it, too, could betray her. But it was not to be denied now.

"Keep your voice down. The others – "

"Bugger the others! Their marriages are the charades – the *modi operandi*, if you like. None of them ever had anything of what you and I had – "

That, too, was true. But it was also true that Elizabeth and Anne and Grace did not have to share their husbands with other women; Patrick, Michael and Marlowe were as incapable of adultery as they were of bestiality; dull husbands though they might be, they believed in loyalty and devotion. "Perhaps they have more. At least they know they can trust each other."

She punctured him with that. He threw up his hands, then went and leaned on the sill and stared out through the windows at the dark night. Without looking round he said, "So what do you want? A divorce?"

"That's the last thing I want," she said, and that brought him round to face her. "Not just because of religion – though I believe in the Church's forbidding of it. Till death us do part – I made a vow and I'll keep it."

"That's all medieval – "

"Perhaps. You'll probably also think my other reason for not giving you a divorce is medieval."

"What's that?"

She had stopped stroking her hair and she looked down at the brush in her hand. It was a silver-backed brush embossed with the Aidan coat of arms and had belonged originally to the first Marchioness. Harry's father had had new bristles put in and had given it to her when Harry had become the heir-apparent; the Marquess had made some dreadful punning joke about hair, but she could not remember now what it had been. But she had been touched by the thought of the gift when he told her it had been his wife's and before her his mother's and so on back through the Aidan women to the first Marchioness: it had been his first practical demonstration that he thought of her as an Aidan. She held up the brush and the light from her dressing-table lamp caught the coat of arms.

"I am part of that now. You made me a part of it, but not only you – I belong to it because of the children. I'm the link between you and Richard – by accident or design or whatever you like to call it, Richard is there to succeed to the title because of me. There are people outside this house who would laugh at me for what I'm saying, but I know that you feel as strongly as your father about continuing the line. And so do I. You've – *infected* me, I suppose is the word. Or converted me. Whatever it is, I'm part of it." She tapped the coat of arms. Then, being a woman in a man's domain, she added tartly, "Though, looking at it, no one may ever know."

It was an argument he could not, immediately, counter. All he could manage was the banal jeer, "You mean you want to go on being the Countess?"

She looked at him more with disappointment than anger. "You know better than that."

He nodded. "I'm sorry. I didn't mean it – " He turned back to the window, stared out into the darkness of the future. "God, I could weep!"

"It might do you good," she said, feeling the trembling inside her. If he had come to her at that moment and flung his arms round her, she might have surrendered; she did not know, either then or later, but because his back was turned he never saw the look that swept her face and he did not cross the few yards of carpet that separated them. She stood up and moved towards her bathroom. "You'd better go."

At the door he said, "How do we act then? I mean in front of the others? And the children?"

"As I imagine hundreds, perhaps thousands, of other couples do. I don't think for one moment that England is full of happily married couples. We'll be respectful towards each other, I suppose is what one would call it."

"Respectful! Christ, what a middle-class word!"

He went out, slamming the door behind him. She went into the bathroom, turned on the taps and stood there staring at herself in the mirrored wall, till the steam rose and her image slowly disappeared like that of a ghost.

3

And so the *modus operandi*, a term they never used again, got under way. The staff, as much part of the household as the family, immediately knew that a new atmosphere and a new arrangement prevailed, not only between His Lordship and his wife but between His Lordship and his father. The Marquess set up his quarters in the far wing of the house, and though the Ministry of Works protested at this denial to the public of some treasured historical rooms, the Marquess took no notice of it and acted in the same high-handed manner as had the original occupant of the wing, the first earl. Time had not, in certain respects, altered the manners of the Aidans.

The Marquess took to drink, though not so much as to become an alcoholic. Constance took to works of charity and in time was mentioned in the Birthday Honours for them. Harry took to women, though he did not become addicted to them; and he was careful to see that they were never mentioned, at least by him. But early in the spring of 1957, six months after their break-up, Constance had a yearning for a reconciliation, and then she learned he was having an affair with Rosemary Kidman. The information came to her through one of those accidents that well-meaning friends or relatives unwittingly manage to contrive.

Grace Truscott was the youngest of the Aidan sisters and was too placid and lazy to be a mischief-maker. She was three years older than Harry and, if she had not allowed herself to put on so much weight, would have still been stunningly beautiful. Winter and summer she wore tweed suits and what she called sensible shoes, but twice a year, as a concession to fashion, she came up to London from Norfolk and had her hair done at one of the fashionable salons in Knightsbridge, emerging looking like a cross between *Vogue* and the *Horse and Hound Yearbook*. If Constance was in town she always went down to the Eaton Square house to have lunch with her.

"Were you down at the Kidmans' last week for Newmarket?" Grace asked, smothering salmon in mayonnaise as if she were no more than skin and bones instead of a too-healthy one hundred and fifty pounds. "I saw Harry in the betting ring with Rosemary. But I couldn't see you and Ian."

"I was down at the Hall," said Constance, trying to stay calm; but the stab of jealousy had not been dulled by time. "And I understand Ian is still at the embassy in Tunis."

Grace, whatever she had done to her figure, had not allowed the placid rural life to dull her sensitivity. She paused with a laden fork half-way to her mouth. "Sorry. Am I talking out of school or something?"

"It's all right, Grace. Sooner or later you'd have known. I mean about Harry and me. Though I didn't know about Rosemary."

"Things aren't what they used to be with you and Harry? Is it that little trollop's fault?"

Constance shook her head. "I think she's just an incidental."

"An incidental? Good God, how can you be so calm and casual about a damned little whore who's sleeping with your husband? Or don't you care?" She put the forkful of food into her mouth, swallowed it. "I'm sorry I said that, m'dear. Obviously you care. What a swine of a man! My own brother, too. Does Father know? He does? That, on top of the row they had over Suez – " She pushed her plate away from her. "I'm sorry, darling. I couldn't eat a thing."

Constance had never felt close to Grace, though she had always been able to tolerate her more easily than she could Elizabeth or Anne. But now she put her hand on Grace's, suddenly glad of another woman to whom she could talk. "Darling, that salmon is Jackson's best. I don't think it should be wasted because of Harry. Men should never be allowed to ruin a woman's appetite."

"Marlowe, unfortunately, encourages my appetite." She looked down at the spread of her lap and hips; then she shrugged and pulled her plate towards her again. "I know Marlowe is a dull man, but he's a dear and he'll never cause me heartburn. I'll only bring that on myself," she said, and pushed the forkful of salmon into her mouth. A few minutes later she said, "What are you going to do? I mean about Harry?"

"We're going to go on living together but go our own ways. It's what I gather is called a modern marriage."

"Pig's foot!" said Grace, who had acquired some rural exclamations, though her language was not quite so explicit as that of the labourers on the Truscott estate. "What are you going to do for love? You know what I mean."

"Do without. Plenty of women do."

"I suppose so. But you look as if you would have enjoyed it too much to give it up just like that." She saw Constance's smile of surprise. "Have I shocked you? Of course I haven't. You're not one of those namby-pamby prigs. If you do take a lover, m'dear, be careful you don't choose someone like Harry. Women usually tend to make the same mistake twice. Some do it three and four times. I think we are masochists at heart."

Constance was shaking her head in mild amazement at this side of Grace she had never suspected. "What sort of life do you country bumpkins live down there in Norfolk?"

Grace finished her salmon, began on her salad. "The same as you slickers up here in London. Only a little more basic."

Several months later Constance found herself on a charity committee for abandoned children with Rosemary Kidman. They had not seen each other for over a year and Rosemary swooped on her with a sabre smile and a kiss that Constance for a moment feared might turn into a bite at her jugular. "Dearest heart! You look simply *marvellous*! Where did you get that absolutely divine suit?"

It was a suit that was three years old and Constance was sure that Rosemary knew it. But this, after all, was a charity committee and she did her best to keep in the spirit of the meeting. "You look well, too, darling. How's Ian?"

Rosemary shrugged, concerned for the abandoned children of London but not for an abandoned husband in Tunis. "Still handing out lemonade to the Wogs, I suppose. It was all he was good for. How's Harry?"

"I thought you'd know," said Constance with her best charity worker's smile.

Rosemary's own smile froze, as if her teeth were suddenly being pulled. Then she blinked and said, "I didn't think you knew. I'm sorry, darling, *really*. I wouldn't have hurt you for worlds. It just *happened* – "

"It usually does with Harry. Just happens, I mean."

"You mean there have been *others*? Oh my God, and I never

suspected! Oh *darling* – " Rosemary was all charity and a mile wide now and Constance knew that if the meeting did not begin within the next ten seconds she was going to brain the bitch with her handbag. "If it's any consolation to you I haven't seen him in *months*, literally. And I wouldn't – not *ever* again. Darling, I feel absolutely *awful* – "

Two seconds more and she would have felt worse than awful; but she was saved by the chairman's bell. Constance took her place at the committee table and soon was listening to the reasons for broken homes and what should be done to alleviate the distress of the innocent victims, while down the table Rosemary watched her with the wary eye of a cuckoo that had landed in very much the wrong nest.

So time went on and Constance managed to do without, as Grace had described it, love, you know what I mean. It was not easy: passion for her had been as natural as smiling or speaking. Once or twice she tried masturbation, the first time since she had been a schoolgirl, but it did not work: it offended her moral sense and her own hand was not as welcome as a man's. Occasionally she found herself looking at men other than Harry, sometimes at a dinner-party, sometimes when she danced with them at a charity ball; but her innate sense of orderliness stopped her from going further, she did not want her life to become even more of a mess than it was. She was not interested in young men and the older men she looked at were invariably married; her sense of honour, a virtue usually considered only a male prerogative, would not allow her to intrude on another marriage; though her own was a wreck, she still believed in it as an institution, the *only* institution for two people in love. She knew Joe Banning was in love with her and, when Harry was not available for balls or parties that she had to attend as part of her duties as the Countess of Bayard, she asked him to escort her. He knew that she and Harry no longer slept together, but he made only oblique references to it and always lightly; and though sometimes when he brought her home he was close to being drunk he never once became difficult. She managed him as easily as she managed everyone and everything else, and though sometimes she knew she was being unfair to him she sensed, correctly, that he would be more hurt if she told him that she did not want to see him again.

So she lived a life of continence, something her Church told her was good for the soul, and never once felt one step closer to heaven. But she did not become dried up and bitter and she did not become a prig. She still kept her sense of humour and her charm and never loved anyone but Harry.

But Harry, as time went by, had other women and had grown careless about whether his affairs with them were known or not. In his obstinacy he had become indifferent to her jealousy, and so the gap between them widened.

4

Harry, for his part, did not enjoy the first year of the separation. The affair with Rosemary did *just happen*: he seemed to have a fatal talent for accidental adultery. He met her by chance at a Foreign Office reception for a visiting Tunisian delegation; he had not known she had left Kidman and come back from Tunis. They went out to dinner afterwards and from there down to the Kidman house near Newmarket. The affair lasted two days: they went to bed and demonstrated to each other how much they had learned since their fumblings together years before, went to the races and backed the winner of the One Thousand Guineas, went back to London and after that saw each other only occasionally at cocktail parties and receptions where they waved to each other across the room or met and exchanged greetings like an M.P. and one of his constituents. Ian Kidman came back from Tunis, resigned from the Foreign Service, was reconciled to Rosemary, turned to politics, won a safe Tory seat in a by-election, and came to Westminster.

Often, when Harry was in the house with Constance, either at Aidan Hall or at Eaton Square, he was tempted to get out of bed and go into her bedroom. Once he did get as far as her bedroom door, but found it locked and he had too much stubborn pride to knock and ask her to let him in. He went back to his own room, seething with frustration and stupid anger, and the next day began his next affair with the wife of a Liberal M.P. who was too liberal for the good of her husband.

Some of the Right-wingers in the Party never forgave him for his desertion at the time of Suez, but most Members had too much on their minds to worry about the past and soon his sin was forgotten. Gradually he began to speak again in the House, never contentiously but always intelligently and to the point; the Front Bench once more took to turning their heads and nodding appreciatively at him, and he was on his way back. So long as one stayed in the Commons one was never really dead politically.

When Macmillan called the 1959 elections there was a movement, led by Colonel Bedley, to look for a new Tory candidate for Bayard, but a counter-movement, led by Joe Banning, proved too strong and Harry was nominated again and returned, though with a reduced majority. The Marquess disqualified himself from discussion on the nomination, but never congratulated Harry on being returned.

Sid Lucas came to England to report to the board of Anglo-Qaran, and while he was in London Harry proposed him as a replacement for one of the octogenarian directors who had finally decided to retire. The main opposition came from the Marquess: "He's a field man, not a management chap. What does he know about our end of things?"

"Probably a bit more than some of us know about his end of it." Harry and his father were at opposite ends of the long board table; the other directors sat between them, aware now of the coolness between father and son. Harry guessed his father's objection to Sid Lucas stemmed from two reasons, that Lucas at the time of Suez had suggested the course which Harry himself had advocated, and that Lucas was Harry's nominee and therefore to be opposed on that ground alone. But he was not going to bow down to his father. "His contract out there is almost up and I know he wants to retire to England to live. We'd be foolish to allow ourselves to be deprived of his knowledge and experience."

The Marquess continued his argument against Sid Lucas, but when it came to the vote Lucas was elected to be proposed to the shareholders at the next general meeting. The directors, long content to allow the Marquess his own way, were now realising that the man who would eventually succeed him was more in tune with the changing world in which the company's fortunes lay. The Marquess took his defeat gracefully and Harry just as gracefully accepted his own victory. But the rift between them was not lessened and the Marquess went back to the country convinced that Harry's nomination of Sid Lucas had been a deliberate provocation of him personally.

Harry took Lucas to lunch at White's. Lucas looked around the dining-room and asked, "Would a Wild Colonial Boy be allowed to join this select band of Englishmen?"

"I don't see why not, so long as you didn't insist on being a Wild Colonial. What do you think of being a director of Anglo-Qaran?"

"I'm grateful to you for putting me up. But I'm getting married, you know. To that Jewish girl I mentioned. Won't that disqualify me?"

"I brought that up at the meeting and persuaded them it shouldn't make any difference. The founder of Shell was a Jew. The Arabs never make any propaganda out of it. Hazza is a practical man. He knows directors' wives have no vote on the board of Anglo-Qaran."

"Speaking of wives, how's Lady Bayard? She would always have my vote. You are one of the luckiest men I know."

"I'll tell her," said Harry, not blinking an eye.

But he never did and they went on playing their charade of man and wife to the public world. Their private world, like some of the more personal possessions of the Aidan family that had been kept from the Ministry of Works, was locked away in the cupboards of their minds, rarely brought out and then only reluctantly.

In 1962 Prime Minister Macmillan, a gentleman with a gentleman's lack of squeamishness, chopped off the heads of seven of his Cabinet Ministers with remarkable suddenness; one or two of them, still unable to believe they had been relieved of their power, ran around for days after with the reflex actions of beheaded roosters looking for the dungheap from which they had been evicted. Other men were called to replace them and among them was Harry, who was given a junior and entirely new post, Minister for Overseas Development.

His father congratulated him politely and coldly. "It's a start. You are fortunate they have forgiven you so soon."

"I don't know that they have forgiven me – not the diehards like you. But Macmillan is pragmatic, to use a fashionable word, and he chooses his horses on their form, not their colours."

Harry maintained his form, quietly rebuilding his reputation while at the same time not treading on more toes than was necessary. The new post meant occasional trips abroad and contacts in London with overseas governments, and he increased his knowledge of world affairs against the day when he might be moved back to the Foreign Office as Minister of State or even, and who knew what might happen under Macmillan, as Foreign Secretary.

Pollution, instead of water, began to flow under the bridge, though the world at large had not yet discovered it. The sixties went by in fits and starts and the older generation, observing events that they could not comprehend, fell into fits that sometimes proved fatal. The Tories were at last put out of power and Labour came in, Harry being returned once again but with a still further reduced majority.

The Pill was marketed and countless women took to it with delight and relief; for the first time in history men got *out* of bed for a rest.

The young became a force, a nation without boundaries, and tried to take over the world with rhetoric, a weapon that history, a subject they greeted with derision, could have taught them was no more effective than a Roman candle. A sub-species of the young, called pop stars, came and went with the same bewildering speed as the strobe lights that lit their performances, all of them with names just as bewildering to the older generation who remembered identifiable durables such as Benny Goodman, Bing Crosby and Vera Lynn: the Nice and Them did not catch the imagination and soon, with the Balls, went the way of all flesh. The Beatles caught everyone's imagination and some of the older generation, deciding their music was an elixir, began to frug and twist to it with a reckless abandon that, according to National Health figures, increased the incidence of slipped discs to a point where it was feared that half the population of Britain might crawl out of the sixties bent half-way to the ground on National Health crutches.

English soccer players took to kissing and hugging each other on the field, leaving foreigners to ponder if *le vice anglais* had at last seeped down to the working class; one wondered where the penalty area was in sodomy. Long hair for men came in and short back and sides went out for all but retired admirals and generals and long-term prisoners. Something called Unisex began to appear on the streets, and a letter from Colonel Bedley appeared in *The Times* asking if the Empire was going to be inherited by hermaphrodites, though the increasing figures for rape suggested there were still enough flagpoles being raised to keep the Union Jack flying.

Harry became friendly with Ian Kidman and each week they played squash together, though by tacit agreement they never entertained each other in their homes and Harry sometimes wondered if Kidman knew of the two-day affair with Rosemary. Kidman began quietly in Parliament, but he soon was recognised outside for his work among orphan children and his efforts to improve the conditions of homes that housed them.

Harry saw his own children growing up, but always, it seemed, at a distance. It was his own fault: Constance did not keep them away from him. All three of the children went off to school as they grew old enough, and each time they came home on holiday he found it more difficult to talk to them than on their previous visit. Constrained in his dialogue with their mother, he was constrained in his approach to them. He was repeating the error his own father had made with him and yet he could not correct it. Fired again by political ambition he had allowed his personal life to drift to a degree where he sometimes found himself a stranger in the house to which he was the heir-apparent.

In the late summer of 1966 he came back from a month in Australia, where he had gone out to look at the family's cattle properties in Queensland and the Kimberleys district of Western Australia. His father had been taking less and less interest in Aidan Estates, the family company, rarely coming up to London now and spending more and more time at Aidan Hall. On his way home Harry stopped off in Canada and looked at the family's timber and mining interests there. Until the past year he had concentrated almost solely on the Aidan investment and influence in Anglo-Qaran. Money in itself did not interest him, but for the first time he appreciated just how much the family was worth. He arrived back determined to talk again to his father about the eventual and inevitable matter of death duties. He had made several attempts to discuss them, but had always been brusquely brushed aside. It was not easy to discuss with an estranged father the possibility and consequences of the latter's death.

He got down to the Hall on a Wednesday afternoon, a public visiting day. The parking area had been laid down to the east of the Hall and to-day was half-full of coaches and cars. Lambert had come up to London Airport in the Bentley to pick him up, and as they drove up the drive, past the occasional groups picnicking on the grass verge or over by the lake shore, Harry could see the stiffening of Lambert's neck that showed he was growing angry about something.

"What's the matter, Lambert?"

"I'm no snob, sir," said Lambert, who wasn't, "but the place isn't the same since you let the public in. Most of 'em appreciate it, that they do, but there's too many of 'em treat it like it was Battersea Gardens and all."

Harry nodded, turning his head to look at some children who, ignored by their parents, were throwing sticks at the ducks on the lake. "I'm afraid it's one of the prices we have to pay for an egalitarian society. All I can say in their defence is that some of my ancestors behaved no better."

"That's as may be, sir. But this country doesn't have much left but its history and we oughta respect it. And that," he nodded up at the Hall as he took the car round it towards the stables and garages at the rear, "is part of it." He stopped the car, got out and opened the rear door for Harry. He grinned, his anger gone. "Perhaps we should shoot the lot of 'em, sir."

"That's the Welsh in you," said Harry. "We English are subtler than that."

"It's true, sir," said Lambert, still grinning. "Don't we Welsh know it and all."

Then Harry saw his three children and John Ferguson standing by an M.G. sports car outside one of the garages, all watching him with the same careful curiosity as the small group of visitors who had just come round into the yard from the gardens. For one moment of fantasy he had the feeling that he was pinned in space, an exhibit to be scrutinised just as much as the furniture, paintings and other treasures inside the house. But then he knew his children would also be looking at him with suspicion, as if they suspected he might also be a fake.

"New car?" he asked as he approached them.

"Fourth-hand." Richard was now seventeen, a lanky handsome boy with a sullen intensity already developed enough for Harry to find it almost impenetrable. "Grandfather bought it for me."

"For your birthday?" As soon as he said it he knew he had made a mistake: a proper father should know the birthdays of his children.

"No. Just a present." Richard turned back to the car, began to tinker with the engine. "My birthday is in November."

Harry looked at Robin and Matthew. "What did Grandfather give you?"

"Nothing so far. He's letting us choose." Robin was fifteen, and looking at her, tanned and beautiful in the bright hot sun, Harry felt a catch in his throat. It was difficult to believe that he had had some part in the creation of this lovely creature and suddenly he felt afraid for her. He knew with what an eye other men would look at her: a philanderer, he told himself, should never have a daughter, not if he wants no worries. "Anything up to two hundred pounds."

"He must be in a benevolent mood," said Harry, looking at Matthew, thirteen, stolid and self-contained. "That's more than I've ever given you."

"Yes," said Matthew.

I find it easier to talk to my chauffeur than to my children. He turned to John Ferguson, trying to struggle out of this quicksand of total unconcern for himself. "How's university, John?"

That had been another of his sins of omission. He had promised to help the boy to get into a university, had a dim memory of the conversation they had had by the lake ten years ago; but it had proved to be a politician's promise, one that he had forgotten till it was too late. By the time he had remembered to ask what he could do for the boy on leaving school, John was already enrolled at Bristol University and the Marquess, against the law, was supplementing his State grant to allow him to live away from home in more than reasonable comfort. Harry, who normally never saw what other men wore, saw now that the young man was as smartly dressed as the Marquess himself might be, a country gentleman at leisure.

"I finished this summer, sir." Another score against me, Harry thought. How to win friends and influence your children. But John Ferguson was kinder than Harry's children had been, he did not allow him to drown in the quicksand: "Time soon goes by. I'm like you, sir, it's hard to believe it's all gone so quickly."

"Yes." Harry nodded, grateful for the thrown rope. He lied, trying to stay on the warm side of his butler's son: "Somehow I thought you had another year to go. What are you going to do now?"

"I'm going out to Kenya."

Harry winced. Another barb; so the boy was not a friend after all. But then what would John know about Janet Buck? "What are you going to do out there?"

"Teach. I want to have a look at the world before I decide to settle down."

"You'll settle down in England?"

John shrugged. He was not insolent, but he had none of his father's deference. "I don't know. I may find a place I like better."

"That shouldn't be difficult." Richard's head came up from under the bonnet of the car.

What do those bloody Jesuits teach these days? "I don't think you're doing too badly," said Harry mildly, not wanting an argument, looking at the cars, the stables and garages, the house itself. "I can't remember seeing anything quite like this out in Kenya."

"At least the atmosphere out there isn't stultifying." Stultifying: an everyday word for a seventeen-year-old.

"Do you find the atmosphere in England stultifying?" Harry looked at Robin and Matthew.

"Oh, Richard's just a Marxist," said Robin.

"*Just* a Marxist? Does Grandfather know that?"

"I'd tell him if he asked me," said Richard; spanner in hand he looked aggressive enough to man the barricades against all the marquesses and earls of England. "So far he hasn't."

"So far he probably doesn't suspect. He gets the vapours if even a Right-wing Socialist speaks to him. But don't be alarmed," Harry added with mock hastiness, "I shan't give you away. Perhaps you and I should have a talk about how Labour is sending the country to the dogs. We may be closer in our views than you think. Are you a Marxist too?" he asked Matthew.

Matthew shook his head. "I'm not interested in politics. But if ever Richard and his crowd tried to take over our place, I'd drive one of the tractors right over them."

"He would, too," said Richard. "He's slightly to the right of Genghis Khan in his outlook."

Harry sighed. "I'm afraid you're going to take over eventually, anyway. There's another Communist in the Lords now. If he's still alive when you become the Marquess, you can sit together and swap hammers and sickles."

Robin laughed and both Matthew and John Ferguson grinned. Richard was a serious young man, but he was not entirely without a sense of humour. He grinned at his father, the first hint of a thaw in their relationship in months. "I couldn't think of a better place to take over than the Lords. All that red leather on the benches – we shan't even have to change the colour scheme."

Harry wanted to stay with them, but he was a politician as well as a father: he knew the right moment to leave. Always leave 'em laughing: it was a maxim for politicians and comedians but especially for fathers trying to establish some sort of rapport with their children. "I'd better go in. Is your mother home?"

"Yes." All the smiles but that of John died on the young faces; the lines of battle were drawn again. "She's in her room, dodging the visitors."

And dodging me: you don't need to say it. He began his retreat. "Come and see me, John, before you go. I'd like to talk to you about Kenya."

"Yes, sir," said John non-committally, as if he recognised another empty gesture by a politician.

The four of them watched Harry go, their faces once more masks of cautious scrutiny. As he came up on to the back terrace he saw his father come out of a door and head towards the gardens. Usually the Marquess stayed out of sight when the Hall was open to visitors, but to-day perhaps the sun had lured him out of his haven. Or perhaps he just wanted to look at the *hoipolloi*, as they would want to look at him: species are equally interesting to each other.

Harry abruptly swung to the right and caught up with his father. The Marquess looked sideways at him, but did not miss a step nor utter a word. Harry fell into step beside him. "Father, I'd like a word with you this evening – "

"What about?"

Why do I bother to come home here? But the answer loomed beside him: the Hall would always be home, no matter how little welcome he got here. "I really think we'll have to start thinking about the possibility of death duties."

They passed a small group of visitors who looked at them with frank smiling approval, like visitors to a zoo looking at two well-behaved llamas which, contrary to expectations, had not spat at them. Harry ignored them, but the Marquess gave them a stiff but polite bow of his head. As they left the group behind the Marquess said, "It doesn't hurt to acknowledge them. I thought you'd know that, as a politician."

"I have a mental block about them. I don't see them unless I have to step over them."

His father said nothing, but gave him a sharp look. Watched now by other groups, they went down into the gardens, crunched their way along the gravel paths, passed through the ornamental gates at the far end and came out on to the stretch of parkland that ran down to what was called the Home Farm. It was the largest of the four Aidan farms and a mixed herd of dairy and beef cattle was raised on it. In the far distance, beyond the stone wall that marked the boundary between the park and the farm, Harry could see some of the dairy herd, black-and-white Friesians, moving slowly through the golden haze of the afternoon like drifting pieces of charred paper. A hay-binder was working far

217

down in the park and the sound carried lazily, somehow adding to the somnolence of the afternoon, up to the two men as they began to walk along the edge of the long grass.

They walked in silence, Harry, by instinct, leaving it to his father to continue the conversation. Though the Marquess had grown noticeably older in the last few years, his hair now quite grey, he still walked very straight, quite frequently, as now, with his hands clasped behind his back. He had become increasingly disenchanted with life, especially after Labour had been returned in 1964 and again this year; he rarely attended the House of Lords these days, though he still continued as chairman of the local Conservative Association. His world, he knew, had gone forever and he had surrendered to the melancholy that had attacked him. His sole interest now were the estate farms: he had become what he had once despised in the Lords, a backwoodsman.

"I've done it," he said at last.

"Done what?"

"Fixed up about the death duties. It's all made over to you. Did it last year while you were out in Jamaica buying that place of yours. All you have to do now is keep me alive for another four years."

"Whittaker never told me – "

"I told him not to." Whittaker was the family solicitor, an elderly man who knew who buttered his bread and never, *never*, disclosed a client's confidence, especially one whose account was as lucrative as that of the Marquess. "We fixed it all up between us. It's no concern of yours till you inherit all this."

"You could have spared my ulcers." Harry was struggling to contain his anger; the old man was still treating him as if he were some irresponsible child. A hundred yards away a family group that had been picnicking in the grass had stood up and were looking up towards them; the father of the family was adjusting the zoom lens on his movie camera as he zeroed in on the ruling class. Make sure you know who's the star of your film, Harry thought, and who is just the extra. "The estate isn't just your concern – "

"You're wrong," said the Marquess, stopping dead and turning to face him. "It is. Until I die it is solely my concern, to do with as I damn well please. When I'm gone, it will be your turn. Until then – " He turned, began to walk back towards the house. He had lost control of his son in politics and business, but *here*, the Aidan of the Aidans, he was still in charge. His words trailed back over his shoulder like a banner: "There is me and only me!"

## 5

And that was the *modus operandi* while the sixties went out and the seventies came in. The Marquess's world disappeared even farther into the fog of history. The Permissive Society made its appearance. Four-letter words became part of the cocktail chatter, their effect made slightly ludicrous by the dogged nonchalance with which they were used. Nudity became obligatory on stage, screen and television; an actor was cast as much for his parts as for a part; bosoms bounced across the screen like loose saddlebags on John Wayne's horse in old-style Westerns. A Children's Charter was proclaimed, making parents obsolete. The Marquess retreated more and more to his Chivas Regal, which remained constant in quality and helped colour the memories of the past.

John Ferguson went off to Kenya and later to other parts of the world, and Harry lost track of and interest in him. Richard went up to Oxford and the gap between him and his father grew wider. Robin went to finishing school in Switzerland and came back more beautiful than ever and a snob as well, something she had not been when she went. Matthew quietly went his own way at school and when he came back on holidays was the only one of the three children to treat the Hall as *home*. Constance went off to occasional holidays at St Moritz and the South of France and at home continued her charity work. She also continued to act out the charade as the serene and happy Countess of the Earl of Bayard and, at last, slowly began to wither inside.

Harry went out to Australia on two quick business trips and was appalled at what he saw of the effects of the long drought there. In another way he was also appalled at the collapse of the mining boom, a situation that he was certain, with the English businessman's usual arrogance towards the efforts of ex-Colonials, could have been avoided in the hands of more experienced men, such as the English. On his return to England he tried to discuss the situation with his father, realising there was a danger of the Aidan interests being caught short if there should be a sudden emergency, but the Marquess had retreated too far from a world that no longer interested him. Harry beat his own retreat into pride and the Aidans, father and son, continued their silent war. But each, in the black shank of an occasional sleepless night, sometimes wondered where the future lay and how the past and present could have been altered.

Then the eighth Marquess died and was succeeded by the ninth.

# Eight

"If the noble Lord Bishop will forgive me – " Lord Wardle nodded towards the opposite benches where the Bishop of Acton, the only Lord Spiritual present, lolled comfortably, one arm spread along the red leather back of the bench and legs crossed negligently " – economics is the absolution of the exploiter. Anything the exploiter may do in the interests of making money is forgiven as being good economics. In a free enterprise society there are few Christians but many churchgoers – well, perhaps not as many as the Lord Bishop would like – "

Go to it, Wardle, thought Harry, they'd have been on your neck as soon as you opened your mouth over in the Commons. But here in the House of Lords the noble lords did no more than shake their heads and only a few did even that. This was a club, the best in London as its members claimed, and polite acceptance of all points of view, no matter how extreme, was an unwritten rule. A lunatic peer, since lunacy did not legally bar the taking of a seat in the House, could rise up and, so long as he did not become physically violent, be listened to with patience and forebearance. Bert Wardle, who had come into the Lords as a life peer in 1960 and in his maiden speech had stated it was his ambition to see the House of Lords abolished, had gradually been won over and now welcomed the graciousness that allowed him to make speeches that would have caused uproar in the Commons.

"I feel a treasonable bastard," Wardle had said to Harry that afternoon as they sat in the bar of the House. He had sat back in his chair, easing his artificial leg out at an angle. The leg had been one of the causes of his elevation to the peerage: it had proved too much of a handicap to him as a working politician in the Commons. He still lived in the semi-detached house in North London where he had lived for the past twenty-five years, and he subsisted on the pension his trade union had given him as one of its retired officers and on the £19.50 he received for attendance at the House three days a week. He and Harry had a drink together once a week and Harry had noticed that Wardle was meticulous about picking up the bill when it was his turn.

Harry, on his part, was just as meticulous that they had only one drink each. He was only conscious of his own money when he noticed the scarcity of someone else's.

"I come in here and the bloody atmosphere gets to me," said Wardle, sipping his Scotch and grinning. "I know in my heart the place is an anachronism, that it has no teeth to do anything really effective in the way of government – not like the American Senate, for instance. But somehow, after half an hour in here, I find I'd hate to see the place wiped out. It has its uses, I tell myself."

"Considered advice," Harry said. "Or it's supposed to be. Some of the chaps from the shires aren't much good, but then they don't come in often enough to worry us."

"M'lord." An attendant had appeared silently behind them. He was a tall man in his fifties, grey-haired, distinguished with that mixed air of deference and bland arrogance that only the best butlers could achieve. They all look like cousins of Ferguson, Harry thought. "The Earl of Bayard has telephoned to say he will be coming to the House this afternoon."

Harry thanked him and the attendant glided away. Wardle looked after him. "You know, when I first came here those blokes frightened me more than you chaps."

Harry wondered what sort of symbiotic class quirk had made Wardle divide the attendants and peers into *blokes* and *chaps*; perhaps he had done it without realising it, a slip that illustrated the growing dichotomy in his own personality. "They're all right. At least they aren't the snobs that some butlers are, or some of the shop assistants in Harrods."

They got up and began to move back to the Chamber. "Your son's coming in this afternoon?"

"He always telephones me before he puts in an appearance. I never know whether it's out of politeness to his father or whether he hopes I'll disappear out of the way. He's like you, y'know – dedicated to the abolition of this place."

"You should get us together," said Wardle with a grin. "I'll try and hold him off till they carry me out of here on a bier. Though when he takes his place in here, perhaps he'll change his mind."

"I hope so."

Now, as Wardle droned on, having got off his introductory *bon mots*, Harry looked across at Richard. His son, dressed in corduroy sports jacket and grey flannels, sat on the steps of the

Throne, occasionally glancing across at Harry and twisting his lips in a slight smile. The custom that insisted that the heir-apparent to a peerage must sit on the steps of the Throne rather than in the gallery was one that embarrassed and annoyed some young men; but Richard, with the perversity of a rebel, took his place like a silent court jester and listened to the debates with a mocking smile that would have been better if it had been hidden in the gallery. Harry, with dry amusement, cursed the unavoidable that would have him dead before he could learn what would happen when Richard's time came to take the Carmel seat in the House.

Harry himself had been in the House two months. He had gone through the required process of proving that he was his father's rightful heir and that he was not a bastard offspring; then the Duke of Swindon and Lord Parrish, two of his father's friends, had volunteered to act as his sponsors when he came to the House to claim his seat. The Carmels' red robe and black cocked hat, which had been worn by four generations, had been sent to the dry cleaners and Harry had been given some perfunctory coaching by the old Duke of Swindon on what would be expected of him when he was introduced to the House.

"Nothing to it, dear boy," the Duke had mumbled. "Had to go through more damned nonsense when I joined the Masons. Only thing to watch is when you retreat – damned robe gets under one's feet. Can't have you falling on your bum."

Harry and Constance had driven to Westminster from the house in Eaton Square. Richard had declined to come down from Oxford, but Robin and Matthew had accompanied their parents in the sixteen-year-old Bentley that Harry had inherited along with the rest of his father's estate.

"Isn't it time we got a new car?" Robin was sitting between her parents and Matthew was up front with Lambert. "There's more room in the new Bentleys."

"This one is perfectly all right." Constance had gradually, over the years, learned the value of the aristocratic whim for being ostentatiously unostentatious. It set them apart from the newly rich, those whose money had been made only in the last fifty or hundred years; it was a subtle statement that, having been able to afford the best, there was no need to go on re-stating the evidence year after year; they were not ashamed, if they were ever capable of such an emotion, to drive a sixteen-year-old car or wear a ten-year-old pair of shoes, as Harry was wearing to-day.

Inconspicuous consumption had not always been a characteristic of the upper classes: extravagant follies such as Aidan Hall and other stately homes were evidence of the fact that the old rich had once been very *nouveau*. But since the end of World War II, as if the aristocracy felt their guarantee of being left alone was by being inconspicuous, those of them who still had wealth had developed a tendency to play it down. Constance herself played the game, though she had never discussed it with Harry, not even when they had been close enough for such discussions. She knew he would have denied, and believed his denial, that he was capable of even bothering to make the effort to hide their wealth. "We'll get another car when this one no longer goes. It gives you no trouble, does it, Lambert?"

"No, ma'am," said Lambert, and winked at Matthew. He and the boy had been discussing cars only that morning and he had remarked that he wouldn't mind driving one of the new Rolls or Bentleys. "This one always gets us there.'

"And that's all one wants," said Constance.

Robin rolled her eyes resignedly. "I notice you got a new outfit for the occasion."

"That was my suggestion," said Harry. "If I had to get dressed up I thought it only fair your mother should also be able to catch the eye. And I must say she does."

Constance smiled, pleased that she had violated the rule of inconspicuous consumption and splurged on the Givenchy suit and the Aage Thaarup hat. Harry would not change into his robes until they reached the House and for the time being she was able, as Harry had said, to catch the eye of anyone who cared to look. Harry had paid her his compliment without any effort to be gallant in front of the children and that also pleased her.

Harry had been piqued at Richard's refusal to come down to Westminster for the ceremony, but he had never mentioned it to his son. He had felt a warm thrill as he had waited outside the Chamber to be called: this formal introduction to his fellow peers was, in a way, the only manifestation of the abstracts of his heritage. The doors were flung wide by two ushers and the Gentleman Usher of the Black Rod, a retired general who took his job seriously, had then led the way into the Chamber. Behind him had come the Senior Herald, Garter King of Arms, his heavily-gilded tabard glinting in the light coming through the upper windows, his hand resting lightly on the sword at his side, his medieval martial appearance only spoiled by the horn-rimmed spectacles he wore, a knight in bifocals. Then Harry had entered, flanked by Swindon and Parrish, all three of them wearing their robes and carrying their black cocked hats. Harry's Letters Patent were read out to the Lord Chancellor, then Harry took the scroll himself and stepped forward to show it to the Lord Chancellor. The Garter King of Arms raised his baton, Harry and his sponsors raised their hats and bowed three times to the Lord Chancellor and he in turn raised his own hat. Then the Lord Chancellor had shaken hands with Harry and there was a loud murmur of "Hear, hear" from the surrounding benches. Harry was officially the Marquess of Carmel and entitled to take his seat in the House of Lords.

And one day, my boy, you'll be doing it too. Harry, now on this May afternoon, looked across at Richard. The boy smiled slightly and, as if reading his father's thoughts, shook his head. Abruptly angry with his son, Harry got up and made his way out of the Chamber. As he came out into the Central Lobby Michael Argus stopped him.

"The P.M. wants to see us for a few minutes. About the New York trip." Argus, grown beefy in middle age, had a face that always reminded Harry of a fist. He had sleek grey hair and his thick body was always impeccably tailored. He was in the Shadow Cabinet, specialising in defence, but it was common knowledge that he had his eye on a higher post. Harry disliked him intensely, but so far had hidden his dislike; with his own interest in the Foreign Office, he did not want to be caught out in any display of petty rivalry. He nodded affably to Argus's suggestion: "Shall we go over now? He's at Number Ten."

As they crossed the Lobby Harry saw Richard hurrying out. He half-raised a hand towards the boy, but at that moment a convoy of constituents, sighting their local M.P., Ian Kidman, swung past in the line of sight. By the time they had run aground on Kidman, who greeted them all with that facial spasm known as the voters' receipt, Richard had gone out the door. Harry gave a small shrug, caught Kidman's eye and gave him a sympathetic smile, and followed Argus towards the exit.

"Shall we walk?"

"Better take my car," said Argus. "One never has time to walk anywhere these days. Go, go, go all the time – that's what the youngsters say, isn't it?"

Though he had been born in Lancashire, public schooling in the South and his own snobbery had at one time removed all traces of Northern accent from Argus's tongue. But over the past few years, with regional accents becoming fashionable with the advent of television personalities from elsewhere but Oxbridge and pop star philosophers who spoke as if through a mouthful of thick Liverpool fog, he occasionally let a word slip with a Northern tinge to it. Harry noticed he had said "yoongsters."

They came to Argus's car and the chauffeur jumped out to open the door for them. It was a regal red Rolls-Royce with the upholstery done in the same shade of red leather as was on the benches in the Lords. But Harry knew that Argus had no ambition towards the Lords; he wanted to stay in the Commons, to stay in the running as a possible future Prime Minister. They drove out of the Yard, the policeman on duty stepping out into the roadway to stop all traffic for them. Out of the corner of his eye Harry saw a van driver jerk his thumb at them, but Argus saw nothing but his own image reflected in the glass partition that separated them from the chauffeur.

Brewster was waiting for them in his study at Number Ten. It was the first time Harry had seen the Prime Minister in his official residence and the first thing that struck him was how much at home Brewster seemed. Even just a few years' occupancy had made it as much home for him as Aidan Hall was for Harry. Permanency for a politician was a relative term.

"This committee – " said Brewster, lighting his pipe and leaning back comfortably in his chair. "Don't forget you're only *observers* – the permanent delegation will be doing all the spade-work in the Council."

"Why two peers?" Harry asked. "Wardle and myself?"

"Tone," said Brewster with a grin. "Cachet. There are still delegates from certain countries who think anyone with a title is a direct emissary from God. You'll be balanced by Michael and the other three chaps from the Commons – two of ours, one of yours and one from the Libs. Bert Wardle will be chairman. I've made it an all-party affair because I want to give the impression that we in Britain are united in our attitude towards this current crisis in the Middle East. We don't want a repetition of Suez. We aren't united," he said, raising a hand as Argus opened his mouth, "I know. But we don't have to advertise it."

"Do you think I should be on the committee?" Harry was not enthusiastic about the trip to New York for the coming debate on the new Middle East situation. There were still a lot of probate inquiries about the estate to be answered and he was afraid that in his absence some wrong answers might be involuntarily given. When Charles Wolfe had first nominated him for the committee he had protested. "Being chairman of Anglo-Qaran, I'm not going to suggest much impartiality."

The Israelis had discovered oil in a section of the Sinai they had captured from the Egyptians in the 1967 war. In the peace talks that had been going on at the time the Egyptians had agreed to relinquish their demands for the return of that particular section of the Sinai; it had been regarded as a useless region of desert rocks and sand and not worth fighting over. But when the Israelis had discovered oil there, in quantities that suggested a major field lay hidden under the worthless surface, the Egyptians, pressed by the other Arab countries, had had second thoughts. If the field *was* a major one, to rival those of Kuwait or Bahrein or Saudi Arabia, then the Arabs would lose a good deal of their political and commercial leverage on the European countries.

"No, we need you, Harry," said Brewster. "I particularly asked for you. You have the contacts, better than any of the others have. You can do some ferreting out for us behind the scenes. The Israelis may make a jibe or two about your Arab connections, but they won't be too loud or too spiteful. They're realists."

Harry let his argument slide. He knew that impartiality was almost rarer than virginity these days and to be excused from the committee on those grounds might only produce jibes from those closer to home who thought hereditary peers did not do as much work as life peers.

"You have your American contacts, Michael." Brewster turned to Argus. "Keep your eye out for any hint of a switch in their attitude – I'm told they're beginning to waver. They're getting very sensitive about world opinion – I've just heard there's another demonstration on outside their embassy this afternoon."

"On a Tuesday?" Argus raised his heavy eyebrows, a trick that his television consultant had advised him to cultivate. There were several politicians with the same thick eyebrows and the same trick and Harry, watching them occasionally on The Box, sometimes thought he was watching a performance by jumping caterpillars. "A little early in the week, isn't it? They're still getting over Sunday's bash."

"We're not dealing with amateurs any more, the ones out for a bit of weekend sport. They have professionals now who can organise a demonstration at the drop of a hat seven days a week."

"Should lock the lot of them up," snapped Argus. "There's too much bloody democracy."

"Try that for size in one of your editorials," said Brewster. "It should guarantee me another twenty years here."

While the Prime Minister and Argus had been talking Harry had been looking around. The study was no different from a dozen others he had seen and less grandiose than some; the one at Aidan Hall was far bigger and better. Number Ten itself was often a disappointment to people when they saw it for the first time, a modest house from the front that did not compare favourably with the residences of Prime Ministers in other countries. The house had been rebuilt internally in recent years, but for over two hundred years, from the time of its first occupant, Robert Walpole, few improvements had been made to increase the comfort and conveniences due to the country's leader. Harry had not been able to believe it when he had first been told that, until the advent of Churchill, there had been only one telephone in the house and that in the butler's pantry. But whatever the lack of grandeur, this, he knew, was the real seat of power. Democracy was on show over at the Commons, but here in this house, in this room and in the Cabinet Room downstairs, the decisions were made that ruled the country.

He looked back at Brewster, half-hidden behind a smoke-screen as he puffed on his pipe. The big mahogany desk in front of the Prime Minister was neatly stacked on one side with red dispatch boxes: they could contain papers on a national crisis, but Brewster would get to them in his own time: he had a reputation for never being hurried but never being caught behind schedule. Harry, watching the big man so comfortable in his ambience and knowledge of power, felt a sudden envy. The feeling startled him: it was the first time in his life he could recall being envious of anyone.

In the years he had been in the Commons his ambition, though dimmed, had never really died. It had flared again briefly while John Kennedy had been in the White House; not only the young men of America had been fired by the glow that had come out of the new Camelot. When the Earl of Home had decided to relinquish his title and become plain (well, nearly plain) Sir Alec Douglas-Home, the dream had been seen to be a possibility: as long as an aristocrat was prepared to show he was just like them, at least in status if not in pedigree, the British public was quite willing to have him as their Prime Minister.

It was Brewster himself who had remarked to Harry at the time that there would always be at least fifty per cent of the British voters who cherished the idea of being governed by their betters.

"So-called betters," he had added. "You should try your luck, Harry, if ever the opportunity presents itself. You'd have as much going for you as Alec. I'm told the ladies like you. In the polling booth, I mean."

Harry had ignored that last crack. His affairs were always short-lived, but at that time they had become too numerous for all of them to remain secret; he was discreet, but he had learned that some of his women had not been. Women held more influence in the Conservative Party at the local level than they did in the other parties; a cabal of men had chosen Alec Douglas-Home to succeed Macmillan, but that had been because the women were not sufficiently enthusiastic about any of the alternatives. But he knew that they would not choose a philanderer and from then on his affairs had become even more discreet. The only danger had been that invariably the affairs had been with women of Conservative, if not conservative, inclination. Somehow or other he had not managed to meet any promiscuous Labour or Communist women; even the wife of the Liberal M.P. had been a turncoat Tory anyway. He had done nothing actively to promote himself as a future candidate for Prime Minister, aware all the time that he would never be considered until he had relinquished all claims to his place in the Lords, but he had also done nothing to spoil his chances of inclusion in the next Tory Cabinet when it next got back into power. It was only one step up from Cabinet Minister to Prime Minister, though some steps were much higher than others. This room where Brewster now sat was the end of the climb, the ultimate realisation of the dream. And he could not help envying the man who had achieved it.

"You look disappointed, Harry," said Brewster, putting down his pipe. "Don't you want to go to New York?"

"Of course," Harry lied. "I was thinking of something else."

"Have they chosen someone to replace you in the by-election down at Bayard?"

"I believe so." Since his elevation to the Lords he had taken no interest in the local Association. The chairmanship had been offered to him, more, he knew, as a compliment to his father than to himself, but he had declined it. "They're waiting on the writ to be issued."

"I wonder what's holding it up?" said Brewster, but a gleam

in his eye suggested he *did* know. He stood up as one of the two doors of the study opened and a Private Secretary poked his head in. "The American Ambassador is downstairs, sir. It's time for his appointment."

"How does he look? Bruised, battered and bewildered?"

"Slightly so, sir. He's come straight from Grosvenor Square. I gather it is a bit ugly up there."

The Private Secretary went out and Brewster shook hands with Harry and Argus. "The Americans really should learn to send only professionals to these posts. They send us these rich ranchers from Texas and the poor blighters can't understand why anyone should hate America. But all these demonstrations may tempt the Yanks to retreat into isolation. We don't want that. We need their Sixth Fleet in the Mediterranean. How could we raise the money to make up a fleet to replace them before the Russians got there?"

"You could abolish the National Health," said Argus. "There is enough money goes into that to equip a new fleet."

"And have me pay for my own duodenal ulcer? Good luck, Michael. Don't be too feudalistic in New York – they may ask you to run for Mayor. Good luck, Harry. I think I know what you were thinking of a moment ago. You've left it too late, I'm afraid."

Harry followed Argus out of the study, wondering how Brewster had guessed what was on his mind. Had his envy been so plain on his face? As they went down the stairs and along the red carpeted hall to the front door, past the bust of Disraeli with its knowing look of what went on in the minds of all men who entered this house, Argus said, "What did the P.M. mean? About your leaving it too late?"

Harry thought it was impertinent of Argus to ask, but all he said was, "It's a private joke. Something to do with local ambitions."

Argus looked blank for a moment, then nodded shrewdly. "Of course. You're both from the West Country. You're a race apart down there, aren't you?"

"Yes," said Harry. "Something like you Northerners."

"Excepting, if you'll forgive my saying so, we have more drive up in the North. Can I give you a lift somewhere?"

"No, you drive. I'll walk."

He went off down the short street and out into Whitehall, fuming with irritation, frustration and, yes, envy. He knew he would never really be happy till he was master in that house in the cul-de-sac that was Downing Street. But he knew only too well that his own life had become a cul-de-sac, one from which there seemed no escape, not even back through the open end into the past.

2

"We are due at the Australian High Commissioner's at six-thirty. Had you forgotten?"

Constance, face made up and hair done, but in a negligee, came into the study after knocking lightly on the door. Harry, in his shirtsleeves and with his shoes off, looked up from the letter he was reading.

"Blast! Do we have to go? I suppose so." He held up the letter. "From Joe Banning. The Probate chaps have paid him another visit. Listen to this: *I am now such a consummate liar that I sometimes feel tempted to embroider our tale. But I think we are winning. Still, I thought I'd better warn you they have not yet given up the scent . . .*" He got up, put a match to the letter and dropped it into an ash-tray. "Don't look so worried. Naturally they are going to be suspicious with Father dying so close to the deadline. But all we have to do is give them the same simple answer we've given them all along. Father was caught in the storm out in the park, over-taxed himself getting back to the house, had one seizure on Friday night and the final one on Tuesday night."

"I just wish we had not had to have Joe and Ferguson perjure themselves."

She regretted now that she had agreed to the conspiracy to defraud the Probate Office; she had been so shocked by her father-in-law's death she had agreed to Harry's suggestion with-out really comprehending the enormity of her commitment. Scrupulously honest even in small matters, she would sometimes jolt awake just as she was dropping off to sleep at night, amazed at the audacity of the deception she was involved in. She was worse than any one of the Great Train Robbery gang; but the joke was a sour one, a whistling in the dark. Harry had continually assured her they were safe, but she still had moments of cold fear that they would be found out and she and Harry, Ferguson and Joe Banning would all be carted off to gaol. Each month when she went to confession she was tortured by her secret; the nuns and the priests still had their hold on her. Was defrauding the government of tax a mortal sin? It was no consolation to be told by Harry that the Vatican itself did not believe in paying taxes.

"They haven't perjured themselves – they haven't been put on oath." Harry himself had no conscience on the matter. There had been no improvement in the Aidan estate liquidity; the position, if anything, had got worse; even Anglo-Qaran's shares, with the latest flare-up in the Middle East, had dropped slightly. There had been several moments of uneasiness when the Probate officers had returned to ask Ferguson additional questions, but the butler had acquitted himself as well as if he had been speaking the truth. As soon as Probate was settled, he must reward Ferguson with something appropriate, though he was not sure what to give him. In the circumstances Constance was not the one to ask for advice on such gifts.

"Joe seems to feel he has perjured himself. Perhaps he feels the Hippocratic oath is some sort of basis for telling the truth."

"If it is, then ninety-nine per cent of the world's doctors are in a constant state of perjury."

"It's a pity you don't have more faith in mankind. I hope you manage to hide your cynicism on this trip to New York. The United Nations is supposed to be a forum for ideals."

Why am I always on the defensive with her? he thought. But he was too tired for argument and he turned it aside with flattery. "You look very attractive to-night. Is that a new hair style?"

"It was when I got it two weeks ago." But there was no tartness in her voice and she smiled. "Harry, compliments to one's wife should be paid weekly, like the milk bill. It's the only way of not getting behind."

"I'll remember it. Well, I'd better bath and get dressed."

"Black tie, don't forget. We're going on afterwards to a film premiere."

"How did we get into that?"

"It's one of my charities – I happen to be chairman of the committee. *Madame* Chairman – it always makes me feel bi-sexual." He picked up his shoes and followed her out of the study. "You don't *have* to come to the premiere. I can always make another excuse for you."

"No, I'll come." He had noted the *another*. "I'm sorry I forgot – I should pay more attention to your charities."

"You don't have to. I enjoy them." They helped to fill in the days and part of the nights: but she didn't say that. "I don't know that you would. They are just for the welfare of ordinary people and your sights never seem to get down that low."

"No, I'll come," he insisted, feeling suddenly generous towards her. They had long ceased to be lovers, but time, the need for convenience and their own sophistication brought moments when they were almost, but not quite, the best of friends. "You put up with my calls on you. This thing at Wolfe's, for instance. From now on each week I'll not only pay you compliments, I'll ask for a report on how your charities are going."

They had reached the door of her bedroom. One hand on the doorknob, she turned back to him. "Don't mock me, Harry. I can stand anything but that."

"My love – " The words slipped out: on his tongue they tasted like another mockery: he used them with every woman with whom he became intimate. He saw the flash of pain in her eyes and at once he said, "I'm sorry. That sounded flippant."

"Yes," she said, and thought, But there was a time when it was all I wanted to hear. But the echoes no longer held the bitter sweetness they once had. Too many years of silence had gone by, the echoes were now too faint.

"I wasn't mocking you. I should show interest in what you are doing and I shall. What is this one to-night for?"

"The International Refugees. It's particularly for the Palestine refugees. What's the matter?"

He smiled wryly. "For the next two weeks, as far as the Middle East is concerned, I am supposed to be snow-white in impartiality. You couldn't organise something to-morrow night for some Jewish refugees?"

"I'm sorry – truly. This was organised months ago. Perhaps you had better not come to-night – "

"Do you mind? I'll drop you at the cinema, then come home. I'm feeling rather tired, anyway. You see, if I had taken an interest in your good works, in people, we could have avoided this."

She shook her head. "No, we couldn't. I'm sorry, Harry, but I'm not going to let politics run my good works – if you like to call them that. Faith, hope and charity are virtues, not pawns in some game."

"You're inconsistent. Where were your principles when I asked you to help me defraud the tax men?" He smiled to show he meant the question in good humour.

"In your pocket." She smiled too, then went into her bedroom and closed the door.

Touché, he thought. He went down to his own bedroom and Ferguson came out of the bathroom. Here in the town house the butler doubled as valet, though he did none of the menial tasks such as cleaning shoes: those things were left to the junior maid. There had been a time a couple of years back when they had had difficulty in finding a full staff for the Hall; with so many well-paying jobs in industry no one wanted to go into service. But over the past few months, with rising unemployment, some of those who had left had drifted back to work at the Hall. Two of the gardeners at the Hall were coloured, a fact that sometimes had stopped Harry's father in his tracks as if he had come on some rare specimen of rhododendron.

"Your bath is ready, sir. And Lady Robin asked me to lay this out for you. It is a present from her. Her remark was that she wanted you to be, if you will forgive the expression, a little more *with it*."

The gift was a pink evening shirt with a frilled white lace front and frilled lace cuffs. Harry looked at it with a cocked eyebrow. "What do you think of it?"

Ferguson at odd moments had a habit of clasping his hands beneath his plump stomach and rotating his thumbs. He did that now, standing beside the bed and gazing down at the shirt. He was not as tall as Harry, but he had such an *unbending* look that he gave the impression of being taller than he was. "I should not wear it myself, sir."

"I know that. But your daughter didn't give it to you as a present. Does your son give you presents that make a demand on you?"

"My son and I do not exchange presents. We do not seem able to exchange even opinions." Ferguson took one last look at the shirt, shook his head, then turned back to Harry. "But that, I gather, is the way it is with the modern generation."

"I'm afraid so. Well, I'll wear the shirt some other night and we'll see if it bridges the generation gap. But you'd better warn my wife."

"Her Ladyship has already seen it. I'm afraid she approves of it."

"We can't win, Ferguson."

Harry had his bath, dressed in a dark lounge suit, then went and knocked on Constance's door. She came out dressed in an ice-blue gown that showed off her figure and went well with her fair skin and blonde hair. He could see the shadow between her

234

breasts that were still remarkably full and firm and he had a sudden feeling of desire for her. But it wasn't love; and sex, he knew, was over forever between them. He was about to compliment her on her appearance when a jumble of loud voices came up the stairwell.

He went to the top of the stairs and looked down. "What's going on?"

"You'd better come down, Daddy." Robin stood in the hall with Richard and a young man whom Harry, looking down on, did not recognise. "There's been some trouble."

Harry heard Constance draw in her breath beside him. She lifted the hem of her dress, he took her hand and they went downstairs: making an entrance, he thought, just as we used to in the old days. By the time they had reached the entrance hall the three young people had gone into the drawing-room. As he crossed the hall from the stairs Harry caught a glimpse of Ferguson standing at the back of the hall by the door that led to the kitchen. There was a look of anger and shock on his plump face that made him look human and vulnerable, two weaknesses Harry had never suspected in him.

"Daddy. Mummy." Robin came towards her parents as they entered the room. "You remember John."

"Good evening, sir. Ma'am." John Ferguson had been kneeling beside Richard, who was sitting on a couch with his head in his hands. He stood up, keeping one hand on Richard's shoulder. Harry understood at once the expression he had seen on the elder Ferguson's face out in the hall: John appeared to be acting as if he were too much one of the family. "Richard has been hurt."

There was a small gasp from Constance as she crossed to Richard. He took his hands away from his head and Harry saw the blood on his fingers. "It's nothing serious. Just a lump."

"How did it happen?" Constance gestured to Robin, who understood what was wanted and went out to get a bowl of water and some plaster. "That's on the *top* of your head. Who hit you?"

Very observant, Harry thought admiringly. And added an observation of his own: "You were in that demonstration up at the American Embassy, weren't you?"

He saw John Ferguson open his eyes in surprise, but Richard, surprised at nothing his father did or knew, said, "Yes. I was hit by a policeman."

"Well, that's part of the risk you take."

"I'm not complaining."

Robin came back into the room, Ferguson on her heels with a bowl of warm water and a towel over his arm. The butler ignored his son. He put the bowl down on a side table, careful to put a newspaper beneath it, then, still without speaking, began to wash the blood from the cut on Richard's head. The latter winced once, but said nothing, just glanced up past the butler's belly and winked at John.

"Why did the policeman hit you?" asked Constance.

"We were getting a bit too close to the steps of the Embassy," said Robin. "They were – "

"*We?* When did you become a demonstrator?"

"I'm not one. I bumped into Richard as I was coming out of Asprey's – "

"An old meeting-house for demonstrators." Harry, so far, was not perturbed by what had happened. For the past two years Richard had been in demonstrations, against the war in Vietnam, against apartheid, against, it seemed to Harry, anything that the Establishment tolerated. "Are Asprey's making silver Ban the Bomb buttons?"

"I'm not that out of date, Father," said Robin coldly. "You may be facetious, but this is serious. I mean for Richard. He hit the policeman back and the chap fell under a police horse and was trodden on. They were carrying him away to an ambulance when we left."

Harry looked at John. "Were you in on this?"

"No, sir." Ferguson senior had paused for a moment, glancing quickly at his son. Then he went back to cleaning Richard's wound as John went on, "I was up there as – an observer, if you like. I saw the copper hit Richard – he was a bit over-zealous, the policeman, I mean – and I moved in when I saw the policeman go down under the horse."

"If it hadn't been for John," said Robin, "we should not have got out of there. The other policemen were trying to get at Richard to arrest him, but they couldn't get through the crowd. They looked pretty furious."

Constance, who had sat down in a chair opposite Richard, looked up at Harry. "It's starting to sound serious. Perhaps he should go and report to the police? If someone recognised him and they came for him, it would look much worse than if he had voluntarily given himself up."

"Why should anyone have recognised me?" Richard felt the

piece of plaster that Ferguson had put on his cut. "I'm not giving myself up to the fuzz. God, they'd be on me like a ton of bricks. They never accept any excuses if one of their own is clobbered. A copper can do no wrong – that's the definition of British justice."

"All right," said Harry sharply, "cut out the propaganda. This isn't Speaker's Corner. Did the policeman look as if he was seriously hurt?"

"I don't know, I didn't look. I was too dazed from the crack he gave me."

Harry looked at John and the latter nodded. "I think he was pretty badly hurt. But if it will relieve your mind, sir – it wasn't really Richard's fault. You know what it's like at a demonstration – "

"I don't," said Harry dryly.

John paused for a moment, then he smiled. "No, I suppose not. Tory M.P.s and peers don't go in for demonstrations, do they?"

"That's enough!" Ferguson senior looked ready to grab his son by the collar and run him out of the room, but Harry waved a placating hand.

"Go on, John. Tell us what happened."

John glanced at his father, who was tight-lipped and red-faced, then went on, "It all got a bit willing. Richard was in the front of the crowd and he and Robin were being pushed forward by those at the back. The police got a bit hot under the collar – I don't think one could blame them, I think most of them would rather not be called upon for that sort of job – "

"They enjoy it," said Richard.

John shook his head. "You're wrong, mate – "

"Don't call His Lordship *mate*!" The butler and the father both sprang out of Ferguson senior at once.

Harry contained the sigh of irritation that welled up in him. "Ferguson, for the moment we'll forget status. It's unimportant in the situation we're in just now. Go on, John."

Ferguson senior, with difficulty, subsided as his son went on: "As I said, the policeman who hit Richard, a young chap, was a bit over-zealous. He clobbered Richard with his truncheon and Richard swung back at him. It was a reflex action – I'd have done it myself. The policeman seemed to jerk back to avoid Richard's swing and the next moment he was under the police horse that was right behind him. I'd swear to that in court, if it gets that far."

"Let's hope it doesn't," said Constance. "I still think Richard should go along to the police and apologise – "

"Apologise! Mother, I'm not apologising. We were there for a perfectly legitimate purpose, to show our feelings about the Yanks' interference in the Middle East – "

"Oh, for Christ's sake cut it out!" Harry snapped. "If the police had not been there to stop you, you'd have gone on into the Embassy and broken up the place. You may think you're going there for a perfectly legitimate purpose, but mobs don't have any intelligence, once they get out of hand they're not interested in anything legitimate."

"Will this affect you?" Constance said quietly.

Harry shrugged. "It won't help me."

"What's it got to do with you?" Richard demanded.

"I'm leaving for New York the day after to-morrow as a member of an observer committee to listen to the Middle East debate at the U.N. I've backed out of going to a Palestine refugees' thing to-night with your mother because, as I told her, I'm supposed to show snow-white impartiality for the next week or two. Having a son charged with trying to storm the American Embassy won't help me with the Americans while I'm there. We are trying to have them be impartial too."

"I'm sorry, then," said Richard, not very graciously. "If I'd known that I might have stayed in the background. But you never tell us anything – "

"Your mother knew. I didn't think you would be interested." Harry wanted to say more, but this was not the time, not with the Fergusons in the room with them. "The point is, do we take the risk that you were not recognised and just sit it out? Or do you want to abide by your mother's principles and go and give yourself up?"

Constance gave him a sharp look when he mentioned her principles. He caught her glance and smiled and shook his head; he had not meant it as a snide remark. But he had done it deliberately: he knew Richard would much less rather go against his mother than against himself.

"Put like that – " Richard said. Then the front-door bell rang.

They all waited in silence while Ferguson went out into the hall. When he came back his voice was as steady as if he were announcing an expected guest: "There is a police constable wishes to see you, m'lord."

"Show him in." Harry raised his voice slightly: "He may have saved us a trip."

The policeman came into the room, cap under his arm, a young curly-haired man whom Harry knew by sight though not by name. "My name is Rayton, m'lord – you have seen me on the beat here. I'm from the local station, down in Gerald Road. There are some inquiries I've been asked to make regarding your son, the Earl of Bayard."

"My son was just coming along to see you. He called in here first to have a cut on his head attended to. He has told me about the unfortunate accident that occurred to the police officer up at Grosvenor Square."

"Yes, sir, it was unfortunate." Constable Rayton shot a glance at Richard that did not promise much sympathy. "I understand the officer is still unconscious. I'd like His Lordship to accompany me to the station."

"He'll be glad to. I'll come too, if I may." Constance had stood up and come to stand beside him; he pressed her arm reassuringly. "You had better ring Wolfe and tell him we shan't be there. Stay with your mother, Robin."

"Shall I come, sir?" John Ferguson asked. "Just in case – "

"I don't think you'll be needed just yet. We'll call on you when the time comes. Do you have a car, Constable?"

"Yes, m'lord. We always provide transport." Constable Rayton was polite but not deferential: he was conceding nothing to young lords who went around bashing up coppers. "Not everyone has their own."

You young whippersnapper, Harry thought: get your licks in while you can. But he just nodded, and he and Richard silently followed the constable out to the police car waiting at the kerb. It was less than a minute's drive to the police station, almost literally just round the corner. As they got out of the car Harry said, "We'll manage to make it back on foot, I think, Constable."

"I'm sure you could, m'lord," said Constable Rayton. "But I don't think the matter is going to end here."

All right, Harry thought, give up trying to score off them. They have us over a barrel. Richard had said nothing from the moment the constable had made his entrance, but Harry could feel the tension in his son, the explosion of nerves that was likely to happen at any moment. He looked at the boy and shook his head warningly. He had the feeling that Richard did not want him here, yet he could also feel that the boy was now uncertain and a little afraid.

They went into the station and were ushered into a small room in which there was a table-desk, three chairs and a steel filing-cabinet; on the wall was a calendar hanging from a coloured photograph of Monte Carlo. Other pictures had been torn off as the months had been torn off: Harry wondered what other exotic places had teased the sergeant who stood up behind the desk as they came in.

"Evening, m'lord." The sergeant was a middle-aged man with prematurely grey hair; his name was Baxter and Harry knew he had been in this station a long time. He was not likely to treat someone with a title as any different from anyone else, but at least he was not likely to be prejudiced. He would not have been kept in this station so long if he were: there were too many people with titles and influence living in the area. "We understand Lord Bayard is in a spot of bother. West End Central, that's the station covers Grosvenor Square and up around there, they've been on to us. Seems a newspaper photographer recognised His Lordship when the – er – incident occurred."

"I'm surprised the newspapers haven't been on to us."

"Oh, they will be, m'lord. Matter of fact, there are some of them outside now. We have 'em – er – detained in a back room. Just wanted to get the formalities over first with a minimum of fuss."

"What are the formalities, Sergeant?" Harry, a man of title, had always been meticulously polite about other men's titles, especially when they had been earned, not inherited.

"We have to book His Lordship – not charge him, just book him for our records – then we'll ship him up to Savile Row, to West End Central. He'll be charged there. That is, of course, if they are going to charge him," he added, but there was no doubt in his voice that there would be a charge. Harry had begun to notice the increasing change in the sergeant's attitude. He might have no prejudice against peers, but he was as prejudiced as the young constable against anyone who went around bashing up policemen. The change in him was evident as soon as he spoke directly to Richard: "Name and address?"

"Richard John Aidan. 200 Eaton Square, S.W.1." Richard's voice was firm enough but very soft.

"Richard John Aidan – " The sergeant, bent over the book in which he was writing, looked up. "Earl of Bayard – right?"

"If you like."

The sergeant, still bent over his book, did not look up this time. "It's not what *I* like. The records call for full identification. Age?"

"Twenty-two." Harry could see Richard's jaw tightening; the boy knew he was in for a rough time. "Colour, white."

"Really?" The sergeant glanced at Harry, but made no comment: then he looked back at Richard. "Were you in Grosvenor Square outside the American Embassy at approximately six p.m. this evening?"

Richard hesitated, then nodded. "Yes. Do I have to answer all your questions here?"

"No." No *sir*, no *m'lord*. Titles might be necessary for full identification in the records, but the sergeant had fully identified Richard in his own mind: the rich peer's son who had nothing better to do than go around putting coppers in hospital. "But you've given us all the information we really need. That will be all, m'lord," he said to Harry. "Now we are going to send him up to West End Central in one of our vans."

"I'd like to go with him, if I may."

"Certainly, sir. I'd just like to warn you, though, if you are with your son when he gets outside here, you're likely to wind up with your photo in the newspapers."

"It will be the first time we've been together in the newspapers since his christening." Constable Rayton had taken Richard out of the room. The two older men, the peer and the policeman, were alone for a moment. "Do you have any sons his age, Sergeant?"

"No, sir. Just a daughter. I'd get a good lawyer for your son, just in case."

"Thank you, Sergeant. And also for the record – I don't condone what my son is alleged to have done this evening."

"I'm sure you don't, sir. That would be the end of the system, wouldn't it?"

Harry smiled: he and Richard had not scored a point since they had got out of the car outside the station. "Unfortunately, that is what my son seems to be working for. But that is *not* for the record. Now we'd better go and get those photographers off your doorstep."

Richard put his hand over his face as they went out of the police station into the barrage of camera flashes, but Harry kept his head up and stared straight ahead. A reporter tried to step in front of them, but Constable Rayton trod on his instep and he lurched back with a four-letter cry of pain.

"No language, sir," said Constable Rayton, "or we shall have to book you. This way, m'lord," he said to Harry; he had already pushed Richard into the police van without a word to him. "Mind the step. It's a bit higher than into a Rolls or a Bentley."

Too much sardonic wit: education was corrupting the police force.

### 3

"Thank you," said Richard as they settled back in the taxi outside the police station in Savile Row. The photographers who had followed them here from Gerald Road had been joined by others; as they had come out of the police station the street had exploded with flashes as if there had been an upheaval of broken power lines. Several reporters, ballpoints at the ready, had attempted to ambush them, but Harry, learning from Constable Rayton, had put his foot down on two convenient insteps, had apologised and was in the taxi with Richard, the door closed and the taxi moving, before the ambushers had recovered.

"That's just the beginning. Thank me if we manage to get you off."

The atmosphere at West End Central had been as chilly as that of a morgue in which one of the supposed corpses had murdered the morgue attendant. The policeman in hospital, still unconscious, had been from this station; it had been impossible for the sergeant booking Richard to be impersonal, though he had done his best. Harry's presence had only been tolerated, not welcomed; he was just one of a dozen parents there to attempt to help their sons and daughters who had been arrested at the demonstration. Richard had known all the others who had been arrested; they were almost like a social club booking into an hotel for the night. They had been laughing and chatting, unconcerned even for their parents' concern, when Richard had been brought in; then abruptly their gaiety had fallen from them as they realised one of them was in real trouble. The boys had patted Richard on the back and one or two of the girls had kissed him, but his morose reaction to the gestures had shown that he knew he needed more support than pats and kisses. He and Harry had been taken into a side room, away from the hubbub of the others who had been charged, and there he had listened stiffly and silently to the charges that had been read out to him. He had politely declined to answer any questions, had expressed his regret that the policeman had been hurt, and had shown a dignity that he had never before shown in front of his father. Whatever he still thought of the fuzz, it did not show in his attitude now. Harry had stood surety for him and he had been released till he was to appear before a magistrate's court on Friday.

"Now if you wouldn't mind phoning for a taxi, Sergeant – " Harry had said.

The sergeant snapped the charge-book shut with a bang. "You will find plenty of taxis cruising around this time of evening, m'lord."

"I'm sure we shall, Sergeant. I'm also sure that my son and I, walking down the street surrounded by a horde of photographers, will create a traffic hazard that your men will then be called upon to disperse."

The sergeant bit back what he thought of the arrogant claim to privilege of some peers and even Richard looked at his father with an expression that suggested they should get out of the police station with a minimum of fuss. But Harry was not just interested in privileges: he knew what he asked was sensible. "I'll pay for the call, Sergeant."

"That won't be necessary," the sergeant snapped, and picked up the phone. "Sid, will you call a taxi for Lord Carmel? I *know*, Sid. We're all busy, busy as hell. Just call the taxi." He put down the phone. "Sid, Police Constable Murray, is Constable Roper's best friend, the man in hospital. They grew up together. But he'll get you your taxi."

"Thank you, Sergeant."

Now they were in the taxi turning out of Savile Row into Burlington Street. Richard looked back through the rear window and said, "Some of them look as if they're following us."

"They're sure to. You're going to have your name and probably your picture in the newspapers several times before this blows over. If it blows over."

"What will happen if that Constable – Roper? – dies?"

"If the magistrate or judge, depending on how far it goes, believes John Ferguson's testimony as a witness, that it was an accident, then you'll probably get off. But whether you get off or not, you have made some enemies in the police. They won't forget what caused their colleague's death."

"Oh Jesus!" Richard suddenly went limp, put his head back and closed his eyes. "It was all so harmless to begin with – "

"So many things are," said Harry, remembering the disasters of his own life that had started out in much the same way.

"I'm sorry for Mother – "

"So am I. But you should have thought of that sooner."

At that Richard lifted his head. "When did *you* last think of her?" He had found a way to distract himself from his own predicament. "Don't be so damned hypocritical. I think you're just here with me to protect your own image – "

I should hit the young bastard; but it is too late for that. And, of course, he is half-right. "Think what you like. But when you get into court – I'll be in New York, so I shan't be able to help –"

"You never can."

Another score: in family politics I seem to be a born loser. "I did not plan to be away at this time. You should choose your times better for getting into trouble." He couldn't help the asperity in his voice; but there was another note there, of disappointment. He sighed, relaxed, though he did not sprawl back as limply as his son had. "We're not getting anywhere going on with this acrimony. We have to stand together – " An echo stirred faintly in his ear: his own father talking to him at the time of Suez. How many other fathers in the generations of Aidans had said the same thing to their sons? "I'd like to be here, but it is too late at this stage for me to withdraw from this delegation. But I'll stand further surety for you if it's called for. I'd suggest that in the meantime you visit the hospital and see how the policeman is progressing. If he should die, make a point of going to see his wife or whoever his nearest relative is – go to his parents anyway. Do it discreetly, so it won't look as if you're courting sympathy through the newspapers."

"Protect my image, you mean?"

Harry kept his voice patient and cool. "That's exactly what I mean. But I also mean that a bit of Christian charity wouldn't go amiss just for your own education. You've had four years at Oxford, but it seems to me that your education is only really going to start in the next couple of days. You're against all the privileges that – forgive the term – our class is supposed to have. Now will be your opportunity to see how far you can go without them."

"I've already been handicapped by them. I'd have done better to-night if I'd been plain Jack Smith."

"Perhaps. It might have depended on how much respect the police had for plain Jack Smith's father. The real test will come when you get into court. Titles may not mean much any more, but most magistrates are middle-class men and the middle class have an in-built respect for men of property."

"Jesus Christ!" Richard shook his head in disgust. "You sound so bloody smug and pompous."

"One always does when one is pointing out the facts of life," said Harry. "My father sounded exactly the same to me. Here we are. Do you have your key? As soon as we pull up, bolt straight inside. I'll handle the reporters."

The two cars carrying the newspapermen arrived only moments after the taxi pulled up; but Richard had already escaped inside. Harry paid off the taxi-driver, then waited for the newspapermen to scramble out on to the pavement and surround him. "My son will not be available for interviews, gentlemen. He, and I too, regret deeply that a policeman has been hurt in the course of his duty. Whatever else is to be said will be said in court on Friday."

"Will you be there, sir, to stand by your son?"

"I shall certainly be standing by him, but not physically. On Friday I shall be in New York attending a United Nations session."

"How does your son feel about that, sir?"

Ask me how I feel about that question. But he held rein on his temper and said coldly, "One of the reasons for my son's being in Grosvenor Square is his belief in the United Nations."

"Does that mean you are also in favour of the U.N. and against the United States on the Middle East oil crisis?"

I must be tired or I'm losing my grip: the tongue really slipped up there. "It means nothing of the sort," he said, and tried to sound convincing. "I am going to the U.N. with a committee of observers. You will learn of our opinions in due course, just as you will learn of the decision of my son's case in court. Good night, gentlemen."

He let himself into the house, a photographer letting off a last flash like a small boy throwing the last derisory firecracker. He closed the door behind him and leaned back against it for a moment, feeling suddenly very tired and a little sick. He looked down at his hand on the doorknob and was surprised to see the fingers trembling. He had been under a great deal of pressure, emotionally, politically and commercially, since his father had died; but he had not noticed any build-up of exhaustion, though this afternoon he had felt unaccountably tired. He heard a foot-step and when he looked up towards the door into the drawing-room he had to adjust his gaze: even his eyes were out of focus. Constance stood there, a sudden look of concern on her face.

"Are you all right?"

He nodded, straightening up. "Just letting my temper simmer down. Newspaper reporters are not my favourite voters."

"I suppose it will be in all the papers to-morrow?"

"Everything but the *Daily Racing Form*. Just let's hope the policeman doesn't die during the night."

"Richard told me he's still unconscious. I'll pray hard to-night."

"Do that." He had no faith in the efficacy of prayer; but nothing else the Aidans might do could help the policeman recover. He followed Constance into the drawing-room. Richard, a glass of beer in his hand, Robin and John Ferguson were there. "Well, that's it for the time being. From now till Friday we all assume what the commentators call a low profile."

"Except for me this evening," said Constance. "I still have to go to my premiere. I can't back out, not this late."

"Are you going with her, Daddy?" Robin asked.

"Your mother understands why I can't." He sank down in a chair, glad of its support.

"Then I'll go," said Richard, putting down his glass with a thump. He was nervous and restless and abrasive, not wanting to be left alone to ponder on how the future would treat him. He was afraid, for himself and for the young policeman possibly dying in the hospital where he had been taken; he was not a callous nor a selfish boy and all at once his causes and beliefs were nothing beside the two fates that an accident had, at any rate for the present, inextricably locked together. "I'll even get myself all dressed up in the approved gear – "

"I appreciate the offer," said Constance. "Especially the sacrifice of wearing a dinner jacket. But your father is right – we should assume a low profile. Especially you. If you went to the premiere with me to-night everyone would see it in the wrong light – the policeman still unconscious in the hospital and you at the cinema enjoying yourself. I'll be perfectly all right by myself."

"Don't think me presumptuous, ma'am," said John Ferguson, "but perhaps I – ?"

Then Ferguson came into the room, a drink on a tray for Harry. The latter suspected the butler had been poised outside in the hall for the proper moment to make his entrance: it was too fortuitous, too much on cue. Ferguson, always aware of the priorities, held the tray out to Harry and the latter took his drink. Only then did the butler say, "May I speak privately with my son, m'lord?"

"You may, Ferguson. But if it is to reprimand him for his presumptuousness – *don't*. I don't think he was being presumptuous and I'm sure Her Ladyship thinks the same. He was only being gallant and helpful. And to-night, God knows, we appreciate both gestures. Lady Robin will accompany her mother." He looked at Robin, who was abruptly ashamed that she had not thought to volunteer. "You'd better get dressed. Your mother can't be late."

Robin hurried out of the room and Ferguson, with a glance at his son, retreated to the back of the house. John Ferguson hesitated, then excused himself and moved to follow his father. Harry stood up, put a hand on the young man's arm.

"I shall have our solicitor here first thing in the morning. He can take your statement on what happened this evening outside the Embassy and then advise us what to do. And thank you – not least for your offer to escort Richard's mother."

"It was presumptuous of me, sir," said John. "But I still think of myself as one of the family – even after all these years." He looked at Richard. "Let me know when you want to go up to the hospital. I'll go with you."

Then he went out of the room and Constance said, "If he does think of himself as one of the family, then I think we should be proud of him. He is a gentleman."

"I could have told you that years ago," said Richard.

Constance tried to hide her disappointment in her son but failed. All the time he had been up at Oxford she had done her best to understand his espousal of radical causes; in the circumstances of the times and his attitude towards his father, she supposed she could expect little else. He had never brought any of his radical friends home to either Eaton Square or Aidan Hall, but on her occasional visits to Oxford she had met some of them and been aware of their only slightly disguised contempt for her position if not for her. The young, she had thought, had never had any talent for hiding their opinions of their betters; then had chided herself for thinking something that sounded like one of Harry's opinions of the young. It seemed that she had really become an Aidan, while her son was doing his best to forget that he was one. But that was not the reason for her disappointment in him this evening.

"Then perhaps you had better take some lessons from him. You sound to me as if you are going to spend the next few days wallowing in self-pity. That won't help you when you get into court on Friday."

She swung round and left the room, her gown hissing as the folds of silk rubbed against each other. Richard stared after her, his face falling apart in shock. He's chosen the hardest route of all to take, Harry thought, and he knows in his heart already that he's never going to make it. He doesn't have the guts his mother has and, if I have any at all, he doesn't have as much as me. All he has is spite and, as his grandfather found out, that's not enough.

Harry took off his jacket, pulled off his tie. He suddenly realised he had a free evening ahead of him, given him by his son's misfortunes. "I don't feel like a heavy dinner. I'm going to have a tray brought up to me in the study. Do you want to join me?"

"I don't feel hungry – " Then Richard looked at his father and seemed to realise he was being offered more than an invitation to share some food. Not ungraciously, putting the cracked pieces of himself together again, he said, "No, I think I'd better have something. I may need all the strength I can muster."

"Better fed than dead," said Harry, then screwed up his mouth. "Sorry. That was tasteless and a bad joke, besides."

Richard suddenly smiled. "It's all right, Father. We all try too hard when we're trying to be nice. I may sound just as corny before the evening is over. Now what would you like? I'll tell Ferguson."


4

Richard and John Ferguson went up to the hospital at ten o'clock and while they were there the injured policeman regained consciousness. By morning he was out of danger and when Harry left for New York Richard's situation was not as serious as it might have been. The charges were still being pursued, but a newspaper reporter, contradicting Harry's opinion of the breed, had come forward to corroborate John Ferguson's story that the incident had been an accident.

"Things look a little brighter," said Constance. "I'll phone you at your hotel as soon as we know the verdict."

"Will you go to the court?" Harry asked.

"I think I should, don't you? I'll go with Robin. Don't worry, we can take care of whatever eventuates."

"I'm sure you can. I just hope Richard can."

"At least this has brought you and him a little closer together."

Harry nodded. "I must try and see that it stays that way." He paused, looked at her obliquely. "Has it brought us any closer?"

"We're still his parents," she said non-committedly, but he felt, rather than saw, the slight stiffening in her. "You look tired. Take care of yourself in New York. I'm told it's a very exhausting city."

Then the Daimler from Whitehall called for him and he went out and got into it and rode out to the airport, wondering how a city could possibly be more exhausting than the folk dance that was domestic relations.

# Nine

But New York *was* an exhausting city. And somehow it suggested that it was an impermanent one: there seemed to be as much demolition going on as there was construction: wreckers and builders passed each other like shift workers clocking off and on. *Build for the Future!* shouted a fifty-foot high sign, while beside it the wrecker's huge iron ball swung back and forth against the walls of a perfectly good building like the pendulum of time itself. Change was progress and progress was change: yesterday had to be eliminated.

"But there is a growing revolt against that," said Lynne Erickson. "A lot of us want to remember yesterday, even the yesterday of our parents. My folks are going to-night to see a musical comedy that *their* folks saw years ago. I saw it last week and loved every minute of it."

Lynne Erickson was the nicest aspect of New York so far, Harry thought. Serendipity had been a favourite word with sophisticated Americans a few years ago, he remembered, when the hippie cult had been flourishing and accidents of joyful discovery had been the innocent pleasures that Americans were then seeking; it was not a word one heard so much these days, not now when Americans, pessimistic and afraid, were past the point of depending upon accident for their relief and escape; their retreat to nostalgia for the past was as planned as anything they had once programmed for the future. Lynne Erickson might be just another good-looking girl to any male New Yorker, but to Harry she had been an accident of joyful discovery, one who had begun to take the grit out of his increasingly irritated opinion of New York.

He had met her his second evening in the city, at the reception given at the Hotel Pierre by one of the African delegations. He had already learned that the most lavish receptions were usually given by the countries that could least afford them, paupers who believed that insolvency was a ridiculous myth that was to be disproved. With *coups* part of the way of life back home, with the possibility of every mail carrying notice of their recall, the delegates of such countries could hardly be blamed for enjoying their own largesse. The biggest eaters and drinkers at such receptions were often the hosts themselves. Yesterday, for them, was quite frequently already disappearing forever: nostalgia did not come on their expense account.

"Why shouldn't I retract what I said?" a cocoa-coloured man was saying; his government had been toppled that morning and he was waiting anxiously for the next cables from home. "I'd rather eat my own words than someone else's."

"If all her wealth was between her legs," the South American bachelor was saying, "she'd be bankrupt. I've never met such a frigid woman. No wonder her husband has turned queer."

"The Americans, of course," the Frenchman was saying, "mistake advertising slogans for a foreign policy."

"The French, of course," the American was saying, "mistake *je ne sais quoi* for a philosophy."

The united bitchiness of the United Nations floated around the room like cigarette smoke. Harry, sipping a martini that tasted like dry ice, wondered again at how much was achieved at functions such as these. But from similar gatherings in London he knew their value: quips, gossip and rumours were among the threads that made up the fabric of politics, domestic or international. An expert observer of the effects, he had quickly felt at home at this reception. But as at home he did not contribute to the gossip or rumours and so far he had not had the opportunity to crack a quip.

"Lord Carmel – " Andrew Mbuli was a muscular middle-aged Kenyan now running to fat; his ex-BBC voice had the added richness of an African drum. "I understand you were out in Kenya when we were having a spot of bother with the Mau Mau. Why have you never paid us another visit?"

"Too many memories," said Harry; but to be honest, he had almost forgotten Janet Buck now. "Tragic ones."

"I understand. We all have memories we'd like to forget. Including my friend here, Mr Roth."

It's all been stage-managed, Harry thought as he turned to greet Mordecai Roth. Yet he knew that at this level the world was no larger than at parish level: sooner or later he would meet Roth again. But coincidence did not have to be so cruelly cynical.

"Lord Carmel and I are old – acquaintances." Roth had reached the verge of being an old man, yet Harry knew he must be barely sixty. The few wisps of hair along his temples were white and he had the bent back of a man stooped under too many disappointments. By now he should have been an ambassador, but Harry guessed he must be still no more than one of the second-level officials in the Israeli delegation. "How is Lady Carmel? I have seen her picture several times in English magazines. You are very fortunate to have such a beautiful wife."

"Thank you." Do you ask a cuckolded husband how *his* wife is? Harry took a chance: "How is Mrs Roth?"

"Very well. She is over there." Harry looked in the direction of Roth's nod, hoping desperately that he was not also going to have to face Deborah Roth. But he could not see her and Roth said, "The grey-haired woman in the blue dress."

It was not Deborah, and Harry, wondering what had happened to her but too careful now to ask, looked back at Roth. "She looks charming."

"I'd introduce you, but unfortunately she is anti-British. Wives are a handicap sometimes for a diplomat, don't you think?"

"Mine is," said Mbuli, "but it wouldn't be worth my life to have her hear me say it. She cannot understand why honesty is not the best policy." He smiled, all shining teeth and high-lighted cheeks. "That's your English Protestant missionaries, Lord Carmel. Myself, I was educated by the Jesuits."

"So was I."

"Then you understand." They both looked at Roth, the Jew who did not understand.

Then someone from a costume ball, all blue silk burnous and diamond rings, joined them in a puff of words. "I am sure to lose face with the Algerians and the Iraqi," said Sheikh Hazza, "but I simply *must* greet my old friend, Lord Harry, even when he is in the company of an Israeli and – where do you come from, my dear sir?"

"A neutral country," said Mbuli, still smiling broadly. "I'll leave you to exchange greetings while I go over and exchange insults with my South African friends."

He moved off and Hazza looked after him. "How innocent and cheerful they all are. Just like big piccaninnies."

"You'd still be selling them as slaves if you had your way," said Harry.

"No need," said Hazza, spreading hands that carried a fortune on the fingers. "We have our oil. Do you think we'd be here, with all this fuss, if it were just a matter of a few slaves?"

Roth's thin face twisted with disgust. Ignoring Hazza, nodding only at Harry, he turned his back on both of them and slipped away into the crowd, lost almost at once, as if crowds were his natural habitat. Hazza did not bother to look after him. "Jews are too sensitive. They think history is a personal insult to them."

"Perhaps they've been insulted more than most," said Harry mildly, though he was glad that Roth had gone. He became aware of a tall, good-looking girl standing just behind Hazza and looking at them with a frank amused stare. She smiled at him when he glanced at her and he smiled back, wondering who she was. Then he looked back at Hazza. "How do you think the debate will go?"

Hazza shrugged. He had put on a lot of weight over the years and no longer suggested the desert hawk he had once been. He had advanced self-indulgence from a vice into a depravity and his hook-nosed, heavy-jowled face was now revoltingly ugly. He breathed heavily and he kept putting his hand to his mouth as if troubled by indigestion; his long blue robes hid a lot but they could not disguise the big swell of his belly. Harry wondered when he had last been on a horse or a camel or whether he now did all his hunting in an air-conditioned Rolls or Cadillac.

"It will just be a lot of words, Harry. The biggest factor for all of us, the Arabs and the Israelis, are the Palestine guerrillas. And they are not represented here. So what is the use of all the words?"

"Their territory, if they still had it, is a long way from the Sinai."

Hazza leaned closer, dropped his voice. "They are putting pressure on me and the other oil sheikhs, just as much as they are on Hussein. They are running out of money and they know if the Israelis develop their oil find, then Israel will have all the money in the world to buy what it wants. The Egyptians are fed up with the war and so are the Saudi Arabians. Frankly, I couldn't care less about the Palestinians – but don't quote me. They pay me periodic visits in Qaran, asking for more money, but I am getting tired of it. I never minded supporting Nasser – I liked him very much – but these chaps – " He shook his head. "I owe them nothing."

"Don't be too rude to them. We don't want them blowing up everything we have in Qaran."

The tall blonde was still standing behind Hazza, still gazing at the two men. She smiled again at Harry when he looked over

Hazza's shoulder at her, but neither turned her head nor moved away, just continued to stand there like some frankly inquisitive schoolgirl. Hazza caught Harry's look, turned his own head and exclaimed, "Miss Erickson! Oh, my manners! I keep forgetting that in America women are first-class citizens. In my country they come behind with the donkeys and the camels."

"You just make my job harder, Sheikh Hazza," the girl said. "Especially if I have to sell your image to some of our newspaper-women. Not to mention Women's Lib."

"Do not mention it!" Hazza threw up a jewelled hand in horror; then he turned the hand into a gesture of introduction. "My very dear friend the Marquess of Carmel. My public relations lady Miss Lynne Erickson. The Prophet must be turning in his grave at such a thought!"

Lynne Erickson gave Harry her hand, firm and surprisingly small considering her height. She was almost as tall as he, built on lines that suggested to Harry that she should be stuck on the prow of a ship baring her bosom to the breeze. Her hair was very blonde and was offset by her tanned skin; her very appearance did nothing for the image of Hazza, only pointed up how epicene and unhealthy he looked; Harry could imagine her doing half an hour's brisk exercise every morning before breakfast. He returned her handshake perfunctorily, dismissing her as a probably aggressive career Valkyrie.

"I don't think the Marquess approves either," she said to Hazza.

"Hah, that is just the English understatement. Take no notice – or still better, take care. Lord Harry has a dreadful reputation with the ladies in London. Englishwomen are the most forward in the world these days, my student spies tell me, and Lord Harry takes advantage of every one of them."

"You are going to be kicked right up the date palm in a moment," said Harry, abruptly irritated by the gossipy Qaranian; Arabs loved gossip, he knew, but Hazza by now should be above bazaar talk. "If you want to stay my very dear friend – "

Hazza held up an apologetic hand; diamonds winked – humbly? "A thousand apologies, Harry. A joke, that was all. Miss Erickson, stay with Lord Carmel and repair my image with him. I see the Saudi Arabians are waving to me. Ah, if only I had their riches!" He sighed, belched behind his hand and wafted away.

"Is he a faggot?" asked Miss Erickson.

Harry looked at her with new interest. "Is that the way American girls talk about their bosses?"

"When you're in public relations you don't only have to know what you have to sell, you also have to know what you have to hide. I'm not sure about him. One minute he's pinching my ass and the next he's floating around as if he's dancing the lead in *Swan Lake.*"

"I suspect he is double-gaited. He's certainly more effeminate than when I first met him years ago."

"It's pretty common with them, isn't it? Homosexuality, I mean. I seem to remember a famous poem about a young boy on the other side of the river – 'he has a bottom like a peach, but alas I cannot swim.' Though maybe I shouldn't say too much. God knows, there are enough faggots in New York to start a brush fire, if you'll forgive the pun."

"Do they stand on the bank of the East River and pine for what they see on the other side?"

"I don't think there are too many peach-assed boys in Brooklyn. How did we get into this conversation?"

"I don't know. But let's get out of it. I don't usually spend my time with women discussing homosexuality."

"What do you discuss with them?" She looked at him quizzically. "You are quite human, you know. Americans have some idea that foreign aristocrats – " she pronounced it a-*rist*-ocrats, a word that was itself foreign to Harry " – are something like Martians. Little pointed heads, their own dialect and a way of making love that's different from us ordinary earthlings."

"The pointed heads bit is wrong. I don't know about the other idiosyncrasies." He decided to be a little rude because he was making no headway with her by being polite: "Do you ever do any public relations for yourself or do you just ignore your own image?"

"Score one for you," she said, and though she smiled he had the feeling he had hurt her and he was instantly sorry. "I gave up trying to sell myself years ago, when I was at high school. I was this size when I was fourteen. I was cheer-leader at our high school football games – imagine a cheer-leader who was bigger than some of the players. Twice I was tackled by the opposing team by mistake – " Then she stopped smiling. "Am I talking too much? That's another of my faults."

"One of my faults is being unnecessarily rude. A woman once told me I had a ruthless indifference for other people's feelings."

Never on such short acquaintance with anyone had he been as candid about himself as this. But Miss Erickson invited such candour; he felt he owed it to her now she had revealed how vulnerable she was. "I apologise. If Hazza will release you, perhaps you'll have dinner with me?"

Hazza, hand resting on the arm of a young Greek, was only too willing to let Miss Erickson go. "My dear young friend here is going to introduce me to the pleasures of Greek food. Enjoy yourself, Miss Erickson. Tell Lord Carmel what a marvellous boss I am and convince him that the Arab cause is the only honest one."

As Harry escorted Lynne Erickson out of the room he saw Roth and his wife watching them, the husband with cynical contempt, the wife with a cold hard stare that was unreadable: had Roth told her of how he had been cuckolded and by whom?

At the door Bert Wardle, leaning heavily on his stick, nodded at Harry. "Escaping, eh? Wish I could. This damned standing around kills me. I've already met Miss Erickson. Pity she doesn't have someone better than Hazza to sell. Have you heard the news? George Brewster has sprung an election for the first week in July. He announced it in the Commons to-night. Doesn't give your crowd much time to get their message across. If they have one," he added with a sly grin.

Six months ago Harry knew he would have a felt a lift of excitement at the news; elections were always the most exhilarating part of politics. But it was all behind him now and all he felt was a momentary stab of regret. "How are Argus and the others taking it?"

"Already writing their campaign speeches, I imagine." Wardle smiled at Lynne. "Lord Carmel and I don't have to worry any more. We're just interested spectators with free seats. Well, I'll give this shivoo another five minutes, then I'm buzzing off too. You don't get decorated for attending things like this, not unless you are in the Foreign Office."

As they waited for a taxi on the front steps of the Pierre, Harry said, "Would you mind if we went back to my hotel first?" He saw her eyebrow go up and he shook his head. "Nothing like that. I have to telephone my wife."

"Well, that sounds safe enough. I've never been out with a guy before who checks in first with his wife. Is that an aristocratic code of behaviour?"

He was liking her more and more. "It's what they mean by *noblesse oblige*."

The committee was staying at the St Regis. When he picked up his key at the desk there was a note with it to say that Lady Carmel had called from London at 6.30 p.m. Up in his room Lynne went into the bathroom while he picked up the phone and asked to be put through to London. The operator rang back as Lynne came out of the bathroom: a long-distance call across three thousand miles of ocean took no longer than the call of nature: the world had become one vast suburb. Lynne gestured silently to ask if he wanted her to go back into the bathroom while he talked to his wife. He shook his head and smiled at the thought of such a retreat to allow him privacy. Then Constance was on the line.

"Harry? I rang earlier – "

"I got the message. I was at a reception. How did Richard go?"

"He was very lucky. The magistrate believed John and the newspaper reporter and he let Richard off the serious charge. But he fined him ten pounds and costs for creating a disturbance."

"Did you go to the court?"

"Yes, both Robin and I. We're in the *Evening Standard* to-night, looking like a couple of well-dressed whores who have just been fined for soliciting – we're coming out of the court with four or five policemen escorting us. They didn't bother to run a picture of Richard. He thinks it's a great joke."

"How is he, other than jokey?"

"I think he's learned a lesson. Whether it will be permanent remains to be seen. He and Robin and John have gone out to dinner to-night to celebrate."

"What are you doing?"

"I'm in bed, reading. What are you?"

He glanced across at Lynne, who was sitting in a chair turning over the pages of a magazine in which she obviously had not the slightest interest. "I'm just about to go out to dinner."

"Alone?"

She sounds like a *real* wife. "No, with Bert Wardle. You heard the news about George Brewster calling an election?"

"There's another piece of news you would *not* have heard – John Ferguson is putting himself up for nomination as the Labour candidate for Bayard. Ferguson, *our* Ferguson, is on the verge of a stroke."

"I'm just glad Father is not alive. He'd have *had* a stroke."

"How do you feel about it?"

"I don't know. Anyone would be better than Bedley – he'll get the Conservative nomination. I suppose if John were accepted as the Labour man and by some fluke managed to get in, well, in a way he'd be keeping the seat in the family, wouldn't he?"

"I'll try that theory on Ferguson, but somehow I don't think it will have much effect. Good night. Enjoy New York. Some day I may take a trip there."

"Do that, my love," he said, and hung up, thinking how easier it was to feel closer to your wife at three thousand miles than when you lived in the same house as her.

Lynne put down the magazine. "Why did you lie to her when you said you were going out to dinner with Lord Wardle? I hope you're not going to turn out to be a sonofabitch like so many others I've met in this game. I'm not a home-wrecker, though I may look big enough – "

"Please don't keep harping on your size," he said irritably. "If I hadn't thought you were attractive I shouldn't have asked you to dinner. Unattractive women bore me."

Lynne shook her head in mock wonder. "Well, I've had some oblique compliments – Okay. I'm attractive. But that still does not alter the situation. In certain ways I am just as much a lady as Lady Carmel – "

"I'm sure you are. But if the situation worries you, I'll put you in the picture. I don't think I should have bothered if you hadn't brought it up, but you seem to call for more candour than any woman I've ever met."

"It's my wide-eyed American morality stance."

They got to know each other a little more that evening at dinner and again on the Saturday. The day itself was spent at a barbecue luncheon at the Long Island estate of one of the American delegates. They drove back to Manhattan in the chauffeur-driven Lincoln he had rented. Although he had been in New York only three days he was already tired of being constantly in the company of Wardle, Argus and the other three M.P.s. He made his contribution to the meetings they held and he was conscientious in his observation of the debates and the informal gatherings he had attended; but he had never been a group man, not even when on committees in the Commons, and he saw no reason to start being one now. Wardle and the others might think he was stand-offish, but what they thought did not worry him. He preferred Lynne's company and he was going to have it.

They went out to dinner again on the Saturday night and Sunday she took him on a tour of Manhattan, showing him her city, the one that she was proud of but for which she wept – "They are killing it."

"Who are *they*?"

"Us, I suppose I really mean. We're all to blame. It could have been a beautiful city, Nineveh on the Hudson. All other great cities, London, Paris, Rome, at least reached their prime. But New York never has and never will."

"Perhaps you're too pessimistic."

"Maybe. I hope so. But if it ever fulfils its promise, it won't be in my time."

She took him back to her apartment in the East Seventies for supper and later they went to bed, as naturally as old lovers. Naked, she was beautiful, voluptuous and graceful; it was as if by taking her clothes off she had discarded the yardstick that made her feel big and ungainly beside other women. She was the first woman he had had since his father's death and he was hungry for her; uninhibited, she denied him nothing of herself. Afterwards they lay exhausted, she lying in the crook of his arm while her gentle hand traced patterns on his chest and stomach.

"Is that what they call *droit de seigneur*? No wonder medieval women never complained when the baron sent down for a bit."

He wished she would not keep referring to his class. "My love – "

Her hand paused and he felt the pressure of her nails. But her voice was only mildly reproachful. "Don't call me that – that's what you called your wife. I'm not your love, Harry. Not yet, maybe never. You don't have to flatter me with lies."

He had not felt so at ease and content with a woman since – well, since the early days with Constance. When he had called her *my love* it had been like an old sweet taste on his tongue: for the moment at any rate he had meant it. For the first time in years he felt the *need* of love; and she was the happy accident, occurring at just the right time. "Whatever else I may say or do, there will be no lies. I promise you that."

"We *could* fall in love," she mused, lying on her back now and staring up at the dimly lit ceiling. Somewhere outside in the dying city a siren wailed, a cry of pain, but here in the comfort of his arms and in this bed she felt the beginnings of a happiness

259

that she had always felt would elude her. But experience had taught her to be cautious and even the improbability of the situation made her slightly incredulous. Lynne Erickson, the Swede from Brooklyn Heights, in bed and in love with an English lord and a rich man to boot. Pinch my ass and tell me I'm not dreaming. "But let's do it gradually. Just so's we can be sure."

2

Two days later the debate in the United Nations was adjourned while the main parties went home for further instructions from their governments – "that means," Harry said, "that someone from the other side has come up with an argument they don't have an answer to."

"I was going home anyway," said Michael Argus. "More important things are happening at home."

"You don't have to worry," said Bert Wardle. "Your majority is safe. You could spend the entire election campaign on the beach at St Tropez and that bunch of fossils in your constituency would return you. I envy you chaps. I always had to battle for every bloody vote I got. But not any more, eh, Harry? We can just sit back now and watch."

"Yes," said Harry, once again feeling the pang of regret and, despite what Wardle had said, recognising the same regret in the other man's eyes.

Lynne did not go out to the airport but came up to his room at the hotel. "One thing public relations teaches you is to be discreet in your private relations. When will you be back?"

"I don't know. A lot will depend on what happens in our elections. Is there any chance of your coming to London?"

She shook her head ruefully. "Not unless I can persuade Hazza he needs a good image in London too. I'm a working girl – I have to stay where the work is."

He hesitated, then said, "I could find something for you to do."

She shook her head again, emphatically this time. "No, Harry. I pay my own way till I've worked out everything about you and me."

He took her in his arms, kissed her gently. "You are the best thing that has happened to me in years. It can't end here, in this room, as if we were just a couple of strangers having an affair on a holiday."

She kissed his nose. "I love that broken nose." Then she pressed her face into his neck and muttered, "I love you, blast you."

"I'll come back," he said, determined now to do so. "If I can't get them to send me over on government business, I'll find a reason to come over on private business. I have interests up in Canada – they can bring me over. Then afterwards we can fly down to Jamaica, I have a cottage at Montego Bay – "

"I'll have my bag packed ready and waiting," she said, but in her heart she knew this was good-bye. She drew back and he saw that she had been weeping, something that surprised and pleased him: it was a long time since a woman had wept over him. "Will you tell your wife about me?"

"No," he said, wondering, like a man, why women always wanted to spoil a moment. "After all these years she's not interested in whom I'm in love with."

"Are you kidding?" she said, wondering, like a woman, why men could so often spoil a moment by brutal frankness. "Never mind, go on believing that, Harry, if it relieves your mind." She stood back, appraising him. "I must say I've been out with better-dressed men. Isn't it time you got yourself a new suit?"

"Whom are you in love with – the man or his tailor?"

She smiled and kissed him again. "Good-bye, darling."

"Good-bye, my love."

On the plane, when they were settled in their seats and had been served drinks, Bert Wardle, seated next to Harry, said, "Did you enjoy New York, then?"

Harry nodded, cautious as ever about his privacy. "But I shouldn't want to live there."

"Me, neither. It was a place I used to dream about when I was young. I almost went there in the late Thirties. Who knows, I might've been a tycoon by now."

"Not you. You're too much of a political animal. A senator, perhaps. Why didn't you go?"

"The wife wouldn't leave London. In the end they're the ones who really run your life, aren't they?" He buried his face in his drink, then lowered the glass and looked sideways at Harry. "Sorry. That wasn't a very politic remark, was it, for a political animal."

"No," said Harry, and turned away and looked out the window, wondering how, after all these years, Constance would react to his approach to her for a divorce. He felt suddenly tired, exhausted not just by New York and Lynne but by an accumulation of loneliness.

Richard, alone in the Bentley, met him at the airport. "There are government cars here, if you want to go back with the others?"

"A week of them has been enough," Harry said, getting into the front seat beside his son. "It was thoughtful of you to rescue me from them."

"That wasn't the real reason, Father." Richard was not a good driver, not concentrating, cutting in front of other drivers who saluted him with an abuse of horns; but Harry, though his feet were braced hard against the floorboards, said nothing. "I just wanted to – well, thank you. And explain that even though I'm not going to change my political views, I'm still grateful."

"I'd never respect any political views that were changed only out of gratitude. Are you still a Marxist?"

Richard smiled, and with the intensity dropping from his face he looked remarkably attractive. Harry wondered if he was interested in girls and if he was having the same success with them as he himself had had at that age. Richard took the car up on to the M4 motorway, pulling into the speeding traffic without a backward glance. He may be all for the proletariat, Harry thought, but not when he's behind the wheel of a car.

"I never really was a Marxist, not one hundred per cent. But I suppose Grandfather would have thought just as bad as one – I'm a Socialist. I'm going to work for John Ferguson in his campaign." He took his eyes off the road to look at his father. "Will you mind?"

"I'll mind more if you drive us up the back of that lorry in front." He waited till Richard had braked sharply, swung out into another lane in front of another car coming up behind them; then he went on, doing his best to sound calm and unconcerned. "It's a free country – "

"You don't really believe that, do you?" Richard said; then raised a hand from the wheel, allowing the car to drift within inches of the lorry they were passing. "No, go on. We'll talk about that some other time."

"If you are going to work for John, do him a favour and don't write his speeches for him. If you try to tell a rural Englishman, whether he's a Tory, Socialist or Communist, that he doesn't live in a free country, he'll run you out of it. They're a different breed from what you've known up at Oxford. No, I shan't mind your working for John. All I ask is that you don't get too personal about the last incumbent. For all my faults, I don't think I did a bad job. Personally, I think you'll be backing a loser. Bayard just isn't Socialist territory."

"You may be in for a surprise. You're in for another surprise too." Richard drove carefully for a hundred yards, concentrating on the road as if reluctant now to say anything further. Harry waited patiently, wondering what news about Constance was about to surprise him: how ironic it would be if she were going to ask him for a divorce. At last Richard said, "Robin has gone overboard for John. And he's that way about her. You never saw such a couple of lovebirds."

It was Harry's turn to ride in silence. Then he said, "It's a bit sudden, isn't it?"

"In a way it's not – they've known each other for years." Richard went back to perusing the road ahead; he seemed to be a more careful driver when he had something on his mind. "I gather you and Mother didn't waste any time."

Was that some sort of criticism? But all he said was, "How does your mother feel about it?"

"I can't work out whether she is speechless or approving. The only one with any definite reaction is Ferguson."

"Who, I take it, is speechless?"

"And apoplectic. He's not speaking to John at all. A Socialist running for his master's late seat and going out with the master's daughter into the bargain – he can't understand what sort of monster he's sired." He took his eyes off the road again but this time fortunately they were in the clear on the motorway and had the road to themselves. "You haven't said how you feel about it. You're not going to be stuffy, are you?"

"I can hardly afford to be, can I?" It was time he tried some candour on his son.

"No," said Richard, equally candid.

It was the first moment of complete truth between them since Richard had been a child and each of them was aware of it. They looked at each other, neither of them smiling but each of them acknowledging that they had reached a new level of understanding. Richard speeded up the car and Harry shoved his feet harder into the floorboards.

When they drew up outside the house in Eaton Square Ferguson opened the front door. Imperturbable as ever, he welcomed Harry back and took his bags. "A pleasant trip, m'lord, I hope."

Richard had taken the car round to the mews garage, so master and butler were alone for a moment. "I understand you haven't had a very pleasant week, Ferguson."

"No, sir. I assume you have been informed of the dreadful state of affairs that has arisen. I don't know what your father would have said – "

"Ferguson, come upstairs to the study."

As they came to the landing outside her bedroom Constance opened her door. Harry took her hand and pressed it. "I'll see you in a moment. Ferguson and I are going to have a talk."

Ferguson had gone on ahead to the study. Constance waited till he had disappeared, then said quietly, "Be diplomatic."

"Are you in favour of what's happened?"

"I'm in favour and I think you will be when you see Robin. I've never seen her so happy."

"Then that's all there is to be considered."

He went on into the study and closed the door. Ferguson stood in front of the fireplace, both hands leaning on the mantelpiece, eyes staring down into the empty grate. It was the first time Harry could remember ever seeing the butler lacking in composure; he looked like a man whose life had been shattered. But as he heard the door close Ferguson straightened up, regaining his composure at once.

"Sorry, m'lord."

"Would you care for a drink?"

Ferguson hesitated, then nodded. "Allow me." He poured two Scotches, careful that his own drink was slightly smaller than Harry's. "Your health, m'lord."

Harry knew that Ferguson would not accept an invitation to sit down, so he sat himself on the arm of the leather chair where he usually relaxed. He sipped his drink, only now beginning to see the situation that lay ahead. "Things look like becoming a bit awkward, don't they?"

"To say the least. I know my son – and perhaps Lady Robin – will think I'm stuffy and old-fashioned for opposing what's going on – I presume we are talking about *them*, m'lord, and not the election business?"

"I think you and I can bear up under the strain of John's running for election. He will be defeated, so that situation will be out of the way in a couple of months." He took another drink.

"Ferguson, I have to tell you this. My wife does not object to the relationship. She thinks my daughter's happiness is more important than any awkwardness that might result between you and me. And her, of course."

"I should have to resign, m'lord."

Harry sighed. "I see your point. But you enjoy working for the family, don't you?"

"It has been my whole life. I can't imagine myself doing anything else." So his life had indeed been shattered.

"How old are you?"

"Fifty-eight, m'lord."

"You have another good ten years in you." Then he smiled. "Sorry. I'm talking as if you're some sort of bloodstock."

"Good butlers are, m'lord." Ferguson smiled and Harry looked up at him in surprise; the butler's flashes of humour were as rare as his gaffes. Suddenly he felt more confident that the situation could be resolved.

"We'll play it by ear for the time being, Ferguson. The romance may die of its own accord – if it does, then we can soon put it at the back of our minds. If it doesn't – " He sighed again, took another drink. "Then perhaps we shall have to bow to youth and progress."

"I never imagined the Permissive Society would go this far."

"You're sounding stuffy again."

Ferguson returned his smile. "Sorry, m'lord. Life does get complicated even in the best of households, doesn't it?"

Harry stood up, put down his glass. He wondered how much more lay behind the butler's blithe remark. "It certainly does. Have the estate tax men been to see you again?"

Ferguson picked up Harry's glass, put it with his own empty one on a tray; he was once more the good and tidy servant, no longer a possible in-law. "They came to see me and Mrs Dibbens last week, m'lord, down at the Hall. They are persistent, but I think we foiled them."

"I suppose *foiled* is the right word. I'm sorry I've embroiled you in this. Perhaps it wasn't worth having you and Mrs Dibbens perjure yourselves."

"They haven't put us on oath."

"If they do – ?"

"We'll face that when we come to it, m'lord. As we'll face the other situation if we have to."

He went out and two minutes later Constance came in. With

her was Robin, looking subdued but – yes, he had to admit it – happier than he had seen her since she was a child.

She kissed him, then put her hand in his, something she had not done in years. "Are you angry?"

"Why should I be?" His surprise was genuine. My God, they know so little of me. "All your mother and I want is for you to be happy – "

"I am – fantastically happy."

"So am I," said Constance.

The happiness of the two women, *his* women no matter how tenuous his hold on them, communicated itself to him. Oh Christ, why did I spoil it all? Suddenly he felt the exhaustion come back on him, but he did his best not to show it. "Let's all go out to dinner. Is John in town?"

"No, he's down at Bayard – he goes before the Labour committee to-night to see if he gets their nomination."

"I thought you'd be with him."

"I advised her against it," said Constance. "There might be some committee members who would resent her relationship with him. They might think she was some sort of fifth columnist planted by the devious Tories."

Harry looked at Robin. "Take notice of your mother. She is a first-rate political wife."

Then he glanced at Constance and saw her face stiffen. But she did not add to the sudden awkward moment. "We'll have a family night out – Richard can come too. It's a pity Matthew isn't home from school – I can't remember when we all last went out together."

Neither could Harry nor Robin, but neither of them said anything. The new mood that enclosed them was too delicate and they were all treading carefully. "We'll go to the Mirabelle," said Constance.

Robin looked at her father. "Is that too extravagant?"

"Where did I get this reputation for parsimony?"

"Well, you never spend any money on yourself. Isn't it time you got yourself a new suit?"

The world was full of echoes. And to-morrow, when he was less exhausted, when the brittle happiness of to-night did not have to be so protected, he would tell Constance about Lynne and ask for a divorce. "I'll buy myself a new wardrobe to-morrow. And to-night no expense will be spared. Within reason, that is," he added with a grin.

"I love you both," said Robin, and then, as if embarrassed by her emotion, went quickly out of the study.

"I hope it works out for her," said Constance. "How did you get on with Ferguson?"

"We adjourned the discussion pending further developments."

"You make the romance sound like a parliamentary works project. She's not one of your constituents."

"I don't have constituents any more."

She had been about to go out of the room, but she turned back at that. "Do you regret it? I mean, with the election coming up?"

He pondered a moment, then said honestly, "I think I do. But there is nothing I can do about it, so it's best to forget it. Perhaps I should try to get John to run as a Conservative. That way I'd get a vicarious pleasure out of it."

She shook her head. "You might turn out like your father. And I don't think John will ever let anyone run him."

"I didn't let Father run me."

"No. I sometimes wonder if you would both have been happier if you had given in to him."

"No. I'm my own man."

"Yes," she said slowly. "Indeed you are."

3

The dinner was a great success, neither expense nor conviviality being spared. Both Harry and Constance gave more of themselves than they had in years; Robin and Richard responded to the same degree. On their way out of the restaurant Ian Kidman rose from a table where he and Rosemary were entertaining another couple and crossed to them.

"Someone's birthday? It's disgraceful the way some people carry on in restaurants." He greeted them all and they in turn waved to Rosemary who threw them kisses as if they were her fan club. "Harry, could I see you first thing to-morrow morning?"

"I was going to sleep till noon." Harry was fighting the exhaustion that now weighed on him like an illness.

"If you don't mind, I'd like to see you at nine. I'll come to your house. It's very important, Harry."

In the car on the way home Richard said, "I wonder what's so important? Perhaps they're going to ask you to renounce the title and run against John."

"Whatever it is, I'm not going to let it spoil my sleep."

He was awakened at nine o'clock the next morning by Ferguson to say that Mr Kidman was downstairs. Still only half-awake he went down in his dressing-gown to see what urgency had brought Kidman to the house at this hour of the day.

Kidman, dapper even at such an hour, was in the drawing-room looking out at the sunlit square. "Could we talk somewhere else but here? I wouldn't want us disturbed."

Puzzled, Harry led him back upstairs to the study where Kidman said, "I'm sorry about all this cloak-and-dagger stuff. But we have to keep it secret."

"For Christ's sake, Ian, get it out!" His tiredness only increased his irritation.

"Sorry, old boy." Kidman took a deep breath, like a schoolboy about to begin a recitation he had been rehearsing all night: "Harry, Charles Wolfe has cancer. The doctors give him no more than six months to live."

His tired mind was functioning only in low gear. "So?"

"Don't you see? Good God – " It was Kidman's turn to be irritated. He ran a nervous hand over his tanned bald scalp. "Harry, it opens up a whole list of possibilities. First, if it gets out that he's dying it could spoil our chances at the elections – no one is going to vote for a party run by a dying man. Two, if he announces his retirement now, that doesn't give us time to elect a new leader before the elections – not one whom we could impress on the voters as being the man they need. Three – "

He was wide awake now. "Three, if it's kept quiet that he's dying, if he leads the Party into the election and wins it – How many know he has cancer?"

"Four of us at the most. Plus his doctor and his housekeeper – though I'm not even sure about her. We're all sworn to secrecy."

"He hasn't told all the Shadow Cabinet?"

"No. He doesn't want any scrambling for succession while we're concentrating on winning the elections. He'll go through with the campaign, keeping his speeches down to a minimum – actually, he looks healthy, though his energy runs out quickly now – and win or lose, he'll retire a month after the elections are over."

"It's not very fair to the country, is it?"

"I know that. But this is when one has to weigh patriotism against politics – and doesn't the latter always weigh most? We have to get Brewster and his crowd out and ourselves in. There are better leaders in the Party than Charles and once we nominate one of them, the rest of the country will soon accept him. What's the matter?"

"I'm trying to think of a better leader than Charles."

"We've already thought of one," said Kidman. "You."

That's the moment to be offered the chance to fulfil your life's ambition: your mind only half-functioning, your head aching, your stomach sour from last night's food and champagne. He sat down in his leather chair, felt the trembling in his hands and shoved them into the pockets of his dressing-gown. "I'm in no mood for jokes."

"I'm not joking. I'm the spokesman for – " He named three of the Shadow Ministers. "Each of them would like the job, but they all know they'd never get enough support. The alternative is Michael Argus."

"Christ, not him!"

"You'd be surprised how popular he is among the constituencies. They think he's got the drive the Party needs. We happen to think he'd make an awful balls of it."

"What makes you think I wouldn't?"

"I could give you a dozen reasons, but I don't have to sell you to yourself. Believe me, dear chap, we have gone into this very carefully. We found out about Charles's illness only two days ago and we've had four meetings since then."

"What about yourself for the job?"

Kidman's hand had been resting on his knee; it tightened, then instantly relaxed. "To be honest, my name comes up in discussion. But I don't think I'd have enough support – not to lick Argus. We have to pick a chap whose odds are as good as those of Argus."

"Does Charles know you intend approaching me?"

"Yes. And he's in favour of you."

His hands had stopped trembling, but he kept them in his pockets. He would not allow himself to succumb to dreaming of the future; he took on the role of devil's advocate against himself: "You've overlooked a couple of things. One, I'm a peer – "

"You'd have to give up the title, we know that."

"Two, the old days of the top families in the Party choosing the leader are over."

"We've thought of all that. My dear chap, none of us belongs to an old family – my grandfather was an inspector for a gas company. You would not be wangled into this job by the Cecils or any of those – we'd see you were elected to it by the constituencies."

"I'm too much of a dark horse – I'm not even in the Shadow Cabinet – "

"Charles will fix that as soon as the elections are over. You will be given the F.O. – either as Minister or as Shadow Minister, depending on whether we get in or not. He's authorised me to tell you that, though you mustn't breathe a word of it to anyone. Officially he is going to remain impartial. He says he wants to die and lie in peace, not be accused of having engineered his succession."

"And not be blamed if I should turn out a failure."

Kidman's face lit up. "So you'll let us nominate you when the time comes?"

"I'll have to think about it. There's the title for one thing – Richard doesn't want it – "

"I suppose you must care about that, it's being such an old one. I must confess it wouldn't worry me if I were offered the P.M.'s job. But then, being plain Mr Kidman is no honour – "

"Don't talk rot. I didn't become another person, the Marquess of Carmel, when I came into the title. It's not the title itself, it's what it implies in the minds of most people. There's no one thinks Alec Douglas-Home is any less of an aristocrat because he gave up being the Earl of Home."

"The ladies in the Party still love an aristocrat."

"The ladies in the Party are going to vote Tory anyway, whether the leader is an aristocrat or someone from the gas company. It's the floaters I'm afraid of, the I-don't-knows from the opinion polls. There are a lot of people in this country to-day who think blue blood is no better than piss. They respect a title more if it's a new one, if they think the man has earned it – "

"I don't think that's entirely true. We're still basically a nation of snobs and the older the title, the more respectable it is. Nobody bothers to find out how the title was earned – if it *was* earned. Some of the oldest were bought. But nobody bothers about that – but they'll sneer at the chap who buys his peerage with cash contributions to a party, though he's no worse than some of those who bought theirs from the kings. I don't think you'll have to worry about being the ex-Marquess of Carmel. Your biggest handicap is going to be that you're rich and everyone knows it."

"I've never advertised it."

"My dear chap, it's being advertised all the time. If your

money was buried in something like insurance, no one would know how much you were worth. But the *Financial Times* runs your name at least once a month as the chairman of one of your companies. Every time Robin is mentioned in the gossip columns it is as the daughter of the *rich* Lord Carmel – they described Richard last week when he was in court as the *rich* peer's son. Our newspapers love to describe people as rich – sometimes I think they're more concerned with money than the Americans are. If there is anyone in Britain who doesn't know you're rich, he must be blind, deaf and in solitary confinement."

"Do you think the ordinary man in the street resents it?"

"Resent – I don't know. Envy, yes. And perhaps resent it too. With the rising unemployment I notice among the kids I work amongst that they talk more about *them* than they used to."

"Do they include you among *them*?"

"I don't know. But they'd certainly include you. You'd be a better bet if you were an impoverished ex-peer, but rich or poor, we still think you're the man we want."

He took Kidman downstairs, showed him out, then went back upstairs to his bedroom. He got back into bed and lay staring at the clear sky outside. Still exhausted, he was in no fit condition to consider the future; but he knew he would not be able to sleep till he had at least put his mind into some sort of order. For too long ambition had been just cold ashes; the fire would have to be stoked again. He needed someone to consult and suddenly he regretted his father was no longer alive.

There was a knock at the bedroom door and Constance, in dressing-gown, face clean of make-up, came in. "I thought you'd like to know John got the Labour nomination last night. So he's the official candidate."

"Robin must be standing on her head with joy."

"She's already left for Bayard. Ferguson just told me Ian Kidman has come and gone. What on earth did he want at this hour? Or am I not supposed to know?"

He looked at her and all at once knew she was another factor that had to be fitted into the future. To-day he had been going to talk to her about divorce; but that was out of the question now and he was glad he had not had time to mention it; at least he had saved possibly hurting her. If it were known that he was seeking a divorce from his wife of twenty-seven years to marry a much younger woman, and an American at that, the ladies in the constituencies would have nothing to do with him. He put divorce and Lynne at the back of his mind.

"I think you have to know." She had never been one to gossip; the information could be trusted with her. "Charles Wolfe is dying of cancer."

"I'm sorry to hear it." She waited patiently and intelligently for him to go on. She has a political mind, he thought admiringly, she can already see the possibilities opening up.

"He's going to go ahead and lead the Party through the elections. Then, win or lose, he'll retire. They'll want a new leader who, if we win the election, will also be the P.M."

"And they want you to be one of the candidates?"

She had been standing at the foot of his bed, but now she moved across and sat down in a chair by the window. The curtains had been drawn back and the sunlight was streaming in, reflected softly by the gold coverlet on the bed and the deep bronze carpet. She wore a yellow silk gown, so that she fitted into the colour scheme of the room, was part of the warm intimacy of it. They were a husband and wife discussing house-keeping, but on a national scale.

"What are your chances? I presume you're interested in standing for the job."

"I've never really lost the ambition," he said frankly. "But it would make a difference to your life, a great difference. I haven't the right to ask you to put up with it."

"No, you haven't," she said just as frankly. She looked out the window, musing. "Being the P.M.'s wife doesn't even make you First Lady, does it? You're always Number Three or Four to the Royals. At least in America the President's wife is Number One. So what are the compensations – at least for me? The money isn't anything. There'd be a few more privileges, but not really much more than we have or can afford now. It seems to me it would be just damned hard work, a lot of public exposure, hours and hours of entertaining or being entertained by bores, English and foreign, and all for what? So I could stand in the background wearing a fixed smile while you stood out in front of Number 10 or wherever we'd be, waving to the crowds and getting all the cheers."

"There'd be an occasional bit of abuse – you wouldn't get that."

"That's a negative sort of reward. It's enough to make one want to join Women's Lib." Then she looked back at him. "But

I'll accept it if you really want to be Prime Minister. Do you?"

His mind was perfectly clear for such a perfectly clear question. "Yes."

"You'll have to give up the title."

"I'll have to renounce it for my lifetime, but on my death Richard could then reassume it."

"If he wanted it, you mean. Well, it doesn't really matter, I suppose. Once I might have been impressed by having a title – I can't remember how I felt when I first became the Countess. It means nothing now. The Marchioness of Carmel – when I see the name in a newspaper it is always someone else, someone I know vaguely. How will you get into the Commons? They'll have to find a seat for you."

"If they wanted me as leader they'd arrange for someone to resign a safe seat, there'd be a by-election and I'd step in and take over."

"Poor chap – I mean the man who has to resign the safe seat."

"They'll give him a title and put him in the Lords."

"Musical chairs. I'll never get rid of my disgust for what goes on in politics."

"Just don't let it show." Then he said carefully, "So you won't mind if I try for the job?"

She got up, moved to the door: he recognised the movement, all their discussions ended with her making the exit line. But he did not mind this morning: he was too tired to try to score off her. He was just grateful that she had been so amenable and once again was glad he had not had time to mention divorce.

"It would have pleased your father," she said. "I'll do it for him."

She went out on that, but it was an exit line that had no bite for him. If his getting the Prime Ministership should please the ghost of his father, then it would please him too; he had never felt bitter towards his father, despite all the years of estrangement, and any gratification the old man might feel, if he was capable of feeling anything where he was now, was something he would not begrudge. He knew in his heart there would be no gratification for Constance; conceit for status had never been one of her sins. If she had any sins at all, he thought ruefully. She would go along with him in his pursuit of the Prime Ministership, not to assist him in his ambition but because she would feel that, if the Party called on him, then he was the man for the job and it was her

273

duty to help him in every way she could. It would not be duty to the Party; he knew she was apolitical when it came to party labels; in her all doctrines were mixed and she was an amalgam of Tory, Socialist and Liberal. She would act out of a sense of duty to the State, a more highly moral sense than his own. She had always had a stricter feeling for duty than himself; her attitude towards their marriage was evidence of that. He knew that if he committed himself it would not be out of a sense of duty or patriotism or Party loyalty: it would be personal ambition and, to a lesser degree, the conviction that he could do the job as well as anyone else and better than his main rival, Michael Argus. He looked out at the trees in the square. Other leaders had lived here in the square, Metternich, Baldwin, Neville Chamberlain: he wondered what their feelings had been when they had been offered the chance at power.

Then he thought of Lynne and he felt a sudden regret that their relationship would have to end. Even in his mind he would not call it an affair; in the few short days he had known her it had gone beyond that. He had come back to England determined to have Constance divorce him, or if she would not grant him one, then go before the courts and under the new law ask for a divorce on the grounds that the marriage had irretrievably broken down. Once he had gone that far he knew she would not have contested it. He knew how painful it would have been for her and the prospect had pained him; but he was close to fifty and he wanted to spend the rest of his life in the warm, amusing companionship of the American girl he had come to love in so short a time.

But all that was out of the question now. Another woman would accuse him of ruthless indifference; but he knew it would be useless to explain. It would be better not to tell her anything of what had been in his mind towards her; he had not mentioned marriage to her, so she would not be expecting it. If he did not get the leader's job, *then* would be the time to talk of divorce and marrying again. He turned over and went to sleep and in his dream there was a woman's voice, faintly familiar, that kept saying sonofabitch, sonofabitch, sonofabitch.

4

Over the next week he went about his affairs as he had done in the weeks before he went to New York. He phoned John Ferguson in the country to congratulate him on his nomination and was surprised at his own genuine feeling of pleasure. He wrote his report on his observations of the debate in the United Nations as far as it had gone; he made a speech in the Lords on Russia's presence in the Middle East that had enough pungency to get him space in several newspapers; he chaired a board meeting of Anglo-Qaran and another of Aidan Estates. He had Kidman and the other two Shadow Ministers, Loudon and Chambers, who knew of Wolfe's illness, to luncheon and broad plans were laid for their strategy when the time came for him, as Kidman described it, to throw his title in the ring. And every day he phoned Lynne in New York, but always from his office, never from home.

"It's wonderful to hear your voice, darling," she said, "but I've been hoping for a letter."

"I never write letters to women in public relations."

"Well, it doesn't matter. I'm taping everything you're saying."

"I never know when you're kidding and when you're not."

"Darling – " Her voice softened. "Do you really think I'd gossip to anyone about you and me? I treasure what you and I have got – "

Why can't I be as truthful with her as she is with me? "So do I." For the time being, anyway.

"I love you – "

What will you think of me in a few weeks' time when I telephone you – not write you, most certainly not write you – that everything has to finish between us? "I'll call you again tomorrow. Good-bye, my love."

Then Charles Wolfe asked him to come and see him – "I shouldn't mention it to anyone, Harry. Just a *tête à tête* between you and me."

Wolfe lived in a solid Victorian house in Kensington that matched his physical looks and his political reputation: a cross between a Masonic temple and a small bank, it suggested that it would hide less corruption than a good many churches. Harry walked from Eaton Square to Wolfe's house, strolling through the early summer sunshine, feeling very much alive and cheerful, and only when he stood outside Wolfe's front door did he remember that he was visiting a dying man, that this was the last summer Charles Wolfe would ever see.

"Don't look like a mourner," Wolfe said as the maid showed

Harry into a study as untidy and cluttered as an old attic. "There will be time enough for that later."

Harry had composed his face into an expression of silent sympathy, but he relaxed at once when he saw that Wolfe had come to terms with his coming death and had put it aside like a bill paid and receipted. Wolfe produced two whiskies, lowered himself into a deep chair opposite Harry and looked at his guest. He said nothing for a full minute while Harry, undisturbed by the scrutiny, held his glass in his hand and gazed patiently back at his host. At last Wolfe nodded.

"I've looked at you before, Harry, but never, I confess, as leader material. I'd hear the women whispering about, about your looks, and that's the way I'd look at you – to see what you had that got their ovaries churning. You have it, whatever it is – it always helps to have some of it on the election platform. And particularly on that bloody invention, television. They tell me my image on telly is enough to get a party political broadcast an X certificate." He chuckled into his drink, took a deep swig, then settled down to business. "Do you think you could handle my job when I'm gone?"

"I think so. One would never know till one was in it."

"That's true. Power not only corrupts, it can crush. It's pulverised a lot of men who, when they got it, found they were too small and weak to handle it. You and I know one or two," he added, and his eyes glimmered like a spiteful schoolboy's. "But I don't think it would make a mess of you. Though you might make a mess of *it*, that's the unknown factor." He took another sip of his drink. "It goes against all my political thinking to consider you for this job, y'know. I swore after the cabal put Alec Douglas-Home in, I'd fight tooth and nail to prevent another aristocrat from being top man. The trouble is, there's no one else to stop Argus. And he'd be a disaster for the Party and the country. He's anti-black, anti-trades union – he's anti-so-many of the facts of life to-day I sometimes wonder if he's real, if he's not a ghost sent to haunt us for past sins. And that damned conceit of his!" Another drink. "Not that you're any shrinking violet. But your sort get away with it better."

"So far I haven't heard one positive talent I have in my favour, except what I suppose you call my sex appeal. And I don't intend to promote that."

"What would your general policy be if you got the job?"

"As P.M. or as Opposition Leader?"

"Let's be optimistic. As P.M."

Harry was silent for a while, but one couldn't answer a question like that in a minute or two. "I wouldn't know what my policy would be till I'd seen yours."

Wolfe nodded approvingly, eyes gleaming shrewdly. "You'd have at least half the Party at your throat if you tried a new broom as soon as you got into office. Whether you like my policies or not, I'd advise you to stick to them for at least a year, till you've worn off the shine of being a new boy. There are still one or two old 'uns who haven't forgiven you for Suez – "

"Have you?"

"Forgiven? I don't know. I've never yet worked out whether forgiveness has any part in politics. No, I probably haven't. But eventually I saw the pragmatism of your argument."

"I don't know whether you know it, but our family motto is *Realism above all else*. It's no rallying cry, it doesn't have the ring of 'For England, Harry and Saint George' about it. But when you come down to it, I don't know a better motto for a politician."

"For a politician, yes, maybe. For a P.M. or an Opposition Leader? Not on your life. Give 'em dreams, Harry – but always have your excuses ready why you can't deliver." Wolfe got up, poured himself a second whisky, then leaned back against a bookcase where books were scattered haphazardly, like the shrapnel of his life. Bound copies of *Hansard*, political biographies and memoirs, Burke, Pitt, Disraeli's novels: he stood against his natural background. He, like Bert Wardle, was a political animal: he could not have done anything else.

"I'll tell you the truth, Harry. The day I resign I'm going to hate the guts of you or whoever gets the nomination. I'm sixty-three and when I went up to Oxford at eighteen and joined the Union I started dreaming of some day being P.M. Forty-five years, and I may make it in July and it'll all amount to nothing." He shut his eyes, but Harry could not know whether it was the pain of his disease or the pain of disappointment that suddenly took the colour out of the big pink face. Then he opened his eyes again, took a long drink from his glass. "I can't give you my blessing, Harry, or hint in any way at all that you should be the one to take over. I have to play it as if I'm going to hold on, either as P.M. or as Opposition Leader, till Kingdom Come. Which is, I suppose, what in effect I'll be doing. My own

Kingdom Come. Good luck, Harry. I don't think we should be seen together till after the elections."

Harry took the soft, outstretched hand. "Why did we never really get on together?"

"It wasn't you. It was all those privileged blighters you're descended from. I didn't resent their privileges, just the use they made of them. Bloody old Derby, for instance. No one in Lancashire or Cheshire ever lifted a finger politically without getting the nod from him. And your family, or some of them, were just as bad in the West Country. It was envy, pure and simple. Not of your money or your titles or the fact that you could name an ancestor who might have had a beer with Henry the Fifth. No, the power they had, that was what prickled me. Of course it doesn't mean a damn any more, but when I first came into politics our Party was still run by no more than half a dozen families. And mine wasn't one of them."

That's the circle, Harry thought. A week ago I was envying Tom Brewster for the power he had, yet his family has had even less privileges than Wolfe's. But Wolfe had too little time left for him to want to discuss such paradoxes: politics was full of them and always would be. "Well, if I make it, it won't be because of the family. Nor even because of my record – what there is of it. I'll get the job on promises and nothing more."

"What other man ever got anywhere in politics on anything else?" Then Wolfe shook his head. "No, this late in the piece I should not be cynical. Now is the time to sound like a shining idealist – I know there won't be time to prove a failure."

"As nice a piece of shining cynicism as I've heard from you. Good-bye, Charles. Take care."

"I'll do that," said Wolfe, face sobering. "I'll do my best to win the election for you and bequeath you the Prime Ministership. And I hope you can hang o· to it for years to come."

Harry left him and walked back to Eaton Square, through sunlight that he no longer felt. The thought of death still chilled him; he was afraid of nothing in life but the end of it. Perhaps, he thought, I should start talking to God again. But by the time he reached home his possible reconversion to religion had faded like a New Year resolution and he was feeling warm again with the sunshine and the prospect that lay ahead of him. Heaven and Downing Street did not equate and he knew, given the choice, which one he would immediately choose.

Late Friday afternoon he went down to the country. Constance had been down at the Hall all week; her presence was evident by the masses of flowers in every room; he had a dim memory that his mother had had the same trait of bringing as much of the garden as she could into the house with her. He was standing out on the side terrace when she came out to him.

"Robin is bringing John to dinner to-night. She rang last night and asked me if it was all right."

"Is that wise? Ferguson isn't going to enjoy serving his own son at our table."

"He can take the night off. It was that or a fearful row with Robin. Love, it appears, has made her as much a rebel as Richard."

She went back inside and Harry stood for a while on the terrace, savouring the peace and quiet of the early evening. He might not have much opportunity for such peace and quiet after the next few months; Baldwin had been the last P.M. who had been able to enjoy the comfortable routine of a country gentleman. He had no desire to be another Baldwin, but there might be times in the future when he would yearn for the peace that surrounded him now. On an impulse he went down the steps and walked down to the beech grove that hid the family mausoleum.

He had not been down here since his father had been buried; he did not believe in pilgrimages to the dead. Last autumn's leaves had been cleaned out of the grove and his footsteps made no sound on the soft earth as he walked down through the trees to the big marble-faced tomb. By tradition only the head of the family was buried here; wives and children were buried in the Catholic graveyard down in Bayard. The mausoleum had been built for himself by the eleventh earl and he had been a man optimistic that the family would survive for a long time: the tomb now held nine bronze caskets and there was room for eleven more. It had been designed by Hawksmoor, with his taste for Doric columns; it had always reminded Harry of a shrunken Greek temple. But its proportions were exquisite and if one cared to be surrounded by beauty when one was dead, then this monument in its grove of sheltering trees would satisfy most men. Harry stood in front of it, not so much communing with the spirits of the dead men inside it as mindless in their presence. He was well aware of his heritage; but dead men did not speak to him nor he to them. If their words were not written down, their deeds not commemorated by what they had built, they were shades to him and nothing more. Of all the men in the tomb only two touched him and only then because he had known them, his father and his grandfather. But now he was here he had nothing to say to them.

He went back to the house. As he climbed the steps he saw Mrs Dibbens standing on the terrace watching Ferguson's small car disappearing down the drive.

"Ferguson has gone off, m'lord. To Roebuck, to the cinema. He explained the circumstances to me. O'Brien will serve at dinner."

"I'm afraid Ferguson is eventually going to have to accept the circumstances. It's a changing world, Mrs Dibbens."

"Indeed it is." She smoothed down the front of her black dress, as if bringing to his attention the mourning she wore for the past. "Nothing personal against young John – but it *is* a pity."

"You'd vote for the past if that was on the ballot paper in the election?"

"Indeed I would, sir. The future never lives up to its promise, does it?"

If I become P.M., I'll have to make sure the *Daily Mirror* and the other Labour papers don't get down here to interview you. "That's a very pessimistic attitude."

"It's the way I feel, m'lord. One can't go against one's feelings."

How wrong you are, Mrs D. "No, I suppose not. Hello, who's this?"

"It sounds as if it might be Dr Banning, that car of his. He's coming to dinner too. Didn't you know, sir?" She looked at him slyly.

"I'd forgotten." I know whose side you are on, you bitch, but I'm not going to let you score points off me. "Tell O'Brien to bring some drinks out on to the side terrace."

Joe Banning pulled up the Lagonda, got out and came up the steps. "Still goes like a dream."

"It must be almost as old as I am."

"Not quite. But I look after it, give it better care than I do my patients."

"Is it on National Health?"

The two men grinned at each other and they walked round to the side terrace. As always Harry felt glad of Banning's company; they might not always agree, but Joe was the one consistent factor in the whole web of his relationship with other people. More consistent even than Constance.

"How have you been?" Banning asked as he took his first drink of the day. Lately he had been trying to cut down on his drinking, but it was proving a tough battle.

"I'll come to you for a check-up. I've had odd moments of exhaustion lately that I've never had before."

"It's that life you lead," said Banning casually, but his expression was hidden behind his hand as he raised his glass. "You should settle down."

"I was thinking of it – " Harry said, then went no further. How ironic it would be if Joe examined him, found something drastically wrong and told him he had only six months to live. Had Charles Wolfe got his death sentence in such an unexpected way, after a routine medical check-up?

"Can you come to-morrow morning? Make it ten o'clock – I have to be over at Shepton Mallet by one o'clock for a vintage car rally." He took another swallow of his drink. "I gather we're having a very democratic dinner-party to-night. Tell you the truth, I think I'll vote for young John. Anyone but that bloody Bedley. Do you think you'll get in? I mean the Party as a whole."

Harry shrugged. "I don't know. Brewster is so damned popular personally, he could carry Labour back on his own."

"That's more than can be said for Wolfe – he's been a loser too long. You need someone else."

"Who, for instance?"

It was Banning's turn to shrug. "God knows. There's no one in the Party I'd personally vote for."

How about me? But all he said was, "Well, you think Labour will waltz back in?"

"Afraid so. But it won't matter to you, you're doing all right. Ah, here comes my favourite patient. She always looks so damned healthy and beautiful. I don't know how she does it."

He stood up, turning away from Harry to greet Constance as she came out on to the terrace. Harry felt a momentary stab of annoyance (of jealousy? But that was ridiculous, he told himself); then he stood up and poured more drinks from the tray O'Brien had brought out. He poured himself a particularly strong whisky, slopping it into the glass; then he looked at it, had second thoughts and switched his own and Banning's glass. He owed it to Robin to stay clear-headed to-night.

"That's a stiff one," said Banning as he took his drink. "Are you trying to get rid of me early?"

'Make that one last, Joe." Constance spoke to him like a wife; he nodded humbly. "You drink far too much."

"Don't nag him," said Harry, and was annoyed at his own irritation. "That's one of the joys of a bachelor's life, not having a nagging wife."

"One of the few joys," said Banning, but his broad smile looked a little forced.

Robin and John Ferguson arrived an hour later and all of them went into dinner immediately. As they sat down in the dining-parlour Robin said, "We've been out campaigning all afternoon, knocking on doors and introducing John. We can't afford to waste any time. So many people don't know him."

"Did they know you?" Harry asked.

"Did they!" John exclaimed. "I'm just depressed by the whole exercise. Everyone who answered their doorbell gushed at her and said how they'd seen her picture in the newspapers and magazines. What hope does a Socialist stand in an electorate where even the workers read *Vogue*?"

But when he looked across the table at Robin it was plain that he loved her; and her love for him was just as evident. Constance made no secret of the pleasure she felt at the romance; and even Banning, like a favourite uncle, beamed with approval on the two young people. Harry, aware of the absent older Ferguson's feelings, was the only one with reservations. But it warmed him to see how happy Robin was and he felt an aching regret that so much of her life had gone by that he had failed to share. His children had now all grown up and sixteen years of their lives were almost a mystery to him.

When it was time for coffee to be served he said, "Would you take Joe and Robin into the library? John and I want a few words with each other."

"If he doesn't treat you gently," Constance said to John, "the appeals committee will be in the library."

"Thank you, ma'am."

When they were alone, coffee and brandy on the table between them, Harry said, "Forgive the old-fashioned phrase, but what are your intentions towards Robin?"

"I haven't proposed yet, sir." John stirred sugar into his coffee, watching it carefully as if it were some sort of chemical experiment. "But if I manage to win the seat, then I shall propose. If that's all right with you and Lady Bayard." He looked up.

"I'm sure it will be all right with both of us. I don't need to ask if it will be all right with Robin – I'm afraid she may propose to you."

John grinned. "She already has, sir. I turned her down – temporarily."

"What happens if you don't win the election?"

John went back to examining his coffee. "That's the problem. To be honest, I don't quite see Robin fitting in as the wife of an L.S.E. lecturer. She also wouldn't fit in if I were running for a seat in some working class district and were elected. But out here – "

"She'd be in her natural background."

"If anything, I could be the misfit around here. Naturally," he grinned again, "I'm trying to tell the voters that Colonel Bedley is the misfit."

"Between you and me, he is. He gives you your best chance of winning. If we had another candidate I don't think you'd stand a chance."

"I'm glad of it. I didn't want to have to attack your record. Actually, if there have to be Tories I wish they were all like you. That's a backhanded, cheeky compliment, I suppose, but I mean it, sir. But I can go flat out against Bedley."

"If I had still been the M.P., would you have run against me?"

John hesitated, then shook his head. "No, sir. I respect you and – though he doesn't seem to appreciate it – I respect my father. It would have been just too intolerable to have run against you."

"Why did you choose this seat anyway? Why not try for another one?"

"It was not my idea. It was George Brewster's."

"The old bastard!" Harry laughed aloud.

"When the P.M. himself sends for you and tells you he'd like to see a Labour man returned from the district where he was born and suggests you, what do you say? I thought it was some student from L.S.E. playing a joke on me when I got the message to go to Downing Street."

"Well, we'll have to see you're elected – to please both Robin and Brewster. But I don't see how I can help you. If I endorsed you, the voters would say I was doing it for my daughter's sake." And what would they say up at Central Office, especially when after the elections my name comes up as the possible Party

Leader? He stood up. "Anyhow, you have our blessing, John. All you have to do is be elected."

"I shan't mention it to Her Ladyship now, but you might thank her for me."

"I'll do that. In the meantime you'd better think of another way of addressing her. I don't think you can go on calling your mother-in-law ma'am or Her Ladyship."

"Force of habit, sir."

"It's all going to be a bit complicated. I never thought I'd have my butler related to me by marriage."

## 5

Next morning, sharp at ten, he was at Banning's surgery in Bayard. The doctor went over him carefully, but at the same time said, "I can't check everything. For a thorough check you should go into a clinic for a day or two."

"Only if you find something that warrants further examination. How do I rate so far?"

"Your blood pressure is too high. That was a splendid dinner last night. I like that boy. You going to let him marry Robin? Breathe deeply."

"Do you want me to answer questions or breathe deeply?"

"Both."

"Yes, I am – going – to – let – him – marry her."

"Good. Your commonsense, I mean. But not you."

"What does that mean?"

"I think you need to take better care of yourself. Had any pains in your chest lately?"

"I've never had any pains in my chest."

"Not even an emotional one?"

"What does *that* mean?"

"Lie back. I haven't finished yet." Then later, as he was washing his hands and Harry was getting dressed, he said, "You'd better ease up for a couple of months. Take a holiday, go out to that cottage of yours at Montego Bay. I don't think you have any serious heart condition. But you're on the verge of a bad case of exhaustion. And if you have any incipient heart condition, that could aggravate it."

"I can't take time off. Not for a couple of months."

"Well, just see you don't knock yourself out. Take it easy, don't concern yourself with anything that's going to keep you awake at night making decisions. Rest is what you want and perhaps that will be sufficient to do the trick. Well, I have to be going, got to be over at Shepton Mallet for that rally. Just as well you're not caught up in these elections, running for office again. I'd have to be galloping everywhere with you like a bloody nursemaid. Give my love to Constance."

# Ten

The Tories, unexpectedly, were returned as the government and John Ferguson, even more unexpectedly, was returned as the Member for Bayard. During his campaign he had not lived at Aidan Hall but at the Coach and Horses pub in the village, considering it not political sense to live right in the heart of enemy territory while he was knocking on doors trying to persuade people it was time they got rid of the Tories. The morning after the poll was declared he came up to the Hall. He spent half an hour with his father, then came in, with Robin holding his hand, to see Harry and Constance.

"How did your father take the catastrophe?" asked Harry.

"He still thinks I'm a heretic."

"How about Colonel Bedley?"

"He thinks I'm the same, I'm afraid." John looked cautiously at the two people who, in the circumstances, were now his future in-laws. "The important thing is, how do you and Lady Carmel feel?"

Harry looked at Constance, allowed her to answer. She moved forward, kissed Robin on the cheek, did the same with John. "You have our blessing. I shouldn't worry about Colonel Bedley – the voters who put you in are the ones with commonsense. As for your father – well, when I first came here I didn't belong." She looked at Harry. "That's right, isn't it?"

"You belonged as far as I was concerned. And Roger, too. But yes – Father and some of the others took some time to adjust."

"You see?" Robin said to her mother. "You *absolutely* belong here now! Anyone can adjust if they put their mind to it. Ferguson will accept the situation – "

"Not if you go on referring to him as Ferguson," said Harry. "He is going to be your father-in-law, not the butler. It is going to be more difficult for you and John than it was for your mother."

"There's always a *modus operandi*," said Robin innocently. Harry and Constance did not look at each other but each felt the sudden tensing in the other. "You should know that, Daddy. Isn't that how everything works in politics?"

John grinned. "Already she has the right cynical outlook. She'll make a great politician's wife." Then he stopped smiling. "Don't worry, sir. We'll work it out with my father."

"Have you thought of when you'd like to be married?" asked Constance.

"The end of September," said Robin. "It would be nice to be married and go away on our honeymoon before Parliament reconvenes, but John thinks he'll have too much on his mind between now and then to be a good lover."

Constance kept her eyebrows in place. "You had better get that split in your personality sorted out early, John."

"Always ask for steak and oysters at all political dinners," said Harry.

John was grinning, feeling at ease with these two people of whom, when he was a child, he had stood in awe. But Robin had told him of the strained relationship that had existed between her parents for so many years and he wondered what each of them did for love. Steak and oysters would be no answer for them.

Later, when John and Robin had gone, Constance said, "Are you pleased or not?"

"I think so. But it isn't going to be as easy for Robin as she thinks."

"It wasn't easy for me when I first came here." It isn't easy for me now: but that was a corollary she left him to guess at. Then she changed the subject: "What about you? You're going to have to make a decision or two now we've got back in. Has Charles Wolfe telephoned you yet?"

"It's too soon. He won't be thinking about his Cabinet till the weekend. And there's no certainty he'll resign. Now he's in power he may hope for a miraculous recovery and stick it out. Westminster has managed just as many miracles as Lourdes."

"If he does, will he send for you?"

"I doubt it. He's never liked the idea of a Cabinet Minister in the Lords – he thinks they should all be in the Commons, where the elected reps can have a go at them. He might consider me for Lord Privy Seal, but I don't fancy that. I'll need a stronger post in Cabinet than that if I'm going to beat Argus for P.M."

"What would you like?"

"The Foreign Office would be the ideal springboard. But Ian Kidman's in the running for that. He'd have to refuse it, take his chances on my getting it and then going on to be P.M., and then trust me to hand it over to him when I'm in Downing Street."

"Would you hand it over to him?"

"What makes you think I wouldn't?"

"I didn't say that."

"No, but you hinted I might have second thoughts."

"Did I? I wasn't aware that I had hinted anything. You must have a conscience. Is Ian doing the wrong thing by backing you for P.M.?"

"You'd better ask him."

She smiled. "No, I don't think I'll do that. I don't want to be accused of meddling. But if you become P.M. I think that might be my price – that you'll honour your promise to Ian to make him Foreign Secretary."

"I made him no promises."

"All right then, you're promising me. I don't know that Rosemary is any advertisement for the country, but I think Ian would make a very good Foreign Secretary. Anyhow, that's my price if I'm to play the gracious lady at Number Ten."

"What if I refused?"

She smiled again. "You won't. You'll sell your soul, if you have one, to get into Downing Street. Harry, my love – " She stopped, the smile still on her face but frozen for the moment like a grimace. "Sorry. A slip of the tongue."

Now was the time to try for a new start. The image of Lynne fell into his mind for just an instant, a slip of the memory, but resolutely (ruthlessly? he wondered) he shut his mind against it. "Could we bury the hatchet and start all over again?"

He might have thought she showed no reaction had he not been watching her so carefully. Her face was still holding the smile, though it had relaxed a little; but he saw the tightening of her hand before she slipped it into the pocket of the yellow silk dress she wore. She let the smile die slowly, but he could not read it: he could not tell whether it was mocking, sad or no more than polite. She turned away from him and began to walk across the Great Hall where they had been standing. She did not appear to know where she was heading. He hesitated, then followed her.

They came into the State Gallery, the long wide hall that ran from one wing to the other and separated the Great Hall from the Blue Ballroom. It was a part of the house that had never appealed to Harry; its proportions never suggested that it had anything to do with family life. Fifty yards long and fifteen yards wide, it might have been designed as a rococo gymnasium. Verrio had painted the ceiling with pastoral athletes vaulting over rocks and bushes in pursuit of equally athletic wenches; Grinling Gibbons had festooned the end walls and the inside wall with wood carvings of high-jumping cherubim; Mortlake tapestries showed wrestlers so muscular that they appeared to be bursting apart the threads that sketched them. The outside wall, interspersed with deep windows, had been re-panelled only a year ago when the old panelling had been found to be ravaged by the death-watch beetle. The new panelling had been instantly aged, the patina of centuries squeezed out of a pressure-pack, and the casual eye could hardly notice the difference. Chairs, tables and sofas were grouped against the inside wall and a long golden rope ran down the middle of the gallery as a barricade against the sticky fingers and questing bums of the public who tramped through here on visiting days. The parquet floor was uncarpeted and every footfall sounded like the hoofbeat of a horse. The gallery was a huge sound chamber and there were signs hung on the rope that suggested *Quiet: Please*. Just the place to attempt a reconciliation with one's wife.

"Well, do you think it's worth trying?" he said.

"No."

"Why not?" His voice sounded like a shout in the long, high-ceilinged hall.

"Drop your voice."

There was an echo on his memory: "Why? Do we have to worry about the bloody servants again?" But he dropped his voice, whispered hoarsely. "Why did you bring me in here?"

"You came of your own accord. Ah, Harry – " She moved across to one of the windows, leaned her haunches against the sill, folded her arms and looked at him. "You see? We're snapping at each other already over small things."

"I'm sorry. I'm just on edge, that's all. It's not every day a man proposes to his wife for the second time."

"It's not a proposal. It's a political proposition – log-rolling, isn't that what it's called? You do something for me, I'll do something for you. Do you honestly think that's enough on which to re-start a marriage?"

"Put like that – "

"How else would you put it?" Her own voice rose, but she quickly cut it off; one of the maids had poked her head in a door at the far end of the gallery, then hastily withdrawn it. "I don't think you've really given a thought to what a reconciliation would mean. It was just something off the top of your head – like a parliamentary interjection."

"For Christ's sake, can't we drop the political analogies? We're not going to live our lives in the Commons – " His throat was strained with trying to keep his voice down. He looked down the long room, saw Ferguson come to the far door, then discreetly retire. "Let's get out of here."

"It doesn't matter where we go, I'm not going to give you an answer right away. You don't deserve one. You sprang this question on yourself as well as on me. I don't think it would work. You can't build a marriage, even a reconstructed one, on expediency."

"Expediency has kept us together all these years – why can't we improve on it?"

"You surely don't call what we've had these last sixteen years a *marriage*? No, you'll have to offer me better than that. I'll be the Prime Minister's wife, if you get to Downing Street, but I shan't be *your* wife. I'd never know when I was going to be dropped in a Cabinet re-shuffle."

One part of him had to admire her: she knew exactly how to score off him. But he was still angry and (yes) disappointed as she walked away from him down the gallery, her heels clicking like distant revolver shots, the sound getting fainter and fainter as she retreated. He leaned back on the window-sill she had vacated, watched her disappear through the door at the far end.

He could feel the trembling in his hands again and once more he felt the exhaustion that had drained him a couple of months ago. He had taken Joe Banning's advice and done his best to relax and take things easily; he thought he had succeeded but evidently he was more emotionally exhausted than he had been physically. Without his being aware of it there had been an erosion of his emotions and suddenly he was afraid. What if he should react like this when some political crisis arose? What reserves would he have to fall back on? Then, blindingly, he knew why he had proposed that they should take up again as man and wife. He needed someone to lean on, a confession he had never made even to himself, and she had been the natural one to turn to. His proposal *had* been one of expediency: she had guessed correctly, but for the wrong reason. But then he had not known the right reason himself.

He straightened up, walked slowly down the gallery towards

the opposite end from where Constance had disappeared, past the woven wrestlers battling each other in bouts that none could ever win: time and age would defeat them all as the threads rotted and the scenes fell apart. He felt as insubstantial as any of the wrestlers; he, too, was coming apart at the seams. He had thought of himself as young middle-aged, but time, the death-watch beetle in the human frame, had begun its work. When the call came from Charles Wolfe he hoped he would have a façade that would hide the sudden doubts that had infected him. But confidence, unfortunately, did not come pressure-packed.

2

He waited in an agony of impatience all weekend, doing his emotions no good at all. Then on the Monday morning Wolfe phoned from London, asked him to come up and see him on the Tuesday. He went up by train, caught a taxi out to Wolfe's house. As soon as he saw Wolfe he felt depressed: the older man looked years younger, the miracle had worked. "Come in, dear boy, come in! Whisky? Oh, come on, I know it's only ten-thirty, but I haven't celebrated our win with you yet!"

"You look well."

"Never felt better in my life!" Wolfe sat back in his chair, stared hard at Harry, then abruptly laughed. "Relax, dear boy. I'm still going to retire. I went to see my doctor yesterday, had some more X-rays taken." He shook his head. "No improvement. Cancer is above politics. A pity."

"What do I say? That I'm pleased or sorry?"

"Be both, Harry. I've swallowed enough disappointment over the past twenty-four hours to be beyond any more bitterness. It's been almost cathartic in a way. I've never felt so damned charitable and magnanimous. I'm beginning to think I'm destined for sainthood."

"Saint Wolfe – it has a nice ambiguous sound to it."

"That's the spirit – don't be maudlin about me." He looked shrewdly at Harry again. "Do you think you'll be up to it? Got any fears or doubts?"

Funny you should ask . . . "No."

"I'll be frank with you. I've already had Michael Argus in. I have to have him in my Cabinet – he's going to be Home Secretary."

"Is that wise? He's so Right Wing he'll want to bring back

some of Palmerston's principles. You know how he feels about immigration, for instance. He makes Enoch Powell sound like Martin Luther King."

"I know. But he asked for the job and I had to give it to him – he said he wouldn't take any other, except the F.O. If I'd left him out of the Cabinet altogether, then announced my retirement, I'd have been accused of trying to prejudice his chances of succeeding me. Whatever happens, we must give the appearance that the Party is just one big happy band of brothers. We've been out of power too long to come back and start looking like a party torn apart by squabbles for the top job. If you succeed me, you can move him into another job before he does any harm as Home Secretary."

"Thank you. Any other dirty work you're leaving me?"

Wolfe smiled. "Dear boy, you have to pay *some* price for what I'm giving you."

"What are you giving me? You haven't mentioned it yet."

"The Foreign Office. All right?"

He rarely drank during the day and certainly never in midmorning. But now he took a long swallow of his whisky, hoping it would settle the excitement that had gripped him. "Thanks, Charles. That puts me back in the public eye."

"Not for long, I'm afraid. I'm announcing my Cabinet tomorrow and two weeks from then I'm announcing my sudden and totally unexpected retirement. The doctor doesn't think I can delay it any longer. He says the pain is going to get progressively worse from now on – right now I'm not feeling it so much because I'm drugged with euphoria, or so he says. But you'll have no more than two weeks in which to re-establish yourself. It may not be enough, but I can't do any better for you."

"What can I do in that time?" He felt suddenly depressed again.

"You'll have to make the most of what you can at the Security Council meeting in New York. I think the drill would be for me to announce that you are going over to lead our delegation to the Arab-Israeli meeting in person."

"One never makes a reputation at a single meeting of the U.N. – not unless one starts banging his desk with one's shoe, as Krushchev did. And I don't think I could come at that."

"I'll have a word with some of our friends in Fleet Street. The B.B.C. will do something on you – I think you should come over very well on television. There will be some pieces for the gossip columns on your butler's son winning your old seat."

"He is also going to be my son-in-law."

"Really? Then *that* should make you a household word – at least for a week or two."

"What a way to become P.M.! Through the gossip columns!"

"Never sneer at your entry into power – the public will only remember your exit. I'm thinking of posthumously publishing a book of political advice." Wolfe had been laughing and chuckling, enjoying himself immensely; then all at once he heaved a great sigh, his face went grey and he seemed to fold into his chair like a deflated dummy figure. "Sorry, Harry. I think I've had enough."

Harry stood up quickly. "Shall I call your housekeeper?"

"She can't help. I'll be all right. I just sit here and after a while it goes. Jesus – !" He put his hand over his eyes and Harry stood awkwardly and uselessly by till at last the older man looked up. "When your time comes, die quickly. It's so much more dignified."

"I'll try and remember that."

A little while later he left Wolfe and caught a taxi to Eaton Square. As he entered the house the phone rang: it was Ian Kidman. "Have you seen Charles? Good. Congratulations on the F.O. He told me he was giving it to you."

"I promise you, if I make it to Number Ten the F.O. is yours. What have you got now?"

"Defence. Not exactly my tea-bag, I'm the greatest coward the Scots Fusiliers ever commissioned, but I'll try and not get us into any wars, at least until you can rescue me and the country both. Thanks for the offer of the F.O. I'd like it, but I shan't hold you to any promise. I don't expect any payback because I'm nominating you for Number Ten. I just think you are the best man."

Harry in the next week did his best to implant that idea in the minds of the British reading and viewing public. It was his first experience of the power of public relations and while the Press Officer of the Foreign Office worked on establishing the image of his new chief, Harry could not help thinking what sort of image Lynne would have created for him. He was not looking forward to seeing her again and telling her what he would have to tell her.

He resigned from the board of Anglo-Qaran and his nomination

293

of Sid Lucas to replace him as chairman was accepted. Richard and Matthew both wrote him congratulatory notes on his appointment as Foreign Secretary and he replied to them, two short notes that were as difficult for him to compose as any official Notes he would have to write: foreign Ministries were sometimes easier to communicate with than one's sons. It seemed somehow indicative of the lack of cohesion in the family that at this time, the most important point of his political career, the two boys should be out of the country. With school and university closed for the summer, Matthew had gone to Iceland on a bird-watching expedition and Richard had gone to Munich for a conference on the Future of the Radical Progressive in the Common Market: to Harry both journeys were for an equally esoteric purpose. He felt closer to Richard than he had in years, despite their difference in political outlook; but Matthew was still a distant boy, someone he had not yet succeeded in touching. It came to him with a jolt that he was repeating with Matthew the relationship he had had with his own father. He made a resolution that Matthew would not finish up as the neglected second son.

Constance came up from the country to Eaton Square to play her part in establishing the image. The more gossipy columnists declared that they were without doubt the handsomest couple to grace the British political scene in years: Camelot, one suggested, had at last come back from Washington to its proper landscape. Constance created her own impression, one that helped Harry immensely.

"I'm grateful," he said to her the night before he was to leave for New York and the reconvened Security Council meeting. "My one worry is that I may finish up being known as Mister the Marchioness of Carmel."

"No better than you deserve," said Joe Banning. He had come up to London on business that day and had come round to Eaton Square for dinner. He had not been abroad in almost ten years and he was planning his first Continental holiday in that time; he had turned sixty and he had decided he must get out of the West Country for at least a month a year before they buried him there for good. He looked at his principal patient, or anyway the one who owed him the most. He wondered what the National Health refund would be on a bill for five million. "How have you been sleeping lately?"

"Why, what's the matter with him?" Constance asked in surprise.

"He could do with a long rest rather than taking on this job."

"I've been taking things easy as you told me," Harry said. "But after all, I'm only just about to turn fifty. I'm just a youth in political age. I haven't got one foot in the grave yet."

"You're fooling yourself," said Banning. "With to-day's pressures no one over the age of thirty-five is young any more. You still haven't answered my question – have you been sleeping well?"

"No," Harry admitted. "But would you expect me to? I've had a lot on my mind. I knew I was in the running for the F.O. job."

"Pity you can't take a couple of weeks off and come over to Wiesbaden with me. The Germans are holding a car rally. We could see that, then just cruise around. I've got a locum in to look after my practice and I'm going over there at the weekend."

"You should have insisted that he went away before this." Constance looked at Harry with some concern. He had never had a serious illness in all the years of their marriage and it had never occurred to her that a day might come when he would suddenly collapse; she was aware of her own approaching age but somehow she still thought of Harry as being the same as when their marriage had broken up. She had not seen him without his clothes in sixteen years and there had been no change in what was visible of him: his hair was still thick and dark, his jawline was still firm, his figure trim. It was as if she had frozen him in her mind as the man she had loved without reservation.

"I'll take care," Harry said, wishing to get off the subject. He looked up at Ferguson as the latter put the cheese platter down in front of him. "You can put Ferguson in charge of me."

"Does that mean you wish me to go to New York with you, sir?"

"It would be a good idea," said Banning. "What's the matter, Ferguson? You look much sicker than His Lordship."

"The thought horrifies me, sir. New York, I read, is now a worse jungle than Darkest Africa ever was. But of course, m'lord, if you wish – ?"

"I wouldn't ask you to make such a sacrifice," said Harry. "No, I'll go to bed early each night, take a sleeping pill – "

"I don't believe in pills," said Banning. "Natural rest, that's the only thing." He spread some Camembert on a cracker, paying careful attention to what he was doing. "Constance should go with you to look after you."

295

Harry's and Constance's quick glances met down the length of the dining table. Even Ferguson was suddenly still; then he turned and went at his usual unhurried pace out of the room. Waiting for him to go gave Harry time to decide his attitude. He did not want Constance in the same city with him while he told Lynne their affair was over.

"You're welcome to come, but honestly I don't think you'd enjoy it."

"I don't think I should either," she said, hiding her mixed feelings almost perfectly. "Anyhow, I have two committee meetings early next week that I can't miss."

"It was just an idea," said Banning lamely, and bit into his cheese and cracker; his one attempt as a marriage counsellor had fallen flat on its face. "You watch yourself, Constance, see you don't wear yourself out as he has done."

"I still sleep like a top," she said; but that had not been true over the past few weeks.

They were having coffee when Ferguson came in to say Michael Argus was on the phone. Harry took the call upstairs, wondering what Argus wanted. "Harry? Sorry to trouble you, old chap, but might I drop round to see you? Yes, now. It's rather important. I gather you're off to New York to-morrow."

"Can't you tell me what it is over the phone?"

"Rather not, old chap. It'll only take half an hour at the most. I can be there in five minutes." Argus lived in Wilton Terrace. "I've been trying to get you since lunchtime, but you appear to have been locked up with the P.M. all afternoon."

Was there a note of jealousy in Argus's voice? Harry had spent only an hour with Wolfe this afternoon. "All right, come round."

He went downstairs to where Constance and Banning still sat in the dining-room. As he walked in Banning sat back in his chair, withdrawing his hand which had been resting on Constance's. In the pale light of the candles on the table Harry thought Constance looked flushed, but he could not be sure. But he was certain of the sharp feeling that pricked him: he was still jealous of another man's attention to his wife.

His voice rasped as he said, "I have to leave you two alone. Argus is coming round to see me."

Constance stood up. "We'll go up to my sitting-room. I'm afraid I don't enjoy meeting Mr Argus."

"I saw him on television the other night," said Banning. "His ego kept upsetting the colour tones. I've never seen anyone who had such a purple opinion of himself."

Argus arrived exactly five minutes later; he was the sort of man who made a fetish of punctuality, as if he were a one-man railway. "Never keep anyone waiting," he said as Ferguson showed him into the study where Harry was waiting for him. "If a man has to wait on you he has time to change his opinions."

"I'm entirely blank-minded," said Harry. "Because I don't know what you want to see me about."

"Brandy, sir?" Ferguson asked.

"Not for me," said Argus. "I've given up liquor of all kinds. I'm in training."

Harry saw the very slight elevation of one of Ferguson's eyebrows; it was rare that the butler ever allowed himself an expression. Argus, whether he knew it or not, and he probably didn't, had invited himself to a house where his popularity was nil. The butler went out, closing the door, and Harry was left alone with Argus.

"You're wondering why I've pushed myself in on you like this. I'll come straight to the point – we don't beat about the bush up North." Harry's own eyebrows went up at that; he hadn't realised Argus had reverted to being a Northerner. But as he listened he detected the faint re-birth of the Lancashire accent. Argus was also working on his image, was preparing to sell himself on the home market. "Harry, I know the situation, I've had it explained to me by Charlie Wolfe." *Charlie*, Harry noted, not *Charles*. Old mates from the North, though Wolfe came from Derbyshire. "When he retires – terrible business, isn't it? Good man like that going at a time like this."

"Terrible."

"Well, you and I are the front runners for his job. I understand Ian Kidman and a couple of others are backing you. Right?"

"Right."

"Must say I was surprised when I heard it. No offence, but if the newspapers knew the situation I don't think they'd consider you more than a long-shot. I thought Kidman himself would be my strongest competitor."

"What odds would you give on him?"

"Five-to-one at the shortest. I have a lot of support, Harry. The constituencies will go for me and, you know as well as I do, that it's the constituencies these days who elect the leader. Now this is what I want to put to you – "

297

"I was wondering when we were going to stop not beating about the bush."

Argus lowered his brows, the frown marking his face like a dark welt. "I detect a note of hostility. Are we going to be difficult with each other?"

"Get to the point," Harry snapped. "What do you want, Michael?"

Argus drew in his breath and he flushed. But his voice was quite firm as he said, "Don't allow them to put your name forward for the P.M.'s job. I don't think you'll do yourself or the Party any good. Now is the time when we must look what we told the country we were – the United Party, the Party of the Future. If there's an unseemly battle to be Prime Minister – "

"I had no intention of being unseemly."

"All right. But it would create a dissension in the Party that we don't want right now. As for yourself, you'd be throwing away your title for nothing. Or do you *want* to come back into the Commons?" It was evidently a possibility he had not considered before.

"I shouldn't consider that a retrograde step. If it's the only way I can stand for the P.M.'s job, I'll have no hesitation in being plain Mr Aidan."

"Well, that's as may be. The point is, I still don't think you can beat me for the job. You haven't enough time to make an impact. I've recognised what you've been trying to do this past week, but your wife has got more press than you have. That's being brutally honest, but I told you I wasn't going to beat about the bush."

Harry could feel the anger beginning to tremble inside him. "Is that the only reason you came here, to tell me I had no chance of beating you?"

"Partly. But as I said, I'm concerned with the image of the Party – more so than I am with yours, if you don't mind my saying so. What I'm suggesting is this – don't run for the leadership, stay in the Lords and I'll guarantee that you can stay Foreign Secretary for as long as you like. I think that together, you and I and the others already in the Cabinet, I think we can make a team that will inspire confidence."

"And you don't think I'd inspire much confidence as P.M.?"

"Frankly, no. The aristocracy has never suggested drive and initiative, not for three hundred years or so. The ordinary Englishman to-day wants someone he can identify with."

"Someone like you who inherited a couple of million?"

"Money doesn't have anything to do with it. No man in his right senses would hold your money against you – "

Then you and Ian Kidman are reading the ordinary Englishman of to-day differently.

" – it's just that they don't know anything about you and they don't want to buy a pig in a poke."

Harry stood up, his manner cool and distant; only he knew the anger that was threatening to erupt from him at any moment. "I'll take my chances. If you become P.M. I don't want to work with you. If I get the job, you won't be in any Cabinet run by me." He pressed the bell to summon Ferguson. "My butler will show you out."

"I'll show myself out," Argus snarled; the Lancashire accent was thick on his tongue. "But I'm warning you – stay in the Lords. You've had your bloody opportunity!"

He went out of the study, tramping down the stairs with heavy tread; Harry heard him growl something at Ferguson, then a moment later the front door was slammed.

Ferguson appeared at the study door. "The gentleman has left, m'lord. Can I get you something?"

Wordless, Harry shook his head. Ferguson looked hard at him for a moment, his plump face creased with concern; but he said nothing, turned away and went back downstairs. Harry felt the trembling in his hands, the sweat breaking on his face, the tightening in his chest. He sat down suddenly; he was fortunate that the chair was right behind him because he did not look round to check. He leaned his head back, closed his eyes and tried to breathe steadily to settle himself. When he opened his eyes Banning was standing in the doorway.

"You're in no fit condition to go to New York to-morrow."

"I *have* to go." He took out a handkerchief, wiped his face and hands. The tightening in his chest had eased and though he still felt weak he was no longer trembling. "It was just that bastard Argus – "

Banning had taken Harry's wrist, was checking his pulse. "Keep on like this and you're going to give yourself a coronary. Did you have a row with Argus?" Harry nodded. "What about? All right, if it's some political secret, keep it to yourself. But as your doctor I'm telling you you can't go to New York to-morrow. You can send someone else competent enough to do the job, can't you?"

"That's not the point – "

"What is the point, then? Vanity?"

"I can't tell you – not yet. But it's important that I go. You'll understand why in a week or two."

"Any pain in your chest?" Harry shook his head; the tightening had gone. "All right, if you have to go I can't stop you, short of holding a gun at your head. But I'm going with you."

"There's no need for that! I don't want to spoil your holiday in Germany – "

"I'll go next year or the year after. I've never been a particular lover of the Germans, not after Arnhem – "

"Don't bring that up again."

"We all have our traumas. I don't know what yours is yet, but you could be building up to one. I've never seen New York, so now is the opportunity while I can go free of charge. I'll put the cost of the trip on your bill. The National Health won't cover it."

"You have a bloody hide – " Then Harry capitulated; perhaps he was more in need of Joe Banning than he cared to admit. "All right. But our plane leaves at eleven in the morning. That won't give you time to go back to Bayard to pack anything. What about your passport?"

Smugly Banning produced it from his pocket. "A brand new one. I picked it up to-day. You can get one of your chaps to get an American visa for me in a hurry. First thing to-morrow morning I'll go up to Simpsons and order what gear I'll need."

"Will that be on my bill too?"

"Naturally. You owe me something like five million, don't you?"

"You know, you're not going to last as long as me. As soon as I get you in some remote spot I'm going to blow your head off."

"Is that a serious threat or are you only joking?" asked Constance from the doorway.

"We're dreaming up a little sideshow for the U.N.," said Banning, and then told her he was going to New York with Harry.

"Are you telling the truth – there's nothing seriously wrong with him?" Though he was no longer her lover he was, she realised, still a necessary part of her life.

He recognised the genuine concern in her voice and felt suddenly better for it. But there was no time to capitalise on it; their years of separation were not going to be swept away by a few words of sympathy between them. And he still had a woman in New York whom he had to let down gently. "Joe will look after me. It's just a bit of tiredness, that's all."

"Take notice of him, then," she said. "How did it go with Mr Argus?"

"We're on opposite sides," he said cryptically, and caught her understanding glance. He stood up, still feeling a slight weakness in his legs. "You'd better stay the night with us, Joe. I'll telephone my secretary and see that you're on the plane with us to-morrow. But if you don't mind, I'd rather you kept quiet that you're my personal physician. Ostensibly you're an ordinary English voter going over to see how his money is spent at the U.N."

He was in bed when Constance knocked on his door and came into the bedroom. "What did Argus want?"

He told her. "I should have kicked him in the behind and thrown him out of the house. Then I might not have had that silly attack."

"Things aren't going to be easier for you if you become P.M. What happens then? There won't be time for you to take that holiday Joe has suggested."

"I'll be all right. I can't let up now." She was standing beside the bed, her hands clasped in front of her. He half-lifted his own hand to reach out to her, then dropped it back on the coverlet; he did not have the strength to-night not to be affected if she rebuffed him. "If I don't get the leadership, then – "

"Then what?"

"I was going to say I'd retire from politics." He smiled, at himself as much as at her; for a moment they were on an old level of honesty. "But I don't think I could let Argus get away with it. I'd wait around in the hope that he'd prove a failure and be kicked out. And he will be."

"Don't let it keep you awake." She had a sudden urge to lean forward and kiss him, but she could not bring herself to that point of surrender. She smoothed down the coverlet, tucking it in, escaping with a show of housekeeping: "It's time to order some more notepaper. Should I order it with or without the crest?"

"We don't lose the crest even if I give up the title. The main thing will be the address. Downing Street and Chequers or Eaton Square and Aidan Hall. If it's the first two, we get the notepaper free."

"I'll go ahead with the order, then. Just in case."

## 3

"The trouble is with the damned Russians." Sir Nigel Straker was a career diplomat and had been leader of the United Kingdom Mission to the United Nations for four years. He was a plump baldheaded man with a slight lisp, the peculiarly English affectation of not attempting to pronounce his "r's" properly; somehow in his mouth the Wussians sounded like a comic opera threat. But if his tongue was a joke to non-English newspapermen, his mind was not. Behind the jocular smile and the twinkling eyes was an intellect as cold and patient as that of any of the Russians he had to face. "They are backing the guerrillas all the way – what have they to lose? The guerrillas aren't fighting any of the major Arab countries. They're fighting the Jordanians and some of the oil sheikhs on the Oman coast, but the Russians turned their backs on that lot long ago."

"What do the guerrillas want?" Harry asked. "I haven't had a chance to talk to Hazza this time."

"The guerrillas want the oil sheikhs to turn over a percentage of their oil revenue to help finance their fight against Israel."

Harry had never realised the letter "r" was so common in English speech till one had to listen to someone who could not pronounce it. Diplomatic negotiations must sound a little farcical when one side used words that might have come straight from the nursery.

"Do the guerrillas appear openly at the sessions?"

"They are in the observers' gallery every day – the Russians make sure they get seats. That's the only time we see them – they never come to any of the receptions, not even when they are invited. They have become very adept at what I suppose one might call guerrilla diplomacy."

"If Hazza gives in to them, he is going to put the bite on us for bigger royalties. That means that indirectly we'll be financing the war against Israel. The Israelis won't like that."

"That's just another of our problems. If that oil field of the Israelis' down in the Sinai turns out to be as rich as I've heard, then they could supply Britain with all our oil needs and we'd no longer be at the mercy of whoever happens to be in power in the Arab countries." The twinkling eyes did not quite camouflage the loaded question he then put: "I suppose, though, it would make quite a difference to some of our current oil investments?"

"Speaking personally – " Then Harry smiled. "No, I can't speak personally any more. But yes, it would make quite a difference. We have thousands of millions of pounds tied up in Arab countries."

Harry and his team of advisers had arrived in New York yesterday from London. They were staying at the Waldorf, where Harry had also managed to book in Joe Banning. He had had no time last night to see Lynne, but he had phoned her and was due to meet her to-night for a late supper. The first reconvened session of the Security Council was to meet this afternoon, and this morning Harry had come to the offices of the U.N. Mission on the tenth floor of 845 Third Avenue. Beyond the closed windows New York steamed under the August sun, but here in the Ambassador's office Harry felt uncomfortably chilled by the air-conditioning. Straker, in his four years here, had become a New Yorker, at least climatically.

Harry stood up: it would be a relief to escape out into the jungle humidity of Third Avenue. "Who's here for the Israelis?"

"Goldsmid, their Foreign Minister. He has Mordecai Roth as his chief adviser. A very good man, except that he doesn't like the British very much – Roth, I mean. Goldsmid takes a lot of notice of him."

"I knew Roth a long time ago. He was very pro-British in those days."

"One wonders what changes them. But then I was not very pro-American when I came here. Now – " Straker smiled. "I hope you aren't thinking of moving me elsewhere?"

"No." Harry felt he was here under false pretences: what would the F.O. men think if they knew he was using his position as their boss only as a stepping-stone to higher office? "I have no intention of moving you."

The session of the Security Council that afternoon was no more than a re-tuning after the intermission: throats were cleared, opinions readjusted, unexpected harmonies worked out. Harry noticed that Goldsmid, the Israeli Foreign Minister, seemed to speak only from notes supplied to him by Mordecai Roth. The latter never once looked in Harry's direction, not even when Goldsmid on one occasion addressed himself to the British delegation in highly critical tones. When the session broke up, having achieved nothing, Harry detached himself from his advisers and waited till Roth, carrying a briefcase that seemed to drag him down on one side, came out of the conference hall. As soon as he saw Harry he stopped, tried to turn aside, but Harry moved forward and intercepted him.

"Mr Roth, would you care to have a drink with me?"

"No, Lord Carmel. You should see our Foreign Minister. I'm sure he would be glad to meet you."

"I have an appointment to meet him later. But he's new – you are the man who knows the ins and outs of what we're trying to solve. And at least we know each other."

Roth had appeared upset when he had first seen Harry, but now he had regained his composure. "Indeed we do. I shall have no hesitation in warning my Minister that you are not a man to be trusted."

Delegates, secretaries, translators flowed by on either side of them; they were in the middle of a stream that could carry what was overheard far beyond this lobby. "I have no track record at all in my official capacity. That is the only one you should judge me on."

Roth waited till the stream had thinned, till they were standing alone. "I judge you on the Balfour Declaration, as I judge most Englishmen. A man from the Stern Gang, as you people called them, once told me he thought that was one of the great broken promises of history."

"I feel as much shame for that broken promise as do a lot of other Englishmen. Not as much as my brother Roger did," he added honestly, "but then he was always much more pro-Zionist than I was."

Roth's face softened for an instant. "I only wish that he were here in your place."

"It was your people who caused the substitution. This was not my ambition – " Then, in 1946, there had been no ambition at all. The irony of the situation made him smile; but there was a double irony: "If my brother had finished up in this job he might have been able to change the Foreign Office's traditional pro-Arab attitude. You should lay *that* at the door of your terrorists. But then one of the troubles with terrorists is that they are never prepared to recognise their friends. Even the Arab countries are realising that with the Arab terrorists you are now dealing with."

"I have no sympathy for Arab problems with the terrorists."

Harry shrugged, gave up. "You seem to have become as intractable as the Russians, Roth."

"Not as intractable. Just as untrusting, perhaps."

He walked away through the lobby, leaning to one side against the weight of his briefcase; Harry wondered if it contained all his pent-up hatred of Britain, the fat dossier on Albion's perfidies. He also wondered if hatred of himself ever weighed Roth down, though hatred, to be sure, was not always a weight. There were some people who gained strength from it, and Roth seemed to be one of them.

That evening Hazza gave a reception at the Hotel Regency – "Just for a few hundred friends and enemies," he told Harry as the latter, accompanied by Banning, arrived late for the gathering. He had sent his Private Secretary and the other members of the delegation on before him, harvesters sent out for the early gossip among the early arrivals, the juniors from the other delegations. On Banning's orders he himself had lain down for an hour.

"This is a friend of mine. Joe Banning. He is over here as the unofficial, unelected representative of the British taxpayer."

"I am here in the same capacity for the Qaran taxpayer," said Hazza, waving his hands, sprinkling the air with diamond and ruby lights.

"Do you have taxpayers in Qaran?" said Banning.

"Only my brothers. I tax them very heavily, since I am also their soul source of income. Ah, here is Miss Erickson. She speaks so highly of you, Harry, that I wonder whether she works for me or for you as a public relations lady."

Harry had known that if he was going to be here in New York for a week or two it would be inevitable that Banning and Lynne would meet. When Banning had asked if he might come to this reception Harry had not hesitated: better to get the meeting over and done with. If Joe had any questions they would be asked and answered long before they were on the plane for home. When Harry went back to London he did not want his mind occupied by anything but the reason for his return: to run for Prime Minister. All personal problems had to be out of the way by then.

"Nice to see you again, Lord Carmel." Lynne was all circumspection and decorum; she hadn't yet placed who or what Joe Banning might be. "A pity you had to come back during our heat wave."

The conversation for the next few minutes was as innocuous and banal as that of strangers stranded in an airport lounge waiting for a delayed plane. But Banning was an acute diagnos-

305

tician of relationship; a glance, a smile, the emphasis on a single word were symptoms that he recognised as readily as those of some medical condition. He held his glass against his chest, not drinking at all, and behind his bland social smile watched two people who, he guessed, were or had been lovers. Harry, sensitive to Banning's reaction, wondered what the latter thought of Lynne.

Then Hazza took Harry away – "I want you to meet a Swiss arms salesman – his idea of neutrality is to sell to both sides" – and Banning and Lynne were left together. Banning took one sip of his drink and decided one would be enough: Hazza seemed to be serving Arabian Scotch. "You've met Harry before?"

"Harry?" Lynne, five feet nine and one hundred and forty pounds, blinked innocently, Shirley Temple in Cinemascope.

"The chap who just left us. I thought you knew him as well as I do."

"Mr Banning – " Lynne shed the look of innocence; she knew the cobwebs of childhood did not suit her. "Just what is your function with the British delegation?"

"Oh, I'm just a tourist, nothing more. Oh, I'm also a friend of Harry's and his wife."

"You sort of underlined that last word. Mr Banning, don't jump to conclusions about me. I'm a simple girl, too simple sometimes, maybe. But I have simple ethics. I know Harry has a wife and I know what the situation is there. If you're such a friend of theirs, you'd know it too."

"I know it well enough." Banning looked down at his drink, decided against another taste of it. "I find it one of the major disasters of my experience. They were the handsomest, happiest couple one could wish to meet. And then – " He looked up. "I still live in hopes of seeing them come together again."

"Whom are you in love with, Mr Banning? Harry or his wife?"

"You're a blunt young lady."

"Harry said that was part of my charm. Personally, I don't find it one of my more endearing facets, but I'm stuck with it. You haven't answered my question."

"Are you in love with Harry?"

"What are we playing – tit for tat?" They had retreated to a corner of the big reception room, high and dry above the cocktail tide. By some trick of acoustics this corner was quiet; they were able to speak almost in whispers to each other. Lynne, with the dexterity of a practised party juggler, lifted a martini from the tray of a passing waiter, depositing her empty glass on the tray at the same time. She did it without seeming to take her eyes off Banning or losing the rhythm of the conversation. "Don't let's fight, Mr Banning. Yes, I'm in love with Harry. What about you?"

"I like him, but I don't think I find him lovable."

"So you're in love with his wife?"

"Perceptive but not provable. Let's just say I'm concerned for her happiness."

"I'm concerned for mine. I wouldn't do anything to break up a marriage – but if that marriage was already a wreck, with no chance of it being put together again, then I think I'm entitled to take my chances."

"His being Foreign Secretary – that might alter things. Forgive me for saying so, but the English still get a touch of jaundice whenever an American girl marries into our aristocracy. We still think you're all like the millionaires' daughters who came over in Edwardian times trying to buy a title to take back home."

"Unfortunately I don't happen to be a millionaire's daughter – my father is a tugboat captain. And if I took a title back to Brooklyn they'd think I'd married either a pop singer or a TV wrestler. When I marry, I don't want to be sneered at or laughed at. So I'm not going to rush down the aisle if and when Harry asks me. I'm a careful girl when it comes to proposals."

Banning raised his glass to her. "I can't drink this, it tastes like something distilled in a camel's bladder, but take the gesture for the act. I drink to you, Miss Erickson. I'm really sorry we got off on the wrong foot."

"All my life," said Lynne, "that's the foot that's been tripping me up."

Then Harry rejoined them and a little later he and Lynne went out to supper. As he left Banning he said, "I don't have to do any explaining, I take it. You seem to know what the score is."

"*I* do," said Banning. "But do you?"

"Don't be enigmatic, Joe. What's that supposed to mean?"

But Banning turned away from him, smiled at Lynne. "Good luck, Miss Erickson. And watch that wrong foot."

Outside the hotel, while they were waiting for the delegation's official car to appear, Harry said, "Are you and Joe in some sort of conspiracy?"

"Against whom?"

But then the car, a black Rolls-Royce Phantom, appeared and Harry, suddenly embarrassed by his own and her question, changed the subject. "Where do we go to eat? The choice is yours."

Lynne picked up the speaking tube, gave the address of her apartment to the chauffeur beyond the glass partition that separated the front and back seats. She hung the tube back on its hook. "I thought those sort of things went out with Queen Marie of Rumania."

"My grandfather had one in his Rolls. He was a bit of an eccentric – he used to carry on one-way conversations with his chauffeur as if the chap was in another county instead of just in front of him. Why are we going to your place?"

"Because I have the feeling we may be having a one-way conversation. If I'm going to break down and weep, I don't want to do it in public."

Neither of them said anything of any consequence till they were in her apartment and had eaten the supper she prepared. He had told the chauffeur to call back for him at eleven. He was determined not to spend the night with Lynne, but at the same time he was aware that by nominating what time the car was to call for him he was, in effect, disposing of a mistress by timetable. He had never been as ruthlessly indifferent as that before, and in an effort not to sound bluntly cruel he found himself groping for the right words with which to tell her he was not going to see her any more.

At last she said, "Harry, darling, are you trying to be diplomatic or something? You're not much of an advertisement for the Foreign Office."

"I'm not much of an advertisement for anything." They were sitting at opposite ends of the couch in her small living-room, but he made no attempt to move closer to her. He had taken off his jacket but, unusual for him, he felt as constrained as if he were in formal dress. "For English chivalry, an English gentleman, anything you care to name."

"How about an English sonofabitch, for starters?" She kicked off her shoes, tucked her feet up under her, smoothed down her dress; she was much more at home than he, even allowing for her being in her own apartment. He wondered if she had been through this sort of thing before, but in that at least he was too much of an English gentleman to ask.

"What did Joe Banning tell you about me?"

"Nothing. Don't blame Mr Banning for anything." Then, the victim of her own impulsiveness, she said, "Don't blame yourself for anything, Harry. I got into this with my eyes wide open." Then she closed her eyes and beat her fist on her knee. "Oh Jesus, why am I always so honest! I always give away my advantage." She opened her eyes again. "I had the advantage, didn't I? You *are* going to be a sonofabitch, aren't you?"

He reached for her hand and she let him take it. "You may not believe me, but when I left here last time I was seriously thinking of asking my wife for a divorce. And I was going to come back here and ask you to marry me."

"Fine words butter no parsnips – isn't that an old English saying? Nor do they butter up discarded mistresses. But go on."

"Don't be flip with me, Lynne. I'm not finding this easy – "

"Neither am I." She softened her tone, pressed his hand. "All right, I'll listen. What made you change your mind, take my dream away from me?"

"Politics." It was a glib answer, but it was true and he could think of none that would not hurt her more: to tell her that he was sacrificing her to ambition would be more ruthlessly cruel than even he was capable of. He had to add a lie to the truth: "When I left here I thought I was out of politics for good. Now, in the political hierarchy as we run things, I'm Number Three in our Cabinet."

"And a divorced Number Three just wouldn't be welcome in the Cabinet?"

"I don't know that the Cabinet would mind. But the ladies in our constituency parties might."

"I guess so. I remember how it was when Governor Rockefeller divorced his wife to marry Happy. I knew one Daughter of the American Revolution who wanted his name put alongside that of Nathan Hale. I was never quite sure whether she thought Rockefeller was an American traitor or just a traitor to marriage. It would have been interesting – " He felt her fingers tighten on his. "Why am I talking so much?"

He saw the tears start in her eyes and he reached out and drew her towards him, something he had wanted to avoid. He said honestly, savouring the truth because he had been able to spend so little of it to-night, "I was very happy with you, my love." She put her fingers to his lips and for a moment he was puzzled. Then he understood. "Sorry. But you could have been my love, you know, if things had gone differently."

"Not things, Harry. *You.*" She sat up, too honest to hide her face while she told him the truth of himself. "I don't know what you want to be, but you want to be more than just a contented husband, whether to me or your wife back home. Oh, I've seen it happen before. This town is full of men with ambition. For all I know, you may not rest till you're Prime Minister of England. Can a lord be Prime Minister?"

"Not a peer, no." How much of himself had he exposed? Had Joe Banning guessed what was in the wind and said something to her?

"Well, whatever you want to be, I wouldn't fit into it. Will your wife?"

"I don't know. There's no reconciliation, if that's what you mean. You haven't suddenly become second best to her."

"Haven't I? Well, I suppose that's some sort of consolation prize." She slid away from him, stood up. "I think we better say good-bye now. I can feel a touch of bitchiness coming on."

He stood up beside her. Even without her shoes she had only to raise her face slightly to look into his; but she had suddenly become small, weak and vulnerable. To-night she was a cheerless cheer-leader, unable to raise a rallying cry even for herself. He wanted to put his arms round her, but he knew that would be the wrong thing. He put on his jacket, kissed her lightly on the forehead.

"I wish it could have been otherwise."

She said nothing, just shook her head and continued to stare at him. He let himself out of her apartment, went down in the elevator. As he crossed the lobby he looked at his watch: ten o'clock. He had proved himself a sonofabitch ahead of schedule.

He went out into the street, looking for a taxi, and at that moment the chauffeur brought the Rolls into the open space at the kerb. In the back seat was Roy Chadwick, his Private Secretary.

"It was fortunate the chauffeur came back to the reception, sir." Chadwick's voice had a sharp, remonstrative note to it; he was not accustomed to Ministers who went off without telling their Private Secretaries where they could be found. "I've been looking everywhere for you."

"Sorry, Roy." Harry got into the car, sank back into the seat. He could feel the trembling in his hands again and he knew he was in for another restless night. "What's the trouble? Something important?"

"The Prime Minister collapsed and died two hours ago."

# Eleven

"I liked Miss Erickson."

Joe Banning settled into the seat beside Harry, two hours after take-off. He had sat patiently in the rear of the first-class compartment, making no effort to intrude, while Harry had worked on papers with Chadwick and Weldon, the senior Foreign Office official who was travelling with them. The rest of the delegation had been left in New York, with Straker to take charge if the Security Council meeting was hastily reconvened. On the news of Wolfe's death there had been an unexpected request for a week's adjournment from the Qaran delegation and when Britain and the United States had been informed why Hazza wanted the adjournment they had rounded up enough support to arrange it.

"I have to go back, Harry," Hazza had said. "The guerrillas have somehow got down into my territory. I suspect my brothers have helped them. They – the guerrillas – are demanding a million pounds from me to help finance them. Otherwise they threaten to blow up the oilfields."

"What are you going to do?"

"I shan't know till I return home. I may have to ask you for troops."

"We're not going to like that. It will be damned awkward for me personally. It's bad enough as it is – some newspapers have already suggested that the family should sell all its holdings in Anglo-Qaran. I may have to do it – " If he should win the race to Number Ten, he *would* have to sell all the Anglo-Qaran stocks after all; his position as Foreign Secretary owning stock in such a strategic commodity was already too delicate, but as Prime Minister it would be untenable. Other political families had been unworried by such scruples, but those days were past. The profit motive, one of the cornerstones of Empire, had become as suspect as imperialism itself.

"I'll buy your shares and sell them back to you when you want them. But we can't let these chappies get away with this." *Chappies*: the schoolmistress from Cheltenham had not lost her influence on Hazza. "We have to stop them, Harry. My Levies may be enough to deal with them, but if not I'll have to ask you for help. You can warn your Prime Minister, whoever he is going to be."

"I'll do that," said Harry.

Wolfe's death had occurred on the Wednesday night; Harry and the others left New York on a VC-10 at eight o'clock on the Thursday evening. Hazza had gone out that afternoon on a Trucial Airlines Boeing 707, one of only two that the airline owned; the company was a conceit of Hazza's and he had the grace to laugh at himself. Harry wondered if Lynne had been at the airport to see Hazza off. He was just glad that he and Hazza had not been travelling together on this BOAC flight: he did not think he could have borne another farewell to Lynne.

"How serious is it with her?" Banning now asked.

"Is it any of your business?"

"No," said Banning, smiling and accepting another drink from the steward. "But you know what an interfering, inquisitive blighter I am. Have you disappointed her too?"

Harry was silent for a minute or two, his head turned away as he gazed out the window at the escarpment of dark clouds building up in the moonlit sky down south. He had not slept well last night, but in the twenty-four hours since the shock of Wolfe's death he had managed to keep his thoughts coherent and had resolutely allowed no fantasies of the future to take hold of him: there were still some major hurdles between him and Number Ten. Yet, though his mind had been cool and ordered, there had been the thought of Lynne, like a splinter buried deep beneath the skin, that he had not been able to remove. Guilt and love were a suppurating sore.

"Yes," he admitted at last, looking back at Banning. "But don't think I'm indifferent about it."

"Who mentioned indifference? But now you're Foreign Secretary, is it wise to have a girl in every port? It may have been all right at the time of the Congress of Vienna, but not now."

"Are you my doctor or my tutor in morals?"

"Take your pick. The fee is the same."

Then Roy Chadwick, standing in the aisle, leaned over them. He was a very tall, thin young man with a long nose and bright shrewd eyes; he seemed made up of mismatched parts, his body moving languidly and his face and head continually alert for danger signals. "The steward allowed me to look at the passenger manifest, sir. I notice that Mr Roth, of the Israelis, is back in the economy section."

"With his wife?"

"Her name isn't on the manifest."

"What's he doing on this plane? If he was leaving New York I'd have thought he'd have been on an El Al plane for Tel Aviv."

"Would you care to have him up for a drink, sir? Perhaps we could find out – ?"

Harry hesitated a moment. "All right, ask him up."

When Chadwick had gone back to the rear of the plane Banning said, "How do you get on with this Roth? You didn't seem too enthusiastic about young Chadwick's suggestion."

"You're too perceptive. Why don't you go for a stroll out on the wing for a while?"

Then Roth came through into the first-class section with Chadwick. Harry had not really expected the Israeli to accept the invitation, but he hid his surprise and did his best to be both diplomatic and friendly. Banning rose to give up his seat, but Roth waved him down and sat down on the arm of the seat across the aisle. "You must be wondering why I'm bound for London, Lord Carmel. Just a stop-over to talk with our Embassy there. My Minister has sent me home to Tel Aviv – I'll be back in New York for the reconvening of the meeting. Will you be coming back?"

"Of course." Unless by then I'm P.M. "I'm sorry we had to interrupt the discussions."

"I gather the Qaranians welcomed it. I understand Sheikh Hazza is having trouble with the Palestinians. I don't relish ironic humour, but sometimes I can't help smiling when I hear of Arab fighting Arab."

"Please, don't move anyone!"

A burly middle-aged man and a young girl stood in the doorway that led to the rear of the plane. The man, black-haired and dark-faced, nervously waved a pistol at the passengers, then shoved it into the back of the senior steward who was immediately in front of him. The girl, short and petite and very young-looking, a schoolgirl on an escapade, also had a pistol.

"If everyone behaves, no one will be hurt." The man's English, though good, was accented by his native tongue and his own nervousness: his voice sounded girlish, reducing its threatening tone. But there was no mistaking the threat of the pistol in his right hand nor the yellow plastic ball he now produced from his jacket pocket. He held up the yellow ball. "This is a grenade. One stupid move, ladies and gentlemen, and it goes off. And we all die. The thought of dying does not frighten us or our comrades – there are four of them down at the back keeping an eye on your fellow passengers."

He nodded to the girl, who produced a grenade from the handbag slung over her shoulder. She moved up the aisle to the front of the compartment, motioned with her pistol for the senior steward to sit down in one of the vacant seats. Her companion took another look around the compartment, nodded again to the girl, then opened the door to the flight deck and stepped quickly through, leaving the door open behind him.

The young girl spoke for the first time, her English also good, her voice firm and under control. "We are going to Algiers, ladies and gentlemen. There, most of you will be off-loaded – unharmed, if you behave yourselves for the rest of the flight. We are interested in only two of you – it is a pity all the rest of you have to be inconvenienced." She looked at Harry and Roth, who had sunk back into the seat behind him. "Gentlemen like yourselves should never travel on scheduled airlines. You only endanger the general public."

"You're making some contribution to the danger," said Harry, nodding to the pistol and the grenade. There was a dryness in his throat, but so far he was not afraid. He looked at Chadwick. "Were their names on the manifest?"

"He would not have recognised them," said the girl. "We are travelling on Turkish and Pakistani passports – forged, of course. They wouldn't know the difference at Kennedy Airport – we're all just Wogs to them."

"How did you get through the inspection?" Banning's grip on his glass was tight, but that was the only sign he was showing any strain. "We were all frisked, even Lord Carmel."

"That would be giving away secrets that we may have to use again. There are always ways and means of beating technology. That is its weakness – it can always be improved upon."

The other passengers in the section had sat quietly so far; but now one of the women, plump and elderly, began to weep in long gasping sobs. Her husband, thin and bald, his face sun-cancered as if he had spent all his life on a farm, tried to comfort her.

The Arab girl nodded to the senior steward. "Get the lady some water." She moved down the aisle after him. As she passed Chadwick, the latter made a slight move; instantly the pistol came round, finished up inches from his throat. "Don't!" Chadwick slid back into his seat. On the other side of the aisle the

314

elderly woman's gasps quickened and the girl looked at her with concern. Then she looked back angrily at Chadwick. "You are only upsetting the lady. Behave yourself!"

She was like a young trainee teacher on her first day in a kindergarten class; she was not accustomed to as much authority as she had now. She clutched the pistol more tightly, followed the steward down to the galley.

Banning leaned out of his aisle seat and looked up through the open door to the flight deck. "Can't see what's happening up there. Don't you think they're a bit ambitious, hoping for Algiers? Will we have enough fuel?"

"Why not Paris or Madrid, if they want to skip London?" said a man behind Banning.

"Algiers is Arab. They know they'll be safe there." There had been no expression on Roth's face since the hijackers had first appeared other than a look of sad resignation, as if this were no more than another in the long list of disasters that was his life. He looked across at Harry. "I can understand their wanting you. But why me? I'm unimportant."

"And unlucky," said the young girl, coming back up the aisle behind the steward as the latter brought a glass of water for the distressed woman. "We just happened to recognise you, Mr Roth, as you boarded the plane. You are a sort of bonus."

"Bonus for what?"

"We'll tell you that at Algiers." The two front seats on the starboard side were vacant and she settled herself on the arm of the aisle seat, the pistol and the grenade held negligently but in full view, like two warning devices, of all the passengers. She looked at the American farmer and his wife. "How is your wife, sir?"

"She'll be okay, I think." The farmer, dressed in his going-to-meeting best, looked at her with a mixture of puzzlement and fear. "You seem a nice polite young lady – why are you doing something terrible like this?"

The girl pointed the pistol at Roth. "Ask him. He understands history, especially the history of Palestine. His comrades fought the English, just as we are fighting the Israelis. He is a nice polite man too."

"I never carried a gun," said Roth.

The girl shrugged. "Everyone has his own means." She looked back at the American farmer. "You have your own guerrillas, sir, from what I read in your newspapers. Perhaps they, too, are nice polite people."

The farmer shook his head, but said nothing further. Harry had been watching the girl carefully, rarely taking his eyes off the hand that held the grenade. The pistol did not worry him so much; but the grenade could be an accident waiting to happen. He could not see what trigger mechanism it had, but if she should drop it or have it jerked out of her hand by someone attempting something foolishly heroic, then the chances were that it would explode at once.

He caught the girl's eye. "May I say something to the others over the intercom?"

"What do you want to say?" she asked suspiciously.

"I just want to ask everyone to do exactly what you want. If Mr Roth and I are the only ones you are interested in, then that makes us partly responsible for everyone on board."

The girl turned to the open door to the flight deck, said something in Arabic to her unseen colleague. Then she looked back at Harry. "All right, Lord Carmel. But don't try to be a hero. Otherwise – " She held up the grenade.

Harry squeezed out past Banning, went up the aisle and into the flight deck. The navigator and flight engineer were sitting in their seats; both of them looked up at him curiously as he appeared in the doorway. The hijacker was squeezed in behind the captain and the co-pilot; he waved the hand holding the grenade at Harry while he kept the pistol pointed at the captain. He seemed less nervous than when he had been out in the passenger compartment, as if he felt certain now that the hijacking had been a success.

"There's the microphone, Lord Carmel," he said. "Go ahead, give them your friendly advice."

Harry took the microphone from the captain. "Have you sent out any messages yet that we've been hijacked?"

The captain shook his head. "Not yet. He caught me on the hop – "

"We'll name the time when the advice goes out," said the Palestinian. "Give them your message, Lord Carmel, and then go back to your seat."

Harry spoke into the microphone, first introducing himself. "I'm afraid that, inadvertently, I am the cause of your being in this predicament. For reasons I don't yet know of, the people now holding you prisoner want to kidnap me. I therefore ask you to be patient and not attempt to thwart these people – anything foolish, no matter how brave, would only endanger us all."

He handed the microphone back to the captain. "Will we have enough fuel to get us to Algiers?"

"We'll be stretching it." Captain Janney was a short burly man with a red beard and strong resonant voice. "We may finish up as the only VC-10 glider in commission. But the man says he won't settle for anywhere but Algiers, so Algiers it has to be."

Harry left the flight deck and went back to his seat. As he stepped into the compartment he glanced down the long aisle into the economy class section; heads were hung out from seats as the passengers looked up towards him. Suddenly someone in the far end of the plane started to clap; the sound spread up the aircraft and was taken up by the passengers in the first-class section. The clapping was quiet and polite, as if the passengers were not quite sure how he would react; no voice was raised and no one made any attempt to stand up. Harry, unused to spontaneous demonstrations in his favour, for once was embarrassed and felt strangely moved. He had just asked people not to be heroes and was rewarded by being greeted as one. He made an awkward gesture of acknowledgment and got into his seat beside Banning.

"You look surprised," said Banning. "You shouldn't be. People respond to a generous gesture. It's natural with most of them."

"How did you respond to it?"

"I'm grateful." Then he added, the tart note gone from his voice, "I'm worried, too. What the hell do they plan to do with you?"

Harry's irritation with Banning's mordant reception abruptly vanished when he saw the extent of the latter's concern for him. He had not yet considered what the hijackers wanted or what they intended to do with him; the situation, though not uncommon over the past few years, was still so bizarre to anyone experiencing it for the first time that it was difficult to consider it logically. The girl did not look like a murderer; but what did a murderer look like? Harry had seen the Mau Mau murderer of Janet Buck; but some of the worst murderers in history had looked like meek saints. In any event how did one look logically on the possibility of one's own murder? For the first time he became worried.

The next few hours were the longest Harry could remember. Night faded and red daylight bled into the sky as the plane flew into it; the day was going to be fine and clear, a tourist's delight.

Once, as the thought of his death seeped up into his mind like a fragment of a nightmare, he found himself sweating and then his hand on the arm-rest between himself and Banning began to tremble.

Banning looked quickly at him. "You all right?"

Harry nodded, struggling to calm himself. "My imagination just got the better of me."

"I'm staying with you," Banning said abruptly. "Wherever they take you from Algiers, I'll come with you."

"Don't be bloody stupid, Joe. I've told everyone else not to act foolish – "

"They're in a different position. I'm not going to do anything like trying to take the plane away from Her Nibs and her friends. I'm just coming with you to see you don't collapse on them. If you did, they might get nasty and take it out on you for letting them down."

Then the captain's voice, firm and a little too loud, said over the loudspeakers: "We shall be landing at Algiers in ten minutes. Fasten your seat belts and no smoking. And please do exactly as our – our hosts tell you."

The plane banked and Harry saw the jungle of white and beige toy blocks that was Algiers. The sun was already bright and away in the distance he could see the green corduroy of vineyards cloaking the shoulders of some hills. This was a country that had been born out of war, but in the clear hot morning, from this altitude, it looked as if it had never known anything but peace: from the right altitude perhaps all the world looked like Eden. He glanced at the young girl, legs braced against the banking of the plane, and wondered if she had ever flown over Israel and what she thought when she looked down on what to her would always be Palestine, the Eden she had never known.

Then the plane was coming in to land. They touched down beautifully, as if Captain Janney were on a demonstration flight, the reverse thrust was applied to the engines and then they were rolling gently to a stop. A moment, then the engines cut out and the interior of the plane was eerily silent. There was none of the usual bustle and movement of passengers impatient to disembark: everyone sat waiting fearfully yet hopefully.

The girl, all at once showing the strain of the last few hours, addressed Harry and the others. "We are staying out here at the end of the field. The airport authorities know the situation and have promised not to interfere – if they do, we have promised to blow up the plane with everyone in it. Buses are on their way out from the terminal now. You will disembark one at a time and when you reach the bottom of the steps you will continue walking till you are fifty yards from the plane. Stop there and wait for the order to board the buses. It may be some time before your luggage catches up with you – we can only apologise. All we do is ask you to remember that we have done this only in the cause of freeing our country."

It was another five minutes before the first of the passengers got up and moved out of the plane. As each of them passed Harry there was a murmur of sympathy or good luck. As the farmer and his wife went past they stopped for a moment and the woman said, "God will protect you, sir. I have been praying for you." Her husband nodded and the two of them moved on, past the young girl who smiled at them like a stewardess farewelling them but got no reaction. At last the first-class compartment was cleared of all but Harry, Banning, Roth and Chadwick; Weldon had been told to get off with the stewards and stewardesses. There was still movement down in the economy section, but the disembarkation there seemed to be just as orderly as it had been up front. The heat inside the aircraft was stifling and Harry and the other three men were sweating profusely.

Then the flight crew, prodded by their captor, came through from the flight deck. As they did so two more men and a girl, all young and all carrying the standard pistol and grenade, came up from the economy section. The man who had been on the flight deck and was evidently the leader said something in Arabic; at once one of the men went back to the rear of the plane and the second man moved forward to stand inside the exit door, putting on dark glasses against the glare from outside. The leader motioned Captain Janney and his crew to sit down, then addressed himself directly to Harry.

"We are re-fuelling now and we shall take off as soon as it is completed. You and Mr Roth will be coming with us, but your secretary and this gentleman – Dr Banning? – will be getting off."

"Not me," said Banning. His face was red and glistening with sweat and he looked most uncomfortable. "I'm Lord Carmel's personal physician. I think it advisable that I stay with him."

"It's not necessary, Joe – "

"What's the matter with you?" The leader looked at Harry as if the latter had played some trick on him.

"Nothing. I'm all right - "

"I'm staying," said Banning flatly. He wiped his face with his already soaked handkerchief; if anyone looked in need of a doctor, it was he. "Despite what Lord Carmel says, I think it *is* necessary."

The leader looked at the two girls, then shrugged. "All right, if you choose to, Doctor. Another one won't matter – we are already stuck with Mr Roth."

"You don't have to be stuck with me," said Roth mildly. He appeared the least concerned of all of those in the plane, even less on edge than the hijackers. "I'm perfectly willing to relieve you of the responsibility for me."

The leader shook his head, humourless and stern-faced. "No, you may turn out to have a price, Mr Roth – we just don't know how much your country is prepared to pay for its diplomats. Your price, Lord Carmel – " He looked back at Harry, savoured the moment before he put the ticket on him. "Your price is one million pounds. In sterling notes deposited in the Bank of Arabia in Beirut within seventy-two hours. That is the message we want Mr Chadwick to deliver to your government as soon as we release him."

"A *million*?" Harry shook his head. "I'm not worth that much to anyone. Especially my government."

"Your modesty is creditable, Lord Carmel. But you *are* worth a million pounds – you are worth a great deal more, but we aren't greedy. Your government may refuse to pay that much for you, but the money can be raised by your family without too much effort. The selling of even just part of your Anglo-Qaran holdings would bring enough. But the money has to be paid over by your government, whether it raises the money or your family does. It's a matter of protocol with us."

"And if it isn't paid?"

"Then I'm afraid in seventy-*three* hours from now you will be dead."

"What about Mr Roth and Dr Banning and these gentlemen?" Harry nodded at the crew.

"We shall have to decide about Mr Roth. The others will be safe, as long as they behave themselves."

"What about the aircraft?" said Janney. "That's my responsibility."

"Perhaps BOAC would pay us a million pounds for its return?" For the first time the leader's face showed a hint of humour; but it was only fleeting, gone like a wince of pain. "No, you and your plane will be safe, Captain. We have the items we want to sell."

"Items?" Harry looked up sharply. "Plural?"

The leader nodded, at last allowing his face to crack in a smile. "Be patient, Lord Carmel. You will meet your fellow commodity."

He gestured to the girls to keep the prisoners covered, then he disappeared out the front exit door, saying something to the dark-glassed guard as he passed him. Two minutes later he was back, jerking his head at Captain Janney.

"Up front, Captain. The re-fuelling is going ahead without any fuss. The starter truck is outside and we'll take off as soon as the tanks are full."

He and Janney went up into the flight deck again and Harry, ignoring the two girls, looked across at the co-pilot. "Why aren't the Algerians trying to stop them?"

The co-pilot, a craggy-faced blond man in his early thirties, shrugged. "Who knows? Maybe they're sympathetic – they probably are. Also, there are another three blokes down there with guns – God knows where they came from."

"They have been waiting here for us," said the girl who had been in the first-class compartment during the flight. "This is a military operation, properly planned."

Harry looked out the window. The last of the buses with the passengers was pulling away, circling the aircraft and heading for the distant terminal: faces were pressed against the bus windows, blank as balloons. Police and army cars and trucks were ringed round the plane, police and troops congregated in small groups like gatecrashers uncertain whether they wanted to come to this party or not. No police or any army officer had approached the plane and Harry guessed that the hijackers' instructions to the airport control had been specific on that point. As far as he could see from his window, the only vehicles near the plane were the two fuelling trucks and the truck with the air compression starter for the jet engines. Men in overalls occasionally flashed in and out beneath the wing, working hurriedly as if, whether they were sympathetic to the Palestinian cause or not, all they wanted was to get the re-fuelling over and done with as soon as possible. Affluent revolutionaries are like affluent workers, Harry mused: solidarity is a last resort.

"Sir?" Chadwick, long, thin and worried, sweat dripping from the end of his long nose, leaned forward in his seat. "Am I to deliver that message for them?"

Harry nodded. "Deliver it. Our Ambassador has probably already been brought to the airport – he'll get you straight through to London. But tell – tell – " He blinked sweat away from his eyes, trying to remember who the acting Prime Minister was. But it would be no decision for an acting P.M.: it would have to be a full Cabinet decision. "Tell them I leave it to them to judge whether to agree to the demand."

"What about your wife?"

Harry felt, rather than saw, Banning glance at him. "She – she will raise the money, I'm sure. But it is the political angle – " He stopped, all at once careless of the political angle; he was suddenly faced with the thought that he might not see Constance and the children again. But he could not immediately comprehend it: hope filtered the starkness of the thought: he was not condemned to death yet. He sought to distract himself from the prospect by being the politician again: "If I'm the pawn in some political game, then tell Cabinet they have to decide for the larger issue."

"Tell them," the young girl interrupted angrily, "we shall kill him if they don't meet our demands!"

The VC-10 took off twenty minutes later, climbing into the clear, calm sky; at the end of the runway an airport worker waved an energetic farewell, like an envious relative of those aboard. The plane headed south, then turned east. At the last moment the co-pilot and the navigator had been off-loaded with Chadwick; the two men had protested, volunteering to stay with Janney, but the hijackers' leader had been adamant. There were now seven hijackers aboard, all of them young people with the exception of the leader. He and another man were up on the flight deck, while the others sat in the first-class compartment with Harry, Banning and Roth.

Harry looked at his watch as the plane took off; four hours later they were crossing the wide, green-bordered scar in the desert that he recognised as the Nile. Then the Red Sea was beneath them, as desolate-looking as the barren land on either side of it; then the vast streaked and wrinkled page of parchment that he knew must be the Sinai Desert. He looked across at Roth, but the Israeli seemed utterly uninterested in where they were heading and lay with his eyes closed, his seat in the reclining position, a weary traveller who might have made this journey a hundred times. But Banning was sitting forward, leaning across Harry to look out the window.

"Was that the Red Sea we passed over back there? You know this part of the world."

"Not that part of it down below."

"Where do you think they are taking us?"

The young girl, whose name Harry had heard was Arabella, turned round in her seat just across the aisle and in front of them. "We can tell you now. You English were kind enough during the war, *your* war, to build some airfields in the desert south of Iraq."

"Was that the place where they blew up those planes some time ago?" Banning looked at Harry. "Dawson Field, was it?"

"We're not going there," said the girl. "Much farther south, miles from the interfering Hussein and his Bedouins."

"Arabella," said Harry, doing some politicking: this girl might be their only source of information in the next seventy-two hours. "I always thought that was a very English name. I had an ancestress named Arabella."

"It was originally Persian. It means a fair Arabian girl." She blushed and the other girl beside her giggled: it was difficult to believe that the two of them were willing to kill and be killed. But Arabella seemed eager to relieve the strain of the past ten hours; she had put away her pistol and grenade and was now combing her hair. "*Her* name is Miriam. That's very Hebrew, but she won't change it."

"Why should I?" Miriam was a plump, plain girl with a pock-marked face. So many of the plain ones, thought Harry, thinking like a man, marry causes because they can't find a man to marry *them*. "The Jews don't own our names too."

Roth opened his eyes at that, but did not sit up. "Miriam suits you. It means bitter."

The girl stood up in her seat, whirling on him, but he had already closed his eyes again. One of the young men snapped something in Arabic, Miriam snapped at him in return, then she dropped back into her seat. Arabella pouted, put away her comb and sat back. Silence fell once more on the compartment.

Almost an hour later there was a change in the engines' note, then the plane banked and began to descend. They went down, the plane bucking a little as the hot draughts of air came up from the desert below, then the wheels bumped, they were rolling along a rough surface and Janney was obviously having difficulty in holding the plane on a straight line. The reverse thrust was roaring and Harry, glancing quickly around him, saw that the hijackers were all sitting tensed in their seats, suddenly afraid. It was one thing to die bravely as a gesture to a cause, but another to die in an accident that would have no meaning and no result. The observation gave him comfort, small though it was: these people were something less than fanatics.

Then the plane's speed slackened and soon it was rolling to a stop. Banning let out a great sigh of relief, undid his seat belt and looked at Harry. "I think that was pretty close. I just said my first prayer in over twenty years. Not since – " But then he shut up. This was going to be nothing like Arnhem: one glance out the window told him that.

He could see the blue-and-white aircraft, a Boeing 707, standing a hundred yards away on the starboard side of their own plane. Beside it were two large trucks and two Land-Rovers; beyond it stretched the desert, a pale yellow glare that danced, like the effects of multiple vision, in Banning's tired eyes. There seemed no horizon: the sky gradually emerged, a pale *blue* glare, out of the rippling waves of heat. And the heat itself was already turning the interior of the VC-10 into an oven.

In another two minutes everyone, hijackers and prisoners, was out of the plane and standing in the shadow of its wings. A truck had been driven up beside the front exit door and everyone had disembarked by jumping down on to the canvas top of the truck and then swinging down to the ground from there. Banning could now see the other trucks, Land-Rovers and what looked to him like a couple of ancient Bren carriers; in all he counted a dozen vehicles. There seemed to him to be about two hundred guerrillas, all of them heavily armed and all of them wearing Arab head-dress. The hijackers, as soon as they got off the plane, were embraced by those waiting for them; there was a babble of excited cries and several of the guerrillas let off their rifles in an ecstasy of welcome. Harry and the other prisoners stood ignored, like unwanted immigrants in an alien land.

Then the clamour of the welcome died down, the group broke apart and two men, both middle-aged, both wearing dark glasses, came towards Harry and the others. The shorter of the two, a thin handsome man with a moustache, was obviously the leader; a gesture of his hand stopped the second man and he stepped forward alone to speak to Harry. He had a hoarse high voice, as if he had spent all his life exhorting crowds to action. This one, Harry judged, would be a fanatic.

"Welcome, Lord Carmel." The guerrilla leader did not look at the others; they were only extras in the scene. "Our message will have been delivered by now to your government. We have our radios tuned in to the B.B.C., but so far there has been no announcement of what your government intends to do. The B.B.C. says an emergency Cabinet meeting has been called, so at least your government thinks you are worthy of discussion."

"I'm flattered," said Harry, determined to give nothing away till he knew more of this man with whom he was going to be dealing. He was hungry and thirsty and tired, but for the time being he would ask for nothing. He nodded across towards the other aircraft, the name, *Trucial Airlines,* clearly visible on its side. "Is Sheikh Hazza all right?"

The leader's dark glasses flashed as he lifted his head. "Who told you he was here?"

"I guessed. One of your men said you would have another item for sale besides myself. How much are you asking for him? Another million?"

"A good guess. Or do men at your level have a standard market rate on yourselves? Yes, Hazza is over there."

"How did you manage to hijack him?"

"Very simple. His bodyguard had *their* price – one per cent of the million. We are only too willing to pay it."

"But will Hazza's brothers pay the million?"

The leader smiled. "It is less than a week's royalties on Qaran's oil. They'll pay it gladly, just to be rid of him."

"They won't be rid of him – not if you set him free."

The guerrilla shrugged. "So long as we get the money we don't care how dogs like him and his brothers settle their differences. If we don't get the money for him, we blow up the oilfields. It is as simple as that."

"And if you don't get the money for me?"

"I gather you already know the answer to that." He took off his dark glasses, polished them with the end of his head-dress; without his glasses one could see a squint in his right eye that spoiled his handsomeness. For the first time he looked away from Harry, towards Roth. "We asked a price for you, Mr Roth – the release of three of our men held in Gaza by your government. We have already had an answer. I regret to say your government refused to deal with us."

Roth was standing beside one of the wheels of the aircraft. He turned round, put his hands on it and leaned forward as if he were going to be sick. He looked desolated, a man crumpled by the sudden knowledge that his life was worth nothing.

Banning abruptly moved to him, put his hand on Roth's shoulder. "Roth, there's nothing final yet – "

Roth shook his head, not looking up. "My government is as rigid as these men – I know it. I don't blame it – it has to be like that. But – " He looked up, staring straight ahead of him at the giant tyre. "But one somehow never expects it."

There was an awkward silence all around, even the guerrilla leader seeming embarrassed by the situation his message had caused. Then Janney, red beard now a dark auburn with sweat, said, "There's food and drink in the aircraft. We haven't eaten since your – " the word could have been *guerrillas* or *gorillas*: the leader cocked his head but wasn't sure what he had heard " – pulled their caper. How about it?"

The leader hesitated, then nodded. "You will eat here beneath the plane. It will be cooler and we can keep an eye on you."

Janney looked around at the barren landscape; the planes and the vehicles were the only landmark, a metal and canvas oasis. "Where do you think we can go? That's another thing – when you do let us go, how do I get the aircraft off the ground? We'll need a starter motor."

"That's BOAC's problem, I'm afraid. Life is too easy for you Englishmen. You are never prepared for emergencies."

"I'm not English, but we *British* rely on our initiative," said the flight engineer. His name was McArdle and he was a parrot-faced, black-haired man in his early forties with a Glasgow suspicion of everyone born south of the Clyde. "When did you Arabs last invent something?"

There was a stirring in the group standing behind the guerrilla leader; a rifle bolt clicked as a warning. The leader looked at Janney. "We don't need your man any more than we need his insults, Captain. Keep him quiet."

"Okay, Jock," said Janney. "This is a bit more serious than a Rangers-Celtic match. Hop aboard, whip up some grub and something to drink. Beer all round would be the trick, I think?"

Harry and the others nodded. The guerrilla leader stared at them all, then he replaced his dark glasses, whirled in a professional soldier's about-turn, and walked quickly away towards Hazza's aircraft, several of his aides hurrying to keep up with him. The guerrillas broke up into small groups, moved away from the prisoners and sat down beneath the shade of the VC-10's fuselage. Only one man, armed with a sub-machinegun, remained near the prisoners to guard them.

Five minutes later Hazza, accompanied by a guard, came striding across the airfield just as Harry and the others were squatting down to their cold meal. "Salek, he's their leader, tells me I am to join you for luncheon. No alcohol, thank you, Captain. A little lemonade, perhaps. So what do you think, dear Harry?"

Harry, munching on a chicken leg, looked around him. No one was moving in the bleached landscape; even the guerrillas, some of them born to desert like this, had surrendered to the heat. There were no camels or goats or dogs, none of the usual livestock that provided some sort of movement in a desert camp; the loneliness would soon regain its dominion over this endless barren plain. In the utter stillness of sound and movement, highlighted rather than broken by the occasional creaking of the metal of the aircraft above them, Harry felt as if he were in the middle of a nightmare land: the silent scream that would jerk him awake would be heard at any moment. Even more here than on the plane it was difficult to be logical and clear-thinking about a situation more bizarre than anything he had ever imagined. Even Arnhem and the night of the Mau Mau killing in Kenya had somehow been closer to reality.

"I don't think I've ever felt so helpless," he confessed. "You, neither, I suppose. And Roth. Whatever happens to us depends on people who don't even know where we are. Where are we, anyway?"

"I am guessing – " Hazza, back in the desert of his boyhood, before wealth and self-indulgence had spoiled him, seemed less epicene; Harry had the impression that Hazza's fat was turning to muscle even as he looked at him. "I think we are about three or four hundred miles inland from the Gulf and about the same distance south of the Iraqi border. My pilot took a rough sunshot just after we landed, just before they shot him."

Everyone stopped eating. "Why?"

"He was too loyal to me, I'm afraid. As soon as he landed the plane he attacked one of my – forgive the expression – my bodyguards. Has Salek told you of their treachery? What men will do for money!" He waved a hand; Harry noticed that the fingers this morning were bare of rings. "Well, they shot him. And the rest of the crew and my secretary." He looked at Harry. "I had almost made up my mind to ask Miss Erickson to come out to Qaran, just for a short trip. I'm glad I didn't."

Harry felt sick, but he said, "Were there any passengers on the plane?"

"They were all off-loaded at Algiers."

"What's happened to your bodyguard?"

Hazza shrugged. "The Palestinians seemed annoyed at the shooting. They've taken them away somewhere – a truck drove off just before you landed. I hope they have shot them." He seemed unperturbed by what had happened, unaware of the sudden chill he had thrown on Harry and the other men. He threw away a stripped chicken leg, picked up another one fat with meat. "I don't think anyone will be coming to rescue us, Harry. My brothers won't send anyone out to look for me, and we are much too far off the beaten track for the English to send out RAF planes looking for you. It would be beyond the range of all but your biggest bombers. It is not like the good old days, is it, when you had RAF squadrons all throughout here. Iraq, Jordan, Palestine, Egypt. One only appreciates the Empire now that it has gone."

"Yes," said Harry, and wondered what Colonel Bedley would think of *that* change of mind. Then he looked at Roth. "We'd be within range of your planes."

Roth had been listlessly eating a tomato; he seemed already to have given up, to feel that to go on eating was a waste of time and effort: he was going to die anyway. "Salek gave you my government's reply. I told you once before, Lord Carmel – I am unimportant. I certainly don't equal the value of three political prisoners."

"But surely to God *someone* will come looking for us!" Banning exclaimed. He was on his third can of beer, but it was coming out of him as sweat as fast as he poured it into himself. He was seated on the dusty earth, his back against one of the plane's tyres; he fanned himself with one of the magazines McArdle had brought down from the cabin and looked on the point of expiry. "*You* are important, Harry. British Foreign Ministers aren't kidnapped every day – "

"Oh, eventually they'll track us down," Harry said. "But I don't think this is going to be the sort of carnival they had at Dawson Field with those other hijacked planes. I suppose paratroopers could be flown in here from Cyprus."

328

"I suggested that to Salek," said Hazza. "He said that at the first sign of any attempt to rescue us, they'll put a bullet in all of us. Captain Janney and Mr McArdle included."

The afternoon wore on. No one came near them, except when the single guard was changed every two hours; it was as if the guerrillas had decided all the worthwhile talking was going on hundreds or thousands of miles away, in Qaran and London. Languid with the heat and pessimism, the prisoners alternately dozed and lay looking at the giant wing beneath which they sheltered, occasionally moving their positions as the sun pushed the shadow east of them. Once Harry looked at his watch and wondered if the Cabinet was still meeting to consider what it should do about him. He knew the money would be paid; Constance would see to that, even if Treasury refused. But would Argus and the Cabinet think he was worth their surrendering to political blackmail?

"Who's the acting Prime Minister?" Banning asked sleepily.

The prisoners had spread out, as if each man wanted to be alone with his depressed thoughts. But Harry and Banning were lying side by side, their heads pillowed on cushions that McArdle had thrown down from the plane. McArdle seemed to have taken on the job of senior steward; till he had lain himself down to doze off he had been solicitous of everyone's comfort, as if making up for the unpleasantness he had caused by his sarcastic remark to Salek. Janney was the nearest man to Harry and Banning and he was twelve or fifteen yards away.

"Funny, I had a mental block about that when we were at Algiers, when I told Roy Chadwick to tell the P.M., whoever he was, I would leave the decision up to him as to what to do. It would be Michael Argus, I should think – which would be just my luck."

Banning turned his head, his eyes all at once wide awake in his red, sleep-lax face. "What do you mean by that?"

There was no harm now in telling Joe; after all, he was his best friend. But he kept his voice low as he told Banning of the events of the past few weeks, of Wolfe's knowledge of his coming death, though it had come sooner than expected, and of the race for the job of Prime Minister. "I telephoned Ian Kidman Thursday morning, before we left New York. He thought there would be only two nominations for the job, me and Argus."

"You surprise me. I'd never have suspected you'd be a starter."

"Why not? The field is wide open."

Banning twitched, an apology for a shrug. "I don't know. Who knows you – politically, I mean? You're too much of a dark horse. Half the country probably doesn't even know you're Foreign Secretary."

"They would now." Harry smiled lazily to himself. "I spent two weeks before I went to New York trying to get myself in the public eye and sometimes I had the feeling it was all a waste of time – I suppose I really don't care what people think of me." It was a casual admission, but he knew it was true. "But now I am probably better known than any other politician in Britain. If the Cabinet agrees to pay over the ransom I'll go home to hog all the limelight away from Argus. He probably thinks I engineered the whole caper just to thwart him."

"Do you think you'd make a good Prime Minister?"

"I'd make a better one than Argus."

"That wasn't the question."

"Charles Wolfe asked me the same thing. I couldn't tell you until I was in the job. Who'd have thought Truman would make a good President? You grow in the job."

"One would hope so," said Banning cryptically. He sat up as Janney moved towards them, squatted down. "Going somewhere, Captain?"

"That's what I wanted to talk to you about. Sir – " He looked at Harry. "I'm not very keen on sitting here on my arse doing nothing but wait till these jokers make up their minds whether to shoot us or not."

"They won't touch you and McArdle." Harry sat up, tried to sound reassuring. "They're not interested in you."

"That's one of the things I don't like. When they move off, disappear out there – " He waved into the distance: the afternoon had begun to cool and the horizon had appeared, a long unbroken line that marked the very edge of the world: eternity lay beyond it " – why shouldn't they just pop us off before they go? We're British, they don't have any love for us."

"What are you suggesting?" Harry, too, waved into the distance. "There's a bloody lot of nothing out there. They'd be on to us before we'd gone a couple of miles, even at night."

"Not if we went *upwards*," said Janney. The guard had stood up, was watching them suspiciously, and Janney suddenly started to laugh, as if he had been telling Harry and Banning a joke. But through his laughter he kept on talking: "If there's an air compression bottle in the wheel compartment of that 707, I could get the engines on that one started. We used to carry them on our aircraft, but we stopped it about five or six years ago when most airfields got their own starter trucks. But out here some of the smaller airlines might still carry them. I'll find out."

He stood up, stretched his arms, looked out at the surrounding scene, then strolled across to the guard and offered him a cigarette. The latter hesitated, then took one. Janney lit it for him, looked out again at the airfield, then strolled back past Harry and Banning.

"How'm I doing?" He was a basically simple man who met all problems head-on; he fought depression with jokes. "Think I could play James Bond?"

Harry smiled, admiring the man: he was producing the initiative that McArdle had claimed for them. He had not been unmindful of the threat to Janney and McArdle, but they had not occupied his thoughts very much: he had taken it for granted that they would be safe. But now he saw the possibility that none of them was safe and all at once he saw the need for some initiative, if only as an emergency measure if worse came to worst. He knew that he would rather die running with a bullet in his back than sitting still waiting for the bullet in the head. Useless bravery, if it was going to have any meaning at all, needed spectators. To die with chin up, true blue and British, was not going to impress the guerrillas. Death, for them, had become just a habit.

Janney squatted down beside the dozing Hazza, prodded him. Hazza rolled over, then sat up. Janney spoke to him, Hazza nodded, then looked across at Harry and nodded again. Then both men stood up and began to walk up and down in the shade of the wing. The guard, butting his cigarette and putting it in an ammunition pouch, moved closer to them, growling something in Arabic. Hazza replied in Arabic, smiling broadly as he did so; whatever he said, the guard seemed satisfied and went back to standing by one of the wheels of the aircraft. He took out the cigarette, lit it again and lounged against the wheel, boredly watching the prisoners.

Harry got lazily to his feet, went through Janney's act of stretching his arms, then unhurriedly moved across and joined Hazza and Janney. The three men walked slowly up and down in the shade of the wing, talking casually, like politicians or priests mulling over the result of a day's conference. Or like three out-of-works, an oil sheikh, a millionaire peer and an airline captain, discussing what to do to-morrow to fill in their day.

## 3

At sundown the prisoners, including Hazza, were told to get back into the VC-10. Four guards climbed up into the aircraft, bringing their supper with them: bread that smelled as if it had been freshly baked, some tinned meat and some dates. McArdle and Banning prepared the evening meal for Harry and the others, McArdle fussing over it as if this were a goodwill flight aimed at persuading Harry, Roth and Hazza to fly BOAC all the time. It was a cold meal again, but this time it was chicken, ham, asparagus, salad, cheese, petit fours and champagne. The four guards looked on expressionlessly as Janney raised his champagne glass.

"To BOAC – the best of service anywhere, any time. I just hope, gentlemen, it's not your last meal with us."

All their plans had been laid and now they just had to wait their opportunity. Janney, by unspoken vote, had been elected leader of the escape attempt; he was the one who had thought up the idea and he was the one who would have to get the aircraft into the air, if and when they made it to the 707. He had spent the rest of the afternoon moving casually from man to man explaining what was to be done.

"First, we have to get rid of our guards somehow – knock 'em on the head, kill them, I'm afraid, if we have to. Then I have to get over to the 707 on my own – I can get up into it through the electric bay hatch. We're damned lucky the Sheikh's planes still carry the air compression bottles in their wheel compartments. Now none of you are to make a move till you hear the compression bottle go off – a bloody great *whoosh!* that will wake up the Wogs too, if they're asleep. Then you all grab this truck here and get over to the 707 as quick as you can. It'll take me twenty seconds to get the first engine to catch – that'll be Number Three, the inboard starboard one. Then it'll be another two minutes before I've got the inboard port one going and both of them are warmed up enough for us to start taxi-ing. We'll be a sitting duck if they open up on us with any sort of heavy gun, but that's the risk we have to take. It will be four minutes before I've got all four engines going and warmed up enough for take-off – we may have to take some evasive action around the field, but you can taxi a 707 around up to nearly forty miles an hour. It's not going to be easy and it may turn out to be an absolute bloody disaster, but I'd rather try it than lay money on our getting out of here any other way. The pity is, they'll probably blow up the VC-10 just out of spite. I hope BOAC doesn't try to deduct the cost of it from my pay."

When they had eaten each man settled down to wait for zero hour – whenever that might present itself. Two of the guards sat at the front of the compartment and two at the rear; the long hot day did not seem to have affected them too much and they were all much more alert than their prisoners. Harry and the others had been told to sit each on an aisle seat, so that none of them was sitting immediately next to each other. Harry, leaning across towards a window, looked out at the desert dusk. The guerrillas were moving about on the airfield, congealing into large groups that a few minutes later split into fragments that wandered aimlessly till they attached themselves to another group; the waiting was evidently wearing on the nerves of the Palestinians too. Daylight suddenly drained out of the sky and instantly the stars asserted themselves. Even though he was inside the plane Harry could feel and hear the thin, fragile stillness that he knew would be out there beyond the airfield. Out there was perhaps the only true isolation a man might know: deserts and oceans were where man discovered his true self. He wondered what he might discover in himself if he suddenly found himself alone out there in the star-gemmed bowl of the night, with the flat horizon enclosing him like a giant ring, stranded in what astronomers called the circle of vision. The vision of one's self? he wondered; but characteristically did not pursue the fantasy. He did not want to know the truth about himself till the last possible moment, till he was dying. Then would be time enough.

The aircraft began to creak again as its metal cooled in the night air. All the prisoners settled down to an imitation of sleep; pillows and blankets were produced by the admirable McArdle. Janney, nodding affably to the guards, strolled to the front exit door and looked out as if getting a last breath of air before turning in. Then he came back, settled down in the seat across the aisle from Harry.

"Looks like early Lights Out," he said casually. "They don't keep late nights out here."

The guards had allowed no other lights but the tiny reading beams to be switched on in the aircraft. The compartment was soon dark except for one reading beam forward and another at the rear; the two pairs of guards sat beneath them, their eyes occasionally flashing in the shadows cast by their head-dresses. Once one of them checked the magazine in his sub-machinegun, the sound harsh and cold in the silence, but none of them spoke to each other. These men were accustomed to one another's silences: they had probably spent other long hours like this when to break the silence would have meant their possible death. Harry sat up for a moment, looked back and across at Roth; though he was their real enemy, the Palestinians had treated him no differently from any of the others. Roth had said nothing in response to the suggested idea of the escape. All afternoon he had been sunk in his own silence, without hope or optimism. In the gloom of the cabin now, his pale bald head shining faintly like a skull, he already looked a dead man.

The minutes and hours ticked on, eating away at the nerves like rats. The guards were changed every two hours, the new-comers coming on grumbling, the relieved ones going off with suddenly-found light-heartedness that threatened to keep them awake now they were free to sleep.

It was the middle-of-the-night shift, that from twelve to two, before the opportunity came for their move. Two of the guards were dozing, while their companions sat blinking in the slim beams of light above their heads, trying hard to stay awake. Janney quietly eased himself out of his seat, pointing up towards the toilet as the guard up front sat up straight. The guard stood up, stepped out into the aisle and nodded to Janney to move up past him. The captain glanced at Harry, saw the latter was wide awake, then moved up towards the toilet.

As he squeezed past the guard Janney suddenly said, "Now!" He pushed the guard and the man with a startled cry fell back into the lap of his companion. At the same time Harry saw Hazza spring out of his seat and plunge down towards the two guards at the rear. Harry went with him, taking the man on the right of the aisle as Hazza took the man on the left. The next few moments were a confusion of scuffling and cursing, of a gasp of pain and the crash of a gun barrel against a window, which fortunately did not break. Then Janney was holding a gun against the head of the man he had pushed, was saying softly but fiercely, "Tell 'em, Sheikh, I'll blow his bloody head right off if they don't quit!"

Hazza snapped something at the two men across whom he

334

and Harry had flung themselves; the limited space between the seats and in the aisle had given the men no room in which to move. Banning had jumped out of his seat, moving up to join Janney. McArdle had run up towards the exit door, was crouched inside it peering out. The guards abruptly were quiet and still. Hazza eased himself back on his feet, taking the gun of the guard he had attacked. Harry stood up, doing the same with his man. Janney stepped up the aisle, backing away from the man he covered, and Banning reached past him and took the gun from the other guard who, with the first man still sprawled across him, handed it over with a snarl. The whole incident had taken only seconds and, miraculously, had been accomplished before the guards had been able to shout or let off a shot of warning.

"Okay, everything's still quiet outside!" McArdle turned round up at the front door. "Let's go!"

Roth, wrapped in a blanket, had continued to sit unmoving in his seat, had not even turned his head to see what was happening to Harry and Hazza behind him. He had shut his eyes for a moment when Janney had rammed the gun's barrel against the guard's head, but now he was staring at the captain as the latter snapped at him, "You first, Mr Roth. You've looked all along as if you're not interested – but you're going with us, whether you like it or not! I want to make sure you're where the others can push you out on to the truck. Come on!"

For a moment it seemed that Roth was not going to move. Then slowly, like an old man told to get out of the chair in which he had been prepared to die, he let the blanket slip from him and stood up. Painfully, arthritically, he stepped out into the aisle and moved up past Janney. The latter moved in above the two guards, still in their seats, so that Roth could pass behind him. The Israeli had just gone past him when Janney suddenly let out a gasp and stepped back into the aisle with a look of surprise on his face.

Harry saw the knife sticking out of the captain's ribs as he fell backwards into the seat across the aisle. Banning was farther up the aisle, beyond Roth, and for a moment the two guards were free to make a second, if fruitless, attack. Harry saw the second knife come up and he yelled at Roth, "Grab the gun!"

But Roth stood paralysed, head turned as he stared down at the dead Janney. Harry had never moved faster than he did in the next moment. He went up the aisle in a jump rather than a run, drove the butt of his gun straight at the head of the guard nearest him as the man dived towards the gun still held loosely in Janney's lifeless hands. The Palestinian went down with a grunt and Harry swung round, catching the other guard under the jaw with the gun-butt as the man swung at him with his knife. There was the sound of two sickening blows from the rear of the cabin, then Hazza came up the aisle after Harry; behind him the two remaining guards were sprawled in their seats, blood pouring from the face of one of them. Roth still had not moved and he fell forward against the door to the flight deck as Banning pushed him out of the way.

"How is he?" Harry asked as the doctor knelt down beside Janney.

Banning shook his head. "Must've gone right up into his heart. What do we do now?"

They all looked towards McArdle and Harry said, "Can you fly the 707?"

The flight engineer's face looked even sharper and thinner than usual; he shook his head, trying to throw off the shock that had engulfed him. "God Almighty – I don't know! I've only flown light aircraft – weekend stuff. I know the theory – I think I could get the 707 off the ground, but I dunno if I could put it down again!"

"Then we'll have to find out the hard way. We can't stay here!" Harry had been prepared to let Janney be the leader of their escape; his arrogance had never allowed him to commit the stupidity of trying to buck the expert. But he saw at once that McArdle was no leader, nor were Roth and Banning. Hazza was, but when Harry looked back at him the sheikh was nodding, already acknowledging him as the man to take charge. He looked back at McArdle. "How much time will you need to familiarise yourself with the set-up of a 707?"

McArdle looked down at the dead Janney, almost accusingly, as if he felt the captain had deserted them by getting himself killed. "I don't know – five minutes at the least, maybe ten." The Scots accent was thick on his tongue; he wished to God he was safe back home in Glasgow with Jean and the kids. "I couldn't just slide into the pilot's seat cold – "

"All right, then, you have to get across there ahead of us."
"How?"

Harry pulled the head-dress off the nearest unconscious guard. "Here. You'll have to pretend you're one of them on roving picquet. Take one of their guns too."

McArdle took the head-dress, but did not put it on. "It won't work – "

"It's got to! Where's your bloody initiative? Come on, man – you're our only chance! We *have* to go now – " He gestured at the unconscious guards. "If they catch us after this, you'll be the first they'll kill. You're the most expendable."

McArdle looked at the four of them, from man to man: only Roth, the little Jew, seemed to have any look of sympathy for him. He jammed the head-dress on his head, took the gun Hazza handed him. "Talk about bloody privilege!" He stepped to the door. "Don't make a move till you hear the compression bottle go off. Then make it over there as fast as you can. And stay away from the inboard starboard engine or you're likely to get your bloody heads sucked off!"

Then he was gone, slipping down on to the truck, and Harry looked at the others. "I'll drive the truck. The rest of you stay on top of the canopy as planned. Joe, you'll be first up into the 707 – go straight up to the flight deck in case McArdle needs a hand. Here, Roth!" He shoved the bloodstained gun Janney had been holding at Roth, but the Israeli stared at it, then dumbly shook his head. "Take it, for Christ's sake!"

But Roth couldn't lift his hands to take the gun. Harry lifted his arm, shoved the strap of the gun over Roth's shoulder, and let the Israeli's arm drop again. Then he moved up to the door, stood just inside it and looked out.

McArdle, gun slung negligently over his shoulder, head-dress tucked about his face, was just moving out from the shadow of the VC-10. In the bright moonlight the 707 seemed to have retreated; it looked twice as far as it had in daylight. McArdle appeared to be dawdling; the guerrillas would be sure to gun him down before he was half-way to the other aircraft. But then he was half-way there and nothing had happened; then he was twenty-five yards from the plane; finally he was lost within the shadow of the 707's wing and there had not been a single challenge or movement from the guerrilla camp. Harry relaxed his grip on the gun he held, easing the cramp in his fingers and feeling the cold sweat lying in his palms.

"So far, so good," he said hoarsely.

Seven minutes later the blast of the air compression bottle shattered the stillness with an explosion that made them jump. "Now!"

Roth stood in the doorway, making no effort to move; Banning gave him a gentle push, then went out the door immediately after him. Harry looked at Hazza, shaking his head in exasperation, then the two of them left the VC-10, Hazza dropping flat on to the canopy of the truck beside the other two outstretched men, Harry sliding down from it and into the driving cabin. The keys were in the ignition: there were no car thieves in the desert, at least not up till now.

The truck's engine kicked over at once. Harry let in the gears, eased the truck away from the VC-10 and headed for the 707. He cursed the brightness of the moon, but McArdle might need it when, *if*, he got the 707 headed down the runway. The truck sped across the hundred yards of blue-white bone dust. Out of the corner of his eye he saw the first of the guerrillas running across the airfield; then he heard the spatter of fire from above him and saw three of the men fall and the others pull up, then drop flat to the ground. He swung the truck in beneath the open front door of the 707, heard the scuffling of the men above him even before he had switched off the engine. Then he heard the whine of the first engine as it took.

He swung out of the driving cabin, clambered up and was on Hazza's heels as the latter hauled himself up into the plane. He turned round, gun aimed across at the distant vehicles just beginning to move, and only then saw Roth still lying flat on top of the canopy.

"Roth!" But his shout was lost in the rising whine of the engine.

The Israeli did not move, just lay face down on the canvas, one arm drawn up protectingly round his head. Harry looked up and across the airfield and saw the first of the Land-Rovers coming towards them, the Bren carriers moving out behind, all of them looking like huge menacing beasts in the soft glare of moonlight flung up from the blue-white earth. Harry looked back at Roth, cursing him, hating him. Then a memory came back, just a subliminal flash: Roth in the room in the King David Hotel in Jerusalem, pleading with him and Roger to flee for their lives.

He jumped down on to the canopy, grabbed Roth under the arms and lifted him up. The Israeli did not struggle but tried to get to his knees, lifted despairing eyes that stared out of the white skull of his face; then he found strength returning to his legs, straightened up and, with Harry's help, pulled himself up into the plane. Harry flung himself up into the doorway, sprawling on Roth, got to his feet and spun round as the first Land-

Rover skidded to a stop at the rear of the plane. Above the still-rising whine of the engine he heard the burst of gunfire from the rear door as Hazza sprayed the men in the Land-Rover before they had a chance to get out of the vehicle. The Bren carriers swung away in a wide circle, both of them heading for the blind side of the aircraft. The guerrillas' trucks were now speeding across the airfield, dragging drogues of dust behind them; the second Land-Rover had driven on past the plane and Harry lost sight of it. He crouched in the doorway, the gun held ready to shoot as soon as any of the guerrillas came within range. Then suddenly he understood Salek's tactics. These Palestinians were professionals, not part-time amateurs; he and the others had overlooked the fact that these men had been learning the ways of war for over twenty years. The trucks and the Bren carriers were swinging out in a wide circle, speeding to build a ring round the plane before it could get moving.

Then Harry heard the inboard port engine cough, take and start its rising whine. There was a burst of fire from one of the nearer trucks and, glancing down out of the door, he saw the explosions of dust fly up behind one of the wheels. He raised his gun, tucked it into his shoulder and let go at the truck. It had been moving slowly in the same direction as the plane, trying to cut across in front of it; then he saw the driver suddenly fall out of the open cabin and the truck slewed round and came straight at the 707. Harry waited for the crash, then all at once the plane began to move. It rolled forward with agonising slowness, the two inboard engines screaming now as if echoing Harry's angry impatience. Then abruptly the plane changed course and he almost fell out of the doorway; the runaway truck passed under its tail and Hazza, leaning out of the rear door of the plane, pumped a burst into the canopy of the truck before it disappeared from view. Harry stood up, grabbed at the door and dragged it shut.

He ran back down the aisle, met Hazza coming up. "Now all we can do, Harry, is pray to God!"

But prayer was something that required contemplation, at least on his part. He turned and went back up through the plane to the flight deck. Roth was sitting in a seat in the first-class cabin and looked up at him as he passed, but he did not stop. He had nothing to say to the little man now and he had no idea what he would have to say, if and when there was the opportunity. They were still a long way from being safe and perhaps he and Roth might die without ever speaking to each other again. But reconciliation was like prayer, it required time in which he could think.

McArdle was working at switches on the complex instrument board in front of him. Banning was sitting in the co-pilot's seat, his face looking more flushed than usual in the dim red light from the instrument panel. He was manipulating the wheel in front of him, swinging the plane in wide sweeping circles as he forced the guerrillas' trucks back from closing their intended circle.

"Doesn't handle like the Lagonda!" Excitement and adventure had turned him into an elderly schoolboy. "But I'm scaring the bloody daylights out of them!"

McArdle had the two outboard engines going now. "I'll take it!" Banning dropped his hands from the wheel as McArdle took over control. The flight engineer glanced over his shoulder at Harry as the latter leaned in behind him. "I hope to God we've got enough power! I can't take her down to the end of the strip, then bring her back – they might put a couple of trucks right across our path! We've got to take our chances – let's hope if we run out of strip, the desert is hard enough to hold us up!"

He swung the 707 in between two converging trucks, going so close to them that both of them seemed to disappear right under the nose of the plane. He gunned the engines, straightened the plane and headed down the runway. Harry, gripping the back of McArdle's seat, saw the broad white path reeling itself into the plane at seventy, eighty, ninety miles an hour. Then ahead of them he saw the faint line that marked the end of the strip.

"Too soon!" McArdle yelled. "We need another thousand feet!"

The end of the runway flashed by, but there was no noticeable change in the smooth run of the aircraft. Harry, tensed and sweating, stared ahead into the moonlit distance rushing at them: the horizon became their destination, the haven beyond the stones, spiky bushes, thin gutters of sand hurtling by beneath the wheels of the plane as McArdle built up its speed. Then suddenly, even as he looked at it, the horizon began to sink.

"We've made it!" McArdle yelled, and pulled the 707 up into the bright, welcoming exclamation of stars.


4

"I'll head for Qaran," said McArdle. "Ask the Sheikh to come up here. He'll know the coastline when I hit it, can tell me which way to head." Now he was airborne he was much more relaxed, had become even a little arrogant: no matter what the others might be, he was the captain of this aircraft. "When we get over Qaran I don't want anyone up here with me, except the Sheikh. I've never put an aircraft as big as this down on the ground and I don't want anyone looking over my shoulder while I'm doing it for the first time."

"Do you think you can do it?" Harry said.

"You'll know if I can't. There'll be a bloody great bang and after that it'll be every man for himself. Now, would you send the Sheikh up?"

Harry, also relaxed now they were in the air, was amused by the flight engineer's crisp authority; God knew how long he had been waiting for just this opportunity; he had seen London bus drivers drunk with power as they had ploughed their way round Hyde Park Corner. "Good luck, McArdle. Dr Banning and I will see you when we land."

He and Banning went back into the first-class compartment and Hazza went up to the flight deck. As Harry went to sit down he saw Roth seated right at the rear of the compartment. He and Banning glanced at each other and he hesitated, wondering what he should do; did Roth want to be intruded upon at this moment? Roth gave him his answer: he looked up towards Harry, then turned his face away and looked out the window. Harry sat down beside Banning.

"You're not going to say something to the poor blighter?"

"He doesn't want me near him."

"What do you expect him to do? Plead with you to come and speak to him?"

Harry, ready to give in to exhaustion, was suddenly angry. "Christ, will you stop needling me! He doesn't want me as his keeper – "

"A little charity doesn't make you his keeper."

Abruptly Harry stood up, went back down the aisle. He stood above Roth and looked down at the Israeli, unsure of how one expressed charity towards a man who hated you. He felt the stirring of something (of pity? Perhaps. He knew it was not contempt; that would require too much effort, more than his exhausted condition could muster), and he said the first thing that came to his tongue: "I'm sorry, Mordecai."

Roth looked up startled, puzzled. "Sorry? What for?"

Then Harry himself was puzzled. "I don't know." He made a bemused gesture; he felt suddenly at a disadvantage. He struggled for some coherence, trying to sort out quickly in his mind all that had gone wrong between himself and Roth. But all he could say lamely was, "I was a bit brusque with you back there on the airfield."

"Brusque?" Roth repeated the word as if he didn't understand it; then an uncertain smile creased his face. "Harry – " He put up a tentative hand and Harry, equally tentative, took it. Only then did he see that above the trembling smile Roth's eyes were glistening. "You saved my life. Don't be sorry for that."

Harry said nothing. He just nodded, squeezed Roth's hand harder and went back and sat down beside Banning.

"You find that difficult, don't you?" said Banning.

"What?"

"Putting yourself in the other fellow's place." He looked up, as if looking for a steward. "I could do with a drink."

"You probably won't find any hard stuff on this one. Hazza, for all his self-indulgence, sticks to his religious laws. You have something on your mind. Go ahead, say it."

Banning sat back. He drummed his fingers on the arm-rest, sucked in his lips against his teeth. "As a doctor, if one is going to be any good at it, you have to learn to put yourself in the other fellow's place. Medical diagnosis is only half of it – I've cured as many by amateur psychology as I have by handing out prescriptions."

"I'll mention it to the B.M.A. if ever they ask me about you. Go on. You're not meaning to give me a lecture on psychology, are you?"

"I am, in a way. You're going to be a hero when we get back to England, aren't you? Oh, you'll weather that all right – you're not the sort to let a bit of hero-worship go to your head. But it should make you the favourite for the Prime Minister's job. You're still going to run for it?"

"I don't know what's happened in the last twenty-four hours. Argus may already be P.M."

Banning shook his head. "Your supporters wouldn't let it happen – not while you're out here improving your image, as it were. No, when you get off the plane, whichever one takes us back to London, you'll be as near to Prime Minister-elect as dammit. We've been starved for heroes since the end of the war. You'll fill the bill – for the time being."

"What are you trying to tell me? That I'll need a course in psychology if I'm to make a success of it?"

342

Banning stared straight ahead of him for a moment, then. making a hard decision, he turned and faced Harry squarely,

"Harry, I don't think you'd ever make a success of it. It seems to me, and I've known you a long time, that you are incapable of putting yourself in the other fellow's place. You never understood why I funked it at Arnhem – " He held up his hand. "No, let me finish. Arnhem is only a very small part of it. It really has nothing to do with my attitude towards this – this ambition of yours. I only mention it because, in retrospect, that was when I began to realise that you never know, you never *care*, what makes the other chap tick. Roth, for instance – you made no attempt to find out how *he* felt. You'd have sat down here and let the poor bugger sit in misery back there till we land at Qaran and he could be pushed off to Israel. But he and I are only two of hundreds you've never tried to understand. The truth is, Harry, you don't care about *people*. People in general. Your father knew the name of every man who works on your estate and on the farms – I doubt if you could name three or four of them. You've been the Marquess for almost six months now and none of the workers on the farms or any of the villagers have seen you. Your father never missed – once a month he'd visit all the farms, all the villages, let himself be seen so that if anyone had a complaint or a favour to ask, he was available. But not you – you haven't been available to anyone for years, not even your own children."

"That wasn't entirely my fault."

"Perhaps not. But you didn't make any effort to remedy it. Christ, I'm not liking this!" Banning looked away for a moment, then turned back. "A couple of years ago I had to treat young Matthew for a boil on his behind – I suggested to him that he ask you to buy him an air cushion to sit on. It was a joke, but he just said he never asked you for anything, that he wouldn't know how to. That was your son, Harry – a boy who didn't know how to ask his father for something as cheap and simple as an air cushion!"

Harry nodded reluctantly. "I know. I've been aware of it – I keep making resolutions – "

"Resolutions? What I'm talking about, at least towards your children, should come *naturally* to you. But it's not just your kids. Nor me and Roth. It's – well, as I said, it's *people*. I think Britain is in for a long hard road over the next five or ten years, what with the Common Market and the rest of it. And I think it needs a Prime Minister who understands people, who will know how

343

to get the best out of them, who will be charitable towards them if they look like failing, who will not be – *indifferent* towards them. And you're not the man."

*Indifferent*: the word haunted him like a curse. "Kidman and the others think I am."

"Perhaps they are only judging you against Argus. I'm not – I'm judging you purely on your own terms." He held up his hand again. "All right, *my* terms. But I'm one of those who's looking for someone to lead the country, me and millions of others. And you're not good enough, Harry – not for what Britain needs."

"Who is?"

"I don't know. I think Brewster was a good man, though I never voted for him. Wolfe might have been – we'll never know. Argus probably isn't. Kidman himself might be – at least he cares for people, he's shown that. All I know is, I can't let you be Prime Minister."

"*You?*"

Banning nodded. "I'm frightened at what I'm taking on – but I'm putting in a single vote against you as Prime Minister. And mine will be the vote that counts."

"Who do you think will listen to you? Christ, you're out of your mind! Kidman and the others won't take any notice of a country doctor – "

"I shan't say anything to them. I'm telling *you*. You may make a great Foreign Secretary, but I think you'd be a disaster as Prime Minister. And if you persist in going for the job – " Then he stopped, reluctant to play his trump card.

"You'll what?"

"I'll go and see the Probate people, tell them what you and I did to dodge the estate duties. No one would want a Prime Minister who diddled his country out of five million quid. Society is not *that* permissive."

"You bastard!"

Banning nodded. "I suppose I am. Perhaps you can find a place for me somewhere on your coat of arms, as your bar sinister. Believe me, as I said, I'm not liking this. But I *know* I'm right, Harry. You're not the man."

"You'd go to prison too, for conspiracy." But he knew it was an empty threat even as he said it.

"I've thought of that. But it's a price I'd be willing to pay.

Patriotism isn't usually rewarded with two or three years in Wormwood Scrubs, but I'd settle for it."

"When did you make up your mind to tell the country what's good for it?"

"I'm not telling the country – I'm telling you. Whether it goes any further is up to you. I've been making up my mind for the past ten or twelve hours – since you told me what was in the wind. It hasn't been easy. You know me well enough to know I'm not mad for any power – I frighten myself with the conceit that I'm taking the decision on myself. I'd prefer to talk it over with someone, but who is there? If I mentioned it to Kidman and your other supporters, they'd drop you at once. So I have to do it myself, Harry."

Of all the bizarre things that had happened in the past twenty-four hours there had been nothing to equal this: none that had had as much effect on him. He sat slumped in his seat, not looking at Banning. His unseeing eyes focused for a moment and he saw the folder sticking out of the pocket of the seat in front of him: *In Case of Emergency* . . . Reach for the oxygen mask, tighten the seat belt: none of those were any help when the bottom had fallen out of one's about-to-be-achieved ambition. He was still a long way from London and safety, McArdle still had to put the 707 down safely, but he had already begun to think of the effect of his return. He would have returned as a hero; he knew, as certainly as if he could already read the headlines, that he would have won the nomination as P.M. And now the man he had thought of as his closest friend, his only *real* friend, had just shot him down in flames.

Hazza came back from the flight deck. "We have about another hour's flying before we reach Qaran. I am looking forward to the expression on my brothers' faces when I turn up at my palace."

Harry tried to shake his own problem out of his mind. "Aren't you afraid they have already taken over from you?"

"There are too many of them, dear chap. They will still be quarrelling to see who is going to be top man – we are not as civilised in that respect as you British. But when I walk in on them with you, they'll know there's no point in trying any further. You are my insurance policy – the Foreign Secretary of the British Government. What better friend could I have?"

Harry smiled wryly. "I have my uses."

"Of course you do, dear chap. We all do, don't we?" He

smiled at Banning, his ebullient good humour doing nothing to relieve his ugliness. "I'm sure Dr Banning is a most useful chap."

"Sometimes," said Banning, and did his best to return Hazza's smile but only succeeded in looking as ugly as the Qaranian.

An hour later there was a change in the pitch of the engines. Then McArdle's voice came over the intercom, crisp and confident. "This is your captain speaking – " There was a chuckle. "Gentlemen, we're going down now. If any of you know any prayers, I'd suggest a few would be in order. But I'll do my best. Good luck."

Five minutes later he landed the aircraft as gently as if he had been flying 707's all his life. Harry, eyes closed, slumped in his seat, felt that even a crash would have been only an anticlimax to what Banning had told him an hour before.

# Twelve

"Did Harry tell you he's changed his mind about running for P.M. ?" Ian Kidman asked.

"Yes," said Constance, hiding her satisfaction. "But he didn't tell me why."

"Nor me. I gather he'd talked to you first from Qaran, then he rang me. He just said he'd had time to think while he was out there in the desert and decided he didn't want it, that he'd rather stay Foreign Secretary. A pity – I think we had it all wrapped up for him, especially after what happened to him. The country has been starved of heroes since the end of the war – heroes who are also politicians, that is."

"What happens now?"

"They're putting my name up." Kidman looked uncomfortable.

"Don't you want it?"

"Of course. Who wouldn't?" Then he smiled, shook his head. "Well, Harry doesn't, of course. Yes, I'd like it. But I was looking years ahead – not *now*."

"You're only three years younger than Harry."

"Yes, but I haven't been in politics anywhere near as long as he has. The British, especially our Party, don't like politicians thrust at them. Not Ministers, especially Prime Ministers. They prefer them to *emerge*. Lovely word, that – conjures up a picture of us being delivered from some political womb."

"Yours might be a breach birth."

"Just so long as it isn't a miscarriage. Still, I'll try for it. Harry says he will back me and for the next few weeks he is the one everyone will want to listen to, particularly in the constituencies. Here's Michael Argus. I understand he had to have himself hypnotised to come out here to meet Harry."

"Constance," said Argus, arriving in the airport lounge in a blaze of self-importance and a cloud of cigar smoke, "you don't know how relieved and delighted we all are that Harry is safe. You must have been worried out of your wits."

"You, too," said Constance, but nothing in her voice or expression gave her away. "How is Ethel liking Number Ten?"

"Oh, we haven't moved in. One can't be premature." But Argus gave her a sharp side glance before he turned away to greet the couple of dozen others who had come out here to welcome the returning hero. There had been a Reuters man waiting in Qaran when Carmel and the others had landed there; Hazza, the Qaranian, probably looking for British support at the Security Council, had told the story of their escape as if he were some sort of public relations man for Carmel; one would have thought the Duke of Wellington had come back to life. Down on the tarmac Argus could see television trucks, newsreel trucks and what looked like half the working population of Fleet Street. An expert in the value of publicity, he had the feeling he had come all the way out here to Heathrow to witness the demolition of his own hopes. He was just glad Carmel was not arriving by train at Waterloo.

Constance eased away from the politicians, made her way through the crush towards Richard, Robin and Matthew. It had moved her tremendously when each of the children, without warning, had returned to Aidan Hall as soon as the news that Harry had been kidnapped had been announced. There had been no mention as to whether they had come home to comfort her or because they had been worried for their father, or for both reasons: it did not matter, they had proved they were still a family.

"God, did you ever see such a smug, dreary lot!" said Richard, looking over and dismissing the British Cabinet. "What a marvellous opportunity to wipe out the lot of them with a bomb!"

"Not now," said Constance. "I'm wearing a new suit."

"John says," said Robin, so newly converted to Socialism that she was still wearing a Princess Anne-style hat, "John says if the Government was honest it would call another election so the country could show whether it wants to be led by a man it knows, like Tom Brewster, or someone it doesn't know."

"Like who?" said Constance.

"I don't know. Like that awful boor Argus, I suppose. Even Daddy would be better than him."

"Your father is quite happy as Foreign Secretary."

"Not if they go on kidnapping him," said Matthew. Richard and Robin had moved away to stand by the window and he looked at his mother. "Were you terribly worried?"

She hesitated, then nodded. "Terribly."

"So was I."

"Why don't you tell him so when you see him?"

He shook his head. "For the same reason you won't. He won't want to know."

"He might. He wasn't always like he is now."

"No, I suppose not. Or you wouldn't have married him. I've never asked you before – but whatever broke up you and Father?"

She had had the answer for some time, but she had never voiced it before: "Two people – he and I. It takes two to make a marriage – quite often it takes two to ruin one. I used to think your father alone was to blame – no, I shan't tell you why, it's none of your business. But now – " She looked out the window, then back at her eighteen-year-old son: he had something of his grandfather, her father, in him: a talent for sympathetic listening. "I've been chewing on my pride for as long as you've been alive. I think it's time I swallowed it."

Ten minutes later she was standing on the tarmac at the foot of the steps that led up to the aircraft that had brought home Harry, Joe Banning and the BOAC flight engineer. Roy Chadwick, who always reminded her of a melancholy stork but to-day was almost bursting apart with joy, took her arm.

"I think you and the family should go up and meet Lord Carmel in the plane. It will be more private." He gestured at the simmering riot of cameramen ready to boil over. "You don't want your welcoming kiss spread across to-morrow morning's papers."

"Heaven forbid," said Constance, and went up the steps followed by Richard, Robin and Matthew. The aircraft was an RAF V.I.P. plane: there were no other passengers but Harry, Banning and McArdle. Constance hugged Banning, who was closest, shook hands with the sharp-faced man behind him, then moved on down the aisle towards Harry. He smiled at her, but she could see the strain in his face: my God, she thought, he looks ready to weep! Is this my Harry?

She halted a foot from him, lifted her arms, felt herself trembling. "I was so worried – "

Then she was in his arms, weeping uncontrollably against his chest, and he was saying, "My love – "

2

They went down to Aidan Hall direct from the airport – "I'll come back to London the day after to-morrow," he told Ian Kidman. "I want twenty-four hours with my family. When is Charles Wolfe being buried?"

"He was buried yesterday. His family wanted it done quietly – they came down and took the body back to Derbyshire, buried him in the family plot. There's a memorial service the day after to-morrow."

"Then I'll come up for that. In the meantime you can let those who count know that I'm backing you for P.M. I've just had a quick word with Lavenham." Lord Lavenham was the Party chairman and had been one of Harry's supporters. "He's agreed to back you."

"You won't tell me why you've turned it down?"

"I just decided your judgment of me was wrong. I'm not the man for the job."

Kidman shook his head wonderingly. "Forgive me, Harry, but I did misjudge you – I never thought you could be as modest as this. But do you think I'm the man for it?"

"I'd tell you if you weren't. That's what friends are for, aren't they?"

"That sounds like a poor joke. But thanks. I'll try and not let you down."

Driving down in the Bentley Constance said, "Rosemary as mistress of Number Ten – I suppose I'll get used to it."

"Ian still has to get the job. Argus won't give up without a fight, especially now I'm out of the way."

"Why did you decide not to run?"

He glanced warningly at the back of Lambert's neck. "I'll tell you one of these days." But he wondered if he ever would; perhaps Joe Banning wanted the secret kept as much as he himself did. He and Joe had hardly spoken to each other on the flight back from Qaran, but once in front of Constance and the children both of them had kept up the façade of friendship. He just hoped he was not going to be involved in another *modus operandi*. He had just finally been released from one and he was grateful; when he had held Constance in his arms in the plane there had been no thought of lost ambitions, only a deep aching regret for all the wasted years. If ever he was going to overcome his curse of indifference, he had made a start there and then. He pressed her hand now, then looked back at the official car following them. "I suppose Joe is telling Robin and the others how he licked the guerrillas on his own."

"I don't think Joe has ever pretended to be a hero. He never talks about the war, for instance. And he seemed very quiet back at the airport. Were they rather nasty towards him?"

"On the contrary."

"Were they nasty to you?"

He smiled. "No scars at all. Except when they started shooting at us, they were politer than the Opposition in the Commons."

"What about Hazza and Mr Roth?"

"Hazza enjoyed it – he's stronger back in Qaran now than he ever was. Roth – " He was silent for a moment. "He told me to tell you he has never forgiven himself for telephoning you that day – he didn't stop to think what it would do to you."

"Poor man. The mistakes we all make – "

"Yes. But this one has been put right – by all of us." He squeezed her hand again, and she looked at him out of narrowed eyes: he recognised the look, an old one from long ago, and they smiled secretly, each wondering what it was going to be like to make love to the other after all these years. He leaned forward towards the chauffeur. "Can't you go a little faster, Lambert?"

"Sorry, m'lord. But there's been a bit of transmission trouble lately – the car is getting on, sixteen years it is – "

Sixteen years: as old as their *modus operandi*. "Get a new one, Lambert. Order it to-morrow."

"Yes, sir," said the chauffeur, beaming in his driving mirror. "Shall I trade this one in?"

"Give it to my daughter," said Harry. "She can drive to Labour Party meetings in it."

All three of them laughed, Lambert as happy as the two in the back seat. He knew the boss and his wife had at last taken up together again and that made him happier than the thought of having a new car to drive. Life was always better when you were working for a happy family.

An hour later they drove up the long drive to Aidan Hall. Ferguson, John, Mrs Dibbens and the staff were waiting on the terrace at the top of the steps. Harry got out and with Constance on his arm went up towards what was home, had been for centuries and would be, he hoped, long after he was dead and gone.

"Welcome back, m'lord," said Ferguson, more than ever one of the family; behind him John moved across to take Robin's arm. "It has been a worrying time for us all."

"Thank you, Ferguson." He looked around at the Hall staff;

351

to-morrow he would go out and meet the farmworkers. "Thank you all."

Then he turned round, saw Banning still at the foot of the steps, went back down to him. "Aren't you coming in?"

"No," said Banning. "Just let it be your family to-night. I'll come over to-morrow. That is, if I'm invited."

There was no one within yards of them: the two cars had been driven on, everyone else was up on the terrace. "Joe, I may never forgive you for what you've done, but I'll do my best to understand. I can't promise any more than that." He looked up towards the house, to the stone coat of arms above the front doors. The family achievement: perhaps it was no more than to have survived for as long as it had. And how had it achieved it? "Realism above all else. Old Roger, the first one, really knew a good motto when he saw one, didn't he?"

"I'll come over to-morrow," said Banning, and pressed the arm of his friend.

30

12/11